Studies in Lo
Volume 47

Logic Across the University: Foundations and Applications

Proceedings of the Tsinghua Logic Conference, Beijing, 2013

Studies in Logic Series Editor
Dov Gabbay   dov.gabbay@kcl.ac.uk

Logic Across the University: Foundations and Applications

Proceedings of the Tsinghua Logic
Conference, Beijing, 2013

Edited by

Johan van Benthem

and

Fenrong Liu

© Individual author and College Publications 2013.
All rights reserved.

ISBN 978-1-84890-122-3

College Publications
Scientific Director: Dov Gabbay
Managing Director: Jane Spurr

http://www.collegepublications.co.uk

Original cover design by Orchid Creative   www.orchidcreative.co.uk
Printed by Lightning Source, Milton Keynes, UK

Table of Contents

v

Preface

This book was initiated in connection with the conference 'Logic across the University, Foundations and Applications' held at Tsinghua University, Beijing, 14-16 October 2013, an event meant to demonstrate the wide range of logic today to a broad academic audience, while at the same time strengthening the ties between Chinese and international colleagues in the field. While there are particular things about this conference, let us start with what is general, perhaps even eternal, about logic.

On its traditional definition, logic studies the laws of correct inference. That may seem like a small topic, but given that the little spark of reasoning in our minds is what has brought us humans from roaming bands of social hunters in pre-history to the peaks of science today, it is no mean subject. Naturally, then, studying reasoning to its full extent means pursuing interests all across the university. Contemporary logic connects many disciplines and crosses many boundaries, from the humanities to the natural and social sciences. In the process, it has picked up many other core topics than just reasoning per se. Modern logic is also about language, information, computation, and agency.

This book is meant to show the vigor and liveliness of modern logic, including its many interfaces with other fields. We have chosen a few general headings to make this clear. A first group of papers under the heading of 'Logic, philosophy, and mathematics' explores new themes at the familiar interfaces between logic, philosophy, and mathematics, that by now date back some two thousand years. A second group is 'Logic and computation', containing lively samples of what is probably the largest and most active area of logical research today, with the most powerful organized communities. A group of papers on 'Logic, language, and cognition' then explores current interfaces of logic with the empirical study of natural language and cognition, perhaps forbidden territory to those who think logic is purely normative, but in reality, one of the most challenging border regions today. Finally, we have added a fourth group of papers under the heading of 'Logic and social interaction' showing how modern logic is also beginning to connect with the social sciences, and even the restless world of information-driven social phenomena today. We hasten to add that these groupings are mainly for convenience.The border lines can be fuzzy, witness the fact that some papers placed under computation and language are in fact about games, a powerful paradigm for interactive computation today, and thus make connections with game theory, reaching out to economics and the social sciences. And striking out in another direction, another paper under computation connects database theory and the foundations of quantum mechanics, again showing how logic disregards boundaries in the most constructive and fruitful manner.

The unity of a field resides not just in what is being studied, but also in how the practitioners approach it, and in particular, how they interact. In this book, we have chosen a somewhat uncommon format, to give the readers a glimpse of this dynamic. The main authors are speakers at the Tsinghua conference, chosen for their prominent position in their respective fields in China and abroad. Some of their contributions are published papers, represented here by summaries, others are newly written papers for this volume. In addition, we have invited two commentators for each paper (in one case, where we got carried away, even three), usually one from China and one from abroad, representing other perspectives on the area, asking them to raise some points showing the sort of discussion that drives our field. The reader will find a wide variety of, often very painstaking, comments, as well as the authors' brief responses, written at relatively short notice.

Naturally, this is just two rounds in an ongoing discussion, and nothing should be considered the last word in cases of disagreement, the engine of scientific progress. Also, we have largely refrained from editing content or language of these texts, since they offer a direct glimpse of the diversity of temperaments and cultures that communicate so successfully in Academic English, the lingua franca of the scientific community today, that is not owned by any particular group of native speakers. Many of the resulting sequences are very lively, raising serious critical points, as well as pointing out new issues for research, and showing the hidden passions of real debate (only saints enjoy being contradicted, and giving in without a fight). We have added introductions to the special parts of this book, but it would be tedious to summarize what authors and commentators say precisely. In particular, our introductions will not reveal the author's responses. You just see for yourself!

Logic not only connects between disciplines, it also connects between cultures. China has had an ancient logic tradition, but in this book, we are after its encounters with modern logic. There is a fascinating history, starting with the famous Jesuits in the 17th century, of absorbing ideas from logic into the Chinese scholarly and scientific community. (However, significantly, the first Western publication in China was Matteo Ricci's *On Friendship*. One needs to have one's priorities right.) And this process is still continuing apace, ever since the first wave of Chinese intellectual pioneers went abroad in the 19th century. Having a conference at Tsinghua pays tribute to great names in this contact, since it was there that JIN Yuelin brought modern mathematical logic to China in the 1930s. This process is still continuing, and China has many lively and growing logic centers at many universities today. This book hopes to contribute to this trend as well, by staging a number of close encounters between Chinese and international colleagues with shared research interests.

While this book is about academic themes, there is a broader context as well. Logic plays a role in industry today, and indeed, it has a role to play in society at large. Two prominent logicians representing these outward looking trends have been invited to give keynote evening lectures at the Tsinghua event. Vincent Hendricks (University of Copenhagen) will talk about 'Bubbles', and Moshe Vardi (Rice University) about 'From Aristotle to the iPhone'. While this book does not cross these bridges toward society, they are there for those who care about the broader significance of our field.

This book offers substantial fare, but it is just one step in an ongoing conversation between colleagues worldwide. The Tsinghua conference is another step, but we hope that, afterwards, our wider Chinese and international readership will feel inspired to join in. We thank our authors, commentators, and support staff (especially, Chanjuan Liu, Chenwei Shi, and Zhiwei Yang) for making all this happen.

Johan van Benthem and Fenrong Liu
9 August, 2013

Part I

Ever since antiquity, logic has interacted with philosophy, the study of human thought and how it relates to the world, and with mathematics, the language of abstraction and precision for all the serious sciences – leading to the golden foundational age of Frege, Russell, and Gödel. These traditional interfaces are still very much alive, and the pieces in this part exemplify some important new themes.

In 'Modeling justification: beyond the propositional paradigm', *Sergei Artemov* (City University of New York) introduces the program of justification logic, a wide-ranging approach to knowledge with early roots in the provability logic of the 1970s, that merges ideas from proof theory with the usual models for modal logics to obtain a sensitive and versatile account of evidence represented by explicit proof terms. Deep mathematical connections then emerge between systems of justification logic and their counterparts in epistemic logic (arising through projection), while there is also an expanding range of applications of the new perspective, from computation to philosophy and game theory. *Hiroakira Ono* (JAIST Kanazawa) asks whether there could be general algebraic structures in the mathematical foundations of justification logic, and also wants to know about temporal and quantified extensions getting closer to epistemic practice. *Bryan Renne* (University of Amsterdam) points at the yet greater richness of our daily practices of justification as a form of story telling, and also asks how justification logic can allow for a variety of non-omniscient agents with frailties like standard humans.

Wesley Holliday (University of California, Berkeley) presents a unifying logical analysis of the intriguing new accounts of knowledge that philosophers have developed since the 1960s. 'Epistemic closure and epistemic logic I: relevant alternatives and subjunctivism' presents one semantic model of belief and ternary world order for knowledge as truth in all relevant alternatives, and knowledge as successful tracking of the truth. In these models, intriguing 'closure failures' occur for knowledge under classically valid implications, not because of bounded agency, but by the nature of knowledge claims. Holliday proves a completeness theorem that determines what is valid and what is not, shedding new light on philosophical intuitions in the area. *Paul Égré* (Institut Jean-Nicod and New York University) points out the importance of a systematic distinction here between the knowing subject and the agent attributing knowledge to others, as modeled in recent semantics for natural language, while also drawing attention to the role of attention in possessing or claiming knowledge. *Zhaoqing Xu* (Sichuan University, Chengdu) points at the importance of simultaneously modeling agents's beliefs as a crucial companion to knowledge, and asks about richer logics where closure might also be about other conditionals than classical implication, such as those found in counterfactual reasoning.

Hannes Leitgeb (Ludwig Maximilian University of Munich) addresses a major problem in 'The stability theory of belief': the connection between quantitative views of degrees of belief, and the qualitative notion of belief as used by logicians. He then uses Locke's Thesis about beliefs occurring above a certain threshold of plausibility to develop a 'stability theory' based on a mathematical theorem saying that, for each probability measure P, there are P-stable logically strongest believed propositions. Such beliefs are the only way to satisfy the minimal requirements of qualitative logical belief, the Kolmogorov conditions on quantitative belief, and a reinterpreted version of Locke's thesis. *Alexandru Baltag* and *Sonja Smets* (University of Amsterdam) point out analogies between Leitgeb's account and current notions of safe belief that is robust under true new information, and ask for a junction with logics of the information dynamics that provides the setting for the stability. *Hanti Lin* and *Kevin Kelly* (Carnegie Mellon University, Pittsburgh) discuss whether Leitgeb's synchronic analysis can be integrated in an AGM style belief revision theory, and proves a new impossibility result showing that a plausible principle of belief revision has to go: the condition that, if new evidence is consistent with our beliefs, the new belief state consists of just the logical consequences of old beliefs plus new evidence.

While scientific laws seem the ultimate generalities, there is a long-standing view in the philosophy of science that they sometimes only work under normal circumstances (ceteris paribus), making allowing for exceptions essential. In 'Is there such a thing as a *ceteris paribus* law?', *Wei Wang* (Tsinghua University) critically examines some prominent objections against ceteris paribus laws, and argues that this kind of exception-tolerant regularity may be non-eliminable even with scientific terminology, while they also admit of testability, the hallmark of scientific knowledge. *Kohei Kishida* (University of Oxford) wonders whether the testability depends too much on the classical conditionals used in Wang's account, and what would happen if we examine the logical apparatus used more closely. *Alexander Reutlinger* (Ludwig Maximilian University of Munich) wonders whether the paper really comes to grips with the criticized arguments by Earman and others, and their connections to Hume's classic notion of a law. Finally, *Liying Zhang* (Central University of Finance and Economics, Beijing) points out the relevance to ceteris paribus reasoning in science of current developments in non-monotonic logics and the semantics of default expressions in natural language.

Feng Ye (Capital Normal University, Beijing) explains his longstanding research in 'Introduction to a naturalistic philosophy of mathematics'. He develops a strict physicalist scientific ontology with a variety of philosophical and technical considerations. Moreover, he shows how the approach has bite by developing original formal systems that allow for strict forms of nominalism and finitism in a wide range of mathematics. *Hannes Leitgeb* (Ludwig Maximilian University of Munich) wonders whether the given formal systems really do, and even can in principle, be as thoroughly finitist as they need to be? He also wonders why empirically oriented physicalism should be the yardstick for a philosophy of an a priori discipline like mathematics. Likewise, *Difei Xu* (Renmin University of China, Beijing) asks why a naturalist philosopher could not accept some abstractions as entirely natural, given the realities of science. She also worries that an all-out naturalism has no room to explain the normativity of mathematical knowledge, and what then becomes of the status of philosophy.

Modeling Justification: Beyond the Propositional Paradigm[1]

SERGEI N. ARTEMOV

Since Plato, *justification* has been considered a principal element of epistemic analysis that was, until recently, conspicuously absent in formal logical models of knowledge and belief. Justification Logic augments epistemic logic by assertions $t{:}F$ that read

t is a justification for F,

hence incorporating the missing justification component. How should epistemic models be upgraded to model justifications?

We retain a classical understanding of propositions models as subsets of the set W of possible worlds. Formally, given a set Fm of formulas, we define an evaluation

$$* : Fm \mapsto 2^W$$

along with natural constraints respecting the logical/modal connectives. We will write F^* to denote the set of worlds that corresponds to formula F, i.e., the *set of worlds at which F holds*, and $u \Vdash F$ as shorthand for $u \in F^*$.

Note that such a model does not distinguish formulas that specify the same proposition, i.e., which truth values coincide at each world. Other traditional models for modal/epistemic logics, such as topological, neighborhood, and algebraic semantics also do not distinguish syntactically different formulas within the same proposition. We call this class of models *models with propositional precision*. The first message of this note is that **propositional precision in not enough to model justification**.

For example, arithmetical formulas $0 = 0$ and Fermat's Last Theorem FLT, being proven mathematical facts, are true in all possible worlds and hence denote the same proposition. In models with propositional precision, a statement

t is a proof of $0 = 0$

should yield

t is a proof of FLT

---

[1] This summary is based on [1].

which disqualifies the model.

As another example, consider a number n which is a product of two cosmologically large primes, $n = pq$ so the statement 'n is a multiple of p' holds and hence as a proposition is no different from $0 = 0$. An epistemic model with propositional precision would not distinguish statements 't *is a proof of* $0 = 0$' from 't *is a proof that* n *is a multiple of* p'. This renders models with propositional precision not adequate for modeling cryptographic reasoning.

In Justification Logic, there is, in addition to the category of formulas, a category of *justifications* with a new sort of proposition $t{:}F$ stating t *is a justification of* F. For the aforementioned reasons, the possible world semantics of justifications should rely on some construction which breaks out of the propositional paradigm. The analysis of existing approaches to model justifications suggests interpreting justification terms Tm at each world as *sets of formulas*,

$$* : W \times Tm \mapsto 2^{Fm}.$$

We write t_u^* for the interpretation of term t at world u. Each t_u^* is a set of formulas for which t is a justification at u. According to this reading,

$$u \Vdash t{:}F \quad \text{iff} \quad F \in t_u^*\,.$$

Note that whereas propositions in such models are interpreted semantically, as sets of possible worlds, justifications are interpreted syntactically, as sets of formulas. This is a principal feature: a model may treat distinct formulas F and G as equal, i.e., for each possible world u,

$$u \Vdash F \quad \text{iff} \quad u \Vdash G,$$

but still be able to distinguish justification assertions $t{:}F$ and $t{:}G$, e.g.,

$$u \Vdash t{:}F \quad \text{but} \quad u \nVdash t{:}G.$$

In a basic setting, justifications are terms with operations *application* and *sum*, although more elaborate justification logics allow additional operations on justifications. The *application* operation takes justifications s and t and produces a justification $s{\cdot}t$ such that if $s{:}(F \to G)$ and $t{:}F$, then $[s{\cdot}t]{:}G$. Symbolically,

$$s{:}(F \to G) \to (t{:}F \to [s{\cdot}t]{:}G).$$

The second basic operation on justifications is *sum* '+.' If $s : F$, then whatever evidence t may be, the combined evidence $s + t$, as well as $t + s$, remains a justification for F. Operation '+,' given s and t, produces $s + t$, which is a justification for everything justified by s or by t:

$$s{:}F \to [s + t]{:}F \quad \text{and} \quad s{:}F \to [t + s]{:}F.$$

Formally, justification terms, *Tm*, are built from justification variables and constants by means of the operations '·' and '+.' Formulas, *Fm*, are built from propositional variables *Var* and truth constants by the usual Boolean connectives and the rule: if t is a term and F a formula, then $t{:}F$ is a formula.

Basic Logic of Justifications J_0:

Classical propositional axioms and the rule Modus Ponens;

$$s{:}(F \to G) \to (t{:}F \to [s{\cdot}t]{:}G);$$

$$s{:}F \to [s+t]{:}F, \quad s{:}F \to [t+s]{:}F.$$

J_0 is the logic of general (not necessarily factive) justifications for a skeptical agent for whom no formula is justified *a priori*. When we want to assume that an axiom A is justified, we postulate $c_1 : A$ for some justification constant c_1. Furthermore, if we want to assume that this new principle $c_1{:}A$ is also justified, we can postulate $c_2{:}(c_1{:}A)$ for a constant c_2, etc. The set of all assumptions of this kind for a given logic is called a *constant specification*.

Let *CS* be a constant specification. J_{CS} is the logic $J_0 + CS$; which axioms are those of J_0 with the members of *CS*, and the only rule of inference is *Modus Ponens*.

For sets of formulas X and Y, we define

$$X{\cdot}Y = \{F \mid G \to F \in X \text{ and } G \in Y \text{ for some } G\}.$$

Informally, $X{\cdot}Y$ is the result of applying *Modus Ponens* once to all members of X and of Y (in the given order).

The notion of the basic modular model corresponds to the classical logic situation (which formally consists of just one possible world).

DEFINITION 1. A **basic modular model** is an evaluation $*$ which maps propositional variables Var to truth values $\{0, 1\}$ and justification terms Tm to subsets of the set of formulas

$$* : Var \mapsto \{0,1\} \quad \text{and} \quad * : Tm \mapsto 2^{Fm}$$

such that

(1) $s^* {\cdot} t^* \subseteq (s{\cdot}t)^*$ and $s^* \cup t^* \subseteq (s+t)^*$.

As usual, we will write $\Vdash F$ instead of $F^* = 1$. The truth value of formulas is defined inductively and respects Boolean logic, i.e., $\Vdash F \wedge G$ iff $\Vdash F$ and $\Vdash G$; $\Vdash \neg F$ iff $\nVdash F$; $\Vdash t{:}F$ iff $F \in t^*$.

Basic modular models are strongly reminiscent to Mkrtychev models for the Logic of Proofs [5]. Soundness and completeness with respect to basic modular models follow from [1, 2].

THEOREM 2. $J_{CS} \vdash F$ *iff* F *holds in any basic modular model respecting CS.*

This theorem shows, in part, that possible worlds are not really necessary for modeling justification; justification logics J_{CS} are complete with respect to single-world models in which all information about justifications is contained in the evaluation function. However, things change when we want to consider both justifications and beliefs represented by modalities. In this case, the standard Kripke semantics suggests using multiple-world models. The standard semantics of *F is believed at world u* is

F holds at all worlds considered possible at u.

How do justifications fit into this picture? Take a Kripke frame (W, R), where W is a non-empty set of possible worlds, R is a binary 'accessibility' relation on W, and consider an interpretation $*$, which is a mapping of the format

$$* : Var \mapsto 2^W, \quad * : W \times Tm \mapsto 2^{Fm},$$

such that for each world u, it specifies a basic modular model $*_u$. To be precise,

$$p_u^* = 1 \text{ iff } u \in p^*, \quad t_u^* = *(u, t),$$

and closure conditions (1) hold for each $*_u$. A conceptually clean mathematical connection of $*$ and R reflecting the epistemic nature of justifications is given by the principle

a specific reason for F yields believing that F.

This principle has been the cornerstone of the Fitting semantics of justifications (cf. [4]); it has also been widely adopted in logical systems with explicit and implicit knowledge, cf. [3]. Let us formulate this principle in the modular model format. Given $\mathcal{M} = (W, R, *)$, let $\Box_u$ denote a set of formulas

$$\{F \mid v \Vdash F \text{ for all } v \text{ such that } uRv\}.$$

Conceptually, at a given world u, t_u^* reflects 'believing for a reason t,' whereas $\Box_u$ represents believing without providing a specific reason. We say that *justification yields belief* in $\mathcal{M} = (W, R, *)$, if

$$t_u^* \subseteq \Box_u$$

for each justification term t and each $u \in W$. In other words, if t is a justification for F at u, then F is believed in u.

DEFINITION 3. A modular model is $\mathcal{M} = (W, R, *)$ in which
 i) W is a non-empty set of worlds and R is a binary relation of W;
 ii) interpretation $*$ has the format
 $* : Var \mapsto 2^W$,
 $* : W \times Tm \mapsto 2^{Fm}$;
 and $*$ is a basic modular model at each world $u \in W$;
 iii) justification yields belief, i.e., $t_u^* \subseteq \square_u$ for each $t \in Tm$ and $u \in W$.

Modular semantics allows us to model justification and beliefs simultaneously. Consider a logic of justifications and beliefs, KJ_0, in the joint language of J and modal logic K, that contains standard postulates for J_0 and K as well as the connection axiom

$$t{:}F \to \square F$$

(stating syntactically that justification yields belief). As before, *CS* denotes a constant specification and KJ_{CS} a logic $\mathsf{KJ}_0 + CS$. Modular semantics seamlessly extends to this case.

THEOREM 4. $\mathsf{KJ}_{CS} \vdash F$ *iff F holds in any modular model respecting CS.*

So, modular models extend Kripke-style evaluations from the usual *formulas are interpreted as sets of possible worlds* to include *and justifications are interpreted as sets of formulas.*

BIBLIOGRAPHY

[1] S. Artemov. The logic of justification. *The Review of Symbolic Logic*, 1(4):477–513, 2008.
[2] S. Artemov. The ontology of justifications in the logical setting. *Studia Logica* 100:17-30, 2012.
[3] S. Artemov and E. Nogina. Introducing justification into epistemic logic. *Journal of Logic and Computation*, 15(6):1059–1073, 2005.
[4] M. Fitting. The logic of proofs, semantically. *Annals of Pure and Applied Logic*, 132(1):1–25, 2005.
[5] A. Mkrtychev. Models for the Logic of Proofs. In *Logical Foundations of Computer Science 1997*, volume 1234 of *Lecture Notes in Computer Science*, pages 266–275. Springer, 1997.

Comments on Artemov

Hiroakira Ono

1 Justification logic

In the paper "The Logic of Justification", the author surveys recent developments of Justification Logic, to which he has made essential contributions in recent years.

By using *justification terms*, Justification Logic will give us a mathematically solid and philosophically sound understanding of the following postulate:

knowledge is no other than justified true belief.

As shown in the paper, various philosophical problems in epistemic logic can be analyzed in a refined way by using justification terms. At the same time, this approach enables us to introduce a semantics for epistemic logic which is different from the conventional Kripke semantics for epistemic modal logic. A remarkable feature of the semantics is its existential character, while Kripke semantics is of a universal character. The *Realization Theorem* that can be expressed as:

epistemic modal logic = justification logic + forgetful projection

then tells us an important connection between epistemic modal logic and justification logic.

2 Several questions

Here are some technical and general questions which came to my mind when I read both the present paper and the paper "The Ontology of Justifications in the Logical Setting".

Two binary operations are defined over justification terms, i.e. application $\cdot$ and sum $+$. Application is a familiar one as it is used in λ-calculus and combinatory logic, while sum is new to me. It is said that $s + t$ will be understood as the *concatenation* of proofs s and t in the logic of proofs LP. Is there any chance of understanding justification terms algebraically? That is;

(1) Is it irrelevant to consider the identity relation $=$ among justification terms? For example, what will happen if we put some algebraic conditions on $\cdot$ and $+$, like the associativity of $+$?

(2) The definition of basic modular models suggests that from an algebraic point of view $s + t$ behaves like an upperbound of s and t. By introducing a *directed* partial (or quasi) order explicitly over justification terms, can we dispense with the operation $+$?

Another subject which might be interesting is to see how to extend the notion of justification terms in combined systems with epistemic logic, especially in temporal epistemic logic.

(3) How to extend justification logic in order to deal with temporal epistemic logic? Isn't it necessary to define justification terms to be functions of flow of time? There is a remark on belief revision on p. 482 of the extended paper mentioned above, saying that justification terms will be non-monotonic for belief revision. But isn't it necessary to take flow of time into account?

The Realization Theorem is a quite important result, since it will clarify the connection between epistemic modal logic and justification logic. It is pointed out on p. 499 that it is not always the case that any modal logic has a reasonable justification logic counterpart. This would be an inevitable situation. Then I would like to see what the existence of justifications means in modal logic, in general.

(4) Is there any reasonable criterion for modal logics without justification logic counterparts? Or, are there any particular feature of modal logics *with* justification logic counterparts?

The study of justification logic has been successful so that some philosophical problems are nicely solved in the setting of justification logic. It will be natural and at the same time interesting from both a philosophical and a mathematical point of view to consider extending justification logic in various ways. This will be a quite inspiring and promising direction of the study. Related to the author's statements at the end of p. 478, I have the following questions.

(5) Introducing identity and quantification in modal logic is often a matter of philosophical controversy. I am curious to see how a first-order version of justification logic can be developed, and in which way these controversial problems can be resolved within the scope of justification logic.

Also I am particularly interested in how to build justification logic over nonclassical logics, i.e., how to get justification logics that are counterparts of nonclassical epistemic (or modal) logics. For example:

(6) How far can results obtained so far on justification logic in the present paper be extended to intuitionistic epistemic logic? Also, how about intuitionistic linear logic with the exponential ! (which behaves like a S4 modality)?

Comments on Artemov

BRYAN RENNE

ABSTRACT. This note presents some comments and questions related to two papers of Sergei Artemov: "The Logic of Justification" [1] and "The Ontology of Justifications in the Logical Setting" [2]. This note was prepared by invitation for the event *Logic across the University: Foundations and Applications* held at Tsinghua University, 14–16 October 2013, in Beijing, China.

1 Justification and Proof

I have come to think of the Justification Logic (JL) approach to reasoning about justification (or "evidence," broadly construed) as more paradigmatic than dogmatic. By this I mean that the justification-combining operations found in a specific JL system should be viewed as just one way to reason about justification, as opposed to an assertion of "the one true way" for reasoning about justification. In this sense, the main message I get from the JL approach is that we can use in-language syntactical bookkeeping to describe and characterize reasoning in a stepwise fashion, and this can be leveraged to provide a more nuanced formal account of knowledge and belief. In particular, by introducing justifications for the first time as separate logical entities, Justification Logic aims to persuade the logic community to consider the previously missing justification component as a key ingredient in logics of knowledge and belief.

While most JLs are based on the *sum* (i.e., justification aggregation) and *application* (i.e., justification Modus Ponens) operations, we should not be overly serious about this. When it comes to reasoning about justification, many other operations are also of interest, even some that are not logically sound. Indeed, much everyday, "real life" reasoning is not logically sound and, even worse, sometimes logically flawed. But there are ways to recover from flawed reasoning via "backtracking" or "revision," and all of this is worthy of consideration in a general study of formal justification. Examples of operations to consider: various other sound operations (e.g., a justification version of Modus Tollens), nonmonotonic or default operations (e.g., a justification version of "φ normally follows from ψ"), commonsense induction (e.g., a justification version of "if every φ-situation seen so far is also ψ, conclude $\varphi \rightarrow \psi$"), and even fallacies or other logically flawed operations (e.g., a justification version of "from $\varphi \rightarrow \psi$ and ψ, conclude φ"). Of course the unsound operations must come with some means of recovery from error, perhaps in a justification-friendly adaptation of Belief Revision theory, which, in the interest of full disclosure, is part of an ongoing project of mine.

In essence, the general picture I see is the following: justification is in some sense about "telling a story" as to why something is the case, and the quality of the justification has to do with the quality of the story (within a particular context and for a particular purpose). So far most work in JL has focused on extremely well-behaved "stories": Hilbert-style proofs constructed from basic assumptions using proof concatenation (i.e., sum), Modus Ponens (i.e., application), and possibly other logically sound operations on proofs. But there is room for a broader perspective, wherein we focus less on proofs and more on potentially unsound justificatory "stories" that are intended to support certain assertions, though this support may be highly subjective or contextual and is subject to revision upon receipt of additional information. From this perspective, the proof-based JLs studied to date are only a first approximation to a generalized study of subjective justification, the key feature of which is an adaptation of the in-language syntactical bookkeeping mechanisms already developed in the JL literature. Of course this extra syntactical structure comes with a cost: typically easy theorems (e.g., Replacement [6]) become difficult because formulas now include detailed, highly syntax-dependent "stories" asserting subjective justifications, and making sure everything is in order with these often requires nontrivial trickery. Nevertheless, what we gain in exchange is the opportunity for fine-grained analysis of a purported justification— something it seems we cannot do absent extra-modal syntactical baggage—and this sometimes makes it worth the additional effort.

Question 1. What would a Justification Logic-style syntactic bookkeeping mechanism for general justificatory "stories" look like?

Question 2. How might justificatory "stories" be compared in terms of their quality or persuasiveness (within a particular context and for a particular purpose)?

2   Justifications and Omniscience

One of the persistent philosophical complaints about modal logics for knowledge and belief is that these logics attribute "too much" knowledge or belief to the agent.[1] The syntactic bookkeeping mechanisms of Justification Logic suggest one way of addressing this: roughly speaking, larger justifications are needed for more distant conclusions, so explicit knowledge or belief (i.e., knowledge or belief witnessed by a specific justification) comes with a specific "cost" [4, 3, 5, 7, 8]. Nevertheless, there is still another kind of omniscience present: in most JLs, justifications are always "out there," even if they are too large for anyone to possibly know them. The question then becomes one of determining which justifications an

---

[1] I borrow this formulation from Melvin Fitting [5].

agent has in hand and how it is that the agent comes to gain (or lose) justifications as a result of further consideration, receipt of additional information, or simple passage of time.

Question 3. What is the cleanest way to study the justifications of non-omniscient agents, taking into account the fact that the available justifications change over time?

3 Conclusion

Inasmuch as there is structure to the story one gives in favor of an assertion, there is in some sense a "logic" afoot, and the syntactical bookkeeping mechanisms used in Justification Logic can be adapted to provide a fine-grained representation of these "logics." So far the focus has been on the usual things one thinks of when speaking of formal logic: sound axioms and rules of inference, proofs, and the like. But when it comes to everyday justifications, additional flexibility is required. In this short note I have tried to provide the briefest sketch of what this might mean for the study of Justification Logic. My hope is that those who wish to take this approach seriously as a general study of justification will agree with me that the JL literature has only scratched the surface, and so there is a great deal of interesting work still to be done.

Acknowledgement This work is funded by an Innovational Research Incentives Scheme Veni grant from the Netherlands Organisation for Scientific Research (NWO).

BIBLIOGRAPHY

[1] Sergei N. Artemov. The logic of justification. *The Review of Symbolic Logic*, 1(4):477–513, December 2008.

[2] Sergei N. Artemov. The ontology of justifications in the logical setting. *Studia Logica*, 100(1–2):17–30, April 2012. Published online February 2012.

[3] Sergei N. Artemov and Roman Kuznets. Logical omniscience via proof complexity. In Zoltán Ésik, editor, *Computer Science Logic, 20th International Workshop, CSL 2006, 15th Annual Conference of the EACSL, Szeged, Hungary, September 25–29, 2006, Proceedings*, volume 4207 of *Lecture Notes in Computer Science*, pages 135–149. Springer, 2006.

[4] Sergei N. Artemov and Roman Kuznets. Logical omniscience as a computational complexity problem. In Aviad Heifetz, editor, *Theoretical Aspects of Rationality and Knowledge, Proceedings of the Twelfth Conference (TARK 2009)*, pages 14–23, Stanford University, California, July 6–8, 2009. ACM.

[5] Melvin Fitting. A logic of explicit knowledge. In Libor Běhounek and Marta Bílková, editors, *Logica Yearbook 2004*, pages 11–22. Filosofia, Prague, 2005.

[6] Melvin Fitting. A replacement theorem for LP. Technical Report TR–2006002, CUNY Ph.D. Program in Computer Science, March 2006.

[7] Melvin Fitting. Reasoning with justifications. In David Makinson, Jacek Malinowski, and Heinrich Wansing, editors, *Towards Mathematical Philosophy, Papers from the Studia Logica conference Trends in Logic IV*, volume 28 of *Trends in Logic*, chapter 6, pages 107–123. Springer, 2009. Published online November 2008.

[8] Ren-June Wang. Knowledge, time, and the problem of logical omniscience. *Fundamenta Informaticae*, 106(2–4):321–338, 2011.

Response to Ono and Renne

SERGEI N. ARTEMOV

The author extends his sincere thanks to Professor Ono and Doctor Renne for providing excellent comments.

Question 1 by Hiroakira Ono. *Does it make sense to consider the identity relation $=$ among justification terms? For example, what will happen if we put some algebraic conditions on $\cdot$ and $+$, like the associativity of $+$?*

Response. Equality/congruence of proofs is a profound question in mathematical logic, closely related to Hilbert's so-called 24th problem that asks for a criterion of simplicity in mathematical proofs. By its nature, equality of proofs is quite different from equality of formulas: proofs are intrinsically syntactic objects, and their equality depends on the conventions concerning proof-preserving syntactical transformations. A natural source of such basic transformations are λ-term reductions and their internalized versions. Such transformations have been explored for the Logic of Proofs in its original combinatory logic format [3, 27] and in the λ-calculus format [6]. In the latter, the base logic is intuitionistic, equality "$=$" has been explicitly introduced to the language, and the resulting system of intensional λ-calculus enjoys confluence and strong normalization. The Logic of Proofs LP represents all logic principles about proofs in the language with $\cdot$, $+$, and ! that hold for any proof system (i.e., invariant w.r.t. the choice of a proof system). This claim has been supported by the arithmetical completeness theorem for LP [1] which implies that if a principle is not derivable in LP, for example, the commutativity of $+$:

(1) $[s + t]{:}F \rightarrow [t + s]{:}F,$

then there is a legitimate proof system and an arithmetical instance of this principle which does not hold in arithmetic. However, individual proof systems may support some principles which are not, generally speaking, invariant. For example, for the standard proof predicate in arithmetic and $+$ as concatenation of Hilbert-style proofs, the commutativity principle (1) is valid. So, adding new principles to LP amounts to narrowing the class of proof systems under consideration to those for which these new principles hold. This can make sense, e.g., for the principles that are valid for the standard proof predicate. Such a possibility has been explored in [3] in which the so-called symmetric logic of proofs SLP has been offered.

In SLP, natural laws of commutativity and associativity of + hold, as well as distributivity of · and +. This demonstrates a way of acquiring a conventional algebraic structure on proof terms: by assuming principles from outside LP that, however, hold for some "good" proof systems for "good" operations of proofs.

Question 2 by Hiroakira Ono. *The definition of basic modular models suggests that from an algebraic point of view, s + t behaves like an upper bound of s and t. By introducing a directed partial (or quasi) order explicitly over justification terms, can we dispense with the operation +?*

Response. An interesting suggestion. Technically, this can be done by postulating a partial order over justification terms but this possibility seems not to have been explored yet.

Question 3 by Hiroakira Ono. *How to extend justification logic in order to deal with temporal epistemic logic? Isn't it necessary to define justification terms to be functions of flow of time? There is a remark on belief revision on p. 482, saying that justification terms will be non-monotonic for belief revision. But isn't it necessary to take flow of time into account?*

Response. A dynamic component indeed makes modal logic more expressive for purposes of modeling knowledge. Epistemic dynamics can be provided, among others, by epistemic actions, by temporal dimensions, or both. One of the principal features of the Logic of Proofs is that proof terms themselves constitute a dynamic component: the proof term t in a proof assertion $t{:}F$ represents the process of cognition, time, effort, etc. If we assume that proving is a process, and proofs become available from simple/short to more complicated/long, then the complexity/size of a proof becomes a reasonable measure of time, cf. [28]. In this respect, the answer to the question of whether or not terms should be functions of time is negative. In the same way that Brouwer-Heyting-Kolmogorov constructive proofs are not functions of time and yet represent the intended semantics of intuitionistic logic capable of capturing the cognition process, proof/justification terms in Justification Logic have enough structure to represent basic epistemic dynamics. So, justification logic may be regarded as a more complex system than temporal modal logic from which the latter can be obtained by some kind of forgetting procedure, e.g., as in [28]. That being said, we acknowledge the potential utility of introducing a separate time component, e.g., as a synchronization device, into Justification Logic, rendering justification terms functions of time. However, it does not appear necessary for basic justification analysis, even with the belief revision component. The same observation holds for the belief revision setting: justification terms are dynamic and need not be functions of time, cf. a justified belief change paper [10].

Question 4 by Hiroakira Ono. *Is there any reasonable criterion for modal logics* *without justification logic counterparts? Or, are there any particular features of modal logics with justification logic counterparts?*

Response. There is no good answer to this question. In known examples of modal logics "without justification logic counterparts," we speak not about absolute impossibility to realize all theorems of a given logic, but rather about impossibility of a realization that satisfies some additional semantic requirements. For example, the positive introspection principle

$$\Box P \to \Box\Box P,$$

naturally defines an operation "!" (proof checker) on proofs such that for each proof u,

$$u{:}P \to !u{:}u{:}P.$$

The negative introspection principle

$$\neg\Box P \to \Box(\neg\Box P)$$

also has a natural explicit version that introduces a natural operation "?" (negative proof checker):

(2) $\neg u{:}P \to ?u{:}(\neg u{:}P).$

However, (2) is not an arithmetically valid principle. Indeed, since any arithmetical proof is not a proof of infinitely many formulas, for any given arithmetical proof u^* there are infinitely many arithmetical sentences P^* such that u^* is not a proof of P^*. Therefore, if (2) were arithmetically valid, $(?u)^*$ would be a proof of infinitely many arithmetical sentences, which is impossible for standard proof systems. However, if arithmetical soundness is not required, the negative proof checker is a "normal" operation on proof terms, cf. [22, 25]. A similar example is provided by the Löb Principle

$$\Box(\Box P \to P) \to \Box P$$

of Provability Logic. It cannot be realized by explicit proofs (since only S4 theorems can), hence the whole provability logic GL does not have an arithmetically sound justification counterpart. The question of which theorems of GL admit explicit proof realization was answered in [16]. So, no general criterion for a modal logic without a justification counterpart is known.

Question 5 by Hiroakira Ono. *Introducing identity and quantification in modal logic is often a matter of philosophical controversy. I am curious to see how the first-order version justification logic can be developed and in which way these controversial problems can be resolved within the scope of justification logic.*

Response. The first-order logic of proofs was developed in [8], and supplied with the Kripke-style semantics and completeness theorem in [15], though neither of these papers considers first-order equality. The latter discusses first-order quantification in a justification logic setting in detail and successfully resolves issues concerning designation of individual terms in the worlds of a model. A quantifier-free first-order justification logic with equality is considered in [2]. Theorem 9.2 provides a way to resolve the Frege puzzle in an epistemic environment: equality of individual objects alone does not warrant substitutivity, but justified equality does: *For any individual terms f and g, justification variable u, and atomic formula $P(x)$, there is a justification term $s(u)$ such that justification logic proves*

$$u{:}(f = g) \to s(u){:}[P(f) \leftrightarrow P(g)].$$

Question 6 by Hiroakira Ono. *How far can results on justification logic in the present paper be extended to intuitionistic epistemic logic? Also, how about intuitionistic linear logic with the exponential " ! " (which behaves like* S4 *modality)?*

Response. According to the format of the Logic of Proofs, axioms of the underlying logic yield proof constants, and admissible rules of inference lead to operations on proofs. Since, in the intuitionistic logic IPC there are infinitely many independent admissible rules (Visser rules), there are infinitely many primitive operations on proofs. Intuitionistic logic of proofs iLP was found in [7] and proved to be complete with respect to Heyting's Arithmetic HA in [11]. It appears that the Realization Theorem applies to intuitionistic S4/iLP without principal modifications. Assertions $t{:}F$ in Justification Logic can be read as

t is a sufficient resource for F,

which naturally connects this area with Linear Logic. This prompts an interesting research project: to develop a logic of resources in the format of Justification Logic and to embed Linear Logic into it in the same way that IPC embeds into the Logic of Proofs via Gödel translation and Realization Theorem. If successful, this would place Linear Logic with its numerous applications into the broader context of the logic of resources. A corresponding paper is now under review.

Question 1 by Bryan Renne. *What would a Justification Logic-style syntactic bookkeeping mechanism for general justificatory "stories" look like?*

Response. There are many informal and formal ways to interpret assertions $t{:}F$ as justificatory "stories," some are listed in [2]. With regard to formal treatment, the guidelines are provided by the "The Ontology of Justification" paper [5]: *justifications at each world are sets of formulas with operations that satisfy some basic closure conditions.* This is a general framework, which is open to extension. It is up to the user to specify, through formalization, these sets of formulas for each justificatory "story" and define operations on them, in addition to a small basic list, usually consisting of · (aggregation with a deduction step) and + (aggregation without an epistemic action). Such a formalization yields access to a variety of methods of justification bookkeeping: direct observation and deduction (Red Barn and Russell's examples in [2]), logical models [5, 13, 21]; tableaux [14, 18, 23]; proof-theoretical methods [18]; deep inference [14]; arithmetical provability semantics [1, 29]; automated proof search [9]; game-theoretic interpretation [24]; evidence tracking [4]; decision procedures [19, 21]; reduction to the reflected fragment [17].

Questions 2 by Bryan Renne. *How might justificatory ?stories? be compared in terms of their quality or persuasiveness (within a particular context and for a particular purpose)?*

Response. One might think of a persuasiveness order on justifications and/or probabilistic measures. One of the possible approaches to the latter was tried in [20].

Question 3 by Bryan Renne. *What is the cleanest way to study the justifications of non-omniscient agents, taking into account the fact that the available justifications change over time?*

Response. First and foremost: check your methods for non-omniscience. For example, if a system/model does not distinguish formulas which specify the same proposition, i.e., which truth values at each world/node coincide, then such a system/model is not adequate for representing non-omniscient agents. Traditional methods for modal/epistemic logics, such as topological, neighborhood, and algebraic semantics do not distinguish syntactically different formulas within the same proposition and hence cannot alone model justification fairly. For example, arithmetical formulas $0 = 0$ and Fermat's Last Theorem FLT, being proven mathematical facts, are true in all possible worlds and hence denote the same proposition. In a model with propositional precision, statement t *is a proof of* $0 = 0$ should yield t *is a proof of FLT*, which essentially disqualifies the model. So, a proper set of methods to model justification should necessarily contain a syntactic component: Mkrtychev models [21], Fitting models [13], Awareness models [12] which are one-term fragments of Fitting models [26], all contain an evidence function that depends on syntactic formulas, not propositions. Modular models [5] evaluate justifications syntactically as sets of formulas, not propositions. How should we handle changes to the set of available justifications? Perhaps there is no a one-size-fits-all suggestion. In its simplest setting, the structure of proof terms may be regarded as carrying temporal information, i.e., that proofs become available during the process of derivation. In this respect, a temporal epistemic logic has some common features with Justification Logic, e.g., both are refined versions of the underlying modal logic, cf. [28]. In more sophisticated models, availability of justifications at each node could be made a function of time with specific model-dependent constraints.

BIBLIOGRAPHY

[1] Artemov, S.: Explicit provability and constructive semantics. *Bulletin of Symbolic Logic*, 7(1):1–36 (2001)

[2] Artemov, S.: The Logic of Justification. *The Review of Symbolic Logic*, 1(4):477–513 (2008)

[3] Artemov, S.: Symmetric Logic of Proofs. In: *Pillars of Computer Science. Essays Dedicated to Boris (Boaz) Trakhtenbrot on the Occasion of His 85th Birthday*. Lecture Notes in Computer Science, volume 4800, pp. 58–71 (2008)

[4] Artemov, S.: Tracking Evidence. In: *Fields of Logic and Computation: Essays Dedicated to Yuri Gurevich on the Occasion of His 70th Birthday*, Lecture Notes in Computer Science, vol. 6300, Springer-Verlag, Berlin (2010)

[5] Artemov, S.: The ontology of justifications in the logical setting. *Studia Logica* 100:17–30 (2012)

[6] Artemov, S., Bonelli, E.: The Intensional Lambda Calculus. In: *Logical Foundations of Computer Science 2007*, Lecture Notes in Computer Science, Springer, volume 4514, pp. 2–18 (2007)

[7] Artemov, S., Iemhoff, R.: The Basic Intuitionistic Logic of Proofs. *Journal of Symbolic Logic*, 72(2):439–445 (2007)

[8] Artemov, S., Yavorskaya, T.: First-Order Logic of Proofs. Technical Report TR-2011005, CUNY Ph.D. Program in Computer Science (2011)

[9] Bryukhov, Y.: Automatic Proof Search in Logic of Justified Common Knowledge. In *Proc. of the 4th Workshop "Methods for Modalities"* (M4M–05), pp. 187–201, Springer-Verlag (2005)

[10] Bucheli, S., Kuznets, R., Renne, B., Sack, J., Studer, T.: Justified belief change. In: LogKCA-10, *Proceedings of the Second ILCLI International Workshop on Logic and Philosophy of Knowledge, Communication and Action*, pages 135–155. University of the Basque Country Press (2010)

[11] Dashkov, E.: Arithmetical Completeness of the Intuitionistic Logic of Proofs. *Journal of Logic and Computation*, 21(4): 665–682 (2011)

[12] Fagin, R., Halpern, J., Moses, Y., Vardi, M.: *Reasoning about Knowledge*. MIT Press (1995)

[13] Fitting, M.: The logic of proofs, semantically. *Annals of Pure and Applied Logic*, 132(1):1–25 (2005)

[14] Fitting, M.: Prefixed tableaus and nested sequents. *Annals of Pure and Applied Logic*, 163(3):291–313 (2012)

[15] Fitting, M.: Possible World Semantics for First Order LP. *Annals of Pure and Applied Logic* (2013)

[16] Goris, E.: Explicit Proofs in Formal Provability Logic. In: *Logical Foundations of Computer Science 2007*, Lecture Notes in Computer Science, Springer, volume 4514, pp. 241–253 (2007)

[17] Krupski, N.: On the complexity of the reflected logic of proofs. *Theoretical Computer Science*, 357(1):136–142 (2006)

[18] Kurokawa, H.: Tableaux and hypersequents for justification logics. *Annals of Pure and Applied Logic*, 163(7):831–853 (2012)

[19] Kuznets, R.: On the complexity of explicit modal logics. In: *Computer Science Logic 2000*, Lecture Notes in Computer Science, volume 1862, pp. 371–383 (2000)

[20] Milnikel, B.: The Logic of Uncertain Justifications. In: *Logical Foundations of Computer Science 2013*, Lecture Notes in Computer Science, volume 7734, pp. 296–306 (2013)

[21] Mkrtychev, A.: Models for the Logic of Proofs. In: *Logical Foundations of Computer Science 1997*, Lecture Notes in Computer Science, volume 1234, pp. 266–275, Springer (1997)

[22] Pacuit, E.: A Note on Some Explicit Modal Logics. Technical report PP–2006–29, Institute for Logic, Language and Computation, University of Amsterdam (2006)

[23] Renne, B.: Tableaux for the logic of proofs. Technical Report TR-2004001, CUNY Ph.D. Program in Computer Science (2004)

[24] Renne, B.: Propositional games with explicit strategies. *Information and Computation*, 207(10):1015–1043 (2009)

[25] Rubtsova, N.: On Realization of S5-modality by Evidence Terms. *Journal of Logic and Computation*, 16(5):671–684 (1996)
[26] Sedlár, I: Justifications, Awareness and Epistemic Dynamics. In: *Logical Foundations of Computer Science 2013*, Lecture Notes in Computer Science, volume 7734, pp. 307–318 (2013)
[27] Shamkanov, D.: Strong Normalization and Confluence for Reflexive Combinatory Logic. In: *Logic, Language, Information and Computation 2011*, Lecture Notes in Computer Science, volume 6642, pp. 228–238 (2011)
[28] Wang, R.-J.: Knowledge, Time, and the Problem of Logical Omniscience. *Fundamenta Informaticae*, 106(2–4):321–338 (2011)
[29] Yavorskaya, T.: Interacting explicit evidence systems. *Theory of Computing Systems*, 43(2):272–293 (2008)

Epistemic Closure and Epistemic Logic I: Relevant Alternatives and Subjunctivism. A Summary

WESLEY H. HOLLIDAY

ABSTRACT. I summarize the formal framework and some of the main formal results in my "Epistemic Closure and Epistemic Logic I: Relevant Alternatives and Subjunctivism" (*Journal of Philosophical Logic*), adding a few methodological comments on formalization in philosophy.

Some topics in philosophy seem to cry out for formal analysis. The topic of *epistemic closure* in epistemology strikes me as a prime example. Its very name invokes the mathematical idea of a set being closed under a relation. In one of its simplest forms, the question is whether the set of *propositions known by an ideal logician* L must be closed under the relation of *logical consequence* from multiple premises: if $P_1, \ldots, P_n$ are in the set of propositions known by L, and P_{n+1} is a logical consequence of $\{P_1, \ldots, P_n\}$ (in which case L has at least deduced and come to *believe* P_{n+1}), must P_{n+1} be in the set of propositions known by L? Epistemic closure questions like this have lead to "one of the most significant disputes in epistemology over the last forty years" ([11], p.256), as different philosophical theories of knowledge give different answers. Some theories support full closure, while others only support weaker closure principles.

One of the goals of my "Epistemic Closure and Epistemic Logic I: Relevant Alternatives and Subjunctivism" ([7], hereafter 'EC&ELI') was to determine exactly the extent of epistemic closure according to a family of standard theories of knowledge. The theories at the center of the closure debate, versions of the *relevant alternatives* (RA) and *subjunctivist* theories of knowledge, are often presented in a kind of modal picture, distinguishing relevant epistemic alternatives or close counterfactual possibilities from the rest of logical space. Thus, it seemed that a natural way to study the properties of these theories was to formalize them with some kind of models for epistemic modal logic. This was the path followed in EC&ELI, which I will sketch in this summary.

In my view, an application of formal methods in philosophy gains value if:

(1) *it is faithful to the philosophical views being formalized*;
(2) *it can handle concrete examples discussed in the philosophical literature*;

(3) *it goes beyond particular examples to provide a systematic and general view of the topic;*

(4) *it leads to philosophically-relevant discoveries that would be difficult to make by non-formal methods alone.*

Points (1) and (2) address the worry that by formalizing we may "change the subject." Points (3) and (4) address the question, "What do we get out of this?" Of course, there are other features that contribute to the value of an application of formal methods, as well as features that detract from it. Here I will explain the ways in which I think EC&ELI satisfies the four desiderata above.

EC&ELI proposes formalizations of the RA theories of Lewis [13] and Heller [3, 4]; one way of developing the RA theory of Dretske [2], based on Heller; and basic versions of the "subjunctivist" *tracking* theory of Nozick [14] and *safety* theory of Sosa [15]. To formalize the theories of Lewis and (one way of developing) Dretske, I introduce *RA models*, defined below. Traditional semantics for epistemic logic represent knowledge by the elimination of possibilities, but without any explicit distinction between those *relevant* possibilities that an agent must rule out in order to have knowledge vs. those remote, far-fetched or otherwise irrelevant possibilities that can be properly ignored. RA models make this distinction explicit by adding *relevance orderings*.

DEFINITION 1. A *relevant alternatives* (RA) *model* is a tuple $\mathcal{M} = \langle W, \to, \preceq, V \rangle$ where $\langle W, \to, V \rangle$ is an ordinary relational model, with $\to$ at least reflexive, and $\preceq$ assigns to each $w \in W$ a *preorder* (reflexive and transitive) $\preceq_w$ on some $W_w \subseteq W$, for which w is a minimal element. Let $u \prec_w v$ iff $u \preceq_w v$ and not $v \preceq_w u$, and define $\mathrm{Min}_{\preceq_w}(S) = \{ v \in S \cap W_w \mid \text{there is no } u \in S \text{ such that } u \prec_w v \}$.

Intuitively, $w \to v$ means that v is an *uneliminated* epistemic alternative for the agent in w; $u \prec_w v$ means that u is a *more relevant alternative* at w than v is; and $\mathrm{Min}_{\preceq_w}(S)$ is the set of *most relevant alternatives* in S at w. For simplicity, assume that each $\preceq_w$ is well-founded: $S \cap W_w \neq \emptyset$ implies $\mathrm{Min}_{\preceq_w}(S) \neq \emptyset$. (The main results also hold with more general definitions without well-foundedness.)

Now we can define three semantics for the basic epistemic language with formulas $\varphi ::= p \mid \neg\varphi \mid (\varphi \wedge \varphi) \mid K\varphi$: C-semantics for Cartesian, D-semantics for (one way of developing) Dretske [2], and L-semantics for Lewis [13].

DEFINITION 2. Given a well-founded RA model $\mathcal{M} = \langle W, \to, \preceq, V \rangle$ with $w \in W$ and a formula φ, define $\mathcal{M}, w \vDash_x \varphi$ and $\llbracket \varphi \rrbracket_x^{\mathcal{M}} = \{ v \in W \mid \mathcal{M}, v \vDash_x \varphi \}$ as follows (with propositional cases as usual):

$$\mathcal{M}, w \vDash_c K\varphi \quad \text{iff} \quad \forall v \in \overline{\llbracket \varphi \rrbracket_c^{\mathcal{M}}} : w \not\to v;$$

$$\mathcal{M}, w \vDash_d K\varphi \quad \text{iff} \quad \forall v \in \mathrm{Min}_{\preceq_w}(\overline{\llbracket \varphi \rrbracket_d^{\mathcal{M}}}) : w \not\to v;$$

$$\mathcal{M}, w \vDash_l K\varphi \quad \text{iff} \quad \forall v \in \mathrm{Min}_{\preceq_w}(W) \cap \overline{\llbracket \varphi \rrbracket_l^{\mathcal{M}}} : w \not\to v,$$

where $\overline{P} = \{v \in W \mid v \notin P\}$. In C-semantics, for an agent to know φ in world w, all of the $\neg\varphi$-possibilities must be eliminated by the agent in w. In D-semantics, for any φ there is a set $\text{Min}_{\preceq_w}\left(\llbracket\varphi\rrbracket_d^{\mathcal{M}}\right)$ of *most relevant (at w)* $\neg\varphi$-possibilities that the agent must eliminate in order to know φ. Finally, in L-semantics, there is a set of relevant possibilities, $\text{Min}_{\preceq_w}(W)$, such that for any φ, in order to know φ the agent must eliminate the $\neg\varphi$-possibilities *within that set*.

Observe that for D-semantics, the whole relevance ordering $\preceq_w$ matters. As Heller puts it, there are "worlds surrounding the actual world ordered according to how realistic they are, so that those worlds that are more realistic are closer to the actual world than the less realistic ones" [3, 25] with "those that are too far away from the actual world being irrelevant" [4, 199], where how far is "too far" depends on the φ in question. By contrast, L-semantics does not use the ordering of more or less relevant worlds beyond the set $\text{Min}_{\preceq_w}(W)$, taken to represent Lewis's single set of relevant alternatives in the current context (for a formal treatment of the dynamics of *context change*, see [5]).

To formalize the RA theory of Heller [3, 4], the basic *tracking* theory of Nozick [14], and the basic *safety* theory of Sosa [15], I introduce CB models:

DEFINITION 3. A *counterfactual belief* (CB) *model* is a tuple $\mathcal{M} = \langle W, D, \leqslant, V\rangle$ where W, $\leqslant$, and V are defined like W, $\preceq$, and V in Definition 1, and D is a serial binary relation on W.

D is a doxastic accessibility relation, so that wDv means that everything the agent believes in w is true in v. To make the truth clauses for K clearer, I add a belief operator B for the D relation; but the main results will be stated for the epistemic language without B. The preorders $\leqslant_w$ can be thought of either as relevance orderings, as before, or as *similarity orderings* as in Lewis's [12] semantics for counterfactuals. With this setup, we can define three more semantics, formalizing three "subjunctivist" views of knowledge: H-semantics for Heller [4], N-semantics for Nozick [14], and S-semantics for Sosa [15]. There are a number of qualifications to be made here, but I will mention only three, referring to the full paper for the rest. First, in EC&ELI, I only treat the basic versions of the tracking and safety theory (for the versions with "methods" and "bases" of belief, see [6, §2.D]). Second, Heller and Sosa only propose necessary conditions for knowledge, so they may wish to take K here to represent sensitive belief or safe belief, rather than knowledge. Third, I interpret the sensitivity, adherence, and safety conditions along Lewisian counterfactual lines, which is common in the literature, but not the only conceivable interpretation.

DEFINITION 4. Given a well-founded CB model $\mathcal{M} = \langle W, D, \leqslant, V\rangle$ with $w \in W$ and formula φ, define $\mathcal{M}, w \vDash_x \varphi$ and $\llbracket\varphi\rrbracket_x^{\mathcal{M}} = \{v \in W \mid \mathcal{M}, v \vDash_x \varphi\}$ by the following truth condition:

$$\mathcal{M}, w \vDash_x B\varphi \quad \text{iff} \quad \forall v \in W \colon \text{if } wDv \text{ then } \mathcal{M}, v \vDash_x \varphi;$$

$$\mathcal{M}, w \vDash_h K\varphi \quad \text{iff} \quad \mathcal{M}, w \vDash_h B\varphi \text{ and}$$
$$\text{(sensitivity) } \forall v \in \text{Min}_{\leqslant_w}\left(\overline{\llbracket \varphi \rrbracket}_h^{\mathcal{M}}\right) \colon \mathcal{M}, v \nvDash_h B\varphi;$$

$$\mathcal{M}, w \vDash_n K\varphi \quad \text{iff} \quad \mathcal{M}, w \vDash_n B\varphi \text{ and}$$
$$\text{(sensitivity) } \forall v \in \text{Min}_{\leqslant_w}\left(\overline{\llbracket \varphi \rrbracket}_n^{\mathcal{M}}\right) \colon \mathcal{M}, v \nvDash_n B\varphi, \text{ and}$$
$$\text{(adherence) } \forall v \in \text{Min}_{\leqslant_w}\left(\llbracket \varphi \rrbracket_n^{\mathcal{M}}\right) \colon \mathcal{M}, v \vDash_n B\varphi;$$

$$\mathcal{M}, w \vDash_s K\varphi \quad \text{iff} \quad \mathcal{M}, w \vDash_s B\varphi \text{ and}$$
$$\text{(safety) } \forall v \in \text{Min}_{\leqslant_w}\left(\llbracket B\varphi \rrbracket_s^{\mathcal{M}}\right) \colon \mathcal{M}, v \vDash_s \varphi.$$

In H-semantics, an agent knows φ iff she believes φ in the actual world but not in any of the "closest" (most similar or relevant) $\neg\varphi$-worlds. This condition implies both that φ is true in the actual world and, according to Lewis's semantics for counterfactuals, that *if φ were false, the agent would not believe φ*; in this sense, her belief is "sensitive" to the truth of φ. In N-semantics, there is an additional requirement for the agent to know φ, namely that she believes φ in all of the closest φ-worlds; in this sense, her belief is "adherent" to the truth of φ.[1] Finally, in S-semantics, an agent knows φ iff she believes φ and in the closest worlds where she believes φ, φ is true; in this sense, her belief in φ is "safe." As explained in EC&ELI, there are structural similarities between D-semantics and H/N-semantics, and between L-semantics and S-semantics.

By drawing RA and CB models as in Figs. 1 - 2,[2] one can represent many concrete examples from the epistemological literature, ranging from examples meant to challenge epistemic closure to examples meant to challenge the necessity or sufficiency of the subjunctivist conditions for knowledge. Doing so helps make clear what exactly one must assume for the challenge to be successful. Indeed, it may be helpful if epistemologists were to draw such diagrams when presenting putative counterexamples.

C/D/L/H/N/S-semantics formalize views of knowledge that have been discussed extensively in epistemology. Having formalized these views, we can move beyond particular examples to answer general questions. Consider, for instance,

---

[1] *Tracking* is the conjunction of sensitivity and adherence. Nozick used 'variation' for what I call 'sensitivity' and used 'sensitivity' to cover both variation and adherence; but the narrower use of 'sensitivity' is now standard. For discussion of different options for formalizing adherence, see Observation 4.5 of EC&ELI.

[2] The solid arrow in the RA model represents the $\rightarrow$ relation, and the dashed arrows in the CB model represents the D relation; $w_1 \simeq_{w_1} w_2$ indicates that $w_1 \preceq_{w_1} w_2$ and $w_2 \preceq_{w_1} w_1$, and $w_1 \equiv_{w_1} w_2$ indicates that $w_1 \leqslant_{w_1} w_2$ and $w_2 \leqslant_{w_1} w_1$; and just as $\prec_{w_1}$ is the strict part of $\preceq_{w_1}$, $<_{w_1}$ is the strict part of $\leqslant_{w_1}$.

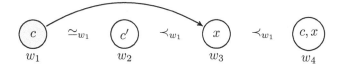

Figure 1. RA model for Example 1 in EC&ELI (partially drawn, reflexive loops omitted)

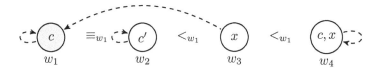

Figure 2. CB model for Example 1 in EC&ELI (partially drawn)

the question of epistemic closure. Given sequences of formulas $\varphi_1, \ldots, \varphi_n$ and $\psi_1, \ldots, \psi_m$, and a propositional conjunction φ_0, let us write

$$\chi_{n,m} := \varphi_0 \wedge K\varphi_1 \wedge \cdots \wedge K\varphi_n \rightarrow K\psi_1 \vee \cdots \vee K\psi_m.$$

Call such a $\chi_{n,m}$ a *closure principle*. It states that if the agent knows each of φ_1 through φ_n (and the world satisfies a non-epistemic φ_0), then the agent knows at least one of ψ_1 through ψ_m. Our question is: which closure principles are *valid*?

Theorem 5 below provides the answer. It is an example of a discovery that we can take back to epistemology, illuminating the closure properties of knowledge according to standard theories, which would be difficult to make without formal investigation. Surprisingly, despite the differences between the RA, tracking, and safety theories of knowledge as formalized by D/H/N/S-semantics, Theorem 5 provides a unifying perspective: *the valid epistemic closure principles are essentially the same for these different theories*. The only twist is with D-semantics over *total* RA models, i.e., RA models in which for all $w \in W$, $\preceq_w$ is a total preorder on W_w, meaning that all possibilities are comparable in relevance. For comparison, I also include C/L-semantics, which fully support closure in the sense that if $\varphi_1 \wedge \cdots \wedge \varphi_n \rightarrow \psi$ is valid, then so is $K\varphi_1 \wedge \cdots \wedge K\varphi_n \rightarrow K\psi$.

The statement of Theorem 5 refers to a "T-unpacked" closure principle, which I will not fully define here (see EC&ELI, §5.2.1). For the first reading of the theorem, think only of "flat" closure principles $\chi_{n,m}$ without nesting of the K operator, which are T-unpacked if φ_0 contains $\varphi_1 \wedge \cdots \wedge \varphi_n$ as a conjunct. A key fact is that any formula in the basic epistemic language is equivalent to a conjunction of T-unpacked closure principles (see §5.2.1), so if we can decide the validity of such principles, then we can decide the validity of any formula.

THEOREM 5 (Closure Theorem). *Let*

$$\chi_{n,m} := \varphi_0 \wedge K\varphi_1 \wedge \cdots \wedge K\varphi_n \rightarrow K\psi_1 \vee \cdots \vee K\psi_m$$

be a T-unpacked closure principle.

1. $\chi_{n,m}$ *is C/L-valid over* relevant alternatives *models iff*

    (a) $\varphi_0 \rightarrow \bot$ *is valid or*

    (b) *for some* $\psi \in \{\psi_1, \ldots, \psi_m\}$,

    $$\varphi_1 \wedge \cdots \wedge \varphi_n \rightarrow \psi \text{ is valid;}$$

2. $\chi_{n,m}$ *is D-valid over* total relevant alternatives *models iff (a) or*

    (c) *for some* $\Phi \subseteq \{\varphi_1, \ldots, \varphi_n\}$ *and nonempty* $\Psi \subseteq \{\psi_1, \ldots, \psi_m\}$,

    $$\bigwedge_{\varphi \in \Phi} \varphi \leftrightarrow \bigwedge_{\psi \in \Psi} \psi \text{ is valid;}$$

3. $\chi_{n,m}$ *is D-valid over* all relevant alternatives *models iff (a) or*

    (d) *for some* $\Phi \subseteq \{\varphi_1, \ldots, \varphi_n\}$ *and* $\psi \in \{\psi_1, \ldots, \psi_m\}$,

    $$\bigwedge_{\varphi \in \Phi} \varphi \leftrightarrow \psi \text{ is valid.}$$

4. $\chi_{n,m}$ *is H/N/S-valid over* counterfactual belief *models if (a) or (d); and a* flat $\chi_{n,m}$ *is H/N/S-valid over such models only if (a) or (d).*[3]

I will discuss some logical aspects of Theorem 5 before its epistemological upshot. Using Theorem 5, we can reduce the validity of any closure principle to the validity of finitely many formulas of lesser modal depth. With a "modal decomposition" result of this form, van Benthem [1] proves completeness of the weakest normal modal logic **K** with respect to relational models (and notes the analogous result for the weakest monotonic modal logic **EM** with respect to monotonic neighborhood models). Like van Benthem's proof, the proof of Theorem 5 in the 'only if' directions assumes that the formulas of lesser modal depth are not valid, from which we infer the existence of models to "glue together" into a countermodel for

---

[3] See §8.2 of EC&ELI on higher-order knowledge for the subjunctivist H/N/S-semantics.

$\chi_{n,m}$. But from there the proof of Theorem 5 requires new techniques. First, since we are dealing with models in which $K\varphi \to \varphi$ is valid, we must use the new idea of T-unpacking. Second, since we are dealing with a hybrid of relational and *ordering* semantics, we cannot simply glue all of the relevant models together at once, as in the basic relational case; instead, we must put them in the right order, which can be done inductively (see §5.2.2).

Using Theorem 5 we obtain the axiomatizations in Corollary 6 below, stated for the axioms and rules in Table 1. **E** is the weakest of the *classical* modal systems with PL, MP, and RE. $\mathbf{ES}_1 \ldots \mathbf{S}_n$ is the extension of **E** with every instance of schemas $S_1 \ldots S_n$. The logic **ECNTX**, which I dub *the logic of ranked relevant alternatives*, appears not to have been previously identified in the literature.

PL. all tautologies

$$\text{MP.} \frac{\varphi \to \psi \quad \varphi}{\psi}$$

T. $K\varphi \to \varphi$ N. $K\top$

$$\text{RE.} \frac{\varphi \leftrightarrow \psi}{K\varphi \leftrightarrow K\psi}$$

M. $K(\varphi \wedge \psi) \to K\varphi \wedge K\psi$

$$\text{RK.} \frac{\varphi_1 \wedge \cdots \wedge \varphi_n \to \psi}{K\varphi_1 \wedge \cdots \wedge K\varphi_n \to K\psi}$$

X. $K(\varphi \wedge \psi) \to K\varphi \vee K\psi$

$$\text{RAT.} \frac{\varphi_1 \wedge \cdots \wedge \varphi_n \leftrightarrow \psi_1 \wedge \cdots \wedge \psi_m}{K\varphi_1 \wedge \cdots \wedge K\varphi_n \to K\psi_1 \vee \cdots \vee K\psi_m}$$

C. $K\varphi \wedge K\psi \to K(\varphi \wedge \psi)$

$$\text{RA.} \frac{\varphi_1 \wedge \cdots \wedge \varphi_n \leftrightarrow \psi}{K\varphi_1 \wedge \cdots \wedge K\varphi_n \to K\psi}$$

Table 1. axiom schemas and rules ($n \geq 0$, $m \geq 1$)

COROLLARY 6 (Soundness and Completeness).

1. **EMCNT** *(equivalently,* **ET** *plus RK, a.k.a.* **KT***) is sound and complete for C/L-semantics over RA models.*

2. **ECNTX** *(equivalently,* **ET** *plus RAT) is sound and complete for D-semantics over* total *RA models.*

3. **ECNT** *(equivalently,* **ET** *plus RA) is sound and complete for D-semantics over RA models.*

4. **ECNT** *is sound (with respect to the full epistemic language) and complete (with respect to the flat fragment) for H/N/S-semantics over CB models (see §8.2 on higher-order knowledge).*[4]

---

[4]Corollary 6.4 gives the answer, for the flat fragment of the epistemic language (without nesting

The epistemological upshot of Theorem 5 is that according to the family of basic RA, tracking, and safety theories of knowledge formalized by D/H/N/S-semantics, an ideal logician is guaranteed to know a logical consequence ψ of what she knows if and only if ψ *is equivalent to a conjunction of propositions she already knows*! By applying this test, one can check that not only is the strong closure principle $K\varphi \wedge K(\varphi \to \psi) \to K\psi$ invalid—which is arguably desirable—but so are very weak and uncontroversial closure principles, such as $K(\varphi \wedge \psi) \to K\varphi$ and $K\varphi \to K(\varphi \vee \psi)$, and even the likes of $K(\varphi \wedge \psi) \to K(\varphi \vee \psi)$ and $K\varphi \wedge K\psi \to K(\varphi \vee \psi)$, while more controversial principles like $K\varphi \wedge K\psi \to K(\varphi \wedge \psi)$ turn out valid. There is much more to be said about Theorem 5 and the ramifications of closure failures for higher-order knowledge (see EC&ELI, §8). But suffice it to say that I take Theorem 5 to be a serious negative result for the RA and subjunctivist theories in question,[5] which motivates the search for new and improved pictures of knowledge in the sequel to EC&ELI.[6]

The last point brings me back to the starting list of virtues in formalization, in order to add one more. An application of formal methods in philosophy also gains value if:

(5) *it leads to the development of new theories that solve philosophical problems.*

To develop such a theory in the case of knowledge is the goal of further work ([10, 8, 6]), for which EC&ELI lays the foundation.

There are many claims in EC&ELI about specific theories and results, summarized in the introduction and conclusion of the paper, but here I will end with some general methodological claims that I think EC&ELI supports:

- While "formal epistemology" often deals with different issues than "traditional" or "mainstream" epistemology deals, EC&ELI shows that a formal approach—in this case, a model-theoretic logical approach—can contribute to our understanding of central issues in traditional epistemology. (For more on epistemic logic and epistemology, see [9].)

- For modal logicians, epistemology represents an area of sophisticated theorizing in which modal-logical tools can help to clarify and systematize parts of the philosophical landscape.

---

of K), to the question posed by van Benthem [1, 153] of what is the epistemic logic of Nozick's [14] notion of knowledge.

[5]There are other problems for C- and L-semantics having to do with skepticism and "vacuous knowledge." See EC&ELI, §3 and [10].

[6]However, we can also take Theorem 5 to be a neutral result about other desirable epistemic properties, viz., the properties of having ruled out the relevant alternatives to a proposition, of having a belief that tracks the truth of a proposition, or of having a safe belief in a proposition. For example, replace the K by a $\square$, reading $\square\varphi$ as "the agent has ruled out the most relevant alternatives to φ," and the newly identified logic **ECNTX**, *the logic of ranked relevant alternatives*, is of independent interest.

- The attempt to apply modal-logical tools to epistemology also benefits modal logic by broadening its scope, bringing interesting new structures and systems under its purview.

- Some of these epistemological investigations may be like NASA space missions, pursued for their intrinsic interest but—as a bonus—also leading to new "technology" usable in other areas. In EC&ELI, proving the main results involved an alternative *modal decomposition* approach to modal completeness theorems, developed in novel ways for new logics. In future work, I plan to show the wider applicability of this approach to other modal logics.

BIBLIOGRAPHY

[1] Johan van Benthem. *Modal Logic for Open Minds*. CSLI Publications, Stanford, CA, 2010.

[2] Fred Dretske. The Pragmatic Dimension of Knowledge. *Philosophical Studies*, 40(3):363–378, 1981.

[3] Mark Heller. Relevant Alternatives. *Philosophical Studies*, 55(1):23–40, 1989.

[4] Mark Heller. Relevant Alternatives and Closure. *Australasian Journal of Philosophy*, 77(2):196–208, 1999.

[5] Wesley H. Holliday. Epistemic Logic, Relevant Alternatives, and the Dynamics of Context. In Daniel Lassiter and Marija Slavkovik, editors, *New Directions in Logic, Language and Computation*, volume 7415 of *Lecture Notes in Computer Science*, pages 109–129. 2012.

[6] Wesley H. Holliday. *Knowing What Follows: Epistemic Closure and Epistemic Logic*. PhD thesis, Stanford University, 2012. Revised Version, ILLC Dissertation Series.

[7] Wesley H. Holliday. Epistemic Closure and Epistemic Logic I: Relevant Alternatives and Subjunctivism. *Journal of Philosophical Logic*, 2013. Forthcoming.

[8] Wesley H. Holliday. Epistemic Closure and Epistemic Logic II: A New Framework for Fallibilism. 2013. Manuscript.

[9] Wesley H. Holliday. Epistemic Logic and Epistemology. In Sven Ove Hansson and Vincent F. Hendricks, editors, *Handbook of Formal Philosophy*. Springer, Dordrecht, 2013. Forthcoming.

[10] Wesley H. Holliday. Fallibilism and Multiple Paths to Knowledge. *Oxford Studies in Epistemology*, 5, 2013. Forthcoming.

[11] Jonathan L. Kvanvig. Closure Principles. *Philosophy Compass*, 1(3):256–267, 2006.

[12] David Lewis. *Counterfactuals*. Basil Blackwell, Oxford, 1973.

[13] David Lewis. Elusive Knowledge. *Australasian Journal of Philosophy*, 74(4):549–567, 1996.

[14] Robert Nozick. *Philosophical Explanations*. Harvard University Press, Cambridge, MA, 1981.

[15] Ernest Sosa. How to Defeat Opposition to Moore. *Noûs*, 33(13):141–153, 1999.

Comments on Holliday

PAUL ÉGRÉ

1. Overview. In "Epistemic Closure and Epistemic Logic I", the first part of
tour-de-force inquiry into the semantics and logic of knowledge attributions, Holl
day formalizes and compares several influential accounts of knowledge and its truth
conditions. Holliday's central result, the Closure Theorem, offers a rigorous and pre
cise syntactic characterization of the demands made by different theories on epistemi
closure. In so doing, Holliday achieves an unprecedented level of rigor and mathema
ical precision in spelling out the predictions of various theories relative to the proble
of epistemic closure. Because the paper raises so many interesting issues, my com
ments will hardly do justice to the richness of its content and results. My goal he
will be more limited, and will essentially concern two aspects of the problem of clo
sure in relation to Holliday's account, namely: i) the question of the relative weigh
of subject factors and attributor factors in shaping the truth conditions of knowledg
attributions, in particular for semantics based on relevant alternatives, and ii) the co
nection between the problem of epistemic closure and the problem of unattended po
sibilities in epistemic logic, in particular with regard to the trilemma that results fro
Holliday's Closure Theorem.

2. Holliday's trilemma. Before we go into these issues, a reminder of the philosop
ical underpinnings of the problem of epistemic closure will be useful. The problem c
epistemic closure is fundamentally a version of the perennial problem of scepticisr
I know that I have hands. My having hands implies that I am not a brain in a va
artfully deceived by an evil scientist or demon into thinking that I have hands. But
do not know, the sceptic points out, that I am not such a brain in a vat. If knowledge
closed under implication, we have a problem. Relying on the force of his observatio
the sceptic concludes, by modus tollens, that I do not know that I have hands aft
all. Holding on to the strength of the main premise, the dogmatist counters, by mod
ponens, that I do know that I am not a brain in a vat after all. For Dretske or Nozic
a third way is to maintain with the dogmatist that I know I have hands, and to der
with the sceptic that I know I am not a brain in a vat, but to resist that knowledge
closed under implication. The rejection of epistemic closure is more costly than w
might have thought at first, and the value of Holliday's contribution is to give an exa
assessment of this cost. One way in which Holliday's result can be interpreted is a
posing a dilemma, or rather, a trilemma, seeing that the second horn of the former su
divides: either we deny epistemic closure, like Dretske or Nozick, but then we end u
denying further closure properties of knowledge that seem intuitively desirable (suc
as the principle $K(\phi \wedge \psi) \rightarrow K\phi$); or we maintain epistemic closure, but then we mu
confront a subordinate dilemma: either we preserve closure by sticking to invariantis
about knowledge (Holliday's C-semantics in the paper), thereby incurring the risk c
scepticism (or dually, of dogmatism); or we adopt a contextualist semantics, of th

ind suggested by Lewis (Holliday's L-semantics), but then we run the risk of making nowledge too easily attainable (what Holliday calls the problem of vacuous knowledge, namely knowledge acquired cheaply by suitably restricting the space of relevant lternatives). Faced with the trilemma, Holliday's inclination appears to be to favor ontextualism about knowledge. I agree that this is a reasonable choice, but before we an assess whether it is the best choice, I think we need to know how compelling the ilemma is, and how much the trilemma depends on the particular framework chosen y Holliday to represent the various theories under discussion.

. Subject vs. attributor The first aspect I would like to question in Holliday's pproach concerns the relative weight of subject factors and attributor factors in the uth conditions of knowledge ascriptions. A very important observation about knowldge attributions is that they involve both a subject about whom we make ascriptions, nd an attributor who ascribes knowledge (see Holliday's emphasis on the distinction g. 9). Knowledge is a function of at least those two perspectives. Prima facie, both erspectives appear to be reflected in Holliday's relevant alternative models, with the istinction between the accessibility relation (whether two worlds are equivalent or ot) and the preorder on worlds (whether one world is more relevant than another or ot). However, Holliday warns us that: "These models represent the epistemic state f an agent from a third-person perspective. We should not assume that anything in ie model is something that the agent has in mind". This is fine, but right away he dds that "Just as the model is not something that the agent has in mind, it is not omething that particular speakers attributing knowledge to the agent have in mind ither.". I agree with Holliday that relevance in particular can be both a subject factor nd an attributor factor, but I end up being confused on which feature of the model presents which perspective. As I see it, relevance is better handled *primarily* as an tributor's factor, simply to reflect the fact that the attributor (who may be an ideal feree, or just some situated agent in a particular community) is the ultimate judge fixing what counts as pragmatically relevant in ascribing knowledge. An abstract 'ay of articulating a relevant alternative semantics would thus be by distinguishing etween the subject's discriminative capacities relative to the whole set of worlds, and ie attributor's assessment of which alternatives are to be considered. In principle, e should probably duplicate perspectives completely, and have, for each subject and tributor, both a respective discrimination set and a respective relevancy set. But one 'ay to keep matters more simple on a first approximation is to put relevance wholly n the side of the attributor, and discrimination on the side of the subject. In particular, ie most basic relevant alternative semantics for knowledge ascriptions can be stated mply by distinguishing, relative to a world w in a set W, the set $R_a(w)$ of the pos- bilities that the attributor a thinks the subject s should entertain, and the set $D_s(w)$ f possibilities that the subject s cannot discriminate from w, irrespective of whether actually entertains those or not:

) $\qquad w \models K_s\phi$ iff for every $w' \in R_a(w)$: if wD_sw', then $w' \models \phi$.

his semantics preserves the principle of epistemic closure, just like the Lewisian L- emantics in Holliday's account, and it can account for the same contrast cases, like the medical diagnosis example discussed by Holliday.[1] Of course, the L-semantics can be seen as one way of specifying this abstract semantics, letting $R_a(w)$ correspond to $Min_{\leq_w}(W)$, and D_s be the arrow relation in Holliday's model. There are two differences however between those ways of articulating a Lewisian semantics: one is the fact that we have attached relevance to the attributor's perspective, and discrimination to the subject's perspective; the second is that Holliday's treatment of relevance gives the notion more structure from the start. In Holliday's framework, alternatives can be *more or less* relevant relative to a world, whereas in the simple-minded semantics presented above, an alternative is simply relevant or not.

4. Attention This brings me to the second aspect I would like to question. The sceptical argument, remember, is that: from $K_s\phi$ (I know I have hands), and $K_s(\phi \to \neg\psi)$ (I know that if I have hands, then I am not a brain in a vat), we must infer $K_s\neg\psi$ (I know I am not a brain in a vat), even for a ψ that is a far-fetched possibility in comparison to ϕ. As talk of far-fetched possibilities suggests, it is quite natural to introduce an ordering on possibilities to capture the difference between ϕ and ψ, the way Holliday does. If we use the semantics laid out above, we have no way of capturing the difference. Nevertheless, I think we can do something similar by distinguishing a second subject factor beside discrimination, namely *attention* to possibilities. The idea would be that, when I claim to know that I have hands, I am failing to attend to the possibility that I am a brain in a vat. Because of that, I may also fail to attend to the implication that, if I have hand, then I am not a brain in a vat. In other words, instead of rejecting epistemic closure, the proposed solution to the sceptical paradox would consist in accepting the validity of closure, but in denying systematic knowledge of the relevant conditional, for cases in which the conditional mentions a possibility that is unattended and properly ignored (see [4]). In [1], truth conditions for knowledge ascriptions are proposed along those lines, building on de Jager's work on unawareness (see [2]). We define an attention and relevance model (for a single subject s and attributor a) as a structure $(W, D_s, R_a, E_s, A_s, V)$, where W is the set of possibilities, D_s is the discrimination relation over W, R_a the relevance relation, E_s a function that to each world associates the possibilities the agents entertains, and A_s a function that to each world associates the sentences that the agent is aware of. E_s and A_s together serve to define the notion of attention to a possibility. Knowledge ascriptions are evaluated as follows, letting $\mathcal{L}(X)$ be the fragment of the whole language consisting of only atoms in X:

(2) $w \models K_s\phi$ iff $\phi \in \mathcal{L}(A_s(w))$ and $\forall w' \in E_s(w) \cup R_a(w)$: if $w' \in D_s(w)$, then $w' \models \phi$

This says that ϕ is known provided the agent is aware of ϕ and ϕ holds in every alternative that is entertained and in every alternative that is considered relevant by the attributor. Now, the important point is that $K_s\phi$ and $K_s(\phi \to \psi)$ will entail $K_s\psi$ under suitable provisos on $\mathcal{L}(A_s)$. This is plausible enough, if we consider that, once I am attending to the proposition that, if I have hands, then I am not a brain in a vat, then I am also attending to the proposition that I am not a brain in a vat. However, it will

---

[1]Let $R_a(w)$, the alternatives the professor considers relevant: in the first context, only include alternative w_c: both A and B would know c; in a second context, let it include also $w_{c'}$, then only A knows c; in a third, if R_a also includes alternative w_x, A does not know c any more.

ossible to have a model in which: $K_s\phi$ is true, but in which both $K_s(\phi \rightarrow \neg\psi)$ and $_s(\neg\psi)$ are false, if indeed ψ is a possibility that is "properly ignored" by the agent, amely both irrelevant and not attended to. On that view of the sceptical scenario, oistemic closure need not be the culprit (the principle can be valid, but need not be ound). Importantly, this solution hinges on the fact that an implication $\phi \rightarrow \psi$ can be niversally true over W without necessarily being attended to, and so, without being atomatically known (in contrast to what a plain L-semantics or a basic semantics like e one in (1) would predict). This means that I can fail to know that I am not a brain a vat, not necessarily for lack of evidence, but for failing to attend to the possibility.

To conclude on this, I wonder if the notion of unattended possibility, in Holliday's levant alternative models, could be captured in terms of a possibility that is so remote the ordering that it is tantamount to being unattended. Or whether inattentiveness ould have to be captured along a distinct dimension.

IBLIOGRAPHY

Aloni M., Egré P. and de Jager, T. (2009). Knowing whether A or B. *Synthese*. DOI 10.1007/s11229-009-9646-1.

De Jager T. (2009). *'Now that you mention it, I wonder...': Awareness, Attention, Assumption*. PhD dissertation, ILLC, Amsterdam.

Holliday W. (2013). Epistemic Closure and Epistemic Logic I: Relevant Alternatives and Subjunctivism. *Journal of Philosophical Logic*.

Lewis D. (1996). Elusive Knowledge. *Australasian Journal of Philosophy* 74(4):549–567.

Comments on Holliday

ZHAOQING XU

Wesley Holliday did an admirable job in "Epistemic Closure and Epistemic Logic I: relevant alternatives and subjunctivism" (I) and its sequel "Epistemic Closure and Epistemic Logic II: A New Framework for Fallibilism" (II). Being a philosophical logician myself, I would not invoke any criticism to the general methodology of formal epistemology, but as far as I see, the full value of formalization would not be appreciated properly without putting I and II together, especially for the value that a formal method "leads to the development of new theories that solve previous philosophical problems". However, for the current purpose and the limit of space, I will confine myself to "Epistemic Closure and Epistemic Logic I", and make only three middle sized comments.

My first comment concerns the concept of belief. Belief plays no role in RA model and being used to eliminate worlds in CB model. My curiosity is what would we get if we add belief in RA model or use the belief relation to constrain the set of relevant alternatives. For the former, the motive reason is that most epistemologists take belief to be a necessary condition for knowledge, although David Lewis has explicitly rejected "knowledge entails belief", I don't remember Fred Dretske did the same rejection. For the latter, Lewis expressed one of his relevance rules in terms of belief in "Elusive knowledge", namely, "the Rule of Belief", which says:

"Next, there is the Rule of Belief. A possibility that the subject believes to obtain is not properly ignored, whether or not he is right to so believe. Neither is one that he ought to believe to obtain - one that evidence and arguments justify him in believing - whether or not he does so believe."([6], p.555)

I agree with Holliday that "if the only point were to add believing φ as a necessary condition for knowing φ, this would not change any of our results about RA knowledge". But the question is whether that is really the only point. For me, it is not. In my opinion, the vacuous problem of Lewisian knowledge is closely related to the way of determining whether a possibility is relevant or not. The worry that Lewis's skeptical knowledge is too easy might come from a misunderstanding. One might overlook the epistemic effort in determining relevant alternatives and take it as automatically obtained. It might be easy to see whether one does believe something to obtain, but it might not be so easy to see whether one ought to believe something to obtain.

Holliday has discussed a variant of Lewis's Rule of Belief in note ∗ on page 10, where he formulates it as "any world v doxastically accessible for the agent in w must be relevant and uneliminated for the agent in w". But for me, this is already the result of combining rule of belief to the idea of using belief to eliminate possible worlds, if we only use belief to constrain relevant alternatives, such a problem would not be raised. Since a doxastically accessible world might still be eliminated, as long as we are not using belief itself as a way of eliminating worlds. The lesson might be understood as CB model is not suitable for Lewisian knowledge, but include the idea of knowledge entails of belief might still be of its own interest. For example, from Lewis's rule of belief, we could conclude that, if φ is true in all relevant alternatives ($P\varphi$), then the subject would believe φ or ought to believe φ ($B\varphi$), i.e., $P\varphi \to B\varphi$ must be valid. And if we denote φ is true in all uneliminated world as $E\varphi$, we would also obtain the validity of $P\varphi \vee E\varphi \to K\varphi$. Details can be found in ([7]). The other direction is not valid in general, but it shows some similarity to Fitting's queer model ([2]) to classical modal logic, which in turn has close connection to the denial of closure and leads to my second comment.

My second comment is on the nature of closure. Mathematically understood, closure is a property of a set under a relation. For epistemic closure, the set is a set of known propositions, and the relation is known implication. From this general perspective, there are two sorts of parameters involved in epistemic closure. The first is the definition of knowledge, and the other is the definition of implication.

Knowledge is commonly interpreted as a universal modal operator, what I mean is the stand translation of $K\varphi$ in first order language is a formula bounded by a universal quantifier, which validate full closure. On the contrast, if we understand knowledge as an existential modality, much less of closure would be valid. My conjecture is that any attempts to drop any closure would make use of the existential quantifier, but this is too sketchy to prove. Holliday did an excellent to characterize the cost of denying closure in various theories of knowledge interpreting knowledge in terms of two universal quantifiers and two existential quantifiers, and show that the order of these quantifiers plays a significant role and give a full description. My concern is whether such observations can be generalized to a definition of knowledge in terms of any number of quantifiers. I have discussed the case of one quantifier. Another way of understanding is to construe the nature of knowledge not in the way of quantifiers, but in the way of conjunction or disjunction, say universal as infinite conjunction, and existential as infinite disjunction. If we have such a spectrum result of the relation between knowledge and closure, epistemologists might be benefited to just see the consequences of their theories just from his starting point.

Without the above hoped spectrum at hand, a Dretskean epistemologist may still find his proper starting point, rather than starting with CB models, instead he might take neighborhood models to be the starting points. Of course, in order to make use of the advantage of more basic semantics, he hand better not define knowledge as universal quantifying of neighborhood (like the one in [1]). Another promising starting point is Fitting's queer models, which validates Dretske's semi-penetrating knowledge (denying validity of K axiom and meanwhile keep epistemic closure under conjunction-elimination and disjunction-introduction) just as neighborhood semantics.

My third comment is closely related the the second one, which concerns the concept of implication. Epistemologists seldom give an exact clarification of the implication involved in epistemic closure, while Holliday rejects epistemic closure even when we replace known implication with knowledge strict implication or known globally implication. What he hasn't discussed is known counterfactual implication. As noted by Kripke, "whatever else we believe about counterfactual logic, surely counterfactuals imply the corresponding material conditionals"([5], p.164). And Timothy Williamson also take counterfactuals as something between material conditionals and strict implication. The result might be more interesting if we take the order for counterfactuals to be the same as relevance order. That is, whether $K\varphi \wedge K(\varphi\square \to \psi) \to K\psi$ is valid in all RA models or CB models? Also, build all epistemic logics on the base of intuitionistic propositional logic might also be an interesting further direction.

BIBLIOGRAPHY

[1] I. Cornelisse. Context dependence of epistemic operators in dynamic evidence logic. Master's thesis, 2011, University of Amsterdam, ILLC.
[2] M. Fitting. *Proof Methods for Modal and Intuitionistic Logics*. Springer.1983.
[3] W. H. Holliday. Epistemic Closure and Epistemic Logic I: Relevant Alternatives and Subjunctivism. *Journal of Philosophical Logic*, 2013. Forthcoming.
[4] W. H. Holliday. Epistemic Closure and Epistemic Logic II: A New Framework for Fallibilism. 2013. Manuscript.
[5] S. A. Kripke. Nozick on knowledge. In *Collected Papers* Vol I. Oxford University Press:2011.
[6] D. Lewis. Elusive knowledge. *Australasian Journal of Philosophy*,1996,74(4):549-567.
[7] Z. Xu. *Knowledge, Evidence and Relevant Alternatives: A Logical Dynamical Approach*. PhD thesis, 2012, Peking University.

Response to Egré and Xu

Wesley H. Holliday

ABSTRACT. In this note, I respond to comments by Paul Egré and Xu Zhao-qing on my "Epistemic Closure and Epistemic Logic I: Relevant Alternatives and Subjunctivism" (*Journal of Philosophical Logic*).

I want to begin by thanking Paul Egré and Xu Zhaoqing for their thoughtful and thought-provoking remarks on my paper, "Epistemic Closure and Epistemic Logic I: Relevant Alternatives and Subjunctivism" ([10], hereafter 'EC&ELI'). It is a privilege to have one's work receive close readings and constructive comments from international colleagues. In the space available here, I will try to address some of the main points raised by Egré and Xu. Throughout I will presuppose familiarity with the summary of EC&ELI in this volume.

Response to Paul Egré

As Egré observes, EC&ELI sets up a trilemma. Each of the theories of knowledge considered in the paper encounters one of the three horns: the Problem of Skepticism (C-semantics), the Problem of Vacuous Knowledge (L/S-semantics), and the Problem of Containment (D/H/N/S-semantics). In EC&ELI, I do not offer a solution to this trilemma, but rather try to systematically investigate the third horn, the Problem of Containment. Given how little I let on in EC&ELI about my own response to the trilemma, it is not unreasonable that Egré infers that "Faced with the trilemma, Holliday's inclination appears to be to favor contextualism" about knowledge attributions (see [8] for more on contextualism). However, for the record I should say that I do not think that contextualism about knowledge attributions is the key to solving the trilemma. Instead, I think the key to solving the trilemma is to replace what I call the Standard Alternatives Picture, in which all of the theories of knowledge in EC&ELI fit as special cases, with a new and improved Multipath Picture of Knowledge [12, 11, 9]. The Multipath Picture of Knowledge is compatible with contextualism; but it is this new picture of *knowledge*, not contextualism about *knowledge attributions*, that in my view solves the trilemma. All of this is discussed in [12, 11], which developed out of [9].

With that clarification made, let me turn to Egré's two main points:

Subject vs. Attributor.

Egré helpfully emphasizes the need to be clear about which aspects of the relevant alternatives (RA) models in EC&ELI depend on the *knowing subject* and which depend on the *person attributing knowledge* to the subject. Egré sketches what he considers a more basic and abstract relevant alternative semantics, which makes explicit a contribution of the subject and a contribution of the attributor. Formally, what Egré labels as (1) is equivalent to the L-semantics of EC&ELI.[1] However, there is a conceptual difference.

In RA models $\mathcal{M} = \langle W, \rightarrow, \preceq, V \rangle$ for L-semantics, the $\rightarrow$ relation represents what possibilities the subject (not the attributor) can discriminate between, so there is no conceptual difference here between L-semantics and the semantics Egré sketches. However, there is a conceptual difference elsewhere: when I presented RA models for L-semantics in EC&ELI, I said that the set $\text{Min}_{\preceq_w}(W)$ of *relevant* worlds at w can depend not only on "attributor factors" but also on "subject factors" (see [5, 30f]), whereas Egré says that (for a first approximation) the set of relevant worlds at w, written as $R(w)$ instead of $\text{Min}_{\preceq_w}(W)$,[2] can only depend on *attributor* factors. (In this sense, Egré's interpretation is less general.)

According to Egré, "relevance is better handled *primarily* as an attributor's factor, simply to reflect the fact that the attributor ... is the ultimate judge in fixing what counts as pragmatically relevant in ascribing knowledge," so "a first approximation is to put relevance wholly on the side of the attributor, and discrimination on the side of the subject." One can certainly interpret the relevance relations in RA models as depending only on the attributor, but doing so would be highly controversial. According to Dretske [6, 7], for example, what possibilities are relevant does not depend at all on the conversational context of the attributor, but rather on objective features of the subject's environment, of which both the subject and attributor might be unaware. In one of his examples, Dretske [6] considers whether a birdwatcher in Wisconsin must eliminate the possibility that the bird he sees on a lake is a Siberian Grebe in order to know that it is a Gadwall. (We are to suppose that a Gadwall cannot be distinguished from a Siberian Grebe unless one sees the underbelly of the bird in flight.) Roughly, Dretske suggests that if, as a matter of fact, there are many Siberian Grebes in the birdwatcher's area, then the bird-

---

[1]This assumes that Egré requires $w \in R_a(w)$, as required for $K\varphi \rightarrow \varphi$ to be valid. However, since Egré takes $R_a(w)$ to be the set of "possibilities that the attributor a thinks the subject s should entertain," it is unclear whether this understanding guarantees that $w \in R_a(w)$ always holds. After all, an attributor in w might not know that w is the actual world and therefore might not think that w is one of the possibilities that s should entertain. Perhaps Egré's idea is that every attributor thinks that the actual world (*de dicto*), whatever it is, should be entertained?

[2]Note that for L-semantics, the rest of the relevance relation $\preceq_w$ beyond $\text{Min}_{\preceq_w}(W)$ does not matter in a fixed context, so there is no substantive difference between $R(w)$ and $\text{Min}_{\preceq_w}(W)$. Yet when it comes to the dynamics of *context change*, the rest of the relevance relation $\preceq_w$ may indeed matter, as discussed in [8].

watcher needs to eliminate the possibility that what he sees is a Siberian Grebe in order for an attributor to truly say of him that he knows the bird is a Gadwall—in that sense, it is a relevant possibility—*even if the attributor is unaware of Siberian Grebes*. Most contextualists also grant that possibilities may be relevant in virtue of these kinds of "subject factors" of which attributors may be unaware. As I say in EC&ELI, "possibilities may be relevant and hence should be included in our model, even if the attributors are not considering them (see [5, 33])."

In any case, I think Egré makes an excellent point about keeping explicit track of the subjects and attributors, who remained nameless in EC&ELI. Since multiple attributors may share the same context, let us keep track of subjects (agents) and contexts (possibly associated with groups of agents). Given a set $\mathcal{S}$ of agent symbols and a set $\mathfrak{C}$ of context symbols, define an extended language:

$$\varphi ::= p \mid \neg\varphi \mid (\varphi \wedge \varphi) \mid K_s^{\mathfrak{c}}\varphi$$

with $s \in \mathcal{S}$ and $\mathfrak{c} \in \mathfrak{C}$, reading '$K_s^{\mathfrak{c}}\varphi$' as "agent s counts as knowing φ relative to context $\mathfrak{c}$." The generalization of RA models for the extended language is just what one would expect: a *multi-agent and multi-context RA model* is a tuple

$$\mathcal{M} = \langle W, \{\rightarrow_s\}_{s \in \mathcal{S}}, \{\preceq^{s,\mathfrak{c}}\}_{s \in \mathcal{S}, \mathfrak{c} \in \mathfrak{C}}, V \rangle,$$

where for each $s \in \mathcal{S}$ and $\mathfrak{c} \in \mathfrak{C}$, $\langle W, \rightarrow_s, \preceq^{s,\mathfrak{c}}, V \rangle$ is an RA model as in EC&ELI. The generalization of C/D/L-semantics from EC&ELI to multi-RA models is also straightforward. For a well-founded[3] multi-RA model $\mathcal{M}$, define $\mathcal{M}, w \vDash_x \varphi$ and $[\![\varphi]\!]_x^{\mathcal{M}} = \{v \in W \mid \mathcal{M}, v \vDash_x \varphi\}$ as follows (for notation, see EC&ELI):

$$\mathcal{M}, w \vDash_c K_s^{\mathfrak{c}}\varphi \quad \text{iff} \quad \forall v \in \overline{[\![\varphi]\!]}_c^{\mathcal{M}} : w \not\rightarrow_s v;$$
$$\mathcal{M}, w \vDash_d K_s^{\mathfrak{c}}\varphi \quad \text{iff} \quad \forall v \in \mathrm{Min}_{\preceq_w^{s,\mathfrak{c}}}\left(\overline{[\![\varphi]\!]}_d^{\mathcal{M}}\right) : w \not\rightarrow_s v;$$
$$\mathcal{M}, w \vDash_l K_s^{\mathfrak{c}}\varphi \quad \text{iff} \quad \forall v \in \mathrm{Min}_{\preceq_w^{s,\mathfrak{c}}}(W) \cap \overline{[\![\varphi]\!]}_l^{\mathcal{M}} : w \not\rightarrow_s v.$$

One can generalize CB models and H/N/S-semantics in the analogous way.

Generalizing to multiple agents and multiple contexts raises interesting new issues, supporting Egré's call to make these aspects of the model explicit:

With multiple agents, we can consider multi-agent closure principles. For example, one can check that the principle $K_s K_u \varphi \rightarrow K_s \varphi$ is not valid according to D/H/N/S-semantics; thus, e.g., according to Nozick's tracking theory, an agent s can know that an agent u knows φ without s knowing φ herself.

With multiple contexts, we can consider cross-context knowledge, as in the so-called Problem of Factivity for contextualism [15, 16, 3, 2], which I cannot resist repeating here:

---

[3] I.e., in which each preorder $\preceq_w^{s,\mathfrak{c}}$ is well-founded as explained in EC&ELI.

According to standard contextualism, as a result of a skeptic's argumentation a contextualist u might find herself in a context $\mathfrak{s}$ relative to which she does not count as knowing ordinary propositions. In such a case, the contextualist might like to respond to the skeptic by claiming that she still counts as knowing those ordinary propositions relative to an ordinary context $\mathfrak{o}$. However, by making such a claim in context $\mathfrak{s}$, she would be claiming something that is *impossible for her to know* in $\mathfrak{s}$, as a simple argument shows. Our initial assumption was $\neg K_u^{\mathfrak{s}} p$. Suppose for reductio that $K_u^{\mathfrak{s}} K_u^{\mathfrak{o}} p$. Relative to any context, knowledge is factive, reflected by the validity of $K_u^{\mathfrak{o}} p \rightarrow p$ according to all of the semantics; and we can assume that the contextualist knows the factivity principle relative to any context, reflected by the validity of $K_u^{\mathfrak{s}}(K_u^{\mathfrak{o}} p \rightarrow p)$ according to all of the semantics. Finally, many contextualists claim that knowledge is closed under known implication relative to any fixed context, reflected by the validity in *L-semantics* of $(K_u^{\mathfrak{c}} \varphi \wedge K_u^{\mathfrak{c}}(\varphi \rightarrow \psi)) \rightarrow K_u^{\mathfrak{c}} \varphi$, an instance of which is

$$(K_u^{\mathfrak{s}} K_u^{\mathfrak{o}} p \wedge K_u^{\mathfrak{s}}(K_u^{\mathfrak{o}} p \rightarrow p)) \rightarrow K_u^{\mathfrak{s}} p.$$

Putting it all together, we derive $K_u^{\mathfrak{s}} p$, reflected by the validity of $K_u^{\mathfrak{s}} K_u^{\mathfrak{o}} \varphi \rightarrow K_u^{\mathfrak{s}} \varphi$ according to L-semantics. But $K_u^{\mathfrak{s}} p$ contradicts our initial assumption of $\neg K_u^{\mathfrak{s}} p$; hence $K_u^{\mathfrak{s}} K_u^{\mathfrak{o}} p$ is impossible. Moreover, it is plausible that the contextualist can follow this derivation and come to know that $K_u^{\mathfrak{s}} K_u^{\mathfrak{o}} p$ is impossible. Hence a weak *norm of assertion*, namely that you should not assert something (in a context) that you know is impossible for you to know (in that context), prohibits the contextualist from replying to the skeptic as she might have liked to in $\mathfrak{s}$. Without going further into this Problem of Factivity and related issues, the argument above suffices to show the interest of the multi-context generalization.

Attention.

Egré's second main point concerns the notion of attention to possibilities (and awareness of sentences). Rather than trying to summarizing Egré's nuanced discussion of attention, I will go directly to my three responses.

First, insofar as an attributor or subject paying attention to a possibility would affect the relevance of that possibility, it would affect the relevance relation $\preceq_w$ in RA models. For example, according to Lewis's [13] *Rule of Attention*, any possibility that the attributors are attending to in a context is relevant in that context, i.e., in $\mathrm{Min}_{\preceq_w}(W)$.[4] On this view, if the attributors attend to additional possibilities, then $\mathrm{Min}_{\preceq_w}(W)$ will enlarge. (For formal modeling of the dynamics of context change, see [8].) One might also hold that a subject s attending to a possibility

---

[4] Although for L-semantics all that matters is whether a possibility is in or out of $\mathrm{Min}_{\preceq_w}(W)$, in D-semantics the rest of the relation $\preceq_w$ matters, and a D-semantic contextualist could take the degree of attention that attributors are paying to a possibility to affect its ranking in the relevance ordering $\preceq_w$.

tends to make that possibility relevant according to $\preceq^s_w$ (whether or not the attributors are attending to it). Indeed, a whole range of views about the relation between attention and relevance are possible and compatible with modeling relevance as in RA models.[5]

Second, on Egré's proposed solution to the skeptical problem, I do not think it is sufficient to solve the problem to say that agents are not paying attention to—and thus don't know—conditionals of the form $p \rightarrow \neg SK$ where p is an ordinary proposition and SK a skeptical hypothesis. It is true that if an agent does not know $p \rightarrow \neg SK$, this blocks *one* skeptical argument for $\neg Kp$, namely the one using $\neg K \neg SK$, $(Kp \wedge K(p \rightarrow \neg SK)) \rightarrow K \neg SK$, and modus tollens. But the essence of the skeptical challenge is this: if an agent cannot eliminate the kind of possibilities described by skeptical hypotheses, how can she know ordinary propositions about the world, given that the skeptical hypotheses are hypotheses according to which she is *radically deceived* about the world? It does not seem to help here to plead ignorance, on the agent's behalf, of some conditionals. If anything, that makes the agent's epistemic situation look even *worse* according to the skeptic, who would otherwise be willing to grant that although the agent doesn't know much at all, at least she knows some obvious conditionals. Of course, there is much more to be said about skepticism (see, e.g., [14, 4]), but there is no room to do so here (see [12]).

Finally, Egré mentions a fascinating paper, [1], that I also cannot discuss here. But I will make one point about the truth clause for knowledge attributions, inspired by [1], that Egré labels as (2) in his comments. According to (2), for an attributor a to truly attribute knowledge of φ to a subject s, it is only required that s has eliminated the $\neg\varphi$-possibilities in the set $E_s(w) \cup R_a(w)$ of possibilities "entertained" by s or "considered relevant" by a. I agree with Dretske, however, that just because a subject isn't entertaining a $\neg\varphi$-possibility and an attributor doesn't consider it relevant, it doesn't follow that the subject need not eliminate that possibility for the attributor to truly attribute knowledge of φ to the subject; consider, e.g., a case in which a patient presents symptoms compatible with several common diseases, but her incompetent medical intern (the subject) and incompetent doctor (attributing knowledge to the intern) only think of one of them and fail to consider the others.

According to Egré's (2) and (1), the fewer possibilities the subject entertains and the attributor considers, the more the attributor can truly say that the subject knows. But just because a close-minded subject and carefree attributor do not entertain or consider possibilities that they *should* does not mean that the attributor can truly attribute knowledge to the subject thanks to their carefreeness and close-

---

[5]I should remind the reader that my own view of knowledge rejects the simple world-ordering picture used in RA models for D- and L-semantics (see [12, 11, 9]), but I am trying to motivate that picture here on its own terms.

mindedness.[6] Knowledge, unlike mere opinion or belief, is an achievement that cannot be acquired so cheaply. A subject's entertainings and an attributor's considerings do not completely determine what counts as relevant; objective aspects of the world, such as the real frequencies of diseases, also affect what possibilities are relevant and must be eliminated for knowledge.

Response to Xu Zhaoqing

Xu helpfully offers three main comments about EC&ELI, as well as a number of interesting side notes. For reasons of space, I focus on the big-picture points.

Knowledge and Belief.

Xu's first point concerns adding *belief* to RA models. In EC&ELI, for simplicity I did not represent an agent's belief in RA models separately from her knowledge (though I did in CB models). To justify this choice, I made two points about belief in relation to RA models. To set them up, suppose that we define *relevant alternatives and belief* (RAB) *models* as tuples $\mathcal{M} = \langle W, \twoheadrightarrow, D, \preceq, V \rangle$, where $\langle W, \twoheadrightarrow, \preceq, V \rangle$ is an RA model and D is a doxastic accessibility relation as in CB models, and redefine the truth clauses as follows:

$$\mathcal{M}, w \vDash_x B\varphi \quad \text{iff} \quad \forall v \in W: \text{if } wDv \text{ then } \mathcal{M}, v \vDash_x \varphi;$$

$$\mathcal{M}, w \vDash_c K\varphi \quad \text{iff} \quad \mathcal{M}, w \vDash_c B\varphi \text{ and } \forall v \in \overline{\llbracket \varphi \rrbracket}_c^{\mathcal{M}}: w \not\twoheadrightarrow v;$$

$$\mathcal{M}, w \vDash_d K\varphi \quad \text{iff} \quad \mathcal{M}, w \vDash_d B\varphi \text{ and } \forall v \in \mathrm{Min}_{\preceq_w}\left(\overline{\llbracket \varphi \rrbracket}_d^{\mathcal{M}}\right): w \not\twoheadrightarrow v;$$

$$\mathcal{M}, w \vDash_l K\varphi \quad \text{iff} \quad \mathcal{M}, w \vDash_l B\varphi \text{ and } \forall v \in \mathrm{Min}_{\preceq_w}(W) \cap \overline{\llbracket \varphi \rrbracket}_l^{\mathcal{M}}: w \not\twoheadrightarrow v.$$

My first point about this in EC&ELI was that the switch from RA to RAB models would not change any of the main results of the paper; the Closure Theorem and the completeness theorem would easily extend to RAB models. My second point was that if we imposed the following constraint on RAB models, then the extra $B\varphi$ requirement for $K\varphi$ would become redundant anyway:

$$\forall w, v \in W: wDv \Rightarrow [v \in \mathrm{Min}_{\preceq_w}(W) \text{ and } w \twoheadrightarrow v],$$

which says that if v is compatible with what the agent believes in w, then v is relevant and uneliminated for the agent in w, ensuring the validity of $K\varphi \to B\varphi$. I will not discuss the plausibility of this constraint, which is a variant of Lewis's [13] *Rule of Belief*. Instead, I will simply say that like Xu, I am curious about the question he raises: are there plausible constraints relating D, $\preceq$, and $\twoheadrightarrow$?

---

[6]The formulation in [1] replaces $E_s(w) \cup R_a(w)$ in (2) with $E(w) \cup S(w)$, where $E(w)$ is the set of possibilities that *are* entertained by the subject and $S(w)$ is the set of possibilities that *should* be entertained by the subject. But it seems to me that there can be possibilities that need to be eliminated for knowledge but that neither are entertained nor *should* be *entertained* by the subject—since "entertaining" possibilities sounds like an intellectual act that takes time, while the agent's perception may quickly eliminate possibilities without so much thought.

Existential Quantification and Neighborhood Models.

Noting that truth clauses for $K\varphi$ usually use universal quantification, Xu points out that in connection with closure it would be interesting to consider truth clauses for $K\varphi$ that involve existential quantification, as in the case of neighborhood semantics, where $K\varphi$ is true at w iff there is a proposition in $N(w)$ that is (a subset of) the set of worlds where φ is true. My short response to Xu's point here is that according to my own positive view of knowledge, the Multipath Picture of Knowledge [12, 11, 9] noted above, the truth clause for $K\varphi$ does have an existential character and the models for knowledge are indeed related to neighborhood models. So my response to Xu's point about existentials and neighborhoods is 'yes' and 'yes'! For the details, see [12, 11, 9].

Stronger Implications.

Finally, Xu raises the issue of stating closure principles with stronger kinds of implication than material implication. In EC&ELI, I noted that not only closure under known implication but also closure under known strict implication, $(K\varphi \land K\Box(\varphi \to \psi)) \to K\psi$, is invalid according to D/H/N/S-semantics. (Note, by contrast, that closure under strict equivalence, $(K\varphi \land K\Box(\varphi \leftrightarrow \psi)) \to K\psi$ is D/H/N/S-valid.) Xu asks whether if we interpret a *counterfactual* $\Box\!\!\to$ using the same preorder used for K, the closure principle $(K\varphi \land K(\varphi \Box\!\!\to \psi)) \to K\psi$ is valid.[7] The answer is 'yes' for C/L-semantics and 'no' for D/H/N/S-semantics; the 'no' answer follows from the invalidity of $(K\varphi \land K\Box(\varphi \to \psi)) \to K\psi$, since the strict implication is stronger than the counterfactual, but to see the reason for D/H/N-semantics, observe that from the facts that the agent has eliminated the minimal $\neg\varphi$-worlds ($K\varphi$) and that the minimal φ-worlds are ψ-worlds ($\varphi \Box\!\!\to \psi$), it does not follow that the agent has eliminated the minimal $\neg\psi$-worlds ($K\psi$). Xu's question underscores an interesting open problem: axiomatize D/H/N/S-semantics for extended languages.

Conclusion

I am grateful to Egré and Xu for raising so many stimulating points, more than I could cover here. I hope, however, that some of the answers—or at least more good questions—arise out of the work that follows EC&ELI [12, 11].

BIBLIOGRAPHY

[1] Maria Aloni, Paul Egré, and Tikitu de Jager. Knowing whether A or B. *Synthese*, 2009. DOI 10.1007/s11229-009-9646-1.

[2] Peter Baumann. Contextualism and the Factivity Problem. *Philosophy and Phenomenological Research*, 76:580–602, 2008.

[3] Elke Brendel. Why contextualists cannot know they are right: self-refuting implications of contextualism. *Acta Analytica*, 20(2):38–55, 2005.

---

[7]I do not think a relation $\preceq_w$ for RA theories should be identified with an ordering for counterfactuals, so I take Xu's question to be mainly about subjunctivist theories.

[4] Keith DeRose. Solving the Skeptical Problem. *The Philosophical Review*, 104(1):1–52, 1995.

[5] Keith DeRose. *The Case for Contextualism: Knowledge, Skepticism, and Context*. Oxford University Press, New York, 2009.

[6] Fred Dretske. The Pragmatic Dimension of Knowledge. *Philosophical Studies*, 40(3):363–378, 1981.

[7] Fred Dretske. Externalism and Modest Contextualism. *Erkenntnis*, 61:173–186, 2004.

[8] Wesley H. Holliday. Epistemic Logic, Relevant Alternatives, and the Dynamics of Context. In Daniel Lassiter and Marija Slavkovik, editors, *New Directions in Logic, Language and Computation*, volume 7415 of *Lecture Notes in Computer Science*, pages 109–129. 2012.

[9] Wesley H. Holliday. *Knowing What Follows: Epistemic Closure and Epistemic Logic*. PhD thesis, Stanford University, 2012. Revised Version, ILLC Dissertation Series.

[10] Wesley H. Holliday. Epistemic Closure and Epistemic Logic I: Relevant Alternatives and Subjunctivism. *Journal of Philosophical Logic*, 2013. Forthcoming.

[11] Wesley H. Holliday. Epistemic Closure and Epistemic Logic II: A New Framework for Fallibilism. 2013. Manuscript.

[12] Wesley H. Holliday. Fallibilism and Multiple Paths to Knowledge. *Oxford Studies in Epistemology*, 5, 2013. Forthcoming.

[13] David Lewis. Elusive Knowledge. *Australasian Journal of Philosophy*, 74(4):549–567, 1996.

[14] Barry Stroud. *The Significance of Philosophical Scepticism*. Oxford University Press, New York, 1984.

[15] Timothy Williamson. Comments on Michael Williams' "Contextualism, Externalism, and Epistemic Standards". *Philosophical Studies*, 103(1):25–33, 2001.

[16] Crispin Wright. Contextualism and Scepticism: Even-Handedness, Factivity, and Surreptitiously Raising Standards. *The Philosophical Quarterly*, 55(219):236–262, 2005.

The Stability Theory of Belief. A Summary[1]

Hannes Leitgeb

What I call the 'stability theory of belief' is a joint theory of rational (all-or-nothing, qualitative) belief and (quantitative) degrees of belief. It starts from what may be a surprising observation: combining the closure of rational belief under conjunction with the so-called Lockean thesis on rational belief does not actually entail a contradiction.

For it seems to be commonly accepted that the following ingredients of what might otherwise be thought an attractive theory of belief and degrees of belief of perfectly rational agents cannot be had simultaneously:

P1 The logic of belief, in particular, the logical closure of belief under conjunction, that is: for all propositions A, B,

$$\text{if } Bel(A) \text{ and } Bel(B) \text{ then } Bel(A \cap B).$$

P2 The axioms of probability for the degree of belief function P.

P3 The Lockean thesis (cf. [2], pp.140f) that governs both Bel and P: for every proposition B, B is believed if and only if the degree of belief in B is not less than r, or briefly,

$$Bel(B) \text{ if and only if } P(B) \geq r,$$

for some threshold r that is greater than $\frac{1}{2}$ and at most 1.

And the reason for rejecting this combination of rationality postulates seems to derive from the following consideration: Either r is set to 1, in which case P3 collapses the Lockean thesis into the trivializing '$Bel(B)$ if and only if $P(B) = 1$', by which all and only propositions of probability 1 are to be believed. But this cannot be right at least as a general requirement on believed propositions: we do seem to believe some propositions B on which we would nevertheless not accept every bet whatsoever, even though we would have to do so according to the standard interpretation of probabilities in terms of betting quotients whenever $P(B) = 1$.

---

[1] This summary is based on a paper that is under review.

Or r is less than 1: which is the more realistic case. But then, if P1–P3 are taken together, they are believed to lead to paradox, as exemplified by the famous Lottery paradox and Preface paradox situations. Most of the classical literature on belief or acceptance thus divides according to which of the postulates above are being preserved and which are given up: for instance, famously, Isaac Levi keeps P1 but rejects P3, while for Henry Kyburg it is just the other way around.

Let us focus now just on the $r < 1$ case: We want to show that what is commonly accepted is not quite right, in the following sense. As it stands, P1–P3 above are *ambiguous* with respect to the position of the 'for some threshold r' quantifier that was used before. Once disambiguated, only in one interpretation it holds that P3 with $r < 1$ cannot be combined consistently with P1 and P2, while in another interpretation this is perfectly possible. Here is why: it is indeed not the case that

there is an $r < 1$, such that for all P (on a finite space of worlds)

assumptions P1–P3 from above are jointly satisfied; however it *is* the case that

for all P (on a finite space of worlds), there is an $r < 1$

such that P1–P3 are jointly satisfied.

Some examples will illustrate what is the issue, the first one of which is lottery-style:

EXAMPLE 1. (1a) Let $r = \frac{999999}{1000001}$. Now consider W to be the set $\{w_1, \ldots, w_{1000000}\}$ of 1000000 possible worlds, and let P be the unique probability measure that is given by $P(\{w_1\}) = P(\{w_2\}) = \ldots = P(\{w_{1000000}\}) = \frac{1}{1000000}$. At the same time, from the Lockean thesis in P3, it would follow that for every $1 \leq i \leq 1000000$, the proposition $W - \{w_i\}$ ought to be believed, as $P(W - \{w_i\}) = \frac{999999}{1000000} \geq \frac{999999}{1000001}$; hence, by P1, the conjunction (that is, intersection) of all these propositions would have to be believed as well; but this conjunction is nothing but the contradictory proposition $\varnothing$, which has probability 0 by P2, and which therefore is not believed according to P3. Contradiction. In a nutshell: for r as being chosen, we can construct a probability measure P, such that P1–P3 do not hold jointly, whatever Bel is like. By the same token, for every $\frac{1}{2} < r < 1$ whatsoever a uniform probability measure can be constructed, such that P1–P3 are not satisfied simultaneously.

(1b) Now let us turn things around: Let W be the set $\{w_1, \ldots, w_{1000000}\}$ and P be given by $P(\{w_1\}) = \ldots = P(\{w_{1000000}\}) = \frac{1}{1000000}$ again. But now set $r = \frac{1000000}{1000001}$: in that case, the only proposition that is to be believed according to the Lockean thesis in P3 is W itself, which has probability 1; trivially, the set of believed propositions is closed under logic then, including closure under conjunction; hence, P1–P3 turn out to be satisfied.

Clearly, in case 1b we were able to avoid the contradiction from 1a by another trivializing method: for given P, by pushing the threshold in the Lockean thesis sufficiently close to 1 so that only the propositions of probability 1 end up be-lieved. While this method allows us to construct for every probability measure (over a finite set of worlds) a suitable threshold less than 1, such that P1–P3 hold jointly, in many cases one can do much better: for one can achieve the same result without triviality, so that at least some proposition of probability *less than 1* ends up believed, too.

This is an example:

EXAMPLE 2. Let $W = \{w_1, \ldots, w_8\}$ be a set of eight possible worlds. Let P be the unique probability measure that is given by: $P(\{w_1\}) = 0.54$, $P(\{w_2\}) = 0.342$, $P(\{w_3\}) = 0.058$, $P(\{w_4\}) = 0.03994$, $P(\{w_5\}) = 0.018$, $P(\{w_6\}) = 0.002$, $P(\{w_7\}) = 0.00006$, $P(\{w_8\}) = 0$. Now consider the following six propositions,

$$\{w_1\}, \{w_1, w_2\}, \{w_1, \ldots, w_4\}, \{w_1, \ldots, w_5\}, \{w_1, \ldots, w_6\}, \{w_1, \ldots, w_7\}$$

only the final one of which has probability 1. Choose any of them, call it B_W, and let Bel be determined uniquely so that B_W is the least or logically strongest proposition that is believed according to Bel; in other words: for all propositions $B \subseteq W$,

$$Bel(B) \text{ if and only if } B_W \subseteq B.$$

Finally, take $r = P(B_W)$ to be the threshold in the Lockean thesis. Then one can show that the so-determined Bel, together with P and r, satisfies all of the postulates P1–P3 from above. Once again, for our given P, there is a threshold r, such that P1–P3 hold simultaneously. But this time, for the first five choices of B_W, there is in fact a proposition of probability less than 1 that is being believed.

Which moral is to be drawn from these examples? *Maybe it is possible to have one's cake and to eat it, too:* to preserve the logic of belief and the axioms of probability while at the same time assuming consistently that the beliefs and degrees of belief of perfectly rational agents relate to each other as expressed by the Lockean thesis even for a threshold of less than 1. The price to be paid for this will be that not any old threshold in the Lockean thesis will do; instead the threshold must be chosen suitably depending on the agent's beliefs and her degree of belief function.

But what does 'choose suitably' mean exactly here? That is: given P, what are the belief sets Bel and thresholds r like which, together with P, satisfy all of P1–P3 above?

In order to spell out the answer we need a probabilistic concept which is closely related to the notion of resiliency introduced by [4]:

DEFINITION 3. With P being a probability measure on the sample space W, we define for all $A \subseteq W$: A is P-stable if and only if for all $B \subseteq W$, such that B is consistent with A and $P(B) > 0$:

$$P(A \mid B) > \frac{1}{2}.$$

So a proposition is stable just in case it is sufficiently probable given any proposition with which it is compatible. In order to get a feel for this definition, consider a consistent proposition A that is P-stable: one of the suitable values of 'B' above is then the total set W of worlds, and hence from P-stability it follows that $P(A|W) = P(A) > \frac{1}{2}$; thus, any consistent P-stable proposition A must have a probability greater than that of its negation. What P-*stability* adds to this is that *this is going to remain to be so* under the supposition of any proposition B that is consistent with A and for which conditional probabilities are defined.

As follows immediately from the axioms of probability, every proposition of probability 1 must be P-stable. Furthermore, the empty proposition is vacuously P-stable. And it might seem that this might actually exhaust the class of consistent P-stable sets, since P-stability might seem pretty restrictive; but in fact things are quite different. One can prove that for almost all P on a finite sample space of worlds (where 'almost all' can be made precise in terms of the geometrical Lebesgue measure) there is a P-stable set of probability less than 1.

The importance of P-stability becomes transparent from the following representation theorem:

THEOREM 4. *Let W be a finite non-empty set, let Bel be a set of subsets of W, (where B_W is the intersection of all members of Bel) and let P assign to each subset of W a number in the interval $[0, 1]$. Then the following two statements are equivalent:*

  I. Bel satisfies P1, P satisfies P2, and P and Bel satisfy the right-to-left direction of P3 with threshold $r = P(B_W) > \frac{1}{2}$.

  II. P satisfies P2, and there is a (uniquely determined) $A \subseteq W$, such that

   – *A is a non-empty P-stable proposition,*

   – *if $P(A) = 1$ then A is the least subset of W with probability 1; and:*

   – *for all $B \subseteq W$:*

     $Bel(B)$ if and only if $A \subseteq B$

  (and hence, $B_W = A$).

This is a universally quantified if-and-only-if statement: its left-hand side I. collects all of our desiderata, if for the moment we restrict ourselves just to the one direction of the Lockean thesis, and if we use $P(B_W)$ as a threshold; and its right hand-side II. says that B_W is P-stable, and if B_W has probability 1 then it is the least proposition of probability 1 (which must always exist for finite W). In other words: Given P and Bel, such that P1, P2, and the right-to-left direction of the Lockean thesis with threshold $P(B_W)$ are satisfied, where B_W is the least believed proposition which exists by P1: then it follows from the theorem above that B_W must be P-stable. And given P and a P-stable proposition (which, if it has probability 1, is the least of that kind): then one can cook up Bel from that P-stable proposition, so that P and Bel satisfy all of the desiderata, and the given P-stable proposition is just the logically strongest believed proposition B_W.

In fact, one can show more: either side of the equivalence statement above actually implies the *full* Lockean thesis, that is, for all propositions B: $Bel(B)$ iff $P(B) \geq P(B_W) > \frac{1}{2}$. This means: one might have thought that one could do just with the right-to-left half of the Lockean thesis, but once one throws in enough of the logic of belief, one always ends up with the full Lockean thesis.

By the theorem above, in a context in which P1 and P2 have already been pre-supposed, we can therefore reformulate postulate P3 from before as follows:

P3 [Reformulated] B_W is P-stable, and if $P(B_W) = 1$ then B_W is the least proposition $A \subseteq W$ with $P(A) = 1$.

Call P1, P2, together with P3—either in its original or in its reformulated form—the 'stability theory of belief'. In neither version should P3 be interpreted neces-sarily as aiming at a reduction of Bel (or B_W) to P: P3 merely expresses a joint constraint or postulate on P and Bel taken together; in order for P and Bel to cohere with each other, P3 should hold; that's all. (This said, the theory is *open* to an interpretation according to which Bel is in fact reducible to P and a contextual cautiousness parameter, which is interpretation given in [3].)

From the theorem above it also follows that if one has complete information about what the P-stable sets for a given probability measure P are like, then one knows exactly how to satisfy P1–P3 from above for this very P: either one chooses a P-stable set of probability less than 1—if there is such—and uses it as B_W; or one chooses the least proposition of probability 1 for that purpose, which is also P-stable. And it turns out that there is a simple algorithm by which one can determine all these P-stable sets for given P easily (which, though, we do not have the space to describe here).

If we apply that algorithm to Example 1 from section 1, the only set B_W so constructed is W itself, which is at the same time the least proposition of probabil-ity 1. This means that the stability theory of belief predicts it to be rational not to believe of any of the 1000000 tickets in a fair lottery that it will be a losing ticket. The corresponding threshold $P(B_W)$ in the Lockean thesis is 1, but one might just as well choose some number that is less than but sufficiently close to 1 instead.

In the case of Example 2, as promised, we get the following stable options for B_W (and corresponding thresholds): $\{w_1, w_2, w_3, w_4, w_5, w_6, w_7\}$ ($r = 1.0$),

$\{w_1, w_2, w_3, w_4, w_5, w_6\}$ $(r = 0.99994)$, $\{w_1, w_2, w_3, w_4, w_5\}$ $(r = 0.99794)$, $\{w_1, w_2, w_3, w_4\}$ $(r = 0.97994)$, $\{w_1, w_2\}$ $(r = 0.882)$, $\{w_1\}$ $(r = 0.54)$. Hence, for example, if $\{w_1, w_2\}$ is taken to be the least believed proposition B_W, then all of P1–P3 are satisfied, and the same holds for $\{w_1, w_2, w_3, w_4\}$; in contrast, neither $\{w_1, w_2, w_3\}$ nor $\{w_1, w_2, w_4\}$ will do. If looked at in terms of thresholds $r = P(B_W)$ to be used in P3, the bravest option would be to use $r = 0.54$ as a threshold, in the sense that it yields the greatest number of believed propositions: all the supersets of $\{w_1\}$. The other extreme is $r = 1$ (or something just a bit smaller than that) which is the most cautious choice: only $\{w_1, w_2, w_3, w_4, w_5, w_6, w_7\}$ and W itself will be believed then by the agent. All the other thresholds lie somewhere in between.

We have taken this example measure P from a paper ([1]) in Bayesian philosophy of science in which P figures as the rational reconstruction of an ideal physicist's degrees of belief concerning the phenomenon of the secular acceleration of the moon: the evidence $E = \{w_5, \ldots, w_8\}$ of the moon accelerating, if taken conjunctively with the relevant part of Newtonian mechanics $T = \{w_1, w_2, w_5, w_8\}$ and a conjunction $H = \{w_1, w_3, w_7, w_8\}$ of auxiliary hypotheses (such as the thesis that tidal friction is negligible) amounts to a proposition ($\{w_8\}$) with probability 0. So given E, either T or H needs to be given up: but which one is it? Since $P(T|E)$ is high while $P(H|E)$ is tiny, Dorling argues that H ought to be given up, and that is exactly what happened historically (for tidal friction is *not* actually negligible). But at the same time [1], pp.179f also admits that for his final qualitative interpretation the precise probabilities should not matter that much—it is a thought experiment after all—and that the scientists themselves would have discussed these matters not in terms of numbers but rather in terms of "finite qualitative probability assignments". However, the Bayesian theory just by itself does not have the conceptual resources to make expressions such as 'qualitative interpretation' or 'qualitative probability assignments' precise and to derive and to assess statements involving them in any systematic manner. That is where a joint theory of quantitative and qualitative belief pays off, such as the stability account: I cannot go into all details here, but the least P-stable set for Dorling's P that is consistent with the evidence E is the proposition $\{w_1, \ldots, w_5\}$: intersecting it with E leads therefore to a plausible proposal for the new logically strongest believed proposition $B_W^{new} = \{w_5\}$ once the evidence has been taking into account (see [3], for the details on the corresponding stability theory of belief *revision*). This means that according to the stability theory, after taking into account the observational data, the ideal astrophysicist at the time should believe precisely the supersets of $\{w_5\}$: she still ought to believe Newtonian mechanics, she takes on board the evidence, but she should also believe the *negation* of the conjunction of the auxiliary hypotheses. In short: $Bel_{new}(T)$, $Bel_{new}(\neg H)$, $Bel_{new}(E)$. Once again, this is exactly what happened in actual history. All of this is consistent with stability theory and with the previous purely probabilistic considerations, since B_W^{new} turns out to be P_{new}-stable again, where $P_{new}(.) = P(.|E)$. Bel_{new} is tantamount to a qualitative interpretation of P_{new} and to a qualitative probability assignment, as intended by Dorling; and Bel_{new} does emerge from a proper theory, not from *ad hoc* conclusions.

Here is one final example, which is a variation of Example 1:

EXAMPLE 5. **Example 1 Reconsidered:**

Let $W = \{w_1, \ldots, w_{1000000}\}$ where each world w_i corresponds to ticket i being drawn in a fair lottery, and let P be given by $P(\{w_1\}) = \ldots = P(\{w_{1000000}\}) = \frac{1}{1000000}$ again. Now introduce the partition

$$\Pi = \{\{w_1\}, \{w_2, \ldots, w_{1000000}\}\},$$

or in other words: the agent is interested only in whether ticket 1 wins or not. Consider the partition cells $\{w_1\}$ and $\{w_2, \ldots, w_{1000000}\}$ as new coarse-grained worlds and Π to be the new set of such worlds. Based on our original P, we can then define a new probability measure P_Π, for which Π serves as its sample space, and where P_Π assigns the following values to subsets of Π: $P_\Pi(\{\{w_1\}\}) = \frac{1}{1000000}$, $P_\Pi(\{\{w_2, \ldots, w_{1000000}\}\}) = \frac{999999}{1000000}$, $P_\Pi(\{\{w_1\}, \{w_2, \ldots, w_{1000000}\}\}) = 1$, and $P_\Pi(\varnothing) = 0$. The new probability for a set X results from applying the original probability measure P to $\cup X$; in particular, $P_\Pi(\{\{w_1\}\}) = P(\{w_1\})$ and $P_\Pi(\{\{w_2, \ldots, w_{1000000}\}\}) = P(\{w_2, \ldots, w_{1000000}\})$.

It turns out then that the corresponding P_Π-stable sets are

$$\{\{w_2, \ldots, w_{1000000}\}\} \text{ and } \{\{w_1\}, \{w_2, \ldots, w_{1000000}\}\},$$

the first one of which has a probability slightly less than 1, while the second one is of probability 1 exactly.

Now let $B_W^\Pi = \{\{w_2, \ldots, w_{1000000}\}\}$ and $r = P(\{\{w_2, \ldots, w_{1000000}\}\})$: then all of P1–P3 are satisfied, and since the negation of the proposition $\{\{w_1\}\}$ is just the proposition $\{\{w_2, \ldots, w_{1000000}\}\}$, the agent believes that ticket 1 will not win relative to Π.

This last example shows how the stability theory can explain why someone might be tempted to believe of a particular ticket, say, ticket 1, in the Lottery situation that it will be a losing ticket: if the underlying space of propositions is set up so that one can only distinguish between ticket 1 winning and ticket 1 losing, then the stability account does in fact allow a perfectly rational agent to believe that ticket 1 will lose.

Our examples also demonstrate what may be regarded as the greatest drawback of the stability theory of belief: the theory entails belief to be strongly context-sensitive (even when P is not sensitive to the context), which seems counterintuitive at least at first glance. Bel is sensitive, first of all, to the choice of threshold in the Lockean thesis, where additionally the range of permissible thresholds is itself sensitive to what the agent's degree of belief function P is like; and, secondly, Bel is sensitive to how the underlying space of possibilities is partitioned, as follows from comparing Examples 1 and 5. It remains to be seen whether this speaks against the theory ultimately, or whether the theory in this way merely illuminates an actually existing feature of rational belief.

Let me conclude this summary by pointing out that P1–P3—combining doxastic logic and the axioms of probability with (the right-to-left direction of the) Lockean thesis—constitute but one possible axiomatization of the theory. Other, provably equivalent, axiomatizations are given by combining the axioms of probability with one of the following: (i) what I call the Humean thesis on belief: for all X, $Bel(X)$ iff for all Y, if not $Bel(\neg Y)$ and $P(Y) > 0$, then $P(X|Y) > r$ (which, one can show, *entails* the logical closure of belief and the relevant instance of the Lockean thesis); (ii) the axioms of standard AGM belief revision, formulated in terms of conditional belief ('$Bel(\cdot|\cdot)$'), and the left-to-right direction of the Lockean thesis for conditional belief; (iii) a thesis to the effect that Bel is the best possible qualitative approximation of P (which I have made precise in different versions which again turn out to be provably equivalent); (iv) a simple qualitative decision theory taken together with standard quantitative Bayesian decision theory and with one central postulate that says that if an action A is believed qualitatively to be desirable, whereas an action B is not believed qualitatively to be desirable, then the quantitatively expected utility of A is indeed greater than the quantitative expected utility of B.

The axiomatization in (ii) has been worked out in detail in [3]. The stability theory of belief that derives from each of these axiomatizations has lots of interesting consequences and applications, and it also relates nicely to some of the empirical work on belief that has been done in recent years; but all of that is the topic of other papers.

BIBLIOGRAPHY

[1]   Dorling, Jon, 1979: "Bayesian Personalism, the Methodology of Scientific Research Programmes, and Duhem's Problem", *Studies in the History and Philosophy of Science Part A* 10/3, 177–187.
[2]   Foley, Richard, 1993: *Working Without a Net*, Oxford: Oxford University Press.
[3]   Leitgeb, Hannes, 2013: "Reducing Belief Simpliciter to Degrees of Belief", *Annals of Pure and Applied Logic* 164/12, 1338-1389.
[4]   Skyrms, Brian, 1977: "Resiliency, Propensities, and Causal Necessity", *The Journal of Philosophy* 74/11, 704–713.

Comments on Leitgeb

ALEXANDRU BALTAG AND SONJA SMETS

Does the *supposed failure* of the Lockean thesis force us to choose between (1) a qualitative account of belief and (2) a quantitative account of degrees of belief? Can we (qualitatively) "believe" propositions to which we assign a credence < 1? In contrast to others who argue either for options (1) or (2) but not both, Leitgeb's stability theory shows how both accounts can reach an alliance when we use a reinterpreted version of the Lockean thesis, in which the belief threshold r is not uniform, but dependent on the probability measure P (cf section 2). In the qualitative account an agent's beliefs are assumed to be consistent and closed under conjunction, so that they can be captured by a normal modal operator in the usual style of doxastic logic, while the quantitative account assumes a subjective probability measure P satisfying Kolmogorov's axioms.

At the center of Leitgeb's "stability theory of belief" lies the idea that for a given probability measure P, there is a unique determined P-stable proposition which is the logically strongest believed proposition. Moreover, it is this P-stable proposition which sets the threshold in the restricted Lokean thesis and regulates the link between qualitative and quantitative beliefs. The main new result of this paper is that beliefs generated by a P-stable proposition are the *only* way to satisfy in the same time the minimal requirements of the qualitative account, the usual (Kolmogorov) conditions on the quantitative account and (the reinterpreted version of) the Lockean thesis (with the threshold dependent on the given probability measure).

"Stability" in this theory refers to a proposition's probability staying above the threshold after learning any new information that is consistent with the given proposition. Leitgeb compares this with Skyrms' notion of resiliency, but in our view Leitgeb's concept is even closer related to the notion of "strong belief" as defined by Battigali and Siniscalchi [6]: indeed, the later concept corresponds to the special case of stability with respect to threshold $= 1$. In fact, the qualitative version of strong belief (studied in the context of Dynamic Epistemic Logic [2, 4]) abstracts away from the probability 1 assumption, thus providing an even better match for Leitgeb's notion of stability.

Another meaningful comparison is with the "stability theory of knowledge", as Hans Rott [10] calls the simplified version of Lehrer's defeasibility theory of knowledge [8], as formalized by Stalnaker [11] (who refers to it as "robust belief") and Baltag and Smets [2] (who call it "safe belief"). This however requires a different form of stability, namely stability after any new *true* information. The two kinds of stability are obviously different, but they are connected in interesting ways, at least in the qualitative setting (of Grove sphere models), as shown in [3]: belief is the closure under logical inference (monotonicity) of strong belief, while defeasible knowledge (safe belief) is the closure under logical inference of *true* strong belief.

The very use of the concept (and term) 'stability' points to a straightforward connection with the research progam on logical dynamics (as pursued mainly but not exclusively in the work on Dynamic Epistemic Logic) [5, 7], and especially with the dynamic perspective on knowledge [1], [2], [3]. The author does not stress the dynamic aspect in this paper, preferring to focus on belief and on a synchronic (rather than diachronic) perspective. However, in the seminal paper [9] from which the current stability theory originates, Leitgeb did connect his account to belief dynamics of a sort, namely to the AGM Belief Revision Theory, formalized using the notion of conditional belief. Conditioning of this kind is what authors in Dynamic Epistemic Logic call "static revision" [2], distinguishing it from truly "dynamic" belief revision (the so-called "update"). The later is the action of revising with new information, action that can be thought of as a transition *between* models: in the quantitative setting, this corresponds to moving to a new probabilistic model, in which the new measure is obtained from the prior by conditioning on the new information; in the qualitative setting, it corresponds to deleting all the worlds that are inconsistent with the new information and keeping (what's left from) the old Grove spheres.

It would be very interesting to look at Leitgeb's stability theory in a dynamic setting of belief updates. This would give a truly diachronic perspective, that could help formalize and address some of the criticisms directed at Leitgeb's account. In particular, the criticism by Kelly and Lin about the "non-commutativity" between revision and Leitgeb's acceptance rule simply amounts to the fact that, although Bayesian conditioning preserves the stability of (what's left of any) old stable sets, it may generate new stable sets (that do not come from any old ones). In particular, the smallest stable set after conditioning might be such a new one, with the consequence that AGM revision is inconsistent with the simplest acceptance rule consistent with the stability theory ("believe whatever is entailed by the smallest stable set"). Of course, there are solutions to this problem, since the stability theory is compatible with many other acceptance rules. In particular, one of the solutions is the "non-reductive" interpretation discussed in this paper (page 13), according to which the stability theory only puts consistency constraints linking the quantitative and the qualitative account, without reducing either to the other (in contrast to the "smallest stable set" rule).

We find this paper's subsequent discussion of the context-dependency of belief (and of thresholds), and its applications to the Lottery and Preface paradoxes, to be truly fascinating (although insufficiently formalized). Varying the "context" (or the set of relevant issues/questions) amounts to choosing a less or more fine-grained partition of the state space and lumping together into the same "state" all the states in a given partition cell. Essentially, the idea is that, while the subjective probability of a given proposition is invariant under such partition-changing (as long as the given proposition is left intact by the change), the threshold for acceptance/belief and (more importantly) the fact of whether the proposition is believed or not (even when threshold stays the same) are *not* invariant under such context-shifts. This gives the announced solution to the Lottery paradox: when the only issue is whether or not a given ticket will win, the agent may well choose a threshold that allows her to believe the ticket is a losing one; but when new issues are added (about whether or not other tickets are winning ones), and in particular when the question is "which ticket is winning?", the context shifts (to a more refined partition) and the previously mentioned belief is lost: in this new context, the agent is *no longer* entitled to believe that the first-chosen ticket (or any other ticket) is a losing one! The solution to the Preface paradox is similar (though the explanation is somewhat less clear and less convincing).

We very much like this solution. It is powerful, illuminating and original, though it also resembles in spirit other known solutions to paradoxes based on exhibiting hidden parameters ("contexts") and context-shifting (e.g. the solution to the Liar paradox by Barwise and Etchemendy). But we think the treatment of context shitfs in this paper is insufficiently formal: the discussion of contexts as "partitions" (intuitively induced by "issues") is a bit vague and consistent with several possible formalizations. We think that a more formalized account would be very useful (and in particular allow for a clearer and more convincing treatment of the Preface).

One obvious way to fully formalize this discussion is again given by the "dynamic" approach: introduce an action of "context-shifting" (or alternative, one off "adding a new issue"), that changes the model. The new context is either given by a partition of the old state space, or indirectly by a set of "issues" or questions (which induces a partition in the obvious way: group together in a cell the states which agree on the answers to all the given questions. Given such a partition, the new model can be defined in at least two different ways. The first (closest to the informal proposal in the paper) is to change the state space: the new states are given by the cells of the partition (and the probability of the new states is given by the old probability of the corresponding cells). A second way (maybe more elegant, from the perspective of Probability Theory) is to keep the state space fixed, but to change the *algebra* (of measurable sets) on which the probability measure is defined. Any partition generates a unique Boolean algebra (the smallest subalgebra of the powerset algebra that contains all the cells of the partition). This second approach has the advantage that the new model is easier comparable to the old one: the new measurable sets (propositions) are of the same type (subsets of the same state space) as the original propositions. In this way, we can meaningfully talk about the same proposition (measurable set) as possibly belonging to two such models, and we can formulate the obvious invariance property: context-changing leaves the probability of any proposition the same (as long as the proposition is still in the new algebra). In contrast, beliefs will typically not stay the same. Such properties (as well as the solution to the Lottery Paradox) can be formally stated in a logical language endowed with dynamic modalities for context-shifting.

Another way to formalize this discussion would be the one matching the above-mentioned "static belief revision". A "static" context-shifting is given, not by changing the model, but by making the belief operator binary, via introducing a second "context" parameter, e.g. $Bel(A|\Pi?)$ (where Π is a partition). This is similar syntactically to conditional belief $Bel(A|B)$ (and will play a similar technical role), but of course its semantics is different: it looks only for stable sets that respect the partition Π, and it returns "true" if A includes such a stable Π-set. The above-mentioned discussion (and the analysis of paradoxes etc) can then be formalized in this setting, without any explicit model-change, similarly to the way in which conditional beliefs were used to talk about belief revision without changing models.

The final application (to formal epistemology, or rather the formalization of issues in Philosophy of Science) is equally interesting. It amounts to a "Lockean" version of Dorling's Bayesian analysis of the historical scientific change induced by the discovery of the secular acceleration of the moon. In itself, the example can of course be perfectly well treated in a purely qualitative theory (using e.g. Grove's spheres or equivalently plausibility models, instead of Bayesian probability). But what Leitgeb's Lockean account adds to this is a principled way to *relate* the two accounts: to extract the qualitative, plausibilistic treatment directly from Dorling's Bayesian account, and thus to formally confirm Dorling's (informal and unproved) claim that "nothing in my final *qualitative* interpretation... will depend on the precise numbers...". This is indeed a beautiful application, and moreover one that points towards the most important promise held by Leitgeb's theory and its elaborations: providing a systematic way to extract simple, intuitively-meaningful qualitative-logical conclusions from the puzzling quantitative complexity of Bayesian statistics.

BIBLIOGRAPHY

[1] A. Baltag. An interview on epistemology. In V.F. Hendrikcs and D. Pritchard, editors, *Epistemology: 5 Questions*. Automatic Press/ VIP, 2008.

[2] A. Baltag and S. Smets. A qualitative theory of dynamic interactive belief revision. In G. Bonanno,

W. van der Hoek, and M. Wooldridge, editors, *Logic and the Foundations of Game and Decision Theory (LOFT 7)*, Texts in Logic and Games, pages 9–58. Amsterdam University Press, 2008.

[3] A. Baltag and S. Smets. Course slides on interactive learning, formal social epistemology and group belief dynamics: logical, probabilistic and game-theoretic models. In *ESSLLI Summer School 2012*. Course material at http://www.vub.ac.be/CLWF/SS/esslli2012lecture3.pdf, 2012.

[4] A. Baltag and S. Smets. Protocols for belief merge: Reaching agreement via communication. *Logic Journal of the IGPL*, 2013.

[5] A. Baltag, H. P. van Ditmarsch, and L. S. Moss. Epistemic logic and information update. In Pieter Adriaans and Johan van Benthem, editors, *Handbook on the Philosophy of Information*. Elsevier, 2008.

[6] P. Battigalli and M. Siniscalchi. Strong belief and forward induction reasoning. *Journal of Economic Theory*, 105:356–391, 2002.

[7] J. van Benthem. *Logical Dynamics of Information and Interaction*. Cambridge Univ. Press, 2011.

[8] K. Lehrer. *Theory of Knowledge*. Westview Press, 2000.

[9] H. Leitgeb. Reducing belief simpliciter to degrees of belief. *Manuscript*, 2010.

[10] H. Rott. Stability, strength and sensitivity: Converting belief into knowledge. *Erkenntnis*, 61:469–493, 2004.

[11] R. Stalnaker. Knowledge, belief and counterfactual reasoning in games. *Economics and Philosophy*, 12:133–163, 1996.

Comments on Leitgeb

HANTI LIN AND KEVIN T. KELLY

Hannes Leitgeb's stability theory of belief provides three *synchronic* constraints on an idealized agent's degrees of belief and the propositions she believes. The theory requires that, for each instant of time, an idealized agent satisfies the following three *synchronic* conditions:

(P1) The set of one's beliefs is consistent and closed under deduction.

(P2) One's degrees of belief satisfy the axioms of probability.

(P3) *Lockean Thesis*: there exists a threshold r with $1/2 < r \leq 1$ such that one's beliefs turn out to be exactly the propositions that have degree of belief at least r.

Given P1 and P2, Leitgeb shows that the Lockean thesis P3 is equivalent to the following condition: every proposition with probability one is believed, and the strongest proposition B_W that one believes is probabilistically *stable* with respect to one's current probability measure P in the sense that, for every proposition E consistent with B_W for which $P(E) > 0$, conditional probability $P(B_W|E)$ is greater than $1/2$.

Given P, there may be more than one stable proposition, and the theory itself does not specify which to believe. Indeed, from what has been said so far, Leitgeb's synchronic theory is compatible with the *generalized odds-threshold* acceptance rules recommended in [6]).[1] Here is a simple geometrical recipe for constructing an odds-threshold rule that satisfies P1-P3 in the case of a ternary partition $\{w_1, w_2, w_3\}$. Recall Leitgeb's figure 2, which we reproduce as figure 1.a below. For each vertex V_i of the triangle, construct the line from V_i that divides the triangle in half. Now mark an arbitrary point P_i on the line that lies above Leitgeb's 1/2 threshold for $\{w_i\}$ (figure 1.b). Draw the straight line that passes through P_i and V_j for each distinct i, j (figure 1.b). The lines so constructed partition the triangle into regions. Label each region with the strongest proposition accepted by a Bayesian credal state in that region in the manner depicted (figure 1.c).[2] The

---

[1] Leitgeb's synchronic theory is also compatible with, say, the defeasibility condition proposed by [2].

[2] Credal states that fall exactly on the boundary of a region may all fall to one side or may all fal to the other.

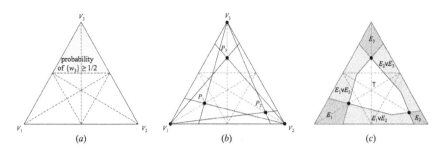

Figure 1. Compatibility of the odds threshold rule with Leitgeb's axioms.

resulting acceptance rule jointly satisfies Leitgeb's P1-P3 along with all of our requirements.[3]

Generalized odds threshold rules have a specific geometrical shape that is not mandated by Leitgeb's synchronic principles P1-P3. The motivation for that shape is based essentially upon *diachronic* considerations, and that is where we begin to disagree with Leitgeb. While we prefer to drop a key principle of the AGM belief revision theory ([1]), Leitgeb expresses a desire to retain it in section 3 of his paper. In this note, we show that, if AGM belief revision is incorporated into Leitgeb's theory, a plausible, diachronic norm of case reasoning must be sacrificed.[4]

Leitgeb interprets his theory as providing nothing more than simultaneous constraints on one's beliefs and degrees of belief. To make that idea explicit, let relation $R(P, Bel)$ mean that it is (synchronically) *permitted* for an agent with probabilistic credal state P to have belief set Bel. Then Leitgeb's thesis reads:

(**Leitgeb's Thesis**) For every P and every Bel, if $R(P, Bel)$, then P and Bel jointly satisfy conditions P1-P3.

Synchronic permission leads naturally to a diachronic constraint on belief revision. Suppose that $R(P, Bel)$ and that, upon receipt of new information E such that $P(E) > 0$, the agent is obligated to update P by conditioning[5] to obtain P_E and revises Bel by a propositional belief revision operator $*$, to obtain new belief state $Bel * E$. Then say that $*$ is *diachronically* admissible for P and Bel if and only if

---

[3] In [6], we likewise assume P1 and P2, but we do not assume P3 unless the underlying partition is binary, so we view the Lockean thesis not as a general norm of rationality, but as a framing effect that arises when one focuses on the partition Q_1 vs. $\neg Q_1$ rather than on the partition Q_1 vs. Q_2 vs. Q_3.

[4] Our argument against AGM in [6]) assumes that the agent must possess a comprehensive *policy* for accepting propositions and for revising them in light of new information—an assumption that Leitgeb does not subscribe to. The impossibility argument in this note is carried out in Leitgeb's much weaker setting.

[5] $P_E(X)$ is defined to be $P(X \wedge E)/P(E)$ if $P(E) > 0$; otherwise it is undefined.

$R(P_E, Bel * E)$, for all E.[6] The most fundamental assumption of the AGM theory of belief revision is *accretiveness*, which requires that, whenever Bel and E are compatible, $Bel * E$ be the set of the logical consequences of $Bel \cup \{E\}$, written $Bel + E$. Since the accretiveness condition holds for all belief revision operators in AGM theory, the assumption that AGM theory is *obligatory* together with the requirement that belief revision be diachronically admissible entails the following requirement, which Leitgeb endorses in section 3 of his paper:

> **(Diachronic Admissibility of Accretive Belief Revision)** For every Bel, every P, and every E such that E is consistent with Bel and that $P(E) > 0$, if $R(P, Bel)$ then $R(P_E, Bel + E)$.

Unfortunately, the two principles Leitgeb endorses—Leitgeb's Thesis and Diachronic Admissibility of Accretive Belief Revision—are jointly incompatible with another plausible, diachronic principle relating acceptance to belief revision. Let $R(P, \Diamond(A))$ say that it is *permitted* for an agent with credal state P to believe A in the following sense:

$$R(P, \Diamond(A)) \quad \text{iff} \quad \exists Bel : R(P, Bel) \wedge Bel(A).$$

Similarly, let $R(P, \Box(A))$ say that it is *required* for an agent with credal state P to believe A in the following sense:

$$R(P, \Box(A)) \quad \text{iff} \quad \forall Bel : R(P, Bel) \to Bel(A).$$

The following is a form of reasoning by cases, the case of learning E and the case of learning its negation:

> **(Diachronic Admissibility of Case Reasoning)** For every probability measure P as an agent's current credal state, every proposition E with $P(E) > 0$, and every A, if $R(P_E, \Box(A))$ and $R(P_{\neg E}, \Diamond(A))$, then $R(P, \Diamond(A))$.

In words, if it required for an agent with probabilistic credal state P to be believe A upon learning E, and if it is permitted for her to believe A upon learning $\neg E$, then it is already permitted for her to believe A prior to learning any new information. For example, for a couch potato who never ventures outdoors, it is obligatory for her to believe that she will not contract poison ivy given that she has genetic immunity against it and it is at least admissible for her to believe that she will not get it otherwise, since she has not traveled outdoors. So it should be admissible for her to believe that she won't get it. Then we have the following impossibility result:

---

[6]In [6], we require that $*$ be a belief revision *policy* defined for all pairs Bel, E, rather than just for a fixed Bel and all E.

THEOREM 1. *Suppose that the set W of possible worlds is finite with cardinality $|W| \geq 3$. Then there is no relation R that satisfies all of the following six conditions:*

1. **Domain of Application.** *For every P and every Bel such that $\mathsf{R}(P, Bel)$, P is a probability measure over W and Bel is a belief set over W that is consistent and deductively closed.*

2. **Probability 1/2 Rule.** *For every probability measure P and every proposition A, if $\mathsf{R}(P, \Diamond(A))$, then $P(A) > 1/2$.*

3. **Probability 1 Rule.** *For every probability measure P and every proposition A, if $P(A) = 1$, then $\mathsf{R}(P, \Box(A))$.*

4. **Non-Skepticism.** *There exist some world $w_i \in W$ and some probability measure P such that $\mathsf{R}(P, \Diamond(\{w_i\}))$ and $P(\{w_i\}) < 1$.*

5. **Diachronic Admissibility of Accretive Belief Revision.**

6. **Diachronic Admissibility of Case Reasoning.**

Acknowledgement

The authors are deeply indebted to Hannes Leitgeb, David Makinson, and Clark Glymour for extensive, probing comments on Hanti Lin's doctoral dissertation. The ruminations in this note were particularly inspired by Hannes Leitgeb's comments.

BIBLIOGRAPHY

[1] Alchourrón, C.E., P. Gärdenfors, and D. Makinson (1985), "On the Logic of Theory Change: Partial Meet Contraction and Revision Functions", *The Journal of Symbolic Logic*, 50: 510-530.

[2] Douven, I. (2002) "A New Solution to the Paradoxes of Rational Acceptability", *British Journal for the Philosophy of Science* 53: 391-410.

[3] Lehrer, K. (1965) "Knowledge, Truth, and Evidence", *Analysis* 25: 168-175.

[4] Leitgeb, H. (2013) "The Stability Theory of Belief", in the current volume.

[5] Levi, I. (1967) *Gambling With Truth: An Essay on Induction and the Aims of Science*, New York: Harper & Row. 2nd ed. Cambridge, Mass.: The MIT Press, 1973.

[6] Lin, H. and Kelly, K. T. (2012) "Propositional Reasoning that Tracks Probabilistic Reasoning", *Journal of Philosophical Logic*, 41(6): 957-981.

[7] Lin, H. (2013) *PhD Dissertation: Probabilistic Reasoning that Tracks Probabilistic Reasoning*, Carnegie Mellon University.

[8] Shoham, Y. (1987) "A Semantical Approach to Nonmonotonic Logics", in M. Ginsberg (ed.) *Readings in Nonmonotonic Reasoning*, Los Altos, CA: Morgan Kauffman.

Appendix: Proof of the Theorem

Suppose for reductio that R satisfies all the six conditions. Since $|W| \geq 3$, let w_1, w_2, w_3 be three distinct possible worlds in W. Denote proposition $\{w_i\}$ by E_i. The probability measure that assigns 1 to E_i is denoted by V_i. Let $\underline{Bel}$ denote the strongest proposition in Bel if it exists.

Let each probability measure P with $P(E_1 \vee E_2 \vee E_3) = 1$ be identified with the triple $(P(E_1), P(E_2), P(E_3))$ of real numbers. Let:

$$P = (x, 0, 1 - x).$$

Prove as follows that $\neg R(P, \Diamond(E_1))$ for every $P = (x, 0, 1 - x)$ with $0 \leq x < 1$.

Case (i): Suppose that $P = (x, 0, 1 - x)$ with $0 \leq x \leq 1/2$ (figure 2.a). It follows immediately from the probability $1/2$ rule (condition 2) that $\neg R(P, \Diamond(E_1))$.

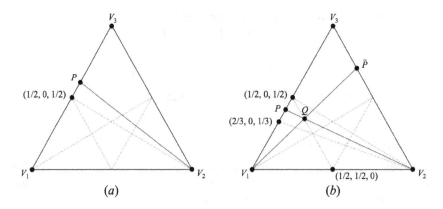

Figure 2. Case (i) in the left, case (ii) in the right

Case (ii): Suppose that $P = (x, 0, 1-x)$ with $1/2 < x \leq 2/3$ (figure 2.b). Suppose for reductio that $R(P, \Diamond(E_1))$. Define Q as the unique probability measure such that $Q(E_1) = 1/2$ and $Q_{\neg E_2} = P$:

$$Q = \left(\frac{1}{2}, \frac{2x - 1}{2x}, \frac{1 - x}{2x} \right).$$

Namely, Q is the intersection of line $\overline{PV_2}$ and the line that passes through $(1/2, 0, 1/2)$ and $(1/2, 1/2, 0)$ (figure 2.b). We have that $R(V_2, \Box(E_2))$, by the probability 1 rule (condition 3). Then, since $Q_{E_2} = V_2$ and $Q_{\neg E_2} = P$ by construction, we have that $R(Q_{E_2}, \Box(E_2))$ and that $R(Q_{\neg E_2}, \Diamond(E_1))$. So, by closure under deduction (condition 1), we have that $R(Q_{E_2}, \Box(E_1 \vee E_2))$ and that $R(Q_{\neg E_2}, \Diamond(E_1 \vee E_2))$. Then, by Diachronic Admissibility of Case Reasoning (condition 6), we have that

$R(Q, \Diamond(E_1 \vee E_2))$. So there exists Bel such that $R(Q, Bel)$ and $Bel(E_1 \vee E_2)$. Argue as follows that $\underline{Bel} = E_1 \vee E_2$. Given the assumed range of x, it is routine to verify that both $Q(E_1)$ and $Q(E_2)$ are no more than $1/2$. So, by the probability $1/2$ rule (condition 2), $\underline{Bel}$ is neither E_1 nor E_2 and, hence, must be $E_1 \vee E_2$. Define $\bar{P}$ as the unique probability measure such that $\bar{P} = Q_{\neg E_1}$:

$$ \bar{P} = \left(0, \frac{2x-1}{x}, \frac{1-x}{x} \right). $$

Namely, $\bar{P}$ is the intersection of line $\overline{V_2 V_3}$ and the line that passes through V_1 and Q. Note that $\neg E_1$ is consistent with $\underline{Bel}$. Then it follows from Diachronic Admissibility of Accretive Belief Revision (condition 5) that $R(Q_{\neg E_1}, Bel + \neg E_1)$. Namely, $R(\bar{P}, Bel')$, where $Bel' = Bel + \neg E_1$. Since $\underline{Bel} = E_1 \vee E_2$, we have that $Bel'(E_2)$. It follows that $R(\bar{P}, \Diamond(E_2))$. But $\bar{P}(E_2) \leq 1/2$, which it is routine verify given the assumed range of x. So the probability $1/2$ rule (condition 2) is violated—contradiction.

Case (iii): Suppose that $P = (x, 0, 1-x)$ with $2/3 < x < 1$. Recall how we define $\bar{P}$ from P in the preceding case; define the map f that sends P to $\bar{P}$ as follows:

$$ \begin{aligned} f(P) &= f(x, 0, 1-x) \\ &= \left(0, \frac{2x-1}{x}, \frac{1-x}{x} \right) \\ &= \bar{P}. \end{aligned} $$

Define the reflection map r with respect to vertices V_1 and V_2 as follows:

$$ r(x_1, x_2, x_3) = (x_2, x_1, x_3). $$

Let $P_1 = P$. Whenever $P_i = (t, 0, 1-t)$ with $2/3 < t < 1$, define:

$$ \begin{aligned} \bar{P}_i &= f(P_i), \\ P_{i+1} &= r^{-1} \circ f \circ r(\bar{P}_i). \end{aligned} $$

Namely, define $\bar{P}_i$ from P_i in the same way that we define $\bar{P}$ from P in case (ii), and define P_{i+1} from $\bar{P}_i$ also in the same way *except* that the roles of V_1 and V_2 are exchanged (figure 3.a). Now proceed to establish what we call the *snaking-up* lemma (cf. figure 3):

LEMMA 2 (SNAKING-UP). *For each* $P_i = (t, 0, 1-t)$ *with* $2/3 < t < 1$, *if* $R(P_i, \Diamond(E_1))$, *then* $R(P_{i+1}, \Diamond(E_1))$.

Suppose that $R(P_i, \Diamond(E_1))$, where $P_i = (t, 0, 1-t)$ with $2/3 < t < 1$. By the same argument as in case (ii) from $R(P, \Diamond(E_1))$ to $R(\bar{P}, \Diamond(E_2))$, we have that

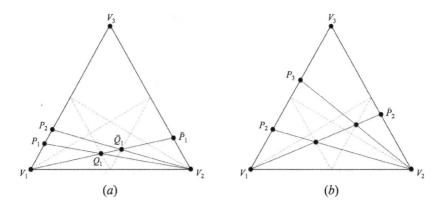

Figure 3. Case (iii)

$R(\bar{P}_i, \Diamond(E_2))$, namely that $R(f(P_i), \Diamond(E_2))$. Now apply the very same argument—except with the roles of E_1 and E_2 exchanged—to the fact that $R(\bar{P}_i, \Diamond(E_2))$. Then we have that $R(r^{-1} \circ f \circ r(\bar{P}_i), \Diamond(E_1))$, namely that $R(P_{i+1}, \Diamond(E_1))$, which completes the proof of the lemma. It is routine to verify that the function g that sends P_i to P_{i+1} can be expressed algebraically as follows:

$$g(t, 0, 1 - t) = \left(\frac{3t - 2}{2t - 1}, 0, \frac{1 - t}{2t - 1} \right).$$

Let S be the following sequence of points on the edge $\overline{V_1 V_3}$: $P_1 (= P), P_2, \ldots, P_i, P_{i+1}, \ldots$. It is routine to verify that, in sequence S, the probability of E_1 is decreasing. It is also routine to verify that, in sequence S, the amount of each decrease in the probability of E_1 is greater than or equal to a fixed number δ determined by $P_1 = (x, 0, 1 - x)$, where:

$$\delta = \left| \frac{3x - 2}{2x - 1} - x \right|$$

$$= \frac{(x - 1)^2}{(x - \frac{1}{2})} > 0,$$

because $2/3 < x < 1$. So the probability of E_1 in sequence S will eventually cease to be greater than $2/3$. Namely, there exists $n \geq 2$ such that:

$$P_n(E_1) \leq 2/3 < P_{n-1}(E_1) \leq P_1(E_1) = x.$$

An example with $n = 3$ is depicted in figure 3: the construction from P_1 to P_2 is in figure 3.a, and the construction from P_2 to P_3 is in figure 3.b. If $R(P, \Diamond(E_1))$, namely if $R(P_1, \Diamond(E_1))$), then it follows from $n - 1$ applications of the snaking-up lemma that $R(P_n, \Diamond(E_1))$, with $P_n(E_1) \leq 2/3$. The conditional just stated has a consequent that contradicts the results either in case (i) or in case (ii), so the antecedent must be false. That is, $\neg R(P, \Diamond(E_1))$, as required.

We have established that $\neg R(P, \Diamond(E_1))$ for every $P = (x, 0, 1 - x)$ with $0 \leq x < 1$. With the help of that result, argue as follows that, for every probability measure P, if $R(P, \Diamond(E_1))$ then $P(E_1) = 1$. Suppose that $R(P, \Diamond(E_1))$. By the probability $1/2$ rule (condition 2), $P(E_1) > 1/2$, so $P_{E_1 \vee E_3}$ exists. Also, there exists Bel such that $R(P, Bel)$ and $Bel(E_1)$. By condition 1, $\underline{Bel} = E_1$. So $E_1 \vee E_3$ is consistent with Bel. By Diachronic Admissibility of Accretive Belief Revision (condition 5), $R(P_{E_1 \vee E_3}, Bel + (E_1 \vee E_3))$. So $R(P_{E_1 \vee E_3}, \Diamond(E_1))$. Since $P_{E_1 \vee E_3} = (x, 0, 1 - x)$ for some x, it follows that $x = 1$, i.e. $P_{E_1 \vee E_3} = (1, 0, 0)$. Hence $P(E_1 \mid E_1 \vee E_3) = 1$. But the choice of w_3 is arbitrary as long as it is distinct fro w_1, so the result generalizes: $P(E_1 \mid E_1 \vee E_j) = 1$ for every possible world $w_j \in W$ distinct from w_1. It follows that $P(E_1) = 1$.

We have established that, whenever $R(P, \Diamond(E_1))$, then $P(E_1) = 1$. But the choice of w_1 is arbitrary. So the result generalizes: for every possible world $w_i \in E$ and every probability measure P, if $R(P, \Diamond(E_i))$ then $P(E_i) = 1$. That contradicts non-skepticism (condition 4), which completes the proof of the theorem.

Response to Baltag and Smets, and Lin and Kelly

HANNES LEITGEB

I would like to thank my commentators for their excellent and extremely helpful remarks.

The comments by Alexandru Baltag and Sonja Smets were so positive that I thought I could be very brief in my reply, and suggest to them instead: why don't we plan on writing an article together on probabilistic stability vs dynamic epistemic logic? You are completely right in pointing to the close relationship between belief according to the stability theory and notions such as strong belief and safe belief in the game-theoretic and the dynamic-epistemic-logical literature; as you say, "It would be very interesting to look at Leitgeb's stability theory in a dynamic setting of belief updates", and we could simply do it jointly!

I also very much second your comments on the topic of context change: in my paper, I presuppose a context (of reasoning and acting) to be given by (i) a set W of possible worlds, which may be interpreted as the set of contextually salient possibilities as seen from the agent's viewpoint, (ii) a partition Π of W, which captures those aspects of the worlds in W that the agent is interested in as far as the present context is concerned (two worlds sharing the same partition cell if they are indistinguishable in terms of all such aspects), and (iii) a contextually determined threshold r that represents how cautious or brave the agent is willing to be regarding her beliefs in that context. But, as you rightly emphasize: I do not formalize context *change* itself—the agent's moving from one context $\langle W, \Pi, r \rangle$ to another context $\langle W', \Pi', r' \rangle$ in the course of her reasoning and acting. It would be important to formalize, and study, how such contexts can be changed rationally; as you suggest, concentrating for the moment just on the second component (ii) from above, this might be done in terms of an operator of the form '$Bel(X|\Pi)$' ("X is believed given partition Π"); or maybe the notion of proposition could be jazzed up along the lines suggested by Steve Yablo's recent work: instead of regarding a proposition as a set X of possible worlds, consider it to be a pair $\langle X, \Pi \rangle$ where Π is a partition of W again and where X is a member of the algebra that is generated by Π; and then an agent might be assumed to believe propositions in this refined sense ("$Bel(\langle X, \Pi \rangle)$"). Either way, the question would be: what rationality postulates do P, and $Bel(\cdot|\Pi)$ or $Bel(\langle \cdot, \Pi \rangle)$, satisfy for *variable* Π over and above static postulates such as P1–P3 from my paper which only tell us something about P and Bel for *fixed* Π? Getting a handle on this would constitute a big step forward, and not just so for formal epistemology and philosophical logic.

Let me turn now to the comments by Hanti Lin and Kevin Kelly, which were equally stimulating though also more critical; here are my responses directed to them.

At the beginning, you observe that your own theory is consistent with my postulates P1–P3. According to your theory, for every two worlds u, v in (say, finite) W there is a threshold $t_{u,v}$, such that for all probability measures P on the power set of W a plausibility ordering between u and v can be determined either by

$$u <_{t_{u,v}} v \text{ iff } P(\{u\}) > t_{u,v} \cdot P(\{v\})$$

or by

$$u <_{t_{u,v}} v \text{ iff } P(\{u\}) \geq t_{u,v} \cdot P(\{v\}).$$

And you illustrate how these thresholds $t_{u,v}$ can be chosen so that P1–P3 hold for all P and for all the Bel that are determined by $<_{t_{u,v}}$ (and where in turn $<_{t_{u,v}}$ is determined from $t_{u,v}$ and P as stated above); which is very nice.

This said, I have two worries: The first one concerns the intended meaning of your assignment of thresholds $t_{u,v}$ to pairs $\langle u, v \rangle$ of worlds. To what kind of mental state is it meant to correspond? It must be a second-order dispositional state that is more basic than the agent's first-order dispositional states given by her degrees of belief and beliefs; a state that is invariant under the agent's learning new evidence in the usual manner, that is, which is invariant under conditionalizing the agent's P on the evidence and revising the agent's Bel in terms of the evidence. Does this assignment of thresholds correspond to some kind of unrevisable a priori state? But then, in order not to bias the agent irrationally for the rest of her life, all of the thresholds $t_{u,v}$ should be set constantly to 1, in which case Bel would always result again from an AGM-like total pre-order of worlds, in conflict with other parts of your 2012 paper (where you impose the additional constraint of "non-opinionation"). Or is the assignment of thresholds revisable? If so, how can it be revised rationally? (Not by Bayesian update or by belief revision, it seems.) Or put more bluntly: if W consists of, say, 8 possible worlds, where do these $8 \cdot 7 = 56$ thresholds $t_{u,v}$ come from? You give some hints at this at the end of your commentaries, but it would be lovely to see this worked out in detail.

Secondly: as far as I can see, satisfying some instance of the Lockean thesis is not optional but mandatory. It is not good enough for a theory of belief to be consistent with it—one should be able to derive some instance of the Lockean thesis from the theory. For, according to all sources, *belief aims at the truth*: for degrees of belief this means that $P(A)$ is the agent's estimate of the truth value of A, and the agent has the following fundamental epistemological aim: pushing

$P(A)$ towards 1 in case A is true, and pushing $P(A)$ towards 0 in case A is false. What this means precisely is worked out in detail in epistemic decision theory. (See, e.g., [4] and [5].) If P is not subject to such a norm, it may well be a probability measure with some other kind of salient interpretation, but, at least as far as I am concerned, it will not be a degree of *belief* function. Similarly for the agent's all-or-nothing beliefs: the agent aims at $Bel(A)$ being satisfied in case A is true, and hence also at $Bel(\neg A)$ being satisfied in case A is false, with the additional qualification that the agent is also permitted to suspend judgement depending on how cautious or brave she aims to be. This being in place, it is difficult to see that how the Lockean thesis could *not* hold; how it could not be the case that $Bel(A)$ if and only if $P(A)$ is high, and hence also $Bel(\neg A)$ if and only if $P(A)$ is low: real-world agents should aim at this being the case, and perfectly rational agents will be such that this is the case.

Here is an example in the terminology of your own theory: assume an assignment of thresholds to pairs of worlds to be given such that P as determined in the diagram yields a strict partial ordering of worlds as depicted in the diagram:

$$w_2: \tfrac{11}{100} \quad w_4: \tfrac{11}{100} \quad w_6: \tfrac{11}{100} \quad w_8: \tfrac{11}{100}$$

$$| \qquad\qquad | \qquad\qquad | \qquad\qquad |$$

$$w_0: \tfrac{8}{100} \quad w_1: \tfrac{12}{100} \quad w_3: \tfrac{12}{100} \quad w_5: \tfrac{12}{100} \quad w_7: \tfrac{12}{100}$$

('|' means $<$; thresholds concerning w_0 are determined so that no $<$-connections emerge for P). This yields, e.g.:

- $Bel(\{w_0, w_1, w_3, w_5, w_7\})$, and $P(\{w_0, w_1, w_3, w_5, w_7\}) = \tfrac{56}{100}$.

- More generally, I have chosen the model "nicely" so that the left-to-right of the Lockean thesis holds unrestrictedly (with threshold $\tfrac{1}{2}$).

- But: it does *not* hold that $Bel(\{w_1, \ldots, w_8\})$, even though it holds that $P(\{w_1, \ldots, w_8\}) = \tfrac{92}{100}$; in other words: although the agent regards $\{w_1, \ldots, w_8\}$ as much more likely to be true than $\{w_0, w_1, w_3, w_5, w_7\}$, the agent does not believe the former to be true while she believes the latter to be true.

That is odd. What kind of disposition makes an agent prefer $\{w_0, w_1, w_3, w_5, w_7\}$ over $\{w_1, \ldots, w_8\}$ when even from the agent's point of view the proposition $\{w_1, \ldots, w_8\}$ is much closer to the truth than the proposition $\{w_0, w_1, w_3, w_5, w_7\}$? And what would make it rational to have such preferences? I could make sense of this perhaps if Bel were not aiming at the truth, or if Bel's aiming at the truth got overridden by some non-epistemic aims of higher priority. But then Bel would not be belief anymore: it would rather be some version of acceptance in the sense discussed by various philosophers of minds, epistemologists, and philosophers of science. (See e.g. [2].) In the case of proper perfectly rational *belief*, the Lockean thesis should be a given; and if so, your (Hanti's and Kevin's) theory collapses into the stability theory that is developed in my paper.

As far as your worries about the AGM postulates or total pre-orders of worlds are concerned (for which I have various justifications, one of which is in terms of a diachronic commutativity principle on belief, acceptance, and probability which can only be satisfied if AGM's characteristic preservation axiom holds): as spelled out in Leitgeb (2013), P-stability^r yields a total pre-order $<$ of worlds so that

$$P(\{u\}) > \sum_{v:\, u < v} \frac{r}{1-r} \cdot P(\{v\}),$$

and in the summary of my present paper as published in this volume, r is simply set to $\frac{1}{2}$, so that the previous condition reduces to the simple sum condition

$$P(\{u\}) > \sum_{v:\, u < v} P(\{v\}).$$

In particular, each (singleton of a) world u that belongs to the logically strongest believed proposition B_W must have a probability greater than the sum of the probabilities of (singletons of) worlds outside of B_W.[1]

And here is now my friendly suggestion to you, Hanti and Kevin: if you insist on presupposing only some strict partial order $<$ on worlds when determining Bel, then you might simply adapt the sum condition from above to such orders:

$$P(\{u\}) > \sum_{v:\, u < v} \frac{r}{1-r} \cdot P(\{v\})$$

where now $<$ would *not* demanded to result from a total pre-order. This is one of the generalizations of my theory that I am very much interested in, and adding a sum condition constraint of this form to your theory would have the advantage of guaranteeing at least the left-to-right direction of the Lockean thesis (with threshold $\frac{1}{2}$)—the direction of the Lockean thesis that Hanti presupposes anyway in the excellent decision-theoretic part of his (Hanti's) PhD thesis. This would thus constitute some nice middle ground between our positions.[2]

Let me conclude with a remark on the very interesting limitative theorem that you prove in your commentary: while the conclusion that you draw from it is to

---

[1] The same sum condition is used by [1] and [7]. But they apply this only to total *partial orders* rather than to total *pre-orders*; the difference is crucial, as only a severely restricted class of probability measures can non-trivially satisfy the sum condition jointly with a total partial order of worlds, while almost all probability measures (on a finite space) can non-trivially satisfy the sum condition jointly with a total pre-order of worlds.

[2] Alternatively, there is a version of my theory in which I start from a *set* of probability measures P instead of just a single such measure; the corresponding theory validates only the system P of preferential reasoning, that is, the system that you prefer over the stronger system P + Rational Monotonicity that corresponds to total pre-orders of worlds on the semantic side.

give up (what you call) Diachronic Admissibility of Accretive Belief Revision, I would propose to give up your constraint of Diachronic Admissibility of Case Reasoning.

Here is an example: Let $W = \{w_1, w_2, w_3, w_4\}$, $A = \{w_1, w_2\}$, $E = \{w_2, w_3\}$ and hence $\neg E = \{w_1, w_4\}$, and let $P(\{w_1\}) = \frac{5}{12}$, $P(\{w_2\}) = \frac{3}{12}$, $P(\{w_3\}) = 0$, $P(\{w_4\}) = \frac{4}{12}$. Then $P_E(A) = 1$, which is why (as we agree, and using your terminology): $R(P_E, \Box(A))$. Furthermore, $P_{\neg E}(A) = \frac{5}{9} > \frac{1}{2}$, which entails (as we agree, I think): $R(P_{\neg E}, \Diamond(A))$. But I would not want to conclude from this that $R(P, \Diamond(A))$, in conflict with Diachronic Admissibility of Case Reasoning. The reason is that (no subset of) A is P-stable: for instance, if A were the agent's logically strongest believed proposition, then there would be a proposition (e.g. $\{w_2, w_3, w_4\}$) that would be a serious possibility from the agent's point of view—as its negation $\{w_1\}$ would not be believed by the agent—but where supposing or learning that proposition would drag down the probability of A below $\frac{1}{2}$ (as $P(A \mid \{w_2, w_3, w_4\}) = \frac{3}{7} < \frac{1}{2}$).

Instead of running trough the various justifications that I have for belief being stable, let me point to the close correspondence with Jackson ([3]) on assertability and Lewis [6] on Jackson on assertability:

> Robustness is an important ingredient in assertability [...] Consider "Either Oswald killed Kennedy or the Warren Commission was incompetent." This is highly assertable even for someone convinced that the Warren Commission was not incompetent [...] The disjunction is... highly assertable for them, because it would still be probable were information to come to hand that refuted one or the other disjunct. The disjunction is robust with respect to the negation of either of its disjuncts taken separately–and just this may make it pointful to assert it. ([3], pp.570f)

> I speak to you (or to my future self, via memory) in the expectation that our belief systems will be much alike, but not exactly alike [...] Maybe you (or I in future) know something that now seems to me improbable. I would like to say something that will be useful even so. [...] Let me say something... that will not need to be given up... even if a certain hypothesis that I now take to be improbable should turn out to be the case. ([6], p.153)

In terms of the previous example: the disjunction $\{w_1\} \vee \{w_2\}$—the proposition A—is not robust enough in order to be asserted; it is not robust with respect to the negation of one of its disjuncts taken separately (just consider conditionalizing on $\neg\{w_1\} = \{w_2, w_3, w_4\}$ again). We should only assert disjunctions that remain to be useful even if the negation of one of its disjuncts comes along as a new supposition or a new piece of evidence. And this holds even when we assert something to our future doxastic self. In other words: when a disjunction is the logically strongest believed proposition (in a context), then it needs to be stable with respect to the agent's degree of belief function. But A in the example above is not P-stable, which is why $R(P, \Diamond(A))$ fails.

BIBLIOGRAPHY

[1] Benferhat, S., D. Dubois, and H. Prade, "Possibilistic and Standard Probabilistic Semantics of Conditional Knowledge", *Journal of Logic and Computation* 9 (1997), 873–895.

[2] Engel, P. (ed.), *Believing and Accepting*, Philosophical Studies Series 83, Dordrecht: Kluwer, 2000.

[3] Jackson, F., "On Assertion and Indicative Conditionals", *The Philosophical Review* 88/4 (1979), 565–589.

[4] Joyce, J.M., "A Non-Pragmatic Vindication of Probabilism", *Philosophy of Science*, 65/4 (1998), 575–603;

[5] Leitgeb, H. and R. Pettigrew, "An Objective Justification of Bayesianism I: Measuring Inaccuracy", *Philosophy of Science*, 77/2 (2010), 201–235.

[6] Lewis, D., "Postscript to 'Probabilities of Conditionals and Conditional Probabilities'", *Philosophical Papers II*, Oxford: Oxford University Press, 1986, pp.152–156.

[7] Snow, P., "Is Intelligent Belief Really Beyond Logic?", in: *Proceedings of the Eleventh International Florida Artificial Intelligence Research Society Conference*, American Association for Artificial Intelligence, 1998, pp.430–434

Is There Such a Thing as a *Ceteris Paribus* Law?

WEI WANG

ABSTRACT. Since there are hardly any universal and exceptionless laws in special sciences, there is more and more philosophical discussion on *Ceteris Paribus* laws and their roles in contemporary sciences. Earman et al. raise several objections to the *Ceteris Paribus* laws. In this paper, I will argue that CP clauses can be ineliminable even with scientific terminology, and that it is also possible to test the contraposition of a CP law, therefore the law itself. Earman's account of differential equations may violate his MRL view of laws of nature. Again, Earman's view of laws of nature may be inconsistent with his supervenience thesis.

1 The Rise of *Ceteris Paribus*

The use of *Ceteris Paribus* (thereafter CP) clauses in philosophy and sciences has a long history. According to [19], the idea of CP in economics can be traced back to William Petty's *Treatise of Taxes and Contributions* (1662), and the concept was firstly used by Alfred Marshall in his *Principles of Economics* (1890).

Most conclusions of economics usually hold "only in the absence of disturbing causes" or "other things being equal". So John Cairnes, in his *Some Leading Principles of Political Economy newly Expounded* (1874), gave a classic example of a CP law: "The rate of wage, other things being equal, varies inversely with the supply of labour".

In the last several years, the topic of CP laws has become an important issue in the philosophy of science, partly because of the discussion of the status of the special sciences. Science is supposed to discover laws of nature, which, according to the standard account, are universal and exceptionless. However, many conclusions of the special sciences are not strict, i.e. universal and exceptionless. How can we still justify their scientific status?

For example, in economics, "When the demand for a product increase, while supply remains constant, the price of the product will increase" just hold if other things being equal. Many interferences, say, irrational behavior or ignorance of customers, could brought about counter results. And in biology, Mendel's law of segregation would not hold if there were certain changes in the initial conditions of the earth or the evolutionary chain. So how can we still insist the scientific legitimacy of the special sciences?

A natural response is to propose that the laws of the special sciences are CP laws and CP laws are as scientifically legitimate as the strict laws. So the special sciences are still scientific even without strict laws.

Some philosophers go further. They suggest CP all the way down – even most of the basic laws of physics are CP laws. So there is no relevant difference between the fundamental and the special sciences. For example, Peter Lipton ([16], p. 155) thinks "Most laws are *ceteris paribus* (CP) laws". And Michael Morreau ([18], p. 163) argues "...hedged laws are the only ones we can hope to find. Laws are commonly supposed to be truths, but interesting generalizations, without some modifier such as *'ceteris paribus'* are by and large false".

So nowadays, there is more and more philosophical discussion on *Ceteris Paribus* laws and their roles in contemporary sciences. Alexander Reutlinger, Gerhard Schurz, and Andreas Hüttemann even published an entry *"Ceteris Paribus* Laws" for *Stanford Encyclopedia of Philosophy* (2011).

The CP clause is usually read as "other things being equal", "there are no inter-ferences", "in the absence of disturbing factors", or something like that. The CP proponents usually deny we can make the CP clause explicit, since there would be an infinite set of conditions in the clause. The CP law can be written as "If CP, all Fs are Gs".

2   *Ceteris Paribus* Lost?

However, the concept of CP is quite debatable in the philosophy of science. John Earman, John Roberts and Sheldon Smith (thereafter ERS) wrote a paper, *Ceteris Paribus Lost?*, which raises, I think, the severest criticism of CP laws. They argue there should be no concept of CP laws in physics in the following 6 ways:

(i) Appeal to examples from physics. It seems to ERS, in order for a "real" CP law to be interesting, CP clause must be ineliminable and the range of this clause can not be made explicit. "Otherwise, the CP clause is merely a function of laziness: Though we *could* eliminate the CP clause in favor of a precise, known conditional, we choose not to do so." ([7], p. 283-284) They argue that CP clauses can be easily eliminated by known conditions if we properly use scientific lan-guage.

(ii) Confusing Hempel's provisos with *ceteris paribus* clauses. ERS think Hempel's central concern "is not the alleged need to save law statements from falsity by hedging them with CP clauses, but rather the problem of applying to a concrete physical system... the conditions of the provisos are conditions for *the validity of the application*, not conditions for *the truth of the law statements of the theory...*" ([7], p. 285) So they would accept Hempel's proviso but reject the CP clauses.

(iii) Confusing laws with differential equations of the evolution type[1]. ERS propose a distinction between a theory consisting of a set of non-hedged laws and an application of a theory that might be hedged (in an easily stateable way). They argue that those examples provided by CP law proponents are just differential equations of evolution type. "But differential equations of evolution type are not laws; rather, they represent Hempel's applications of a theory to a specific case. They are derived using (unhedged) laws along with non-nomic modeling assumptions that fit (often only approximately) the specific case one is modelling. Because they depend on such non-nomic assumptions, they are not laws." ([7], p. 286)

(iv) Early Cartwright on component forces. ERS raise two objections to Nancy Cartwright's local anti-realism about component forces: in many cases they (component forces) are measurable; it is not clear that it follows that something is not occurrent just because it is not measurable. ([7], p. 287) Thus, it seems to me, they propose the local realism of component forces to support their (local) realism about the (component) laws.

(v) Cartwright's argument from Aristotelian natures and experimental method. ERS point out, I think correctly, that Cartwright's primary goal is to argue against a "Humean" view that restricts the ontology of science to the behaviors of physical systems and regularities in those behaviors, and in favor of a broader ontology that includes natures and capacities. In order to emphasize that laws could be linked to (supervene on) behaviors even without the concept of capacity, ERS repeat their supervenience thesis in this part: "One can grant that there is a lot more to being a law of nature than just being a true behavioral regularity, and even grant that what laws state is helpfully understood in terms of capacities, while maintaining that laws (and capacities) must supervene on the behaviors of physical systems." ([7], p. 288)

(vi) The world as a messy place. The CP laws proponents would argue: "The world is an extremely complicated place. Therefore, we just have not good reason to believe that there are any non-trivial contingent regularities that are strictly true throughout space and time." ERS don't think the inference is valid. They acknowledge the premise of this argument is "undeniably true", however denies the conclusion.

ERS also mention two objections to CP laws in other sciences: (1) there seems to be no acceptable account of their semantics; (2) there seems to be no acceptable account how they can be tested. They think the first objection is "not fatal to CP laws", while the latter, untestability of CP laws, is decisive. They argue: "... either this [CP] auxiliary can be stated in a form that allows us to check whether it is true, or it can't. If it can, then the original CP law can be turned into a strict law by

---

[1]As I understand, the acceleration of falling objects equals g could be an example of differential equations. Because it can be deduced from Universal Gravitation Law and some non-nomic assumptions, say, diameter and mass of the earth, and there is no air friction, etc.

substituting the testable auxiliary for the CP clause. If it can't, then the prediction relies on an auxiliary hypothesis that cannot be tested itself. But it is generally, and rightly, presumed that auxiliary hypotheses must be testable in principle if they are to be used in an honest test. Hence, we can't reply on a putative CP law to make any predictions about what will be observed, or about the probability that something will be observed. If we can't do that, then it seems that we can't subject the putative CP law to any kind of empirical test." ([7], p. 293)

So if my reading is correct, ERS' main arguments against CP laws can be summarized as following three theses: (1) CP clauses can be easily eliminated if we properly use the scientific language, i.e. ERS'(i). (2) The CP laws can not be tested, if we can not substitute testable auxiliaries for the CP clauses. (3) So called "CP laws" are just differential equations of evolution type (which hedged on nonnomic assumption), but laws are strict, i.e. ERS'(ii), (iii), (iv) and, perhaps, (vi).

I will argue against the first two theses in section 3. The third is more complicated, would involve a deep understanding of laws of nature, so I would like to discuss it in section 4.

3   Eliminability and Untestability

Is a CP clause eliminable? Here is Lange's example. To state the law of thermal expansion [the change in length of an expanding metal bar is directly proportional to the change in temperature], "one would need to specify not only that no one is hammering the bar on one end, but also that the bar is not encased on four of its six sides in a rigid material that will not yield as the bar is heated, and so on" ([13], p. 234).

ERS think this example is expressed in a language that "purposely avoids terminology from physics". If we use technical terms from physics, the condition can be easily stated: "The 'law' of thermal expansion is rigorously true if there are no external boundary stresses on the bar throughout the process." ([7], p. 284)

But how can we be sure any forces on the metal bar, say gravity by the earth or electric force by electric charges nearby, would not be a stress, which could influence the expansion of the metal bar? Even we agree with ERS'strict terminology, consider the temperature is raised higher than the melting point of the metal, would the length of the metal bar be still be proportional to its temperature? In fact, ERS do not mention the melting temperature at all in their strict or rigorous reconstruction of the thermal expansion law.

ERS give another example "... Kepler's 'law' that planets travel in ellipses is only rigorously true if there is no force on the orbiting body other than the force of gravity from the dominant body and vice versa." ([7], p. 284)

But is "other than" terminology from physics?[2] Even it is, would the ellipse

---

[2]If there is no clear-cut distinction between the language of physics and ordinary language, perhaps because they always overlap, I do not understand why ERS urge "terminology from physics".

law still hold if, say, the mass of the sun is increasing or decreasing because of certain chemical reaction, which is not "force" at all? If we consider all such interference, I am afraid that ERS' rigorous reformulation would finally have to expand infinitely.[3]

Is a CP law testable? ERS mention two common views for testability of CP laws: (1) We can confirm the putative law that CP, all Fs are Gs by finding evidence that in a large and interesting population, F and G are highly positively statistically correlated; (2) We can confirm the hypothesis that "CP, all Fs are Gs" if we find an independent, non-ad-hoc way to explain away every apparent counter-instance, that is, every F that is not a G.

ERS think the former just lend confirmation to the stronger claim that in some broader class of populations, F and G are positively statistically correlated, that would not be a CP law. And the latter is not sufficient. Here is their counter-example: "CP, white substances (or compounds containing hydrogen) are safe for human consumption." Although we can explain away any white substance would be not safe for human by modern biology or medicine, it is not a law at all.

I can agree ERS'two objections are nice, especially the first, but I would like to raise another testability possibility for CP laws – their contrapositions are testable.[4] If we write a CP law "If CP, then L" as "$CP \rightarrow L$", it is logically equivalent to "$\sim L \rightarrow \sim CP$". We can easily get, say, "If not all Fs are Gs, then not CP".

Here I consider two interpretation of CP, "there are no interferences" and "other things being equal". I give the former the logical form $\sim (I1 \lor I2 \lor I3 \ldots)$, Ii refers to different Interferences, which could be a infinite set. The latter can be written as $(E1 \land E2 \land E3 \land)$, here Ei means various Equal conditions, which again could be infinite. So whenever F is not G, there must be at least one interference or one conditional unequal. So experimenters try to find the interference or unequal condition, finally they do – there is a disturbing factor or something unequal.[5] I think that is a confirmation of the CP law!

With regard to ERS's "white substances" counter-example, I think it involve the distinction of genuine laws and accidental generalizations. According to the best knowledge of modern science, we regard "CP, white substances (or compounds containing hydrogen) are safe for human consumption" as an accidental generalization rather than a genuine law, even if it is true. But generalizations of the strict form would face the same problem. "All gold are less than 10^6 Kg" would be true

---

[3]Here, ERS provide two reasons why they are no CP laws: (a) the CP clause is easily eliminable by a known condition, and (b) they are not laws anyway [but differential equations] ([7], p. 284). I would argue against the second claim in Section 4.

[4]I learned the idea from [8]. They raise a contraposition argument against Nancy Carights claim that the fundamental laws do not apply in the real world.

[5]If not, they have to keep going on. But the confirmation of strict laws has the same induction problem.

[6]It seems to me it is possible to work out distinction between a genuine CP law and an accidental

while not regarded as a law of nature. So I do not think ERS'counter-example justify the untestability of CP laws.

4   Differential Equations and Supervenience

ERS argue that those so called "CP laws", say, thermal expansion law or Kepler's law, are just differential equations of evolution type. They are not laws at all. But I think, that claim would be inconsistent with Earman's so called "system approach" ([1], p. 2) to the understanding of laws of nature.

In the discussion on laws of nature, there are mainly two camps in the philosophy of science. David Armstrong, Michael Tooley and Fred Drestke give a necessitarian view. They think a kind of physical or nomic necessity distinguishes the genuine laws from the accidental generalizations. But it is still difficult for them to work out an explicit definition of that necessity. So nowadays J.S. Mill, Frank Ramsey and David Lewis'view (therefore MRL) is more popular. John Earman is in this camp.

MRL think laws are "consequences of those propositions which we should take as axioms if we knew everything and organized it as simply as possible in a deductive system" ([20], p. 38). So "a contingent generalization is a law of nature if and only if it appears as a theorem (or axiom) in each of the true deductive systems that achieves a best combination of simplicity and strength" ([15], p. 73).

Therefore, according to MRL, and Earman, "No sphere of uranium-235 has diameter greater than 100 meters" is a law of nature, while "No sphere of gold has diameter greater than 100 meters" is just an accidental generalization. Because the former belongs to our knowledge system of quantum physics, but the latter does not.

A system approach (MRL and Earman) would acknowledge even theorems (not only axioms) of our deductive system of the modern sciences are laws of nature. The thermal expansion law or Kepler's law are the consequences of modern physics, say, solid mechanics or Universal Gravitation Law. Why Earman insists they are just differential equations of evolution type, not laws of nature? I think his claim in the discussion of the CP law is inconsistent with his point concerning the law of nature.

ERS think there is a distinction between conditions for the truth of a law (CP) and conditions for the validity of its application (Provisos). They accept Hempel's provisos but reject CP clauses. Conceptually, conditions for the truth of a statement may not equal to the conditions for the validity of its application. "The Independence Day of USA is July 4" is true; but is not necessary for me to apply it in my

---

CP generalization by MRL view, discussed in section 4. Here is a simplified example. Consider from a CP laws, $CP1 \rightarrow L1$, an axiom of our best deductive system, with other initial conditions O, we can logically get $CP1 \rightarrow (L1 \wedge O)$. If the consequent imply, say, another law-like sentence L2, $CP1 \rightarrow L2$ could be a theorem from the systems, therefore another CP law.

pursuing a lady. But the situation in fundamental physics is a little bit different.

The fundamental laws of physics are always abstract. It seems there is no direct way for us to justify their truth. Of course, it does not mean those fundamental laws are untestable. We can deduce something, usually with the help of bridge principles (or correspondence sentences), from the abstract laws, combined with non-nomic assumptions or initial conditions. From my point of view, that is a kind of application of the abstract laws to the real situations. Since these derivations are testable, we can confirm (or disconfirm) the truth of the fundamental laws.

So the conditions for the truth of a law are closely related, if not logically equivalent, to their applications. Consider ERS'supervenience thesis. They insist laws must supervene on the behaviors of physical systems. According to *Oxford English Dictionary*, "supervene" means "to come on or occur as something additional or extraneous; to come directly or shortly after something else, either as a consequence of it or in contrast with it; to follow closely upon some other occurrence or condition". So the supervenience thesis should remind us that laws always "come after" the real behaviors.

Here is Cartwright's nice analogy. She regards the relation between laws of nature and real situations as a kind of abstract-concrete relation, like morals and fables. She quotes Leesing's claim, "The general exists only in the particular and can only become graphic (anschauend) in the particular". Consider the moral: The weaker are always prey to the stronger ([3], p. 37-43). We can find the real and concrete situation as described in Lessing's fable "A marten eats the grouse. A fox throttles the marten; the tooth of the wolf, the fox." I think her analogy gives a wonderful example of supervenience: the moral supervene on the fable.[7] In what sense the moral is true? It provides a nice idealization of the real situation, say, the relation between martens, grouses, foxes and wolves. It can be applied to the relation between other animals, perhaps humans and nations too. Suppose in a possible world, there are no animal, human, nation or something like that, the moral need not be true any longer, since there is nothing it can supervene on. Now suppose in another possible world, the electric charge of everything is removed, because of a certain evolution of the universe, while other things (so laws) remain equal. Would the law of electromagnetism force, one of the four fundamental forces in modern physics, still hold? According to Earman's distinction of the conditions for truth and for application, one can argue the law of electromagnetism force still holds however it can not apply any longer. But according to the supervenience thesis, the law can not supervene on any physical behavior since there is no electric charge at all. Again, I am afraid Earman's point on the CP laws is inconsistent

---

[7]Here we need not hold anti-realism about the fundamental laws. Since the moral The weaker are always prey to the stronger can still be true in the idealization of the real situation. Cartwright([3], p. 23) wrote, "Nowadays I think that I was deluded about the enemy: it is not *realism* but *fundamentalism* that we need to combat."

with his understanding of the law of nature.

Perhaps the reason why ERS stick to the strictness of laws is the mathematical tradition, I think that is a kind of residual Platonic idealism, in the modern sciences. According to [12], there are two traditions, mathematical versus experimental, in the development of physical science. Because in ancient Greece, the classical sciences, astronomy, harmonics, mathematics, (geometrical) optics, and statics, are practiced by a single group and participating in a shared mathematical tradition, Kuhn also calls the mathematical tradition as "classical mode". The experimental tradition, since mainly the result of the work of Francis Bacon, is called "Baconian mode". Kuhn thinks "In to the nineteenth century the two clusters, classical and Baconian, remained distinct" ([12], p. 48).[8] Although both traditions can be found in Europe, the center for Baconian work was Britain, and for the mathematical the Continent, especially France. Until the 20th century, the long-standing division between mathematical and experimental tradition has been more and more obscured or even seems to have disappeared. But Kuhn insists "it continues to provide a source of both individual and professional tensions" ([12], p. 64).

Cartwright also mentions the distinction between English and French tradition. She quotes Duhem's two kinds of thinker: the deep but narrow minds of the French, and the broad but shallow minds of the English. Her position is surely the English: "The realist thinks that the creator of the universe worked like a French mathematician. But I think that God has the untidy mind of the English." ([2], p. 19)

Historian of science, Alexander Koyre, proposes that Galileo's scientific revolution is a revival of Platonism ([11]), although Galileo himself prefer to be regarded as "New Archimedes". I am not sure whether it is very crude to name the mathematical tradition as Platonic mode, but it seems to me that the modern sciences, perhaps unconsciously, assume Platonic view.

Plato's concept of an idea is well known. According to him, the idea of human being is perfect – moral, wise, healthy, beautiful, whatever. But since we ordinary humans are just imperfect copies of that idea, there are many problems in a human's world.

Similarly, many thinks the laws of nature are true in an ideal realm. The laws of nature, at least after we discover and confirm them, would be true forever, no matter what actually happens. If real observation does not conform to the derivation of the laws of nature, it is just because the real situation is not as ideal as required. That sounds like the reality becomes a copy, usually imperfect, of the ideality. I wonder whether that is a remnant of Platonic idealism. I do not plan to

---

[8]Galileo and Newton are apparent exceptions. But Kuhn acknowledges only Newton is a real one. Galileo's dominant attitude toward that aspect of science remained within the classical mode. Newton did participate unequivocally in both traditions. His *Principia* lies squarely within the tradition of the classical science, the *Opticks* is in the Baconian ([12], p. 49-50).

argue against Platonic idealism in this paper. But I find there is tension between Platonic view and Humean view (especially the supervenience thesis) of laws of nature. If somebody want to be a coherent Humean, why not reversing the order, to consider the laws of nature as the idealization, so often imperfect, of our real world?

5 Conclusion

Earman et al raise several objections to the *Ceteris Paribus* laws. In this paper, I argued that CP clauses could be ineliminable even with scientific terminology, and that it is also possible to test the contraposition of a CP law, therefore the law itself. Earman's account of differential equations may violate his MRL view of laws of nature. I also suggest it seems there is still a kind of residual Platonic idealism in the modern sciences, perhaps because of the mathematical tradition. Again, the Platonic view of laws of nature may be inconsistent with Earman's supervenience thesis (or perhaps Humean empiricism). If we give up the remnant of Platonic idealism, the concept of CP laws would not be so difficult to accept.

BIBLIOGRAPHY

[1] Carroll, J. ed.: *Readings on Laws of Nature*. University of Pittsburgh Press, Pittsburgh (2004)
[2] Cartwright, N.: *Natures Capacities and Their Measurement*. Oxford University Press, Oxford (1989)
[3] Cartwright, N.: *The Dappled World*. Cambridge University Press, Cambridge (1999)
[4] Cartwright, N.: In Favor of Laws that Are Not *Ceteris Paribus* After All. *Erkenntnis* 57, 435-439 (2002)
[5] Earman, J., Roberts, J.: *Ceteris Paribus*: There is No Problem of Provisos. *Synthese* 118, 439-478 (1999)
[6] Earman, J.: Laws of Nature. In: Y. Balashov and A. Rosenberg ed.: *Philosophy of Science—Contemporary Readings*. Routledge, London& New York (2002)
[7] Earman, J., Roberts, J., Smith, S.:*Ceteris Paribus* Lost. *Erkenntnis* 57, 281-303(2002)
[8] Elgin, M., Sober, M.: Cartwright on Explanation and Idealization. *Erkenntnis* 57, 441-450 (2002)
[9] Giere, R.: *Science without Laws*. the University of Chicago Press, Chicago (1999)
[10] Glymour, C.: A Semantics and Methodology for *Ceteris Paribus* Hypothesis. *Erkenntnis* 57, 395-405 (2002)
[11] Koyre, A.: *Galileo Studies*. Humanities Press, Altantic Highland, NJ (1978)
[12] Kuhn, T.: *Essential Tension*. the University of Chicago Press, Chicago (1977)
[13] Lange, M.: Natural Laws and the problem of Provisos. *Erkenntnis* 38, 233-248 (1993)
[14] Lange, M.: Whos Afraid of Ceteris-Paribus Law? Or: How I Learned to Stop Worrying and Love Them. *Erkenntnis* 57, 407-423 (2002)
[15] Lewis, D.: *Counterfactuals*. Harvard University Press, Cambridge, MA (1973)
[16] Lipton, P.: All Else Being Equal. *Philosophy* 74, 155-168 (1999)
[17] Mitchell, S.: *Ceteris Paribus* An Inadequate Representation for Biological Contingency. *Erkenntnis* 57, 329-350 (2002)
[18] Morreau, M.: Other Thing Being Equal. *Philosophical Studies* 96, 163-182 (1999)
[19] Persky, J.: *Ceteris Paribus*. *Journal of Economic Perspectives* 4, 187 (1990)
[20] Ramsey, F.: *Foundations of Mathematics*. Humanities Press, Altantic Highland, NJ (1978)
[21] Reutlinger A. et al: *Ceteris Paribus* Laws. *Stanford Encyclopedia of Philosophy* (2011) ⟨ http://plato.stanford.edu/entries/ceteris-paribus/⟩
[22] Schurz, G.: *Ceteris Paribus* Laws: Classification and Deconstruction. *Erkenntnis* 57, 351-372 (2002)

[23]   Sphohn W.: Laws, *Ceteris Paribus* Conditions, and the Dynamics of Belief. *Erkenntnis* 57, 373-
       394 (2002)
[24]   Woodward, J.: There is no such thing as a *Ceteris Paribus* Law. *Erkenntnis* 57, 303-328 (2002)

Ceteris Paribus(,) Logic Helps

KOHEI KISHIDA

One of the objections to *ceteris paribus* (CP) laws that Earman, Roberts, and Smith (ERS) [2] raise is that CP laws are not testable, as long as they are genuine and cannot be rewritten into strict laws. In his defense of genuine CP laws from this objection, Professor Wang offers an interesting idea of how a CP law can be tested, and in particular confirmed. In this set of comments I would like to mention a couple of questions that arise from Professor Wang's argument and assumptions.

Let us briefly review how the ERS objection and Professor Wang's defense go regarding the testability of CP laws. CP laws are stated under the form "All other things being equal, L", where L is a strict law providing a universal generalization, such as "All Fs are Gs", or a statistical one, "Every F is a G with probability p". As is clear from their arguments, both ERS and Professor Wang take a CP law as a form of conditional that implies L given that all other things are indeed equal. Very roughly put, the ERS objections goes as follows, where we write CP for "all other things being equal". If the clause CP in an apparently CP law CP, all Fs are Gs has a truth (or assertability) condition C that makes the clause testable, then the law is really a strict law All Fs that are Cs are Gs. So, in a genuine CP law, CP cannot have any such truth condition. Therefore the condition under which we can derive "All Fs are Gs" from the CP law is not testable, and hence the law cannot make testable predictions (even with the help of testable auxiliary hypotheses).

Professor Wang's defense of the testability of CP laws goes as follows. He assumes that a CP law is a conditional of the form

(1) If CP, then L,

and that CP is a conjunction, and presumably an infinitary one, $\bigwedge_{i \in O} E_i$, where O is the set of "all other things" and each E_i means that the particular $i \in O$ of the "other things" is "equal". Moreover, he takes (1) as equivalent to its contrapositive

(2) If $\neg L$, then $\neg CP$.

Then taking an instance of $\neg L$, the antecedent of (2), and showing that it satisfies $\neg E_i$ for some i and therefore $\neg CP$, the consequent of (2), confirms (2), and so confirms the contrapositive (1).

For the purpose of the debate between ERS and Professor Wang, it is an interesting question whether Professor Wang's characterization of the clause $CP =$

$\bigwedge_{i \in O} E_i$ would let an ERS-type argument go, that it would make (1) a strict law since CP could then be turned into a testable clause with a definite truth condition. Answering this question seems to require more reflection upon logical properties of testability of propositions,[1] and in particular ones involving infinitary operations (for instance, infinitary conjunction preserves falsifiablity but not verifiability). A proper logical account of testability may help to expand Professor Wang's idea and to provide a formulation of CP laws that renders CP laws testable, based on the testability of E_i, without turning them into strict laws.[2]

Among the assumptions in Professor Wang's argument, the nature of "If ... then ..." in (1) may be interesting to non-classical logicians. Professor Wang takes (1) as a conditional that is equivalent to its contrapositive (2). Nonetheless, there are many approaches to CP clauses in which (1) is not a classical conditional; rather, the apparent antecedent, CP, is often regarded as a modality that operates on L. For instance, by reading a CP law "CP, L" as "normally L", one may take it as stating that L is true in normal cases (or normal possible worlds, if one prefers it).[3] Or one may take a CP law "CP, all Fs are Gs" as endorsing "If F then G" as a non-monotonic or defeasible conditional.[4] The point is that logicians offer some reasons to read (1) as a non-classical statement that may not be reducible to a classical conditional; if (1) is indeed such a statement, it may fail to be equivalent to its contrapositive (2), or even to have a contrapositive at all. Thus, a logician would invite Professor Wang to the question of whether CP laws are reducible to classical conditionals.

A related issue concerns Professor Wang's argument steps on confirmation. He combines the following two principles:

(3) A (universally generalized) material conditional is confirmed by an instance satisfying both its antecedent and consequent;

(4) Contrapositives are equivalent and so confirmed by the same instances;

and he concludes that it confirms (1) to provide an instance with which its consequent and antecedent both fail. This line of argument immediately reminds us of Hempel's celebrated paradox of the ravens—by the same form of argument, anything non-black that is not a raven is supposed to confirm "All ravens are black", though it is implausible that it does. It is a question worth considering whether (4) applies to CP laws, since, as I mentioned above, they may fail to be equivalent to

---

[1] It may be more proper to call them *topological* properties; see [4].

[2] In fact, in this direction of investigation, Glymour [3] (in response to ERS [2] in the same journal issue) provides a formulation of CP laws that renders them testable in the limit, but that does not turn them into strict laws.

[3] See, e.g., [7] and Sect 8.2 of [5]. See also Sect 4 of [1]. Glymour [3] reads "CP, L" as "normally L", but does not treat "normally" as a modal operator.

[4] See, e.g., [6] and Sect 8.3 of [5].

their contrapositives or even to have contrapositives. Yet, in either case, Professor Wang's idea seems to call our attention to further investigation into exactly how (3), the so-called Nicod principle, can apply to statements involving CP clauses, and, more generally, exactly how such statements can be confirmed.

BIBLIOGRAPHY

[1] J. van Benthem, P. Girard, and O. Roy, "Everything Else Being Equal: A Modal Logic for *Ceteris Paribus* Preferences", *Journal of Philosophical Logic* **38** (2009), 83–125.

[2] J. Earman, J. Roberts, and S. Smith, "*Ceteris Paribus* Lost", *Erkenntnis* **57** (2002), 281–301.

[3] C. Glymour, "A Semantics and Methodology for *Ceteris Paribus* Hypotheses", *Erkenntnis* **57** (2002), 395–405.

[4] K. T. Kelly, *The Logic of Reliable Inquiry*, Oxford University Press, Oxford, 1995.

[5] A. Reutlinger, G. Schurz, and A. Hüttemann, "*Ceteris Paribus* Laws", *The Stanford Encyclopedia of Philosophy* (Spring 2011 Edition), E. N. Zalta, ed.,http://plato.stanford.edu/archives/spr2011/entries/ceteris-paribus/.

[6] A. Silverberg, "Psychological Laws and Non-Monotonic Logic", *Erkenntnis* **44** (1996), 199–224.

[7] W. Spohn, "Laws, *Ceteris Paribus* Conditions, and the Dynamics of Belief", *Erkenntnis* **57** (2002), 373–394.

Ceteris Paribus **Found?**

ALEXANDER REUTLINGER

In his "Is There Such a Thing as a *Ceteris Paribus* Law?", Wang argues for an affirmative answer to this question by way of undermining arguments against the existence of and methodological difficulties connected to *ceteris paribus* laws (henceforth, cp-laws). Wang's main target is John Earman, John T. Roberts and Sheldon Smith's paper *"Ceteris Paribus* Lost" ([4]). Roberts and Smith present – here I agree with Wang– the most forceful arguments against philosophical accounts of cp-laws (see also [2]). Although I agree with Wang that there are such things as cp-laws, that these law statements have truth-conditions, that they can be confirmed by evidence, and that laws of this kind play an important role in scientific practice, I am less convinced of at least three of his objections to Earman, Roberts and Smith. I go through Wang's objections in chronological order and present one challenge and one objection.

1   A Challenge for Wang's Account of Testability

Wang argues that – contra Earman, Roberts and Smith (henceforth, ERS) – cp-laws are indeed testable. His main argument consists in the proposal that a statement of the form "cp, all Fs are G" is testable in the following way:

> [...] whenever F is not G, there must be at least one interference or one condition unequal. So experimenters try to find the interference or unequal condition, finally they do there is a disturbing factor or something unequal. I think that is a confirmation of the CP law! (Wang, pp. 6)

Initially the approach Wang takes looks quite promising. Identifying an interference that lead to a particular counter-instance of the law does not seem to entail a typical problem for cp-clauses, to which Wang refers: for explaining one specific counter-instance, it is not necessary to list infinitely many possible interfering factors. The trouble for Wang's reply is that Pietroski and Rey[6] advocates a similar and elaborated version of this view – the epistemic completer account – and, unfortunately for Wang, Earman and Roberts ([2]), among others, have raised an objection to Pietroski and Rey's view (cf. [8] section 5.2). According to Pietroski and Rey, the cp-law statement "cp, all Fs are G" is non-vacuously true (and therefore empirically testable) iff the following conditions are satisfied:

1. F and G are nomological, not "*grue*-like" predicates.

2. If a counter-instance to "all Fs are G" (that is, an F that is $\neg G$) occurs, then one is committed to explain $\neg G$ by referring to a factor H, which is independent[1] of the cp-law in question. Pietroski and Rey allow two possibilities regarding the independent explanatory force of H: (i) H alone explains $\neg G$, or (ii) H in conjunction with the cp-law explains $\neg G$.

3. It is the case that either (i) $F \wedge G$, or (ii) $F \wedge \neg G$ and conditions 1 and 2 are satisfied (cf. [6], pp. 92).

Condition 2 does the crucial work for establishing that cp-laws are testable. This condition is strinkingly similar to Wang's proposal of how to test cp-laws. However, the critics of Pietroski and Rey's approach have argued that conditions 1-3 are not sufficient for saving special science laws from being vacuously true, because conditions 1-3 permit that F is completely irrelevant for G, and $\neg G$ is still perfectly explained by an independent factor H ([2], pp. 453-454; [9], pp. 366-367). To illustrate the charge of irrelevance, Earman and Roberts present the following counterexample:

> Unfortunately, [Pietroski and Rey's account] is not sufficient for the non-vacuous truth of the cp-law. To see why, let "Fx" stand for "x is spherical", and let "Gy" stand for "$y = x$ and y is electrically conductive". Now, it is highly plausible that for any body that is not electrically conductive, there is some fact about it – namely its molecular structure – that explains its non-conductivity, and that this fact also explains other facts that are logically and causally independent of its non-conductivity – e.g., some of its thermodynamic properties. Thus,[the conditions of Pietroski and Rey's account] appear to be easily satisfied. If Pietroski and Rey's proposal were correct, then it would follow that **ceteris paribus**, all spherical bodies conduct electricity. ([2], pp.453)

Earman and Roberts conclude:

> The general moral of this observation seems to be that it is not enough simply to require, [...] that when cp: $(A \rightarrow B)$, any case of A accompanied by $\sim B$ must be such that there is an independent explanation of $\sim B$. This is because this requirement does not guarantee that A is in any way relevant to B, which surely must be the case if cp: $(A \rightarrow B)$ is a law of nature. ([2], pp. 454)

---

[1] According to [6], pp. 90, $\neg G$ is explained *independently* by referring to H if (a) H is not a logical consequence of $\neg G$ (i.e. logical independence of H), and (b) the explanatory factor H is not an effect of $\neg G$ (i.e. causal independence of H).

The challenge for Wang is to address Earman and Roberts's counterexample and to show that his own approach succeeds in avoiding the charge of irrelevance, unlike Pietroski and Rey's account (cf. [7] for a reply to Earman and Roberts).

2   Objection: Not a Failure of Humean Accounts of Laws

Wang objects to ERS that the distinction between abstract laws of nature and applied "differential equations of the evolution type" ([4], pp. 285-286) – which ERS use to argue against the importance of cp-laws in physics – contradicts the Humean view of laws. This is an original objection. The Humean view of laws consists in the best system account of laws and Humean supervenience about nomic facts (cf. [5]). According to the best system account, the laws are, roughly, the axioms and theorems of the simplest and empirically strongest deductive system "summarizing" the non-nomic facts. Humean supervenience is the claim that nomic facts supervene on non-nomic facts. At least Earman endorses the best system account and Humean supervenience ([1], chapter 5). Wang claims:

> According to Earman's distinction of the conditions for truth and for application, one can argue the law of electromagnetism force still holds however it cannot apply any longer. [...] Again, I am afraid Earman's point on the CP laws is inconsistent with his understanding of the law of nature. (Wang, p. 8)

But why believe that – as Wang suggests – some abstract statement L can be a law according to the best system account, even if L is not applicable? That is, why believe that L is a law even if L cannot be applied to a real physical system by using "provisos" and other "nomicly contingent modeling assumptions" in order to derive a differential equation of the evolution type? An advocate of the BSA might simply argue that a statement L only makes it into the best system if it is simple and empirically strong. The empirical strength of L paradigmatically consists in allowing us to derive (many) differential equations that apply to concrete physical systems (theorems of the best system). Hence, the Humean concludes that the abstract statement L only qualifies as an empirical statement summarizing non-nomic facts if it can be applied to physical systems in the way that ERS suggest. Therefore, it is impossible that, as Wang objects, L belongs to the best system and L is not applicable to concrete physical systems. Surprisingly, Wang seems to concede that the abstract law statements have empirical content only under the conditions that the law can be applied:

> The fundamental laws of physics are always abstract. It seems there is no direct way for us to justify their truth. Of course, it does not mean those fundamental laws are untestable. We can deduce something, usually with the help of bridge principles (or correspondence sentences), from the abstract laws, combined with non-nomic assumptions or initial conditions. (Wang, p. 7)

However, if Wang concedes this much about the empirical content of the abstract laws, it is unclear how some statement L, which cannot be applied to physical systems, could be an axiom or a theorem of the best system, since propositions figuring in the best system are required to be (true) empirical statements. Moreover, proponents of the best system account may hold that the abstract laws are testable only if they can be applied (by, among other things, employing Hempelian provisos) and testability entails Humean Supervenice (cf. [3]). Hence, Wang's objection fails: a Humean account of laws is compatible with distinguishing abstract laws and applied derived equations.

Wang also presents another related concern. He holds that Earman's best system account implies Platonism (in this context, the view that laws are abstract mathematical objects), which is in contradiction with the best system account and Humean supervenience (Wang, pp. 9). However, I do not believe that this is a convincing objection to ERS for the following reasons. *Either* it is the same objection as before (i.e, it is Wang's concern that a statement L might have no empirical content and still count as a law in the framework of the best system account). If that is how to understand Wang's objection from Platonism, I have presented reasons undermining this objection (see above). *Or* the objection rests on a conflation of laws and law statements. The law *statements* figuring in the best system may well be abstract objects qua being propositions (depending on one's preferred account of propositions). However, the laws *themselves* are – for a Humean – regularities (or patterns) in the concrete Humean mosaic of non-nomic facts. Regularities understood this way are not at all abstract (mathematical) objects. Hence, contra Wang, a Humean is not committed to laws being abstract objects.

BIBLIOGRAPHY

[1] Earman, J.: *A Primer on Determinism*. Reidel, Dordrecht (1986)
[2] Earman, J., Roberts, J.: *Ceteris Paribus*: There is No Problem of Provisos. *Synthese* 118, 439-478 (1999)
[3] Earman, J., Roberts, J.: Contact with the Nomic. A Challenge for Deniers of Humean Supervenience about Laws of Nature Part II: The Epistemological Argument for Humean Supervenience. *Philosophy and Phenomenological Research* LXXI: 253-286 (2005)
[4] Earman, J., Roberts, J., Smith, S.: *Ceteris Paribus* Lost. *Erkenntnis* 57, 281-303(2002)
[5] Lewis, D: Humean Supervenience Debugged. In *Papers in Metaphysics and Epistemology*, 224-247, Cambridge University Press, Cambridge (1994)
[6] Pietroski, P. M. and Rey, G.: When Other Things Aren't Equal: Saving *Ceteris Paribus* Laws From Vacuity. *British Journal for the Philosophy of Science* 46 (1):81-110 (1995)
[7] Reutlinger, A.: A Theory of Non-universal Laws. *International Studies in the Philosophy of Science* 25, 97-117 (2011)
[8] Reutlinger A. et al: *Ceteris Paribus* Laws. Stanford Encyclopedia of Philosophy (2011) The Stanford Encyclopedia of Philosophy (Spring 2011 Edition), Edward N. Zalta (ed.), URL = http://plato.stanford.edu/archives/spr2011/entries/ceteris-paribus/
[9] Schurz, G.: *Ceteris Paribus* Laws: Classification and deconstruction. *Erkenntnis* 57, 351-372 (2002)

Ceteris Paribus as Defaults

LIYING ZHANG

In his paper "Is There Such a Thing as a *Ceteris Paribus* Law", Wei Wang defends the view that the existence of *Ceteris Paribus* (CP) laws is reasonable. He argues that CP clauses cannot be eliminated even with scientific terminology.

More specifically, Wang says: "Since there are hardly any universal and exceptionless laws in special sciences, there is more and more philosophical discussions on *Ceteris Paribus* laws and their roles in contemporary sciences". He also discusses the tension between Platonic and Humean views: since our real world is imperfect, we need laws of nature that are imperfect too. His conclusion is: "If we give up the remnant of Platonic idealism, the concept of CP laws would not be so difficult to accept." I agree with this perception of the world. In fact, in recent decades, there have been more and more studies on default reasoning and non-monotonic reasoning, which typically contain generic sentences. The most striking feature of generic sentences is their being tolerant of exceptions. Behind the current trend from classical logic to default or non-monotonic reasoning, and from universal sentences to generic sentences, there is a trend from the Platonic view to a Humean view.

In the paper *"Ceteris Paribus* Lost?" by John Earman, John Roberts and Sheldon Smith (ERS), the authors argued in six ways that there should be no concept of CP laws in physics. Wang summarizes their six theses to three:

(1) CP clauses can easily be eliminated by proper use of the scientific language.

(2) CP laws cannot be tested, if we cannot substitute testable auxiliaries for the CP clauses.

(3) Exampes of "CP laws" are just differential equations of an evolutionary type, but laws are strict.

As to (1), Wang thinks that ERS rigorous reformulation would have to expand infinitely, so CP laws for science cannot be eliminated. For (2), Wang argues that the contraposition of a CP law is testable, whence the law itself is testable too. (Writing a CP law "If CP, then L" as "$CP \rightarrow L$", it is logically equivalent to its contraposition "$\sim L \rightarrow CP$".) As for (3), Wang argues that Earman's account of differential equations may violate his MRL view of laws of nature, and Earman's view of laws of nature may be inconsistent with his supervenience thesis.

I agree with these arguments on the whole, and I think the argument is successful in refuting the ERS's viewpoint. But there are some details I want to supply and some discussion points I would like to raise.

For a start, the author formalizes CP laws as "$CP \rightarrow L$", and there are then two main interpretations of these:

1) $\sim (I1 \vee I2 \vee I3 \ldots) \rightarrow L$ (there are no interferences)

2) $(E1 \wedge E2 \wedge E3 \ldots) \rightarrow L$ (other things being equal)

But how to formalize L? The paper provides no details. But in the philosophy of science, normally, L is formalized as a universal sentence $\forall x (Fx \rightarrow Gx)$.

From the viewpoint of a logician, these formulations leave room for improvement. What are "no interferences" or "other things being equal"? Can we represent these features formally? If we take L as some Ei, or $\sim L$ as some Ii, is there a circle? In fact, in the fields of logic and linguistics, there are many studies that may be helpful for these issues. For example, studies on default reasoning and non-monotonic reasoning focus on generic sentences that tolerate exceptions. Generic sentences can represent laws from different fields, general knowledge, habits, etc. Among these, laws of science are a very important standard kind. So, combining studies of CP laws and studies on default reasoning and non-monotonic reasoning with generics would be a very good thing, both to the philosophy of science and to logic and linguistics.

As I mentioned above, CP laws have two different interpretations:

3) If there are no interferences, then L.

4) If other things being equal, then L.

In "Everything else being equal: A modal logic approach to *ceteris paribus* preferences" ([3]), the authors summarized these two interpretations as:

3)' All other things being normal

4)' All other things being equal,

give a detailed logical account of the equality reading, but for CP as normality, they referred to existing default logics. As I mentioned above, there are indeed a lot of studies on normality as well. Among them, in the paper "Making the right exception" ([5]), Frank Veltman represents a sentence of the form

Ps are normally Q

by a formula
$$\forall c((Px \wedge \neg Ab_{P_x, Q_x} x) \rightarrow Qx)$$

If an object satisfies the formula $Ab_{P_x, Q_x} x$, this means that it behaves abnormally with respect to this rule. In his semantics for these formulas, Frank Veltman improved the method of abnormality from McCarthys circumscription approach in AI ([1]).

In the paper "An analysis of the meaning of generics" ([2]), Yi Mao and Beihai Zhou interpret a generic sentence with subject-predicate structure as

> (normal S) (normally P):
> for any object x, if x is normal S regarding P or not P, then, normally, x is P.

According to this analysis, the formalization of generic sentences is

$$\forall x(N(\lambda x Sx, \lambda x Px)x > Px).$$

By adding predicate and negated predicates as the parameters, both methods can avoid circularities that may bring trouble when we formalize CP law as 1 or 2. Due to a different focus, Veltman's method is simpler, but it did not include modal interpretation, while Mao and Zhou's Method is a little more complex, but it can capture the intensionality of defaults and generics.

Which method is better to represent CP laws? I leave it to the philosophers of science.

BIBLIOGRAPHY

[1] McCarthy, J: Circumscriptiona form of non-monotonic reasoning. *Artificial Intelligence*, vol.13, 27-39 (1980).
[2] Mao, Y. and Zhou, B.: An analysis of the meaning of generics. *Social Sciences in China*, Vol. xxiv, No.3, 126-133 (2003)
[3] Van Benthem, J., Girard, P., Roy, O.: Everything else being equal: A modal logic approach to *ceteris paribus* preferences. *Journal of Philosophical Logic* 38, 83-125 (2009)
[4] Veltman, F.: Default in Update Semantics. *Journal of Philosophical Logic* 25, 221-261 (1996).
[5] Veltman, F.: Making the right exceptions. to appear.

Response to Kishida, Zhang and Reutlinger

WEI WANG

All the three reviewers read my paper carefully and provide nice criticism. Here are my brief answers to their critical comments.

1 Testability of CP statements

Dr. Kohei Kishida and Dr. Alexander Reutlinger point out that my defense of testability of CP laws, appealing to confirmability of their contrapositions, may have the same problem of C. G. Hempel's famous "raven paradox". Raven paradox is a classical paradox of confirmation ([1]): since "All ravens are black" is logical equivalent to "All non-back things are non-ravens", whether a piece of white paper, which may confirm $(x)(\neg Bx \rightarrow \neg Rx)$, can confirm $(x)(Rx \rightarrow Bx)$, which is quite counterintuitive?

However, in the philosophy of science literature, white paper can be regarded as a logical confirmation of "All ravens are black"! It seems implausible only because we know white paper is non-raven, the observation provides no new information. So according to John Vickers, the important lesson of raven paradox, as well as Hempel-Goodman resolution, is that inductive inference is sensitive to background information and context ([6]).

Certainly, I agree with Dr. Kishida, infinitary conjunction preserves falsifiability but not verifiability. In fact, my colleague at Peking University, Prof. Yongping Sun, raised the same problem when I finished the draft. If we take falsifiability a better criterion of testability than confirmability, I must think more about the falsifiability of CP laws.

2 Platonism and Humean accounts of laws

Dr. Reutlinger finds no failure of Humean account of laws in ERS's paper. I would like to make more clarification here: I am arguing John Earman's supervenience thesis is wrong, but it may be inconsistent with his distinction of the conditions for truth and for application. If we accept the distinction, we can tolerate laws without any physical behavior. But that would conflict with the supervenience thesis: those laws supervene on no physical behavior, therefore be empty. So my worry is not the failure of Humean approach, but Earman's two positions, distinction of the conditions for truth and for application and supervenience thesis, seem to be incoherent.

In addition, I do not mean Earman's best system account implies Platonism. In fact, the best system account is a kind of Humean approach to laws of nature. I just think Earman's worry about CP laws might presume a kind of Platonism, which is not *the view that laws are abstract mathematical objects*, as Dr. Reutlinger understand, or a kind of mathematical realism, but that our real situations are just, usually imperfect, copies of ideal laws. According to this version, laws of nature are always true in ideal situation, and whenever we find counterexample, that is because the real situations are not so ideal. I am afraid of this kind of Platonism is inconsistent with Humean account of laws.

3   Logical formulation of CP laws

All the three commentators, especially Dr. Liying Zhang, question the logical forms of CP laws, e.g. "$CP \to L$", which I regarded as the most serious challenge to CP laws. Personally, I am not satisfied with the logical formulation of CP laws in philosophy of science either.

Due to the great role of mathematic logic in philosophy of science, philosophers of science used to form an alliance with logicians, so established the Division of Logic, Methodology and Philosophy. We can also find the name of "logical" in the early schools in philosophy of science, e.g. logical empiricism, logical pragmatism, etc. However, with the rise of the tradition of HPS, history and philosophy of science, after the publication of T. Kuhn's The Structure of Scientific Revolutions, philosophers of science pay more attention to case studies in scientific practices, but less attention to the new development of contemporary logic. So in the philosophical literatures of CP laws, there is very little discussion on its logical formulation.

As Dr. Zhang proposes, there are some logical studies of generic sentences. An important approach is her PhD. supervisor Beihai Zhou and Yi Mao's formalization and interpretation ([2]). Prof. Johan van Benthem and Fenrong Liu, in their email, also recommend default logic ([3], [5]) and their modal logic of preference ([4]).

I am not an expert on those logical fields yet, however I do wish to learn a lot from the recent logical contributions to improve the logical formulation of CP laws. Hopefully, the logical improvement should partly solve some technical difficulties in philosophy of science, e.g. eliminability or untestability of CP statements, and finally deepen our understanding of the important concept of laws of nature in fundamental and special sciences.

BIBLIOGRAPHY

[1] Hempel, C.G.: Studies in the Logic of Confirmation. *Mind* 54,1-26 and 97-121 (1945)
[2] Mao, Y., Zhou, B: An analysis of the meaning of generics. *Social Sciences in China*. Vol. xxiv, No.3, 126-133 (2003)

[3]   Reiter, R.: A logic for default reasoning. *Artificial Intelligence* 13, 81-132 (1980)
[4]   Van Benthem J., Girard P., Roy O.: Everything Else Being Equal: A Modal Logic for *Ceteris Paribus* Preferences. *Journal of Philosophical Logic* 38, 83-125 (2009)
[5]   Veltman F.: Defaults in update semantics. *Journal of Philosophical Logic* 25, 221-261 (1996)
[6]   Vickers, J: The Problem of Induction. *Stanford Encyclopedia of Philosophy* (2012), http://plato.stanford.edu/entries/induction-problem/

Introduction to a Naturalistic Philosophy of Mathematics

FENG YE

This article introduces my research project exploring a nominalistic, strictly finitistic, and truly naturalistic philosophy of mathematics. There are two distinctive features of my approach compared with other contemporary naturalistic and/or nominalistic philosophies of mathematics. First, it is strict finitism, which means a position that does not assume reality of infinity in any sense, any format. Second, it takes physicalism as its philosophical basis, and in particular, it emphasizes physicalism about cognitive subjects and processes. The technical work in this research tries to propose a logical explanation of the applicability of classical mathematics in the sciences consistent with nominalism and strict finitism [29]. The philosophical work starts from methodological naturalism and argues that a coherent naturalist should adopt physicalism about cognitive subjects and that this should imply nominalism and strict finitism in philosophy of mathematics. The research is still in progress, but I hope it has accomplished enough to make it attractive.

I will only briefly compare my approach with other contemporary philosophies of mathematics in this paper. [26] gives the motivation for my approach in the context of contemporary philosophies of mathematics, and [31] and the book [28] (in Chinese) are my comments, from my own naturalistic perspective, on the major figures in philosophy of mathematics in the 20th century, from Frege to Quine.

1 Strict Finitism and Applicability

1.1 Why strict finitism?

I will introduce my argument for nominalism later. Here I will briefly explain why nominalism and naturalism imply strict finitism. First, contemporary physicists never straightly claim that there is infinity in this physical universe. Macroscopically, the universe is believed to be finite; microscopically, physicists so far study only things above the Planck scale (about $10^{-35}m$, $10^{-45}s$ etc.). No physicists are considering things at even lower scales (say, $10^{-350}m$), and a strictly discrete space-time structure is an open option. Now, philosophers cannot know more about this physical world than physicists do, and we do not want our philosophical account of human mathematical practices contingent on the assumption that someday physicists will confirm that there is infinity in this physical world. Therefore, if a philosophical account of mathematics assumes infinity (in any sense, any format, including potential infinity), then it must be infinity either in an abstract world or in a mental or spiritual world. The former could be alleged abstract mathematical objects or mind-independent intensional entities that somehow embody infinity (e.g., Fregean senses in the third realm); the latter could be alleged intensional entities understood as creations of minds (e.g., functions according to the intuitionist Brouwer, which are potentially infinite).

Assuming an abstract world apparently contradicts nominalism. On the other side, methodological naturalism requires philosophers to take seriously the contemporary cognitive scientific claim that human thinking is ultimately neural processes in brains inside this physical world. That is, human thinking does not create any non-physical entities. This then implies that there is no non-physical world of mental things to host infinity (in whatever sense). Therefore, a truly nominalistic and naturalistic account of human mathematical practices should not assume reality of infinity in any format, in any sense. That is, a truly nominalistic and naturalistic account should be strictly finitistic. At least, a strictly finitistic account is preferable, for it is more compatible with our current knowledge in physics and cognitive sciences.

1.2   The problem of applicability and my answer

Under nominalism and strict finitism, the problem of applicability of classical mathematics in the sciences is: how can apparent references to non-existent abstract (and infinite) mathematical entities in the applications help to derive literally true conclusions about strictly finite and discrete physical things above the Planck scale in this universe? My research project intends to offer the following answer:

(1) Mathematical axioms that appear to refer to mathematical entities are not really among the final premises logically implying our scientific conclusions about strictly finite and discrete physical things in the universe; an application of classical mathematics in the sciences can in principle be transformed into a series of logically valid deductions from literally true premises about strictly finite and discrete physical things alone to a literally true conclusion about them.

If this is successful, we will have a logically plain explanation (actually, a straight logical demonstration) of the applicability of classical mathematics in the sciences consistent with nominalism and strict finitism. More specifically, it is well-known that infinite (and continuous) mathematical models are approximations to finite and discrete physical things (above the Planck scale) in scientific applications; the models are not exact representations of physical things. That is what happens when we use semi-Riemannian manifolds to model large-scale space-time structures. However, logically, it is not very clear how such 'approximations' work. An ordinary scientific derivation of a conclusion about physical things from scientific hypotheses and mathematical axioms is usually not a series of logically valid deductions on literally true and exactly true statements, even if we assume that mathematical axioms are literally true of mathematical entities. The hypotheses that connect infinite and continuous mathematical models with finite and discrete physical things are never exactly true. The answer (1) suggests that, by eliminating all apparent references to infinity and mathematical entities extraneous to the physical world, we can eliminate the 'approximations'. We can identify the literally true and exactly true premises in an application and show that they strictly logically imply the conclusion of the application, and all these literally true and exactly true premises turn out to be exclusively about strictly finite and discrete physical things, not including any premises about alleged abstract mathematical entities extraneous to this physical world. This result will come back to support nominalism and strict finitism against realism, for it actually says that the applicability of classical mathematics to this finite physical world is explained not by assuming the literal truth of mathematical axioms about alleged mathematical entities; on the contrary, it is explained by eliminating any apparent references to infinity or anything extraneous to this finite physical world.

However, I must clarify here that this is not to suggest abandoning classical mathematics in scientific applications. Exactly on the contrary, using infinite and continuous models to approximate finite and discrete physical things is an ingenious invention of our scientists. It greatly simplifies our theories and makes them tractable and ultimately possible. My research is merely to make a logical and philosophical point: to get a logically plainer picture about which literally true and exactly true premises logically imply a literally true and exactly true conclusion in an application. Moreover, I do not intend to use this finitistic explanation of applicability alone to refute realism and defend nominalism and strict finitism. I believe that nominalism and strict finitism follow from physicalism about cognitive subjects, which in turn follows from methodological naturalism, as I have mentioned above. This strictly finitistic explanation of applicability is mostly to resolve the logical puzzle of applicability given physicalism. It is to show that the fact that humans have successfully applied classical mathematics in the sciences can fit into physicalism as an overall worldview.

The work done so far for demonstrating (1) is reported in the book *Strict Finitism and the Logic of Mathematical Applications* ([29] and the paper [27]). The strategy for demonstrating (1) is as follows. First, strict finitism is formalized as a formal system **SF**. It is essentially a fragment of quantifier-free primitive recursive arithmetic with the accepted functions restricted to elementary recursive functions. Strict finitism is therefore quantifier-free elementary recursive arithmetic. An application of strict finitism to physical things can be interpreted as a series of valid logical deductions from literally true premises about strictly finite and discrete physical entities to a literally true conclusion about them. That is, the applicability of strict finitism is logically plain and is clearly consistent with nominalism and strict finitism.

Then, to explain the applicability of classical mathematics, I try to show that the applications of classical mathematics in the sciences are *in principle* reducible to the applications of strict finitism. This is achieved by trying to develop sufficiently rich applied mathematics within strict finitism. The book [29] is devoted to this. So far, the book has developed the basics of the following in strict finitism: calculus, ordinary differential equations, metric spaces, complex analysis, Lebesgue integration, (bounded and unbounded) linear operators on Hilbert spaces, and semi-Riemannian geometry. These include mathematical theories needed for the basic applications in classical quantum mechanics and general relativity. After a mathematical theory is developed in strict finitism, the applications of (the classical version of) that theory in the sciences will be reducible to applications of strict finitism. This demonstrates (1) for those applications (of classical mathematics).

1.3   Strict Finitism (SF) as a formal system and mathematics in SF

Here I will explain a little more about the formal system **SF** and about how to develop ordinary mathematics in **SF**. Please consult [29] for the details.

The language of **SF** is the language of quantifier-free typed λ-calculus with constants for base elementary recursive functions and operators for constructing new terms by bounded primitive recursion and bounded minimalization. We use o to denote the numerical type, and for any types $\sigma_1, ..., \sigma_n, \sigma$, there is the type $(\sigma_1, ..., \sigma_n \to \sigma)$ of functionals. Bounded primitive recursion and bounded minimalization can only be used to construct terms of the numerical type. This restriction implies that all terms of the numerical type represent only elementary recursive functions. Well-formed formulas of **SF** are limited to equations between terms of the numerical type and Boolean combinations of these equations. Recall that the language of **SF** is quantifier-free. An equation between two terms of a higher type, for instance, $\lambda xy.(x + y) = \lambda xy.(y + x)$, would implicitly contain a universal quantification. Generality in **SF** can only be achieved by using free variables. That is, instead of $\lambda xy.(x + y) = \lambda xy.(y + x)$, we have to state the commutative law as $x + y = y + x$, which is indeed provable in **SF**. The axioms of **SF** include the axioms of quantifier-free typed λ-calculus and the axioms characterizing base elementary recursive functions, bounded primitive recursion and bounded minimalization. Besides, **SF** contains a (quantifier-free) induction rule.

We can develop Gödel's encoding techniques and define bounded quantifiers within **SF**. These allow developing basic arithmetic in **SF**. We encode rational numbers as numerals. Real numbers are understood as elementary recursive Cauchy sequences of rational numbers, where a sequence of a type σ is understood as a term of the type $(o \to \sigma)$. Functions of real numbers are then terms of even higher types. Since physics theories are accurate only up to some finite precision (above the Planck scale), we expect that elementary recursive real numbers are sufficient for representing all physics quantities.

An ordinary mathematical proposition in classical mathematics uses unbounded quantifiers and is therefore not directly expressible in the language of **SF**. To prove such a proposition in strict finitism, we translate it into a claim stating that some terms in the language of **SF** can be constructed to satisfy a condition, where the condition is represented by a (quantifier-free) formula of **SF** involving the constructed terms. For instance, suppose that $t\,[n]$ is a numerical term in the language of **SF** with a free numerical variable n, and consider the proposition '$\lambda n.t\,[n]$ is a Cauchy sequence of rational numbers'. In classical mathematics this can be stated as

(2) $\forall k > 0 \exists n \forall l, m > n \,(|t\,[l] - t\,[m]| < 1/k).$

$(|t\,[l] - t\,[m]| < 1/k)$ can be formalized as a (quantifier-free) formula of **SF**, using the arithmetic developed in **SF**. Now, in strict finitism, we understand (2) as the claim that we can construct a term $\lambda k.N\,[k]$ and derive the following quantifier-free formula with free variables k, l, m in **SF**:

(3) $k > 0 \wedge l > N\,[k] \wedge m > N\,[k] \to |t\,[l] - t\,[m]| < 1/k.$

More generally, we use Gödel's *Dialectica* interpretation to translate a formula in the language of **SF** augmented with unbounded quantifiers of all types into a formula of the format

(4) $\exists x_1 \ldots \exists x_n \forall y_1 \ldots \forall y_m \varphi\,[x_1, \ldots, x_n, y_1, \ldots, y_m]$

where $\varphi\,[x_1, \ldots, x_n, y_1, \ldots, y_m]$ is a quantifier-free formula of **SF**. This is then understood as the claim that we can construct terms $t_1, \ldots, t_n$ to satisfy the condition $\varphi\,[x_1, \ldots, x_n, y_1, \ldots, y_m]$. Proving the claim in strict finitism means constructing the terms $t_1, \ldots, t_n$ and deriving $\varphi[t_1, \ldots, t_n, y_1, \ldots, y_m]$ in **SF**. We say that the terms $t_1, \ldots, t_n$ witness the claim (4). The proof of (2) by constructing $\lambda k.N\,[k]$ and deriving (3) is such an instance. Another example is to translate (assuming that $\varphi\,[n]$ and $\psi\,[n]$ are quantifier-free)

(5) $\forall n \varphi\,[n] \to \forall m \psi\,[m]$

into

(6) $\exists X \forall m\,(\varphi\,[X\,(m)] \to \psi\,[m])$

according to Gödel's *Dialectica* interpretation. Then, a proof of (5) in strict finitism means constructing a term $\lambda m.t\,[m]$ (for $\exists X$) and deriving in **SF** the formula $\varphi\,[t\,[m]] \to \psi\,[m]$. Such translations are supposed to extract the finitistic contents of quantified statements. In this way, we can state mathematical theorems in much the same format as they are stated in classical mathematics, using quantifiers freely, but we always understand them as claims in strict finitism in the format (4). For instance, in strict finitism, we can prove the existence of solutions to ordinary differential equations under some common conditions. When translated into a claim in the format (4), the proof actually means constructing elementary recursive approximations to the solutions.

Moreover, we can show that, when interpreted this way, unbounded quantifiers and other logical connectives connecting quantified statements satisfy the intuitionistic logical laws. Therefore, we can directly use statements with unbounded quantifiers of various types in an informal proof of a theorem, as long as we use the intuitionistic logic. The terms $t_1, \ldots, t_n$ for witnessing a claim like (4) (as the conclusion of the proof) can be automatically extracted from the informal proof. The result is that developing mathematics in strict finitism is very similar to developing mathematics in Errett Bishop's constructive mathematics ([3]). The only difference is that, in strict finitism, we have to limit recursive constructions to bounded recursions on numerical functions, so that we never go beyond elementary recursive functions, and we have to limit inductions to quantifier-free statements.

We also use terminologies such as 'sets' and 'functions from sets to sets' in our informal presentations of mathematics in strict finitism. This is similar to referring to classes when developing axiomatic set theory in **ZFC**. Recall that, when working with **ZFC**, a 'class' is actually a formula with an indicated free variable (and perhaps other parameters) giving the membership condition of the class. Similarly, whenever we refer to a 'set' in an informal presentation of mathematics in strict finitism, we actually mean a statement $\psi[a]$ with an indicated free variable, giving the membership condition of the set, and another statement $\chi[a, b]$ with two free variables, giving the equality condition for the members of the set. (ψ and χ can be statements with unbounded quantifiers.) Then, '$\langle \psi[a], \chi[a, b] \rangle$ defines a set' can again be translated into a claim in the format (4). In this way, we can refer to subsets of real numbers, sets of continuous functions on a real interval $[a, b]$, and so on and so forth. Similarly, we achieve abstraction by stating theorems as theorem schemas involving arbitrary formulas and terms of appropriate types. For instance, a theorem in the metric space theory is a schematic theorem with two arbitrary formulas giving the membership and equality conditions of the underlying set, together with an arbitrary term giving the distance function. We can state theorems about arbitrary linear spaces or Hilbert spaces in a similar way.

In developing semi-Riemannian geometry, we consider n-dimensional open spheres with rational centers and radius, and we consider an open subset that can be represented as the union of an elementary recursive sequence of such spheres. Since semi-Riemannian manifolds are only approximations to large-scale physical space-time structures in general relativity, we expect that such open subsets are sufficient for representing physically meaningful space-time regions. Then, we can develop some basic topology with such open sets, define basic notions in semi-Riemannian geometry and prove some basic theorems.

1.4 The Conjecture of Finitism

The success of this strategy for explaining applicability depends on the following **Conjecture of Finitism**:

> Strict finitism is in principle sufficient for formulating scientific theories and conducting proofs and calculations in the sciences.

Besides the fact that an impressive part of applied mathematics has been developed within strict finitism, there are other intuitive reasons supporting this conjecture. Firstly, the methods for developing mathematics in strict finitism are general and we expect that much more applied mathematics can be developed similarly. Secondly, infinity (and continuity etc.) is merely an idealization to approximate strictly finite physical things in mathematical applications in the sciences. We intuitively think that infinity should not be strictly logically indispensable, for otherwise the proofs and calculations could be scientifically meaningless because of taking the idealization too literally. Experiences in developing mathematics within strict finitism show that there is a natural connection between the fact that a proof is reducible to strict finitism and the fact that infinity is not taken too literally by the proof. For instance, proving (6) above in **SF** means that to derive $\psi[m]$ for a specific m we do not really need the premise $\varphi[n]$ for all n. We only need $\varphi[X(m)]$ for a known elementary recursive function X. In that sense, a finitistic derivation of $\psi[m]$ does not 'use up the full premise $\forall n \varphi[n]$' or does not 'take the idealization in that premise too literally'. Physically meaningful proofs and calculations should not take the idealization to infinity too literally. Therefore, we expect that they are in principle reducible to strict finitism. See Chap. 3 of [29] for more illustrations on this. Finally, some people may have doubts about the Conjecture of Finitism because of Gödel's Incompleteness Theorem. Section 1.3.2 of [29] contains a response to such doubts.

The status of the Conjecture of Finitism is still open and more work is needed to confirm it, but perhaps what has been done so far already shows that it is plausible and that my strategy for explaining applicability is workable.

2 From Naturalism to Nominalism and Strict Finitism

2.1 From methodological naturalism to physicalism about cognitive subjects

Today, many philosophers accept methodological naturalism, which is usually understood as the view that scientific methods (*vs.* the so-called First Philosophical Method) are our best methods for achieving knowledge. However, there is a trap here. One could unconsciously take methodological naturalism to mean that scientific methods are the best methods for a non-physical 'subject' to know things in a world 'external to the subject'. In this image of cognition, the entire physical world belongs to the external world, and the subject stands against this external world and tries to 'know (or describe, or gain epistemic access to)' this external world. This literally contradicts the naturalistic and scientific image of humans. In the naturalistic image, humans are themselves physical things in the physical world and human cognitive processes are physical processes inside the physical world. A naturalist may start her philosophical thinking with a vague conception of herself as a 'cognitive subject', but after accepting contemporary scientific theories from physics to evolutionary theory and cognitive sciences, she should realize that she herself is just a physical thing and that her cognitive processes are physical processes in this physical universe. Accepting methodological naturalism requires her to abandon any conception of an amorphous subject that is not itself a part of this physical world but somehow stands against an external world that contains the entire physical world (including her own brain). This conception of a subject is a homunculus fallacy. It seriously distorts the scientific and naturalistic image of humans. I will call the scientific and naturalistic image *'physicalism about cognitive subjects'*.

All naturalists reject Cartesian dualism. That is for sure. However, after rejecting Cartesian dualism, one may still unconsciously take the stance of a supernatural subject, or the stance of an amorphous, atomic, non-physical self, instead of consciously treating oneself literally as a composite physical system interacting with its physical environments in this physical world. I believe that this fallacy comes from human egocentric instinct deeply rooted in human nature. That is, one takes scientific methods to be methods for knowing the 'external world', but oneself as a cognitive subject is somehow exempt from scientific scrutinization, and one is unaware of the incoherence here because of an egocentric instinct. This conception of self as an amorphous, non-physical 'subject' facing an 'external world' is a brain's misconception of itself and its relation with its physical environments. It is a *brain's* homunculus fallacy.

2.2   The impacts of physicalism on philosophy

Recognizing this fallacy will profoundly change our view about philosophy, because so many traditional philosophical notions *presuppose* a non-physical cognitive subject. For instance, reference and truth are traditionally understood as relations between a subject's language or ideas and the external world. If a human subject is literally a physical system within the physical world, then reference and truth (at least reference to and truth about physical things) ought to be physical connections between a physical system and its physical environments. Similarly, since a human is a physical (biological) system growing out of an embryo, we can meaningfully talk about what are innate (i.e., determined by genes) *vs.* what are nurtured, or what are innate for an individual *vs.* what are innate for a species, but the traditional notion of apriority presupposes an *a priori* subject that cannot be literally a physical system in the physical world. Besides reference, truth and apriority, many other traditional philosophical notions (e.g., logical validity, onto-logical commitment, objectivity, necessity, knowledge, justification, etc.) face the same problem. They all presuppose a non-physical subject, or at least it is very unclear what they could mean if cognitive subjects are literally human brains as physical things in this physical world. A consistent naturalist should either aban-don these notions or replace them with truly naturalistic notions. It seems that many contemporary naturalists are not aware of this problem.

For philosophy of mathematics, since a human mathematician is a physical system in the physical world, human mathematical practices and applications are physical processes in the physical world, namely, neural activities in human brains and their physical interactions with human physical environments. In particular, we should not conceive of mathematical practices as activities of an amorphous, non-physical subject accessing a world of mathematical entities. Instead, a study of human mathematical practices should be a continuation or extension of cog-nitive sciences, dealing specifically with the mathematical cognitive activities of human brains.

Some people may not like this idea, because it introduces cognitive sciences into philosophy of mathematics. However, if you do not deal with brains, then you have to speculate about the cognitive capacities of a non-physical subject. Dealing with brains may be scientifically very difficult, but it is at least concep-tually clear. You are studying plain physical things no matter how complex they are. You look into brains and build models to simulate how brains work, just as you do in other branches of science. On the other side, philosophical speculations about a non-physical subject and its cognitive abilities (e.g., the ability to 'grasp' mind-independent abstract concepts) are always vague and even inscrutable. They are idle and fruitless so far, and there appears to be no way to improve them and proceed further. Therefore, the naturalistic approach is perhaps more attractive.

Moreover, even if you *are* a Cartesian dualist and believe that there are souls, you still have to admit that human brains can do all the things that cognitive scien-tists say they can do, which include almost everything 'we' can do in mathemat-ical practices. Then, the positive results of a naturalistic study referring to brains should be acceptable to you as well. That is, the naturalistic approach is actu-ally based on minimum assumptions acceptable to all philosophical camps, even including Cartesian dualism. If the naturalistic approach is successful, then you can reconsider why you still believe that there are other things besides the brains. There is also a chance that such researches will in the end discover and clearly identify a place where we have to assume that there are other things (say, souls).

That will be a good outcome too. Therefore, the naturalistic approach always has its positive value for everyone.

Furthermore, since our interests in philosophy of mathematics are only on the philosophical and logical aspects of human mathematical practices, introducing cognitive sciences is only for clarifying the philosophical foundation and setting a clear background for discussing philosophical and logical issues. When we start investigating specific philosophical and logical questions about human mathematical practices, we can rely on highly simplified cognitive models of human brains, and we can abstract away cognitive scientific details as much as we can. My work in this research project does not rely on any specific cognitive scientific theory. Some cognitive scientists are indeed investigating the cognitive scientific aspects of human mathematical practices (e.g., [9]), but they do not discuss the logical and philosophical issues that interest contemporary logicians and philosophers (e.g., the issue of applicability mentioned in the last section).

2.3   The nature of human mathematical practices under physicalism

To describe mathematical practices, the notions of reference, truth, and logical validity are most critical. They must be replaced with naturalistic notions. There are already some attempts to characterize reference and truth as naturalistic relations between concepts and thoughts realized as neural structures in brains and the physical things and states that these concepts and thoughts represent. See [19] and [12] for some surveys and debates. This is called 'naturalizing content' or 'naturalizing representation' in philosophy of mind. There are problems in the current theories for naturalizing representation. [32] and [30] proposes a new approach to solve those problems.

With reference and truth (for physical things) naturalized, the logical validity of a logical inference rule is also naturalized. A logical inference rule is valid, if in all applications of the rule, as a natural regularity, whenever the premises are true (in the naturalized sense), the conclusion is also true. An assertion about the validity of a logical inference rule is thus an assertion about some natural regularity in a special type of natural processes. Note that this is not the so-called 'psychologism' if 'psychologism' means the view that valid logical rules are the rules that are most frequently followed by human brains. A brain can frequently hold false beliefs, that is, thoughts that do not bear the (naturalized) truth connection with physical things in the environment, and a brain can frequently conduct invalid logical inferences on thoughts. Normativity in logical validity comes from normativity in reference and truth, which in turn is naturalized normativity. See [27], Chap. 1 of [29] for more discussions on some subtle points in this naturalistic characterization of logical validity.

Naturalized reference and truth apply only to concepts and thoughts about physical things. Mathematical concepts and thoughts (as neural structures in a brain) do not represent anything. They have other cognitive functions in the cognitive processes in a brain. For instance, the concept '5' in a brain can combine with the concept 'finger' in the brain to form the concept '5-fingers', which does bear the (physical) reference connection with physical things. Similarly, the concept 'point' in a geometry theory in a brain can be translated into a concept directly representing small physical spatial regions, when the brain applies the geometry theory to physical space. Mathematical concepts and thoughts are abstract representational tools inside brains. Brains take valid logical inference rules on thoughts about physical things (in the naturalized sense) and apply them to mathematical thoughts. That is, brains adopt the same logic for mathematical thoughts as that for thoughts about physical things. This has the advantage that when brains apply mathematics, their mathematical inferences on mathematical thoughts can be more straightforwardly translated into valid logical inferences on thoughts about physical things.

Brains also apply the word 'true' to some mathematical thoughts. This may come in several ways. First, brains may abstract a mathematical thought from some true thoughts about physical things with the same pattern. For instance, brains may abstract the thought '5+7=12' from '5-fingers plus 7-fingers are 12-fingers' and other similar true thoughts about physical things. Second, brains adopt the disquotation schema '"p" is true if and only if p' for a mathematical thought p. This schema, together with the logical rules, implies 'either p is true or $\neg p$ is true'. Brains also apply 'true' to all inferential consequences of 'true' premises using logical rules. Third, brains may decide to apply 'true' to a mathematical thought (e.g., the axiom of infinity) because, after some practices of entertaining, manipulating or applying the thought, brains find the thought useful, logically consistent, or simply 'appearing obvious'. These uses of the word 'true' on mathematical thoughts serve important cognitive functions in the cognitive activities in a brain. However, they may give a brain the illusion that some mathematical thoughts are 'true' in exactly the same way as thoughts about physical things are true.

Note that, in this naturalistic description of the cognitive functions of mathematical concepts and thoughts in brains, mathematical concepts and thoughts are as 'meaningful' as anything can be. Having meanings consists in having cognitive functions, and representing things is merely one of many cognitive functions of a cognitive component in a brain. Mathematical concepts and thoughts do not represent anything, but they play more important cognitive functional roles than other concepts and thoughts that represent physical things do. This naturalistic account is similar to formalism in many respects, but I will never say that mathematical thoughts are 'meaningless' or that mathematical theorems are 'literally false' (or 'only vacuously true'). Nor does this account evade the alleged 'true meanings' of mathematical concepts and thoughts, for their 'true meanings' just consist in their true functions in the cognitive activities of a brain. [33] contains more discussions on meaning and 'truth' in mathematical practices.

Understanding, knowledge, experience and intuition in mathematical practices can be explained by referring the cognitive functions of mathematical concepts and thoughts in brains. For instance, having mathematical knowledge on a particular subject matter means having relevant mathematical concepts and thoughts in the brain, having relevant ability to do inferences on those concepts and thoughts, having relevant knowledge about the expected outcomes of such inferences, and finally having relevant knowledge and ability in translating mathematical concepts and thoughts into concepts and thoughts that directly represent physical things in mathematical applications. In particular, understanding, knowledge, experience and intuition in mathematical practices have nothing to do with the alleged abstract mathematical entities. ([33])

The traditional notion of objectivity is understood from the point of view of a non-physical subject. Ideas, sense data and so on belong to the subject and are considered subjective; physical objects or the alleged abstract objects belong to the external world and are considered objective. Under physicalism about cognitive subjects, being objective can only mean being independent of a brain (or all human brains) in some way. [34] discusses several notions of objectivity under physicalism. These naturalistic notions should replace the traditional notion of objectivity. I argue that admitting objectivity of mathematics in the naturalized senses does not imply admitting the existence of abstract entities or mind-independent (actually, brain-independent) concepts. The true basis for people's sharing 'the same concept', successfully communicating their concepts, and following 'the same rule' and so on is the fact that human brains share the same fundamental cognitive architecture and obey the same neural-physiological laws, as well as the fact that they have similar individual developmental environments.

The traditional notion of apriority similarly presupposes an *a priori* subject. Under physicalism, we must replace the traditional notion with a naturalistic notion. For instance, there can be some 'predetermined harmony' between brains and their environments, as a result of evolutionary selections. This will result in some '*a priori*' knowledge for brains, in the sense that all biologically normal human brains will automatically have that knowledge after a normal individual developmental process in a normal environment. In particular, logic seems to be '*a priori*' in this sense. On the other hand, logic and arithmetic are not absolutely universal. As researches in quantum logic suggest, classical logical constants may not be directly applicable to properties of quantum particles. That is, when two propositions p and q attribute two incommensurable properties to a particle, the conjunction $p \wedge q$ is either meaningless, or meaningful but not obeying the distribution law between conjunction and disjunction. Either way, classical logic is not universally applicable. However, human brains are evolved in their interactions with deterministic, medium size physical objects in human immediate environments.

As a consequence, humans can only clearly conceive of objects similar to those deterministic, medium size physical objects. That is, we cannot clearly conceive of quantum particles with wave-particle duality. Classical logic and arithmetic represent some most general features of these deterministic, medium size physical objects. Therefore, classical logic and arithmetic are universal and necessary in a limited sense, namely, in the sense that they are valid for anything clearly conceivable by humans. [35] discusses these issues.

2.4   From physicalism to nominalism and strict finitism

These are some aspects of human mathematical practices under physicalism. All these suggest that physicalism about cognitive subjects should imply nominalism in philosophy of mathematics. [25] argues for this implication. Many traditional arguments against realism in philosophy of mathematics originate from [2]. Typically, they first propose constraints on what cognitive subjects could possibly refer to or have knowledge about, and then they argue that we cannot possibly refer to nor can we have knowledge about abstract entities, because abstract entities do not satisfy the constraints. However, reference and knowledge are complex philosophical notions. Realists can always reject the constraints proposed by the arguments. [25] takes a different route in arguing against realism, by relying on physicalism about cognitive subjects. Here are the basic ideas.

First, a complete naturalistic description of the mathematical practices and applications of a brain will describe neural activities and their physical interactions with other physical things. It will not say which abstract mathematical entities a neural structure (as a mathematical concept) in a brain 'refers to' or 'represents'. (Recall that reference to physical things is a physical connection between brains and other physical things.) One might think that for a brain to refer to (or be committed to) abstract entities is just for the brain to use pronouns, variables and quantifiers in some specific manner. This appears to be exactly what Quine ([17] and[18]) was saying. However, 'the brain is using words in so and so manner' is already a complete naturalistic description of relevant things in the natural world. Redescribing what is happening in the brain as 'the brain is referring to abstract entities' is pointless.

Second, there is a psychological explanation about why people tend to think that they are 'referring to abstract entities'. They take the stance of a non-physical subject standing against an external world, and when a nominal term does not refer to any physical thing in the external world, they tend to think that it refers to an abstract entity in the external world. That is, it is a homunculus fallacy mentioned above. The same homunculus fallacy occurs when one objects, 'you have to use mathematics in describing brains; aren't you already committed to abstract entities?' A brain is practicing and applying mathematics, and in describing that brain I do not say which abstract entities a neural structure in that brain 'refers to'.

Now, my brain is exactly the same sort of physical system as that brain. Therefore, neural structures in my brain do not 'refer to abstract entities' either when I use mathematical concepts to describe that brain. Apparently, here the objector agrees that it seems meaningless to ask which abstract entities a brain as a physical system 'refers to', but she forgets that she is just a brain herself. That is why she feels that she somehow has to be 'committed to abstract entities' by using mathematical terms in describing other brains.

Besides, I argue that confirmation holism and the disquotational conception of reference and truth cannot save abstract entities either. It is again due to a homunculus fallacy that one thinks that holism and disquotational reference and truth can explain how a subject accesses abstract entities. See [25] for the details.

Physicalism about cognitive subjects also implies strict finitism. Human brains and everything in human immediate environments are strictly finite. Brains can have knowledge about things beyond their immediate environments, but it is still a fact that contemporary sciences never claim that there is infinity in this physical universe. A complete naturalistic description of the mathematical practices of a brain does not mention which infinite entities or structures a neural structure in the brain 'refers to'. What we confidently know today is a finite world of physical things above the Planck scale. This world might contain infinity below the Planck scale. However, before scientists confidently assert that, our philosophical account of human mathematical practices had better to be strictly finitistic.

3 Comparisons with Similar and Related Researches

I have mentioned at the beginning that two distinctive features of my approach are physicalism and strict finitism. There are other nominalistic and naturalistic philosophies of mathematics in the contemporary literature. I discuss their flaws and the motivations for my approach in [26]. The basic idea is that, after denying the existence of abstract mathematical entities, nominalists should offer literally truthful accounts of various aspects and features of human mathematical practices, by referring to what really exist in mathematical practices. These aspects and features should especially include those that are taken to be reasons supporting realism by realists. For instance, nominalists should explain the nature of human mathematical knowledge, intuition and experience; they should explain the relationships between the mathematical and the physical, including the apparent, approximate isomorphism between a mathematical structure and a physical structure, which is the reason for the applicability of a mathematical model; they should identify various aspects of objectivity in mathematical practices and show that admitting objectivity does not imply admitting the existence of abstract entities; they should explain the apparent obviousness, apriority, universality and necessity of logic and arithmetic; and finally, they should explain the applicability of classical mathematics in the sciences. Moreover, the nominalistic accounts of these aspects of mathematical practices should be literally true accounts and should respect the possibility that there are only strictly finitely many concrete objects in total in the universe. That is, the nominalistic accounts must be 'realistic' accounts in themselves and must be strictly finitistic accounts. As I have explained in this paper, my research project is devoted to these tasks. However, these are missing or not adequately done in other current nominalistic philosophies of mathematics.

For instance, some current anti-realistic philosophies in the literature (e.g., [13, 14, 15, 16]; [10, 11]; [8]) are not 'realistic enough'. They did not offer literally truthful accounts of the aspects and features of mathematical practices mentioned above. Maddy denies the robust existence of mathematical objects, but she does not offer any alternative description of human mathematical practices by referring to what really and robustly exist. Maddy says that mathematical entities have 'thin' existence and 'thin' existence is not essentially different from non-existence. If that is the case, we had better abandon the mysterious and obscure notion of 'thin existence' and give an account of human mathematical practices by referring to whatever really and robustly exist. Leng says that mathematical entities are 'fictional entities' and scientists use 'fictional entities' to model real things in mathematical applications. However, since 'fictional entities' do not really exist, the claim 'scientists use fictional entities to model real things' is a literally false claim. What is the point of making literally false claims in one's philosophical articles? It only shows that one has not really explained what scientists are doing. Similarly, Hoffman's assertions about the so-called ideal agents are also literally false, since ideal agents do not exist. Such philosophies cannot stand against realists' charge that nominalists are 'intellectually dishonest'. Apparently, we have to refer to human brain activities in a literally truthful account of human mathematical practices, because, under naturalism and nominalism, brains are what really exist in human mathematical practices. There are no non-physical minds, nor are there any non-physical abstract objects or concepts graspable by non-physical minds. There are only brains and their physical environments. That motivates my physicalistic approach.

Besides, fictionalists (e.g., Leng) insist that mathematical models are fictional and do not really exist. However, if infinity is *logically strictly* indispensable for the applications, then there is at least a logical puzzle: Our literally true scientific conclusions about physical things in this finite universe somehow logically indispensably depend on literally false premises about non-existent fictional things. Note that our assertions about fictional frictionless planes or absolutely rigid bodies in physics textbooks are not strictly indispensable. When applying textbook examples, calculations and proofs to real things, one can in principle translate textbook assertions about frictionless planes or absolutely rigid bodies into literally true (though approximately true, perhaps) assertions about real physical planes or bodies that are sufficiently smooth or sufficiently rigid. That is why apparent references to fictional frictionless planes or absolutely rigid bodies are useful for deriving truths about real physical planes or bodies. Similarly, if we can translate our assertions about a mathematical model into literally (and approximately) true assertions about physical things that the model aims to simulate, then we can explain why assertions about the fictional model are useful. That is, if fictional things are *in principle* dispensable in the applications, we can have an explanation of the applicability of fictional things. We can show that referring to fictional things is just a simplified and convenient way to encode assertions about real things, and after eliminating all apparent references to fictional things, we can get literally true premises about real things that logically imply our conclusion about real things. Now, since real things we deal with in the sciences are always finite, dispensability of fictional things at least means in principle dispensability of infinity for the applications. In other words, if infinity is logically, strictly indispensable for the applications, then there will be an irresolvable logical puzzle about why assertions about non-existent fictional models are useful for deriving truths about real things. We will not know which literally true premises are the premises that logically imply our conclusion, and from the logical point of view, we will not know how the truth of our conclusion comes about. Contemporary fictionalists did not try to resolve this puzzle, while resolving the puzzle is a main goal of my project.

Some other current anti-realistic philosophies of mathematics (e.g., [5, 6]; [7]; [4]) are 'too realistic' about infinity. They accept at least potential infinity, which means assuming something beyond this physical universe in case the physical universe is strictly finite and discrete. Therefore, their philosophies are not coherent nominalism. These philosophers might refer to the pragmatic values of assuming potential infinity to justify assuming potential infinity in mathematics, but then this opens the door for a pragmatic justification of the existence of abstract objects. Moreover, the real puzzle of applicability of mathematics is the puzzle about how exactly 'infinite mathematics' is applied to this strictly finite and discrete physical world. Applying the allegedly nominalistic mathematical systems by Field, Hellman, and Chihara in the sciences is still using infinite models to approximate strictly finite and discrete physical phenomena. The puzzle of applicability has not been resolved.

There are other logicians and philosophers pursuing a strictly finitistic foundation of mathematics, for instance, [36] and [23, 24]. My research project differs from theirs in terms of philosophical basis, motivations and goals. I would not say that there are only finitely many numbers or that there is the largest number. If numbers are supposed to be mind-independent abstract objects, then there are just no such things as numbers. I won't say that there are small abstract objects but no arbitrarily large abstract objects. Now, suppose that numbers are our imaginations, not real. Then, why should we stop at imagining finitely numbers? Since they are just our imaginations anyway, why don't we go ahead and imagine that there are infinitely numbers and even imagine that there is a universe of all sets as in **ZFC**? Mathematicians and scientists have been imagining that there are infinitely many numbers for many centuries. Our mathematical imaginations in contemporary classical mathematics obviously look quite coherent and have been very useful in the sciences. Perhaps we can coherently imagine a finite largest imaginable numbers, as van Bendegem has shown. However, there is no good motivation for doing that, unless such imaginations are even more convenient for scientific applications, which should be left for working scientists to decide. In contrast, in my research project, strict finitism as a mathematical formal system is only a logical tool for analyzing why classical mathematics can apply to this finite physical world. I have no intention to offer a new foundation for mathematics or to suggest a new way to practice mathematics. As for strict finitism as a philosophical view, it follows from physicalism. Regarding mathematics, it says only that, since mathematical entities in classical mathematics are our imaginations, not real, philosophical accounts of human mathematical practices should not again refer to mathematical entities as if they were real, and thus philosophical accounts of human mathematical practices have to be strictly finitistic. *Strict finitism is a restriction for philosophers; it not any restriction for working mathematicians and scientists.*

The formal system **SF** for strict finitism developed in my book [29] is closely related to the system $\widehat{\mathbf{T}_0}$ in [1], or the system $\mathbf{T}_0$ defined in [22], with the recursion operator restricted to numerical functions. **SF** is a proper subsystem of these systems because **SF** admits only bounded primitive recursions and thus only elementary recursive functions, not all primitive recursive functions. **SF** is quantifier-free and admits only elementary recursive functions, and therefore it is weaker than the weakest system **RCA**$_0$ studied by most researches in reverse mathematics ([21]). On the other side, I did not consider issues of computational complexity, and the system **SF** is stronger than the systems considered in bounded arithmetic.

Finally, as I have mentioned above, developing ordinary mathematics in **SF** is very similar to developing ordinary mathematics in Bishop's constructive mathematics ([3]). The only essential difference is that we have to limit recursive constructions to ensure that we never go beyond elementary recursive functions. This is to make sure that the numerical bounds for all values appeared in all calculations and proofs can always be explicitly estimated. Then, when we translate the calculations and proofs into inferences on statements about real finite and discrete physical things in this universe, we can examine whether we really get statements literally true of those real finite and discrete physical things.

4   Conclusion

Some philosophers of mathematics may think that strict finitism and physicalism are too radical. I will only urge them to put philosophy of mathematics in a broader context of pursuing a coherent philosophical worldview and to consider the fact that physicalism attracts most philosophers in contemporary philosophy of mind, for good reasons. This research project intends to show that the fact that humans have successfully practiced and applied classical mathematics in the sciences is consistent with a clean, clear and coherent physicalistic worldview, in which the only things that are known to exist are physical things (including brains) above the Planck scale. This worldview is attractive, because bringing in any extra things (e.g., abstract entities or irreducibly mental entities) will threaten its coherence, given our current scientific knowledge. While such a project can never be literally finished, I hope I have already shown that it deserves some attention.

Acknowledgements

I would like to thank Johan van Benthem and Fenrong Liu for their comments and suggestions, which help to improve this article greatly.

BIBLIOGRAPHY

[1] Avigad, J., and S. Feferman. (1998). 'Gödel's functional ("dialetica") interpretation'. In S.R. Buss (ed.). *Handbook of proof theory*, 337–405. Amsterdam: Elsevier.
[2] Benacerraf, P. (1973). 'Mathematical truth'. *Journal of Philosophy* 70, 661–79.
[3] Bishop, B. and D. Bridges (1985). *Constructive Analysis*, Springer-Verlag.
[4] Chihara, C. (2005). 'Nominalism', in Shapiro (2005), 483–514.
[5] Field, H. (1980). *Science without Numbers*, Oxford: Basil Blackwell.
[6] Field, H. (1989). *Realism, Mathematics and Modality*, Oxford: Basil Blackwell.
[7] Hellman, G. (2005). 'Structuralism', in Shapiro (2005), 536–62.
[8] Hoffman, S. (2004). 'Kitcher, Ideal Agents, and Fictionalism', *Philosophia Mathematica (3)* 12, 3-17.
[9] Lakoff, G. and R. Núñez (2000). *Where Mathematics Comes From: How the Embodied Mind Brings Mathematics into Being*, New York: Basic Books.
[10] Leng, M. (2002). 'What's Wrong with Indispensability?', *Synthese* 131, 395-417.
[11] Leng, M. (2005). 'Revolutionary Fictionalism: A Call to Arms', *Philosophia Mathematica (3)* 13, 277-293.
[12] Macdonald, G. and D. Papineau (2006). *Teleosemantics: New Philosophical Essays*, Oxford University Press.
[13] Maddy, P. (1997). *Naturalism in Mathematics*. Oxford: Clarendon Press.
[14] Maddy, P. (2005). 'Three Forms of naturalism', in Shapiro (2005), 437–59.
[15] Maddy, P. (2007). *Second Philosophy: A Naturalistic Method*. Oxford: Oxford University Press.
[16] Maddy, P. (2008). 'How pure mathematics became pure', *The Review of Symbolic Logic* 1, 16-41.
[17] Quine, W. V. (1992). *Pursuit of Truth*. Cambridge: Harvard University Press
[18] Quine, W. V. (1995). *From Stimulus to Science*. Cambridge: Harvard University Press.
[19] Rupert, R. D. (2008). 'Causal theories of mental content', *Philosophy Campus* 3, 353-380.
[20] Shapiro, S. (ed.). (2005). *The Oxford Handbook of Philosophy of Mathematics and Logic*, Oxford: Oxford University Press.
[21] Simpson, S. (2009). *Subsystems of Second Order Arithmetic, 2nd ed.*, Cambridge University Press.
[22] Troelstra, A.S. (1990). 'Introductory note to 1958 and 1972'. In S. Feferman, et al. (eds.). *Kurt Gödel Collected works, volume II*, 217–239. Oxford: Oxford University Press.
[23] Van Bendegem J. P. (1999). 'Why the largest number imaginable is still a finite number.' *Logique et Analyse* 42, 107–126.
[24] Van Bendegem J. P. (2012). 'A Defense of Strict Finitism.' *Constructivist Foundations* 7(2), 141–149.

[25] Ye, F. (2010a). 'Naturalism and abstract entities'. *International Studies in the Philosophy of Science* 24, 129-46.

[26] Ye, F. (2010b). 'What Anti-realism in Philosophy of Mathematics Must Offer', *Synthese* 175, 13-31.

[27] Ye, F. (2010c). 'The applicability of mathematics as a scientific and a logical problem'. *Philosophia Mathematica* 18, 144-65.

[28] Ye, F. (2010d). *Philosophy of Mathematics in the Twentieth Century: A Naturalistic Commentary* (in Chinese), Peking University Press.

[29] Ye, F. (2011a). *Strict finitism and the logic of mathematical applications, Synthese Library, vol.355.* Dordrecht: Springer.

[30] Ye, F. (2011b). 'Naturalized truth and Plantinga's evolutionary argument against naturalism', *International Journal for Philosophy of Religion* 70, 27-46.

[31] Ye, F. (2012). 'Some naturalistic comments on Frege's philosophy of mathematics', *Frontiers of Philosophy in China* 7, 378-403.

[32] Ye, F. (online-a). 'A Structural Theory of Content Naturalization', http://sites.google.com/site/fengye63/. Accessed July 24, 2013

[33] Ye, F. (online-b). 'On what really exist in mathematics'. ibid.

[34] Ye, F. (online-c). 'Naturalism and objectivity in mathematics'. ibid.

[35] Ye, F. (online-d). 'Naturalism and the apriority of logic and arithmetic'. ibid.

[36] Yessenin-Volpin A. (1970). 'The ultra-intuitionistic criticism and the antitraditional program for foundations of mathematics'. In: Kino A., Myhill J. & Vesley R. (eds.) *Intuitionism and proof theory*. North-Holland, Amsterdam: 3–45.

Comments on Ye

HANNES LEITGEB

In his "Introduction to a Naturalistic Philosophy of Mathematics", Feng Ye puts forward several strong theses in the philosophy of mathematics and beyond. I am sympathetic to all of them, and I do not endorse any of them, at the same time. I will concentrate on why I would not want to endorse them.

First of all, Feng Ye defends what he calls the Conjecture of Finitism: "Strict finitism is in principle sufficient for formulating scientific theories and conducting proofs and calculations in the sciences" (section 1.4). In other words: we can do science, at least as far as its deductive aspects are concerned, without assuming the existence of infinitely many objects. Secondly, his philosophical view in general, and his philosophy of mathematics in particular, are nominalistic: he believes that there are no abstract entities. Thirdly, his philosophy and his philosophy of mathematics are also naturalistic: he adheres to a methodological naturalism according to which knowledge is achieved best by means of scientific methods, and where knowledge in this sense includes philosophical knowledge and knowledge about ourselves, which he regards both as special cases of knowledge about physical entities. (He is also a physicalist, about which more below).

In part (I) of my comments I will raise some worries concerning Feng Ye's finitism and nominalism. In part (II) I will do the same with respect to his methodological naturalism. The outcome of (I) will be a challenge, while the outcome of (II) will be a question.

(I) Problems concerning finitism and nominalism

Feng Ye's main argument for his Conjecture of Finitism relies on his formal system SF and its properties: in his monograph from 2011 (which is cited in his paper) he tried to show that much of the mathematics that is to be found in the sciences can actually be reconstructed in the system SF; and since SF does not seem committed to the existence of infinitely many objects, his conjecture of Finitism seems to be supported. My worry concerns precisely this formal system SF, or rather, the manner in which it is developed vis-à-vis Feng Ye's own background assumptions and philosophical aims.

As every formal system, SF is based on syntax: it consists of a vocabulary as well as of a set of grammatical rules by which the set of well-formed expressions in the formal language of the system is defined inductively. Subsequently, a set of axioms and derivation rules is introduced by which some well-formed expressions in the language of the system can be derived as theorems. All of this is standard practice in proof theory, of course. However, Feng Ye's finitism is non-standard, and that leads to my concern: SF's vocabulary is infinite, its set of well-formed expressions is infinite (and there is no finite bound on the length of well-formed expressions), its set of axioms is infinite, its set of rules of inference is infinite, and its set of theorems is infinite. On the other hand, since SF is a scientific theory – or if it is not, what is it (in the eyes of a naturalist such as Feng Ye)? – the Conjecture of Finitism should apply: it should be possible to carry out the relevant applications of SF in science without assuming that any of the sets from before is infinite. So if Feng Ye's conjecture is right, then it should be possible to build the relevant parts of the syntax and proof theory of SF without assuming the existence of infinitely many objects of any kind. That is then my first challenge to Feng Ye: to determine the scientifically relevant parts of the formal system SF without assuming the vocabulary, the set of well-formed expressions, the set of axioms, the set of rules of inference, or the set of theorems to be infinite (and also without the assumption that the length of well-formed expressions is not bounded by any finite number). The following reply by Feng Ye is conceivable: he might say, well, as you know, I do not actually "suggest abandoning classical mathematics in scientific applications. Exactly on the contrary, using infinite and continuous models to approximate finite and discrete physical things is an ingenious invention of our scientists. It greatly simplifies our theories and makes them tractable and ultimately possible" (section 1.2). So maybe Feng Ye could say something like that also about his system SF? It is infinite, yes, but that would only be so in order to approximate and simplify the "real" system, say, SF*, that does capture all the mathematics that is required by science without assuming infinity in its syntax or proof theory? If so, my challenge from above reappears in the following form: to determine this system SF*. And the worry is that this cannot be done in the same manner in which Feng Ye tries to reconstruct other parts of mathematics: for instance, as he explains in section 1.3, he suggests to translate typical mathematical statements as being used in the sciences into ones whose intended interpretation is along the lines of: "one can construct terms t_1, ..., t_n that satisfy conditions". The expression "terms" means here: members of the infinite inductively defined set of well-formed terms of the language of SF. However, a statement of that form could not be the ultimate translation of a theorem of SF into the language of SF*, as there is no such infinite inductively defined set of well-formed terms according to the strictly finitary syntax of SF*.

Just as Feng Ye's finitism causes trouble for what he says and assumes about the formal system SF, also his nominalism does: standard syntax and proof theory maintains that the expressions of formal languages are abstract objects: either they are taken to be abstract sequences of symbols, or abstract natural numbers (via coding), or abstract types of concrete physical inscriptions, or some other kind of abstract entites. But none of this is acceptable to Feng Ye, since his nominalism excludes the existence of abstract entities altogether. This is thus my second challenge: to determine what kind of concrete entities the formal expressions in the language of SF are meant to be. Are they linguistic expressions that we actually write on a blackboard or which we type into a computer? What if someone erases any of them again? What if someone writes two inscriptions on the blackboard that look the same? Combined with the previous challenge, the two challenges amount to: to determine the scientifically relevant parts of the formal system SF without assuming infinitely many objects, and while only assuming the existence of concrete objects.

(II) Problems concerning methodological naturalism

A philosopher who is a methodological naturalist differs from the prototypical natural scientists solely in terms of the questions and problems that he or she is interested in. Other than that, the methodological naturalist is engaged in the same overall scientific endeavor and uses the same methods as the prototypical scientists in order to pursue his or her goals. A methodological naturalist is of course allowed to criticize scientific work occasionally, just as all other scientists are permitted to do; but he or she will never attack a scientific theory merely on the grounds of philosophical presuppositions, unless these presuppositions are themselves justified by scientific theories that are better confirmed than the theory that is attacked.

I take this characterization of methodological naturalism for granted now, and from what he says about this in the abstract and in section 2.1, I do think that Feng Ye embraces this characterization as well. This said, I also wonder if Feng Ye is actually a methodological naturalist in this sense of the word. First of all, as far as I can see, a naturalist of this stripe would have to take physics pretty much at face value: only very, very rarely he or she would dare to correct what the present-day physicists have to say. And of course Feng Ye does not present himself as correcting modern physics either. But then again he writes: "contemporary physicists never straightly claim that there is infinity in this physical universe" (section 1.1), and "it is still a fact that contemporary sciences never claim that there is infinity in this physical universe" (section 2.4). Strictly speaking, this looks false to me: e.g., according to modern physics, space-time is a continuous manifold that consists of infinitely many space-time points. Hence modern physics does claim that there is infinity in the physical universe. End of argument. Or not quite – I suppose that Feng Ye might have a come back that might go like this: well, he might say, taken literally, the physicists might say so; but that's just for their own convenience. Really, we can revise physics, or extract the "real" physics form it, so that the result is not committed anymore to an infinity of space-time points, and we can actually do so without losing anything that we care about scientifically and deductively.[1] But now assume that the physicists do not accept this: no, they say, we have considered your proposal carefully, Feng Ye, but we think our physical theories need to postulate the existence of a space-time continuum, and not just for convenience but because we think that there is such an infinite space-time continuum in the physical world. Would Feng Ye go along with this? If he is a methodological naturalist he should; after all, it's not his job as a naturalized philosopher to correct the specialists about space-time. And at some places in his paper (e.g. in section 1.1. and at the end of section 2.4), it seems that this is exactly what Feng Ye will do: follow the physicists' lead on this matter. If he did not do so, he would let his finitary philosophical instincts overrule our best scientific theories, which would contradict his methodological naturalism.

But now let us move on to other parts of science. Consider mathematics: a thorough methodological naturalist about science in general, including mathematics, will never dare to tell the specialists that they are wrong in claiming the existence of infinitely many mathematical numbers, functions, sets, and the like. After all, it is not his or her job as a methodological naturalist to correct the specialists about numbers, functions, and sets. Back then, when Quine did judge pure mathematics solely in terms of its applicability in the natural sciences, he forgot that mathematicians should count just as much as belonging to the family of science as e.g. physicists do, and as a good naturalist he should have judged pure mathematics by the standards of the world-wide community of mathematicians, and not by his own very much empirically minded philosophical standards. Penelope Maddy made this very clear in her own, much more thorough version of methodological naturalism (her "second philosophy"). In contrast with Quine, she does take seriously what the modern mathematicians tell us about mathematics. Now back to Feng Ye: Will he be a good methodological naturalist and leave it, with only very, very rare exceptions, to the mainstream mathematicians to tell us which numbers, functions, and sets exist? And since they accept that infinitely many numbers, functions, and sets exist (and indeed infinitely many infinite sets), will he go along with their verdict? That is my question to Feng Ye. If he does so, presumably, he will have to give up his nominalism, since it will be difficult to locate all the sets in the cumulative hierarchy somewhere in the physical universe. And also his finitism will seem a bit pointless then: for if he is happy to embrace the infinitary commitments of pure mathematics, why worry any longer about the infinitary commitments of applied mathematics in science? But if he does not go along with the mathematicians' verdict, why not? On what grounds? Is it because

---

[1] Just as an aside: Would Feng Ye also defend the following stronger Conjecture of Finitism: "Strict finitism is in principle sufficient for formulating scientific theories, for conducting proofs and calculations in the sciences, *and for the inductive confirmation of scientific hypotheses*"? This is important, since assumptions of infinity might also play a major role in the inductive systematization of scientific data, much as theoretical terms do, as had been urged a long time ago by Carl Gustav Hempel.

his philosophical views tell him not to do so? That would be the moment at which he would cease to be a proper methodological naturalist. The upshot is: as far as I can see, there is actually a serious tension between Feng Ye's methodological naturalism on the one hand and his nominalism and finitism on the other.

It seems to me, therefore, that Feng Ye should not have committed himself to methodological naturalism in the first place. Instead he might want to restrict himself just to defending a naturalism of a different ontological sort: physicalism—the thesis that everything is physical. Which is not the same as methodological naturalism. Physicalism entails Feng Ye's nominalism, which makes good sense to him, of course; and without the commitment to methodological naturalism, he would be free to reject our mathematicians' verdicts concerning the existence of infinities, since he would not be bound anymore by the methodologically naturalistic obedience to scientific – in this case: mathematical – practice. Indeed, in various parts of the paper that is exactly what he says: he says that he is a physicalist (e.g. in the abstract and in section 1.2). But then again in other parts (e.g. in the abstract again) he speaks of himself as a methodological naturalist, and it is not clear how one can be both at the same time: physicalism is a proper philosophical thesis; mathematics, and science where it is mathematical, contradicts physicalism in this strict sense of the word, since mathematics and mathematical science accept the existence of infinities of abstract entities which are not physical; if someone sticks to the position of physicalism in spite of this, he or she must rate their philosophical inclinations more highly than mathematics or mathematical science; which means that he or she cannot be a methodological naturalist.

Comments on Ye

DIFEI XU

There are two parts in the following pages. The first is primarily on his general naturalized philosophy; the second is on his strict finitism.

1 Naturalized philosophy

Ye's naturalistic philosophy is a radical physicalism, which only admits what exist are only the material things in space and time, or objects that physics study on (we can call these objects are physical objects). As matter of fact, his naturalistic philosophy is ontological naturalism (also called physicalism), which is the strongest form of naturalism. This philosophy is a kind of common sense realistic philosophy, which admits concrete objects, like water, dog, electronic, etc, exist, while abstract objects, like 2, 2, or standard natural number model, don't exist. But how can we tell what is an abstract object? This philosophy turns to the common sense, and claims that material things which exist in space and time of the universe are concrete and they are the only things that exist in the world. I think maybe this strategy using commonsense is a wiser way to shut the traditionally philosophical notions out. Unlike Ye, some philosophers who declare that they themselves are naturalistic in philosophy of mathematics are realistic in philosophy of mathematics. They think besides physical objects, there are also abstract things including mathematical objects, such as sets. These philosophers include Quine, Burgess, and even Maddy. Today many philosophers are naturalists, and it seems that scientific spirit is the spirit of our times. Naturalistic philosophy has many branches, but they has one same principal belief that generalized scientific methods are the best methods we can rely on to know the world and there's no method prior to or better than scientific methods.

In [8], he analyzes many difficulties that these naturalistic philosophies of mathematics face and gives the underlying reason why they face the difficulties. The reason is that the naturalistic philosophers didn't take cognitive subjects as physical objects in the physical world, and forgot that the cognitive practices are only physical processes in the physical world. Although Quine admitted human beings as cognitive subjects were physical denizens in physical world, Ye argues, the reason why Quine admitted that reality of mathematical objects was necessary for mathematical knowledge is that he unconsciously took the cognitive subject as an detached mind standing outside of the world. Ye emphasizes that cognitive subjects are physical objects in the physical world, and he criticizes many other philosophies including other kinds of naturalistic philosophies in this way. But how does he decide that these philosophers take cognitive subjects as detached one outside the physical world? Ye once claimed if you were pondering at the reference of the abstract concepts like "2", you were thinking yourself as a detached subject, trying to project your "inner notions" to the "outer world". I think this argument is not so clear. On one side he claims his philosophy permit brain's introspection, but why this way of introspection is wrong? Sciences also need experiment to testify their theories. I believe Ye doesn't deny the method of "projection". The only thing not appropriate here seems to me is that the object 2 is not in the world! This argument only shows that it only conflicts with the basic assumption that the only things in the world are material things, but not take the cognitive as a detached mind. Maybe we can make the criterion clearer to decide whether a philosopher take the cognitive subject for a detached one outside the world.

Some philosophers, argue that mathematical intuition is also an important source of mathematical knowledge. The natural concepts like "water", is a universal of particulars. Since we permit "capturing" the similarities or same pattern of particulars, why should we deny the reality of some kinds of abstract objects? To deny these objects, in my opinion, a philosoper must use the ontological basic belief of physicalism. For Ye, his strategy to convince the other philosophers that there are only physical objects is to show (1) his philosophy is minimal philosophy, because his ontology only admits reality of physical objects; (2) his philosophy can explain in a clear and coherent way what the other philosophy can explain. Therefore, I think that Ye's philosophy does not starts from mathematical naturalism, but from ontological naturalism instead.

The main task of naturalism is to explain what role the prima facie non-phenomena, like mental representation, play in the physical world. Ye adopts naturalism. He believes epistemic questions should be turned into cognitive questions, since the most reliable methods are scientific methods. The traditionally philosophical notions, such as truth, reference, logical validity, should be naturalized, and its aim is to put these naturalized new notions fit into the world revealed by natural sciences. Ye has been working on naturalized semantics these years. Briefly, Ye argues that he only admits the reference relations between natural concepts, like "dog", in brains (remember all the concepts in brains are neuronal activities, and they are material things) and the material things that the concept represents, because this kind of reference relation is between material things. However the traditional reference relations between the abstract concepts including mathematical concepts and mathematical objects don't exist, because there are no mathematical objects in the world at all. Physicalists only admit the relation between material things. Here

Ye uses ontology of physicalism to deny the traditional reference relation.

Ye emphasizes that Tarski's semantics is a simplified semantics, and it doesn't tell anything about the normality of representation. In modern philosophy of mind, many philosophers are trying to describe the representing relation between concepts in brain and the material things in the naturalistic frames. Of course there are many kinds of frames reflecting different points of view on how the concepts work. For example, some philosophers think that all concepts have structures, while some philosophers don't think that all concepts have structures.

Ye has been trying to provide his own semantics utilizing the new research results of cognitive sciences, methods of semantics and non-classical logics (like dynamic logics or commonsense reasoning). In [11], Ye argues that concepts have structures, so that they can combine together to act as cognitive process. Although the abstract concepts, including mathematical concepts, don't have references in the physical world, they have cognitive functions instead. This research project of naturalized semantics is still going on, and we expect that the results in this research will give us better understanding of our cognition process. However some philosophers are not satisfied with the naturalized semantics even if they are finished someday in detail. They don't mind the semantics is so complicated, although it is. What they are confused about this kind of semantics is why this semantics endow some kind of concepts with reference functions while others not. They think that all concepts have cognitive function, and they are also the results of cognitive process. We know that even these concrete concepts are the results of abstraction of cognitive subjects. Why are they superior to the concepts of mathematics? It seems, to some philosophers, that these kinds of semantics are not only like a fairy tales, but also beg the question. Naturalists provide different naturalizing frames. Since these frames all look like explanatory stories, how can philosophers decide which one is the best?

2   Strict Finitism

Ye's strict finitism is the consequence of physicalism and nominalism. Ye believes that there are finite material things in the physical world, not only in macro scope but also in micro scope. So there is a puzzle: why classical mathematics including infinity can be applied in sciences which only study the finite physical objects? He conjectures that the mathematics applied in science can be reduced to strict finitism mathematics. Ye has showed us that this project is feasible. And Ye gives a logical or mathematical solution to this puzzle, because he thinks this problem is a logical problem. Of course, strict finitism itself doesn't imply nominalosm. There are still philosophical questions like what are natural numbers? Are they independent of cognitive subjects? But with the physicalism as his basic philosophy, Ye gives a way to solve the problems that physicalism should answer, and this solution is coherent with his physicalism. And in turn this coherent answer supports his philosophy. I would say that Ye's technical work is very significant to physicalism. In the field of foundation of mathematics, many mathematicians have been working in reduction ways. They always claim that "a little bit goes a long way". We appreciate their work because they show us how a little bit goes a long way and give us a clear picture of what rest on what. Maybe these mathematicians think they are doing philosophy in mathematical or logic ways. And we've already seen that Ye also solve a philosophical problem in logic or mathematical way. As a matter of fact, many philosophical works are very specialized, like logic and mathematics. Russell said at the beginning of his book "Human Knowledge" as following: "The following pages are addressed, not only or primarily to professional philosophers, but to that much larger public which is interested in philosophical questions without being willing or able to devote more than a limited amount of time to considering them. Descarts, Leibniz, Lock, Berkley, and Hume wrote for a public of this sort, and I think it is unfortunate that during the last hundred and sixty years or so philosophy has come to be regarded as almost as technical as mathematics. Logic, it must be admitted, is technical in the same way as mathematics is, but logic, I maintain, is not part of philosophy. Philosophy properly deals with matters of interest to the general educated public, and loses much of its value if only a few professionals can understand what is said." Here, I'd like to ask some general questions: Does philosophy really lose its value if only a few professional can understand what is said? Today some philosophers are defending for traditional philosophy. For example, George Bealer argues for two principles. One is the thesis of autonomy of philosophy and the other is the thesis of authority of philosophy. Naturalism epistemology also faces many difficulties, such as accounting epistemic normativity. Ye's philosophy also face the general difficulties that naturalism face. How Ye's philosophy cope with these difficulties is what we are expecting to know. As we all know, in philosophy of mathematics, Gödel held the other extreme of philosophy. His famous realism is also coherent. In his correspondence reappeared in [15] (page 8-11) he explained how his philosophy helped him get the outstanding theorems. Here, we want to know how can we decide which philosophy is better for mathematics?

BIBLIOGRAPHY

[1] Bealer, George. a priori knowledge and the scope of philosophy. *Epistemology, An Anthology*, 2nd (2008) Blackwell publishing Ltd, 612-624
[2] Gödel(1964). "Cantor's continuum problem", reappeared in *Kurt Gödel Collected WorksII* (1990), Oxford University Press.
[3] Macdonald, G. and D. Papineau (2006). *Teleosemantics: New Philosophical Essays*, Oxford University Press.
[4] Russell, B. *Human Knowledge Its Scope and Limits*, reprinted 1997, Routledge.
[5] Ye, F. (2010a)."Naturalism and abstract entities". *International Studies in the Philosophy of Science*, 24, 129-46.
[6] Ye, F. (2010b). "What Anti-realism in Philosophy of Mathematics Must Offer", *Synthese*, 175, 13-31.

[7]   Ye, F. (2010c). "The applicability of mathematics as a scientific and a logical problem". *Philosophia Mathematica*, 18, 144-65.
[8]   Ye, F. (2010d). *Philosophy of Mathematics in the Twentieth Century: A Naturalistic Commentary* (in Chinese), Peking University Press.
[9]   Ye, F. (2011a). *Strict Finitism and the Logic of Mathematical Applications*. Synthese Library, vol.355. Dordrecht: Springer.
[10] Ye, F. (2011b). "Naturalized truth and Plantinga's evolutionary argument against naturalism", *International Journal for Philosophy of Religion*, 70, 27-46.
[11] Ye, F. (online-a). A Structural Theory of Content Naturalization, http://sites.google.com/site/fengye63/. Accessed February 15, 2013
[12] Ye, F. (online-b). On what really exist in mathematics. ibid.
[13] Ye, F. (online-c). Naturalism and objectivity in mathematics. ibid.
[14] Ye, F. (online-d). Naturalism and the apriority of logic and arithmetic. ibid.
[15] Wang, H (1974) *From Mathematics to Philosophy*, New York: Humanities Press

Response to Xu and Leitgeb

FENG YE

1   Reply to Difei Xu

First I want to thank Professor Difei Xu for her thoughtful comments. The comments raise many interesting questions. I will respond to the questions and take this chance to clarify my ideas.

If I understand correctly, Xu's first question is, 'why can't a brain "project" its mathematical concepts onto the external as abstract mathematical objects?' The idea, if I understand correctly, is the following: The only difference between 'projecting' mathematical concepts and 'projecting' other concepts representing physical things is that the former results in abstract objects, not physical objects; if you deny the former but allow the latter, then your reason can only be that abstract objects do not really exist in the world; but then you beg the question against mathematical realism in your argument. 'Projecting' mathematical concepts and 'projecting' other concepts representing physical things are on a par. If physicalism allows the latter and admits that it does not assume a mind detached from the world, then the former does not assume any detached mind either.

Now, [5] explains why it is senseless to say that a *brain* 'reifies' (or 'is committed to', or 'posits') abstract objects. It targets Quine's claim that we 'reify' (or 'are committed to', or 'posit') abstract objects. The same strategy can explain why it is senseless to say that a brain is entitled to 'project' its mathematical concepts onto the external as abstract objects.

First, note that 'projecting concepts onto the external by a brain' is only a metaphorical, non-literal manner of speech. A brain cannot really 'project' anything out of its skull. A brain is a system of neurons physically interacting with its physical environments. You certainly do not want to mean by 'projecting' that a brain is throwing its neurons out of its skull.

Second, note that physicalism is *not* the idea that the 'external world', or the 'world external to me', is all physical. If I understand physicalism this way, then I implicitly take myself to be an exception. That is, *only* the world external to me is physical, and *I myself* am a non-physical 'subject'. Physicalism is the idea that everything including myself as a cognitive subject is physical, completely physical, period. There is no exception. It is not that the 'external world' is physical but I am a non-physical 'subject'. The term 'external world' should be meaningless under physicalism. The traditional idea is that my concepts belong to my 'inner world' and I 'project' my concepts onto the 'external world' when I try to know (to gain knowledge about) things in the 'external world'. This idea makes no sense anymore under physicalism, because cognitive subjects and cognitive processes are completely physical things and physical processes within the physical world, namely, brains and neural activities and their interactions with their physical environments. Cognitive processes are *not* accesses to the 'external world' by a 'subject' standing outside the 'external world', where the 'external world' should contain the brain of the 'subject', and then the 'subject' has to be distinct from its brain and has to be non-physical.

Therefore, brains physically interact with their environments; brains do not literally 'project' anything onto the 'external world'. However, that said, I admit that we can say that, *in a sense*, a brain 'projects' its concepts representing physical things. Think of a robot in a lab. The robot first observes some red balls and blue cubes in the lab and creates two concepts RED-BALL and BLUE-CUBE accordingly. Then, by an internal process, the robot creates another concept RED-CUBE and then searches around for what the concept represents, namely, red cubes. You may want to describe this scenario as 'the robot is projecting its concept RED-CUBE onto the external'. That is fine. However, remember that the phrase 'projecting concepts onto the external' here is reducible. It is a loose and figurative way of saying what is really happening in the lab. You can replace it with a literal, more accurate and completely physical description of physical things and events in the lab.

Then, consider how the robot uses its mathematical concept '9'. The concept (as a chunk of bits in the robot's memory chip) participates in controlling the robot's counting actions in counting balls and cubes. There may also be sentences like '9 is an odd number' in the robot's knowledgebase, and the concept '9' also plays some inferential role in the internal inferential processes in the robot's AI system. These are some of the cognitive functions of the concept '9' for the robot. Now, would you say that the robot is 'projecting its concept "9" onto the external as an abstract object'? Or would you, following Quine, say that the robot is 'reifying (or positing, or committed to) the abstract object 9'? That seems pointless. You can have a complete physical description of what is really happening in the lab. In that description, you just describe how the physical components of the robot (including components realizing data and programs) work in the robot and how they physically interact with the robot's physical environment. You do not use figurative words like 'project'. Nor do you use epistemic/semantic terms like 'reify', 'posit', or 'be committed to'. Nor do you use the puzzling philosophical term 'abstract object', which engineers designing the robot will have a hard time to understand. After you have described how the physical components of the robot work, what is the point of adding 'the robot is now projecting its concept "9" onto the external as an abstract object', or 'the robot is now reifying the abstract object 9'? It seems pointless. It adds nothing informative for describing the robot's cognitive activities in the lab. It only introduces vague philosophical locutions such as 'project', 'reify', 'be committed to', and 'abstract object', which engineers designing robots will never use and actually would not understand. Now, for a physicalist, a human is a completely physical thing and is not essentially different from a robot. That is why it is also meaningless to say that a brain 'projects its mathematical concepts onto the external as abstract objects'. Brain neurons just physically interact with each other and interact with other physical things in human environments. After you describe how neurons work in the physical world, it is pointless to add 'now the brain is projecting those neural structures (namely, mathematical concepts in the brain) onto the external as abstract objects'.

Finally, there is a psychological explanation as to why so many people strongly feel that they are entitled to 'project' their mathematical concepts onto the external. My assumption is that people habitually take the stance of a 'subject' facing an 'external world'. When you take that stance, physical objects and abstract objects 'look' on a par to you. Both belong to the 'external world' and are your 'posits'. In other words, you think that you believe in the existence of physical objects in the 'external world' because you 'project' your concepts onto the external. Therefore, you have an equal right to project your mathematical concepts onto the external and believe that there are abstract mathematical objects in the 'external world'. If you seriously treat a human cognitive subject as a physical system, a robot, and seriously describe human cognitive processes as physical processes, as neural activities and their interactions with their physical environments, you will see that you never need to worry about whether a neural structure somehow 'represents abstract objects', or 'reifies abstract objects'. You just describe how that neural structure interacts with other neural structures or other physical things. The notion of 'abstract object' won't even occur to you, just as it never occurs to engineers designing robots. Engineers just deal with data and programs (realized as bits and bytes on silicon chips), the robot's mechanical body and arms, and other physical things in the robot's environment. They never ask 'does this piece of data (e.g., an integer '9', or a floating point '9.0', or a double precision '9.0') represent an abstract object?'

That concludes my clarification regarding Xu's first question. Please see [5] for more details.

Xu's second question, if I understand correctly, is the following. Brains can discern similarities between physical things, or the same pattern of particulars. Why do you deny the existence of a type as an abstract object besides those physical things or particulars? For instance, brains can recognize various tokens of the letter 'a' in various fonts, including fuzzy hand-written tokens, and brains can classify them into tokens of the same letter 'a'. Why do you deny the existence of the letter type 'a' as an abstract entity?

I discussed this question in [7]. The answer is similar to the answer above. We consider some ink marks on paper to be tokens of the letter 'a' because they look sufficiently similar to 'a', and some ink marks on paper are not considered tokens of the letter 'a' even though they look a little like 'a', because we think that they are not sufficiently similar. There is a vague borderline between those that are considered tokens of the letter 'a' and those that are not. What decides this vague borderline? It is the actual pattern recognition processes in human brains. Brains decide which ink marks look similar enough to 'a' to be considered a token of 'a'. To say that they are tokens of the same letter is just to say that the pattern recognition processes in actual human brains happen to classify them and produce the same label 'a' to label them. (If aliens secretly change human brains, it may cause different ink marks to be considered tokens of 'a' by human brains and the borderline for tokens of 'a' will change.) In describing the whole pattern recognition process performed by a brain, you talk about ink marks on paper, neurons in brains doing the pattern recognition, and a label 'a' as a neural circuitry in the brain's memory. You will never mention the alleged letter type 'a' as an abstract object independent of all human brains. Assuming the existence of the letter type 'a' as an abstract object independent of human brains helps nothing in explaining how human brains work in recognizing and classifying ink marks on paper. It is the actual human brains that determine which ink mark is a token of 'a' and which is not. The alleged letter type 'a' as an abstract object cannot determine by itself which ink mark is a token of 'a'. You think that there is a unique abstract object as the type of all tokens of 'a', only because you use the phrase 'the letter "a"' as a singular term. Such linguistic usages are convenient, but linguistic usages are physical events in the physical world. They have nothing to do with the alleged letter type 'a' as an abstract object. The idea of a letter type as an abstract object won't occur to you if you just describe the linguistic usages as physical events. The engineer designing a pattern recognition program for a robot to recognize tokens of 'a' never wonders whether there exists an abstract object as the type of all the ink marks that her pattern recognition program happens to classify into the same and label by the same label 'a'.

You might insist that, as long as a pattern recognition program classifies some ink marks into the same and label them by the same label 'a', there exists an abstract object there, for instance, the class of those ink marks, which can be seen as the letter type. This is related to Xu's third question. The question is whether my argument presupposes ontological nominalism not just methodological naturalism. The argument will be circular if it presupposes that abstract objects cannot exist, and fans of abstract objects will not be convinced by such an argument.

To answer this question I must first clarity that I did not mean to offer any knockdown argument to show that abstract objects just cannot exist. Philosophical arguments are rarely knockdown arguments. All we can do is to weigh pros and cons. For instance, suppose you insist that a neural structure can bear an irreducible reference relation with some abstract objects (that is, neural structures can refer to or represent abstract objects), and suppose you insist that the reference relation between a neural structure and abstract objects is fundamental and does not need further explanations. I cannot say that you are logically inconsistent. You are consistent to hold such a view. You are consistent to insist that as long as a pattern recognition program classifies some ink marks and label them with the same label 'a', there exists a type as an abstract object. However, your view has disadvantages. It is ontologically more extravagant by assuming abstract objects. It assumes more unexplainable primitives, namely, the reference relation between neural structures and abstract objects. Recall that we can in principle explain how a neural structure tracks some physical objects by some natural regularities and thus it can be said to 'refer' to those physical objects. Reference to physical objects is a reducible relation in the physicalistic image of the world; it is not an unexplainable primitive. Moreover, introducing the so-called abstract objects and reference to abstract objects does not add anything new and informative for describing and explaining what is happening within the physical world. Our description of what is happening within the physical world is self-sufficient, without the need to mention any alleged abstract objects. Then, your theory will be like an astronomic theory that assumes that every star and planet in the sky has a guardian angel but the guardian angel never interferes with the natural world or how the stars and planets move. The additions are redundant and serve no good purpose.

Therefore, my line of argument is the following. Accepting methodological naturalism, we philosophers should accept what contemporary scientists confidently tell us. These should especially include what cognitive scientists tell us, that is, that human brains have so and so cognitive abilities. From there, together with other naturalistic views about humans offered by the contemporary sciences, we reach what I called 'physicalism about cognitive subjects', namely, the view that cognitive subjects themselves are completely physical. Then, we see that the idea that a brain is 'reifying abstract objects', or 'committed to abstract objects', or 'projecting its mathematical concepts onto the external as abstract objects' is unclear and gratuitous. It serves no purpose for describing how brains work within the physical world. Therefore, we reach nominalism. Nominalism is certainly not the starting point in my argument.

Indeed, many mathematicians strongly believe that there is an objective abstract world of mathematical entities. Methodological naturalism requires us to take all scientific views seriously. I admit that they should include mathematicians' views, but those mathematicians' Platonism appears to be in conflict with our naturalistic and scientific view about humans. When scientists (and mathematicians) are in a little conflict among themselves, we philosophers have to come and reconcile the conflict. I claim that we should uphold the naturalistic view about humans, because our cognitive scientists, evolutionary theorists and biologists know much more about what humans are and what human thinking processes are than mathematicians do. Besides, not all mathematicians are Platonists. I am willing to say that some mathematicians are so engrossed in their mathematical work that they have illusions about the ontological status of the alleged mathematical world and have illusions about their own thinking processes. I cannot forsake the naturalistic view about humans supported by all other branches of science, from physics to evolutionary theory and cognitive sciences, in favor of a few Platonist mathematicians who do not really care about all other branches of science. Besides, remember that this is just about philosophical interpretations of mathematical practices, not about mathematical practices themselves. Mathematicians can still do their work in their own ways. Mathematicians are experts regarding mathematics, but they are not experts regarding humans or human mathematical thinking processes, nor are they experts about how to reconcile the ontological status of the mathematical world with the naturalistic worldview offered by contemporary sciences.

The fourth question that I want to consider is 'why this semantics endows some kind of concepts with reference functions while others not'. In other words, why do some concepts in a brain represent physical objects but some other concepts, say, mathematical concepts, do not represent anything? The underlying thought, I suppose, is again that if we are entitled to think that some of our concepts represent physical things, why aren't we entitled to think that our mathematical concepts also represent entities?

Well, the answer is: that is exactly what we see when we open a skull and carefully examine how neurons work in the brain and how they interact with other things in the environment. Some concepts as neural structures in the brain bear a special kind of physical connection with some physical objects. For instance, a concept RABBIT as a neural structure in the brain bears a special kind of physical connection with rabbits. We recognize that this special kind of physical connection is what we intuitively called 'representation' or 'reference'. Then, we look at the concept '9', another neural structure in the brain. We do not see that this neural structure bears *the same kind of* physical connection with anything in the environment. Actually, this neural structure '9' bears more complicated connections with physical things in the environment. For instance, it can combine with the neural structure RABBIT to form another concept 9-RABBIT, which then bears the same kind of physical connection, namely, reference, with some physical things in the environment. That is one of the cognitive functions of the concept '9' in the cognitive activities of the brain. The point is that it serves no good purpose to assume a *new* kind of objects, abstract objects, and then to assume a *new, irreducible, and unexplainable* non-physical reference relation between our neural structures and these abstract objects.

Finally, I will respond to Xu's last question, 'which philosophy of mathematics is better for mathematical practices?' Gödel used to imply that the belief in Platonism helped his discovery of his famous theorems. Therefore, the idea is that Platonism might be a better philosophy for mathematical practices.

Now, Platonism might indeed have helped Gödel, but one can also say that formalism helped Cohen in his discovery of his independence proof. Platonism about a specific subject matter, e.g., the universe of sets, may psychologically encourage people to imagine the subject matter as something objective, eternal and of a higher value, and thus it may encourage people to put more enthusiasm on the subject matter. However, it has potential downsides as well. For instance, it may lead to idle, fruitless fantasies, and it may also discourage exploring other subject matters, because it makes people believe that one subject matter is the unique Reality. On the other side, nominalism can be more liberating. Nominalism claims that mathematical worlds are our imaginations. When we apply mathematics to physical things, we use imaginary mathematical models to smooth over extremely complex details in the finite physical world and build simplified and tractable models to represent the physical world. Scientists' ingenuity just lies in constructing appropriate tractable fictional models that smooth over enough details but are still accurate enough. By viewing mathematics this way, nominalism encourages us to explore new kinds of mathematical imaginations for scientific applications, not limited to the set theory in **ZFC**. Since there is no unique mathematical Reality, you can imagine any kind of mathematical worlds whatever you can. On the other side, when applying mathematics to the physical world, it is this finite physical world that constrains our applications. In this way, we can have a more liberal spirit in mathematical practices, while at the same time we do not lose our grasp of the physical reality. That is, nominalism may be better for mathematical practices.

2   Reply to Hannes Leitgeb

First I want to thank Professor Hannes Leitgeb for his comments. Leitgeb raises several concerns, challenges and questions. I will respond to them and also take this chance to clarify my ideas.

Leitgeb's first concern, if I understand correctly, is the following: In presenting my formal system **SF**, I appear to be committed to infinitely many syntactical entities, which contradicts finitism. He puts the challenge this way: 'to determine the scientifically relevant parts of the formal system **SF** without assuming the vocabulary, the set of well-formed expressions, the set of axioms, the set of rules of inference, or the set of theorems to be infinite.'

In my book [6] and in the target introductory paper that Leitgeb gives his comments, I explain why the system **SF** is NOT committed to the existence of infinitely many numerals. The system **SF** is quantifier-free. It never makes any assertion

about 'the entire domain of numerals'. It does not need to assume that there exists
a fixed domain of numerals. It does not even assume that numerals can grow without limit. That is, it does not even assume potential infinity. General laws about
numerals are expressed by free variables as in $x + y = y + x$. Then, whatever
numerals you actually have, you can substitute them for the free variables and get
a sentence. Obviously, the same should apply to the syntax of **SF** itself. When
presenting the syntax of **SF**, I present schemas for constructing terms and formulas. The schemas themselves are completely finite. You can follow the schemas
to construct new terms and formulas whenever you actually need them. That is all
about it. No infinity in any sense, not even potential infinity, is assumed. I certainly
do NOT claim that there are or will be literally infinitely many terms or formulas
in this physical universe, nor do I ever claim that humans can or will construct
arbitrarily long terms. Both are plainly false. 'The scientifically relevant part of
SF' just consists of whatever terms and formulas that we will actually construct
when applying **SF** to analyze scientific theories in this real world, following those
finitely many schemas for constructing terms and formulas. On the other side, we
can *imagine* that there are infinitely many terms (or *imagine* that an ideal human
agent can construct arbitrarily long terms). However, when we do the imagination,
we again use *finite* sentences to express our imaginations. Imagination activities
in a brain are completely finite neural processes within this finite physical world.
Your brain does not literally *create* infinitely many 'fictional entities' when you
imagine that there are infinitely many numerals. All your brain creates are just a
few sentences, for instance, a sentence like 'for each numeral t, St is a numeral'.
That is, your brain again just creates finitely many schemas for constructing new
numerals or finitely many linguistic descriptions of the alleged 'infinitely many
numerals'. Nothing involved is really and literally infinite. That should be very
obvious. Remember, we are finite beings in this finite physical universe.

On the other hand, you will create a lot philosophical troubles if you take infinity here to be real. How can a human brain as a finite system of neurons magically
'refer to', or 'comprehend', or 'grasp', or 'be committed to' (whatever you want to
call it) infinitely many numerals? What is that mysterious semantic and epistemic
connection between a human brain as a finite and concrete physical thing within
this physical universe and the alleged infinitely many numerals as abstract entities?
Are there anything more than the fact that the brain is manipulating sentences like
'for each numeral t, St is a numeral' in the internal inferential processes within
the brain, which are completely finite in themselves? If you think of yourself as
a non-physical mind with a mysterious non-physical cognitive channel or non-
physical cognitive ability to 'access' the world of abstract entities, then you can
perhaps have a coherent view here. But just 'perhaps', because as I have explained
in the target paper, I accept methodological naturalism, which requires us to take
seriously the contemporary cognitive scientific view that it is brain neurons that

are thinking, processing languages, creating concepts, doing logical inferences, planing actions, and so and so forth. It is not any non-physical mind with mysterious cognitive capabilities unexplainable in physical terms. That is, I believe that, in the context of contemporary sciences, methodological naturalism will lead to physicalism about cognitive subjects. Then, if you insist on the literal existence of infinitely many numerals as abstract objects, you will have to explain the epistemic and semantic connection between a *brain* as a finite and concrete physical system of neurons and the alleged infinitely many abstract objects. That is a real trouble, an insurmountable difficulty given what a brain is. You might think that for a brain to 'refer to' or 'be committed to' abstract objects is just for the brain to use pronouns, variables and quantifiers in some manner with some effects in controlling the body actions. This appears to be what Quine was saying. However, 'using pronouns, variables and quantifiers in so and so manner with so and so effects in controlling the body actions' is already a complete description of the physical things and events in the physical world. What is the point of paraphrasing it as 'the brain is now committed to abstract objects!'? What is the point of introducing so-called abstract objects? In my papers I offered a psychological explanation as to why people tend to do that. The idea is that people have a hard time to think of themselves as physical things. They unconsciously take themselves to be non-physical 'subjects' facing an 'external world', which contains all physical objects, including their own brains (and may contain abstract objects as well). They unconsciously take this stance of a supernatural 'subject' even after they openly admit that humans are 'physical denizens of the physical world' (Quine). Please see [5] and the target paper for more details on this.

Leitgeb's second concern is that we ordinarily take expressions of a formal system to be abstract objects, which is not allowed for nominalists. He puts the challenge this way: 'to determine what kind of concrete entities the formal expressions in the language of **SF** are meant to be. Are they linguistic expressions that we actually write on a blackboard or which we type into a computer? What if someone erases any of them again? What if someone writes two inscriptions on the blackboard that look the same?'

Note that this is a question for all nominalists, not just for my version of nominalism. Besides, it is among the first set of questions that a nominalist will naturally try to answer. All contemporary nominalist philosophers (e.g., Field, Hellman, Chihara, Maddy, Leng in philosophy of mathematics and many others in metaphysics) already have their own answers to this question. (Otherwise, the contemporary debates between nominalism and its alternatives will be completely dumb.) That is not to say that I am satisfied with the answers offered by contemporary nominalists. I have explained the answer that I personally favor in [7] and in my reply to Difei Xu above. It is again based on physicalism about cognitive subjects. I believe that physicalism about cognitive subjects helps to show clearly that assuming the existence of a symbol type as an abstract entity is irrelevant and unhelpful for describing and explaining human cognitive activities in recognizing and processing languages. The notion of a type as an abstract entity independent of all brains is a philosophical notion completely alien to the naturalistic descriptions of the cognitive activities of human brains. It only adds, unnecessarily, the philosophical puzzle of how a brain can 'grasp' (or 'comprehend', or 'be committed to', or whatever you want to call) the alleged abstract entities independent of all brains. Here I just want to add a few points.

If all tokens of the letter 'a' in the world are erased but the pattern recognition programs for recognizing 'a' in human brains are intact, then we can still talk about the letter 'a' meaningfully, by which we can still mean concrete tokens that *will* be recognized by human brains as 'a'. Now, suppose that aliens secretly erase all pattern recognition programs for recognizing 'a' from human brains, so that from now on no human brain will recognize current tokens of 'a' in various fonts as tokens of the same letter, will you say that the abstract letter type 'a' still exists? How about the time before humans invented the letter 'a'? Would you say that the abstract letter type 'a' exists independent of time, eternally? Apparently, human brains create the letter 'a'. But how can brains within space-time create abstract entities not in space-time? I teach my dog to recognize some tokens of 'a', but no matter how hard I teach, my dog is not able to recognize some tokens that we human brains recognize as 'a'. I write a pattern recognition program for recognizing 'a' and the program has the same problem. Does my dog or my program therefore create a letter type different from the letter type 'a' created by human brains? Suppose that someone tries to invent a new letter for English. He preaches his new letter to other people, but few accept him. How many English speaking people must accept him before we can say that he really creates a new letter type? What if he forgets about his invention later because he cannot convince others? Or is it the case that whenever you draw a mark on paper and intend it to be a token of a type you successfully create a new type even if no other humans will recognize any tokens of that type and you yourself forget it altogether later?

Therefore, it is not hard to see that many problems similar to the one that Leitgeb raises are actually difficulties for realism about abstract entities, not difficulties for nominalism. This fact shouldn't be anything surprising at all. Open your eyes, you see concrete things (including human brains) in the world. These are real things and they have to be the bases for anything else if there are indeed anything else. A so-called letter type is first of all what you imagine when you close your eyes and imagine a representative for all those tokens that you recognize to be 'the same' in some aspect. If you think that it really exists and is not just your imagination, surely you will have to offer more philosophical explanations to explain what kind of reality it belongs to and how that realm of reality is connected with human brains. You will have to overcome many philosophical difficulties. I am not saying that the problems raised in the last paragraph are unanswerable. For instance, you might hold that, after you draw an ink mark on paper, as soon as you have the mental intention to let the ink mark to be a token of a symbol type, you successfully create a symbol type as an abstract entity, and then that abstract entity exists forever even after you yourself forget your previous intention at the next moment. You might also hold that the existence of a symbol type is a matter of degree. As more and more people start to recognize tokens of a type you invent, the existence of that type becomes more and more solid. Since 'abstract object' is such an obscure notion, you can attribute any exotic feature you like to alleged abstract objects.

But why all the fuss? In contrast, nominalism is a much more innocent philosophical position and will engender much fewer philosophical puzzles. The challenge 'to determine what kind of concrete entities the formal expressions in a language are meant to be' *is* a very realistic challenge. It is a challenge that AI software engineers face everyday when designing their pattern recognition programs for recognizing linguistic tokens. However, it is a technical challenge, to be met by engineering efforts. It contains nothing philosophically or conceptual puzzling, nothing like the puzzles that the notion of 'abstract object' causes. A more interesting question is perhaps, 'why so many philosophers think it obvious that there have to be types as abstract objects?' As I have mentioned above, I offered a psychological explanation in my research project.

Leitgeb's next concern is that I may violate the principle of methodological naturalism. Here, his first question is about my assertion, 'contemporary physicists never straightly claim that there is infinity in this physical universe', which he quotes in his comments. He then says,

> 'Strictly speaking, this looks false to me: e.g., according to modern physics, space-time is a continuous manifold that consists of infinitely many space-time points. Hence modern physics does claim that there is infinity in the physical universe. End of argument.'

He then considers reading my claim as a philosophical reinterpretation of what physicists say, and then he asks,

> 'But now assume that the physicists do not accept this: no, they say, we have considered your proposal carefully, Feng Ye, but we think our physical theories need to postulate the existence of a space-time continuum, and not just for convenience but because we think that there is such an infinite space-time continuum in the physical world. Would Feng Ye go along with this?'

Therefore, he suggests that I violate the principle of methodological naturalism by denying what physicists say.

Apparently, Leitgeb assumes here that, by applying semi-Riemannian manifolds to simulate large-scale space-time structures in general relativity, physicists are saying that physical space-time is a continuum. I thought it is already common knowledge among philosophers of mathematics that this understanding of general relativity is wrong. It is wrong according to physicist or according to standard physics textbooks, not wrong according to any philosophy. No physicist understands general relativity that way. Semi-Riemannian manifolds are meant to simulate space-time structures at a large scale. General relativity, by itself, says nothing about the micro structure of space-time. What physicists are currently doing in theories of quantum gravity is just to make general relativity compatible with discrete space-time. Applying semi-Riemannian manifolds in general relativity is like applying continuous models in fluid dynamics, where we know that fluids are not continuous. Similarly, validity of semi-Riemannian manifolds for modeling large scale space-time structures is relatively independent of the structures of space-time at the micro scale. These are common knowledge among physicists and physics students. These examples of literally false but useful mathematical models in physics are also cited frequently in the literature in philosophy of mathematics.

In the target paper, my next sentence following my assertion above quoted by Leitgeb is

> 'Macroscopically, the universe is believed to be finite; microscopically, physicists so far study only things above the Planck scale (about $10^{-35}m$, $10^{-45}s$ etc.). No physicists are considering things at even lower scales (say, $10^{-350}m$,), and a strictly discrete space-time structure is an open option. Now, philosophers cannot know more about this physical world than physicists do ...'

It is obvious that I am there referring to theories of quantum gravity in contemporary physics. I am reading physicists' claims literally. I am NOT reinterpreting what physicists say. Many physicists 'straightly claim' that the Planck length is perhaps the minimum physically meaningful spatial length and that space-time may be discrete at the Planck scale. For instance, in the introduction chapter to a collection of articles on various approaches to quantum gravity, physicist Rovelli says (referring to the fact that the Planck length is perhaps the minimum physically meaningful spatial length),

> 'A number of considerations of this kind have suggested that space might not be infinitely divisible. It may have a quantum granularity at the Planck scale, analogous to the granularity of the energy in a quantum oscillator. This granularity of space is fully realized in certain Quantum Gravity theories, such as loop Quantum Gravity, and there are hints of it also in string theory.' ([2], p.6)

Both String Theory and Loop Quantum Gravity, the two most popular theories of quantum gravity, suggest that spatial length becomes physically meaningless below the Planck length. Loop Quantum Gravity further explicitly assumes the smallest possible spatial length, area and volume ([3], p.136, Figure 26). Even if you don't like any of these contemporary theories of quantum gravity, it is still a unanimous view among physicists (and actually it is basic knowledge among physics students) that general relativity says little about the micro structure of space-time and that it is an open question whether space-time is a continuum.

One of the goals of my book ([6], Chapter 8) is just to offer a logical explanation about why our physics conclusions in general relativity, especially, a version of Hawking's singularity theorems, can still be valid at the large scale, even if space-time is not continuous at the micro scale. That is, strict finitism allows us to show that physically meaningful statements, proofs and calculations in general relativity can be interpreted as statements, proofs and calculations about physical space-time *only* at the large scale, without being committed to anything at the micro scale. I believe that this exactly reflects working physicists' intuitions about their theory. It is not a philosophical reinterpretation of what physicists believe. In fact, you might even accuse me of being a too naive scientific realist, accepting physicists' naively realistic beliefs without any philosophical reflection. That is fine. If a scientific realist can never straightly claim that there is infinity in this physical world, then, I suppose, an scientific anti-realist will be less likely to claim that there is infinity in this physical world.

Now, even if space-time is discrete at the Planck scale, there might still be infinity in other places or aspects of the physical world. Therefore, I only made the cautious assertion that contemporary physicists never 'straightly claim' that there is infinity in the physical world. I didn't even say 'physicists suspect that there is no infinity in the physical world'. The modest claim is already sufficient to motivate a strictly finitistic approach to philosophy of mathematics for a naturalist and nominalist. Because, if you assume the reality of infinity in your philosophical account of human mathematical practices and applications, then the infinity you assume has to belong to either the abstract world or the non-physical mental world, since you do not really know that there is infinity in this physical world. You cannot know more than physicists do. Then, you have to abandon either nominalism or the naturalistic view about human thinking processes.

Leitgeb's second question regarding methodological naturalism is about my attitude toward mathematicians' realistic beliefs about mathematics. The idea is that mathematicians are also scientists, and methodological naturalism requires us to respect mathematicians' judgements. By denying mathematicians' claim that there are numbers, I may violate the principle of methodological naturalism.

This is one of the common objections to nominalism in philosophy of mathematics. There are already responses to this objection in the literature. My own response to this objection should be clear from the specific strategy that I use to argue for nominalism. As I have explicitly emphasized in the target paper, I start from methodological naturalism and first conclude that methodological naturalism requires us to take seriously contemporary cognitive sciences and to accept physicalism about cognitive subjects. Then, I argue that physicalism about cognitive subjects implies nominalism and strict finitism in philosophy of mathematics. This means that some mathematicians' realistic beliefs about mathematics are in conflict with our scientific and naturalistic view about human thinking processes. I choose to uphold our scientific and naturalistic view about humans and reject the realistic interpretation of human mathematical practices.

I have elaborated this point in my reply to Difei Xu above. Here I want to add a few points. A philosopher of mathematics with solid graduate-level mathematical education from a mathematics department and with real experiences in proving mathematical theorems and developing some (perhaps a little shallow, though, or a little 'formal', to use mathematicians' words) mathematical theories should already have sufficient mathematical knowledge and experiences for investigating *philosophical* issues about the nature of human mathematical practices and about the ontological status of the alleged mathematical world. Mathematical competence beyond that point, for instance, special mathematical talents for proving very difficult mathematical theorems or developing very abstruse mathematical theories, does not appear to be relevant for investigating *philosophical* issues about mathematical practices. Such a philosopher is in fact in a better position to offer *informed opinions* regarding philosophical issues about mathematics than an ordinary working mathematician is, where by 'an ordinary working mathematician' I mean a mathematician who never carefully thought about philosophical questions before and who might even be unaware of the researches in foundations of mathematics. (Note that, among mathematicians who did think a lot about philosophical issues and/or did work on foundations of mathematics, a large proportion actually rejected realism or Platonism.) It could happen that the philosopher is too much influenced by traditional philosophical speculations that are obscure, fruitless, and alien to scientific ways of thinking and that have been forsaken by contemporary scientists. That will certainly be bad for the philosopher. However, suppose that the philosopher actually carefully rejects traditional 'unscientific' philosophical speculations. Suppose that the philosopher is instead influenced by other branches of contemporary science, from physics to evolutionary theory and cognitive sciences, which shapes the physicalistic view about human cognitive subjects. Then, the philosopher should be in a much better position to offer informed opinions regarding philosophical issues about mathematics than an ordinary working mathematician is, where the working mathematician may never care so much about other branches of science and their implications on what humans and human mathematical practices really are. Remember, we are talking about philosophical questions about human mathematical practices. We are considering the ontological status of mathematical entities, such as whether they are fictional or mind-independent. We are exploring a coherent worldview so that facts about human mathematical practices and applications in the sciences can fit into that global worldview (which should include our naturalistic view about humans and human thinking processes, as I repeatedly emphasize). We are NOT proving new mathematical theorems or developing new mathematical theories. Obviously, for investigating those philosophical issues, knowledge of other branches of science, especially, our scientific knowledge about human thinking processes, is much more relevant than any mathematical competence beyond what the philosopher already has. That is, the philosopher is in a much better position to offer informed opinions regarding such philosophical issues about mathematics than an ordinary working mathematician is. Therefore, even before we get into sophisticated philosophical debates in philosophy of mathematics, such as those regarding fictionalism, inference to the best explanation, the consquence of indispensability of mathematics and so on and so forth in the current literature, we can already see that the objection to nominalism citing an ordinary working mathematician's opinion is just uninteresting. It asks the wrong person for informed opinions.

Leitgeb did not comment on my argument from methodological naturalism to physicalism about cognitive subjects and then to nominalism and strict finitism in philosophy of mathematics. Perhaps he believes that no such argument can possibly be valid and can have any value. Perhaps he believes that as long as one accepts methodological naturalism one must automatically accept realism in philosophy of mathematics. I hope my response above can persuade him away from this view. In fact, few contemporary philosophers of mathematics believe that methodological naturalism directly leads to realism in philosophy of mathematics. Leitgeb cites Maddy as an example. Unfortunately, Maddy is a wrong example for Leitgeb's purpose. In her book *Second Philosophy*, Maddy formulates two positions, Thin Realism and Arealism. Arealism straightly claims that mathematical entities do not exist and that set theoretical axioms are not true (or are not the kind of sentences apt for truth evaluation). Thin Realism claims that mathematical entities have 'thin existence', which means that mathematical entities exist just in the sense that mathematicians find it methodologically good to accept relevant axioms stating their existence. 'Thin existence' is supposed to be different from the more robust physical existence. Now, after comparing the two positions, Maddy summarizes as follows, 'In sum, then, I see no significant difference between Arealism and Thin Realism; they are alternative descriptions of the same underlying facts.' ([1], p.390) Therefore, Maddy's position is much closer to anti-realism than it is to realism. We can say that she is essentially an anti-realist about mathematics. Maddy is exactly an instance of those philosophers who deny the implication from methodological naturalism to realism in philosophy of mathematics ([4]).

BIBLIOGRAPHY

[1] Maddy, P. (2007). *Second Philosophy: A Naturlistic Method*. Oxford University Press.

[2] Rovelli, C. (2009). 'Unfinished revolution', in D. Oriti (ed.). *Approaches to Quantum Gravity: Toward a New Understanding of Space, Time, and Matter*, pp.3-12, Cambridge University Press, 2009.

[3] Smolin, L. (2000). *Three Roads to Quantum Gravity*, Basic Books.

[4] Ye, F. (2008). 'Review of Penelope Maddy's Second Philosophy: A Naturalistic Method', *International Studies in the Philosophy of Science, 22*, 227-230.

[5] Ye, F. (2010). 'Naturalism and abstract entities'. *International Studies in the Philosophy of Science*, 24, 129-46.

[6] Ye, F. (2011). *Strict finitism and the logic of mathematical applications. Synthese Library, vol.355.* Dordrecht: Springer.

[7] Ye, F. (2012). 'Some naturalistic comments on Frege's philosophy of mathematics', *Frontiers of Philosophy in China*, 7, 378-403.

[8] Ye, F. (online). 'Naturalism and objectivity in mathematics'. http://sites.google.com/site/fengye63/. Accessed February 15, 2013.

Part II

Logic has fruitfully coexisted with computation ever since the days of Leibniz, but in the 20th century, after the work of pioneers such as Turing, von Neumann, and McCarthy, the fast-evolving interface of logic and computer science has become a major engine of both theoretical and practical innovation, spawning ever new mixtures of logical and computational themes. The resulting wide variety of flourishing specialized fields, ranging from programming language semantics or data base theory to automated deduction or artificial intelligence, probably contains the bulk of logic research today.

In 'Relational databases and Bell's Theorem', *Samson Abramsky* (University of Oxford) explores a surprising connection between the foundations of computation and quantum mechanics. He shows how the fundamental result in physics called Bell's Theorem, that shows the impossibility of removing non-local quantum-mechanical correlations in favor of theories with local variables, links up with questions that arise naturally and have been well-studied in relational database theory. He explores logical structures underlying relevant notions like non-locality and contextuality, and the reasoning common to both areas. *Balder ten Cate* (University of California, Santa Cruz) gives a brief summary of modern database theory, including topics like data exchange, data integration, and multiple schemata, and suggests that Abramsky's logic bridge may be widened. *Jouko Väänämen* (University of Helsinki and University of Amsterdam) explores some general concepts behind Abramsky's analysis, such as contextuality, makes a connection to the current research program of dependence logics, and suggests that there is a general logic behind all this, reaching from physics to computation and social behavior.

Computational techniques are crossing over to social scenarios like games, and in 'Qualitative extensive games with short sight: a more realistic model', *Chanjuan Liu* (Peking University), *Fenrong Liu* (Tsinghua University), and *Su Kaile* (Peking University and Griffith University, Brisbane) consider the recent idea of short-sight games, where players only have a limited horizon for viewing possible moves. They propose a logic-friendly model for these games, bring a number of well-established computational techniques to bear, including $\alpha - \beta$ search from AI, propose new solution algorithms, and link to concrete benchmark examples. *Thomas Ågotnes* (University of Bergen and Southwest University, Chongqing) points out the relevance of current work in AI on agent modeling in problem solving, and offers links with recent work on logics of strategies for bounded agents. *Paolo Turrini* (Imperial College London), one of the original proposers of the sort-sight game model, points out fruitful links to more sophisticated results from modern game theory, and suggests bringing in a deeper analysis of the agents in the scenarios of this paper.

Ramaswamy Ramanujam (National Institute of Mathematical Sciences, Chennai) takes up another basic game-theoretic notion, that of a strategy, in 'Strategies in games: a perspective from logic and computation'. He points out that, moving away from the overly detailed strategies in game theory, specifying properties of generic strategies can be done elegantly by techniques from computer science. This approach relates strategies to automata theory, and results in new logical systems for computation and games that work at just the right level of generality. *Olivier Roy* (Ludwig Maximilian University of Munich) asks how the defeasible character of intuitive strategies, as plans that can be abandoned, can be accommodated in this framework, and also points at the importance of another notion: refraining from, or avoiding actions. *Yde Venema* (University of Amsterdam) asks how local strategies can be composed into global ones on Ramanujam's view, and also makes some suggestions for positioning the logic of strategies in the landscape of existing computational logics, such as the modal μ-calculus.

In 'Refuting random 3CNF Formulas in propositional logic', *Sebastian Müller* (University of Toronto) and *Iddo Tzameret* (Tsinghua University) give a sample of state of the art analysis of propositional proof complexity in terms of average time behavior on random input. This adds a dimension of probabilistic computational information about propositional proof systems beyond what logicians have considered, suggests new takes on how far polynomial time goes in what proof systems can achieve, as well as, in principle, new ways of making proof search more efficient. *Hans de Nivelle* (University of Wroclaw) discusses possible connections with the best available practical satisfiability testing methods, where many approaches are in fruitful competition, and asks what can be learnt from this more theoretical style of analysis. *Xishun Zhao* (Sun Yat-sen University, Guangzhou) asks whether this work can take us toward a better grasp of the P = NP problem, and draws attention to the potential wider thrust of striking new ideas in the paper such as incomplete but still useful proof methods that deliver correct results only with high probability.

Relational Databases and Bell's Theorem. A Summary

SAMSON ABRAMSKY

Our aim in this paper is to point out a surprising formal connection, between two topics which seem on face value to have nothing to do with each other: relational database theory, and the study of non-locality and contextuality in the foundations of quantum mechanics. We shall show that there is a remarkably direct correspondence between central results such as Bell's theorem in the foundations of quantum mechanics, and questions which arise naturally and have been well-studied in relational database theory.[1]

---

[1] The paper is available online http://arxiv.org/abs/ 1208.6416. It will appear in *Festschrift for Peter Buneman.*

Universal relations, data integration, and quantum-consistency

BALDER TEN CATE

In the relational database model (cf. [1]), a database is viewed as consisting of a collection of *relations*, each having a specified set of *attributes*. We can think of a relation as a table, where the attributes are the column names. A *tuple* in the relation then corresponds to a row of the table, and it consists of a value for each attribute of the relation (where the values come from some fixed infinite domain). At the foundation of the relational model lies Codd's relational algebra, whose operations manipulate relations. Two fundamental operations on relations are *projection* and *natural join*. Taking a projection of a relation means dropping some of the attributes, while the natural join of two relations (with possibly different sets of attributes) is the relation over the union of the two sets of atttributes, consisting of all tuples whose corresponding projections belong to the two given relations. Other operations of the relational algebra include *selection, attribute renaming, union,* and *difference* (i.e., *relative complementation*). Codd's completeness theorem states that these operations together capture, in a precise way, the queries definable in first-order logic. Various extensions of the basic relational database have been developed and studied, such as probabilitistic databases (where, in the simplest setup, a probabilistic relation is a map that assigns to every tuple a probability). An elegant general framework was introduced in [3] based on universal algebra: for any commutative semiring $\mathcal{K} = (K, \oplus, \otimes, 0, 1)$, a $\mathcal{K}$-relation (with a given set of attributes) is a function that maps every possible tuple (with given attributes) to an element of K, in such a way that all but finitely many tuples are mapped to 0. The operations of projection and natural join can be naturally generalized to $\mathcal{K}$-relations (projection being defined in terms of the additive operator $\oplus$ and natural join in terms of multiplicative operator $\otimes$). The same holds for other the other operations of the relational algebra, except for the difference operator, which turns out to be less straightforward in this context. In this way, traditional relational databases (under the usual set-based semantics) can be identified with $\mathcal{B}$-relations, where $\mathcal{B} = (\{0, 1\}, \vee, \wedge, 0, 1)$ is the Boolean semiring, and probabilistic databases, as well as several other well-studied extensions of the basic relational model, can be similarly captured by appropriate commutative semirings.

The flat, tabular format in which data is organized in the relational model, in general, cannot to account for all inherent structure in the data. To accomodate for this, the specification of a relational database schema may be accompanied by further integrity constraints. These constraints impose further restrictions on the data stored in the relations of a database. For instance, an integrity constraint may express that a relation is the graph of a function, or that all values appearing in some column of a relation also appear in some column of another relation. Constraints such as these are used in several ways. They allow us to detect inconsistencies in a database, they are useful for query optimization (i.e., they may affect the choice between different execution plans for a given query), and, finally, they are used in schema design.

To illustrate the use of constraints in schema design, consider the following example: in designing a database schema for an application concerning courses and grades, one may determine that the relevant information to be stored is *Course, Teacher, Room, Hour, Student, Grade*. Thus, one may consider storing the data in a single relation R with these six attributes. Natural integrity constraints that one expects to hold in this case include the fact that the values of *Teacher, Room*, and *Hour*, are functionally determined by the value of *Course* in any given tuple (but the same does not hold for the *Student* and *Grade* atttributes). These constraints imply that the relation R can be decomposed into a join of two relations, whose sets of attributes are {*Course, Teacher, Room, Hour*} and {*Course, Student, Grade*}. One natural and well-studied approach to schema design follows precisely the steps we took in this example: first, one specifies a single *"universal relation"* schema, consisting of a single relation with all relevant attributes. Then, one writes down the integrity constraints that are expected to hold for the universal relation. Finally, the universal relation is decomposed into a number of relations, based on the integrity constraints. If a relational database schema was constructed following this design process, we may naturally expect that the actual data stored in the database satisfies a structural condition that is known as the *universal relation assumption* (URA): *there exists a relation (the "universal relation"), of which every relation in the database is a projection.* The URA is a structural property that may or may not hold of a database. Testing whether it holds is NP-complete [4].

Professor Abramsky's paper established a remarkable link between the URA and the concept, in quantum mechanics, of local realism, or, reproducibility in a physical theory based on local hidden variables. Bell's celebrated theorem [2] (as well as subsequent refinements due to Hardy and others), states that quantum mechanics predicts observations that fail to satisfy local realism, and, in this sense, cannot be accounted for by classical theories of mechanics. Such observations, then, corresponds to databases that fail to satisfy the URA. This link between concepts from quantum mechanics and the URA holds not only for standard (set-based) relational databases, but also for probabilistic databases, and in fact can be stated at a more abstract category theoretic level that, in particular, accounts for

$\mathcal{K}$-relations for arbitrary commutative semigroups $\mathcal{K}$.

As discussed above, the traditional uses of database constraints have to do with maintaining data consistency, query optimization, and schema design. Over the past decade, a different set of applications has emerged, where constraints are used not to specify structural properties of relations within a single database schema, but rather to express relationships across different schemas. In particular, for instance, in *data exchange* and in *data integration*, constraints are used to relate elements of two or more database schemas to each other. In the context of these applications, some of the fundamental concepts and techniques studied in the literature on database constraints have found new uses and interpretations. Here, I will briefly discuss data integration [5] (in a very cursory fashion) and explain how (a generalization of) the URA naturally arises in this setting.

In a data integration scenario, we are given a collection of databases (also called *sources*), each of which may adhere to a different schema, as well as an additional "global schema". Furthermore, we are given a set of constraints involving the relations in the source schemas and the relations in the target schema. These constraints may, for example, express that a source relation is a projection of a relation from the global schema, or that a source relation is a join of two relations in the global schema. These constraints are then used to answer queries posed over the global schema, using only the source databases. Formally, if q is a query posed over the global schema, then the *certain answers* of q, with respect to the given source databases, are those tuples that are an answer to q in every *legal global database*, where a legal global database is a database over the global schema that, together with the given source databases, satisfies all the given constraints. Note that, depending on the constraints and the source databases, there may be any number of legal global databases. Potentially, the query could yield a different result when evaluated against these different legal global databases. The certain answers to the query are those tuples that are a correct answer to matter in which legal global database the query is evaluated. The problem of computing certain answers to queries in data integration settings has been extensively studied, for different classes of constraints and queries.

A fundamental decision problem that underlies query answering is the problem of testing whether for the given collection of source databases and the given constraints, a legal global database exists, that is, whether the given collection for source databases is *consistent* with respect to the constraints. If the answer is negative, then, clearly, the concept of certain answers becomes trivial. Now, in the special case where the global schema consists of a single relation and the constraints express that each source relation is a projection of the global relation, the consistency problem precisely asks for the existence of a universal relation. Since, in practice, source data often gives rise to inconsistencies, it is natural to consider refinements of the concept of certain answers, that produce meaningful results even in the presence of inconsistencies [5].

The newly established connection between the universal relation assumption and Bell's theorem suggests another angle at this problem. Even if a data integration setting is inconsistent, it may still be realizable by a quantum mechanical process. Could there be meaningful (and effectively computable) notions of "quantum-consistency" and of "quantum-certain answers" in this context?

BIBLIOGRAPHY

[1] S. Abiteboul, R. Hull, and V. Vianu. *Foundations of Databases*. 1995.

[2] John S. Bell. On the Einstein-Podolsky-Rosen paradox. *Physics*, 1:195–200, 1964.

[3] Todd J. Green, Grigoris Karvounarakis, and Val Tannen. Provenance semirings. In *PODS*, Beijing, China, June 2007.

[4] Peter Honeyman, Richard E. Ladner, and Mihalis Yannakakis. Testing the universal instance assumption. *Inf. Process. Lett.*, 10(1):14–19, 1980.

[5] M. Lenzerini. Data Integration: A Theoretical Perspective. In *ACM Symposium on Principles of Database Systems (PODS)*, pages 233–246, 2002.

Comments on Abramsky

JOUKO VÄÄNÄNEN

Abramsky points out a connection between a non-locality phenomena in quantum physics and a non-locality phenomenon in database theory. This seems surprising as one would not expect—a priori—such a connection to exist. The proof of the relevant quantum theoretic phenomenon turns out in Abramsky's analysis to be a mathematical rather than a physical phenomenon. It is based on the nonexistence, in the mathematical sense, of probability distributions satisfying certain conditions. The probability distributions can even be replaced by 2-valued distributions with the same effect.

A similar mathematical non-existence phenomenon exists in database theory. The phenomenon is twofold. On the one hand there is the decomposition problem: how to decompose a large database into smaller parts, which would form an instance of a database schema with the associated dependencies. On the other hand we may start from an instance of a schema and ask whether there is a universal relation.

Abramsky calls this *contextuality* and suggests that a common *logic of contextuality* is emerging from these examples. We may also ask, is contextuality an even more general phenomenon? If it is, there is all the more reason to take a logical approach to the question. By a logical approach I mean establishing a language, a semantics of the language and an attempt to figure out the logical properties.

The characteristic feature of contextuality, be it quantum physics or database theory, is the presence of several tables of data and the question is whether they arise from just one universal table by projection. This co-existence of several limited "realities" is not an unusual theme in the history of ideas, and neither is the question whether there is one ultimate "truth". This is the case in different areas of humanities for rather obvious reasons, but it exists even in mathematics. For example, in set theory there is discussion about a multiverse position according to which the fact that we have not been able to solve questions such as the Continuum Hypothesis is a consequence of the circumstance that the set theoretical reality of mathematical objects is a multiverse, i.e. a collection of universes, very "close" to each other, some satisfying the Continuum Hypothesis and some not. One may indeed reformulate first order logic so that the semantics is not based on the concept of a model satisfying a sentence, but on the concept of a multiverse model satisfying a sentence [5].

The question of contextuality arises in database theory rather naturally. When dealing with large databases one observes that a certain datum, e.g. the telephone number of a person, occurs repeatedly, and it is more efficient to have a separate small database for telephone numbers. This decomposition of a large database into smaller pieces is possible because of *dependencies*, in our case that of the person functionally determining his or her telephone number. For an ideal result one would also want the components to be *independent* of each other. The decomposition results in an *instance* of a database schema. The question is, given an instance, does it arise from the decomposition of a big "universal" database into smaller ones. Abramsky gives strong evidence to category theory, especially the language of presheaves and sheaves, being a natural framework for studying this kind of issues concerning schemas and relational databases.

The logic generally used to express dependencies in database theory is first order logic or a fragment of it called the *relational calculus*. Individual databases of instances are treated as *relations*. The existence of a *functional dependence* $x \to y$ between attributes x and y in a relation R can be expressed in first order logic as

(1) $\quad \forall x, y, y', \vec{z}, \vec{z'}((R(x, y, \vec{z}) \wedge R(x, y', \vec{z'})) \to y = y')$.

Another kind of dependence is *inclusion*, $x \subseteq y$, expressing the concept that every value of the attribute x in relation R occurs as a value of the attribute y in relation S. This can be expressed in first order logic as

(2) $\quad \forall x, \vec{z}(R(x, \vec{z}) \to \exists \vec{u} S(x, \vec{u}))$.

However, to express that a universal relation exists for binary relations R_1, R_2, R_3 seems in general—not least because of its NP-completeness in the general case— to require existential second order logic:

$$\exists S(\forall x, y(R_1(x, y) \leftrightarrow \exists z S(x, y, z)) \wedge$$
(3) $\quad\quad \forall y, z(R_2(y, z) \leftrightarrow \exists x S(x, y, z)) \wedge$
$$\forall x, z(R_3(x, z) \leftrightarrow \exists y S(x, y, z))).$$

So here we go beyond first order logic. Is existential second order logic the right "logic of contextuality"? As it is, existential second order logic is badly non-axiomatizable: the set of valid sentences is complete Π_2 in the Levy hierarchy, which means that to check the validity of a sentence requires scanning the cumulative hierarchy up to inaccessible cardinals (if they exist) and beyond. In the database context this is not so drastic as databases correspond to finite models and even first order logic itself is non-axiomatizable in finite models.

The approach of *generalized quantifiers*, introduced by A. Mostowski in 1957, is to consider particular second order definable quantifiers, such as "for infinitely many x" or "for uncountably many x", or in finite models "for an even number of x", and add them to first order logic. In many cases this has resulted in fragments of second order logic with nice properties, such as effective axiomatization and countable compactness. Perhaps we can do the same here, not by adopting generalized quantifiers, but by adopting *generalized atomic formulas*. Since we are talking about *logical notions* here, the atomic formulas that are most naturally subject to generalisation are the identities, such as $x^1 = y^1$. Accordingly, let us take an atomic formula

(4) $=(x^1, y^1)$

with (1) as the meaning with the understanding that the variables $x^1, y^1, \vec{z}^1$ denote the attributes of the relation R. Thus a relation R satisfies (4) iff (1) is true. Likewise, let us take an atomic formula

(5) $x^1 \subseteq x^2$

with (2) as the meaning with the understanding that the variables $x^1, \vec{z}^1$ denote the attributes of the relation R and the variables $x^2, \vec{z}^2$ denote the attributes of the relation S. Thus an instance (R, S) satisfies (5) iff (2) is true. If we add these new atomic formulas to first order logic and define the logical operations in a canonical way we get a "schema"-version of what is in the single database (or universal database) case known as *dependence logic* [6], if only (4) is added, *inclusion logic* [1], if only (5) is added, and (essentially) *independence logic* [3], if both are added.

One may reasonably ask, in what way are dependence logic, inclusion logic and independence logic better than mere existential second order logic, of which they all are fragments? The point is that we have more control and can make finer distinctions. For example, inclusion logic has on finite models exactly the expressive power of fixed point logic [2]. Dependence logic formulas can express exactly all second order properties of (single) relations which are closed downwards (as (1) is but (2) is not) [4]. Finally, independence logic can in fact express all existential second order properties of (single) relations [1].

The origin of dependence logic is in game theoretic semantics. There a typical database consists of plays of the semantic game associated with a sentence and a model. If the variables x, y, z take on values during the game, then subsequent plays of the game yield a database with x, y, z as attributes. From this database we can see e.g. whether the player that picked z was playing a strategy (expressed by $=(x, y, z)$), whether the player used partial information (expressed e.g. by $=(x, z)$), whether the player was committed to play only values that the player picking y uses (expressed by $z \subseteq y$) etc. These examples show how the above generalized atoms arise naturally in game theoretic semantics. They arise also naturally in experimental science ("the time of descent is functionally determined by the height but independent of the weight"), social choice (the value of the social welfare function on the choice between a and b is functionally determined by the choices between a and b of the voters), biology (Mendel's Laws), and philosophy (e.g. inquisitive logic). In all these application areas one can see relevant universal relation type questions, even if they may have not been subjected to mathematical study. Laws governing logics arising from atoms such as $=(x, y)$ and $x \subseteq y$ may offer the beginning of a new "logic of contextuality", but this requires development of "multi-relation" versions of these logics.

Let us return to (3), which plays a central role in Abramsky's paper. The condition (3) is an existential second order property of the relations R_1, R_2 and R_3, but over which domain? We can construe it as being over the so-called *active domain*, which occurs in (3) only implicitly. Currently this seems to go beyond the logical apparatus of even the strongest of the above logics, the independence logic. However, we can take it as a new atom, in the spirit of $=(x, y)$ and $x \subseteq y$. So let us adopt a new atom

(6) $\bowtie(x^1 y^1, y^2 z^2, x^3 z^3)$,

with (3) as the meaning with the understanding that the variables x^1, y^1 denote the attributes of the relation R^1, the variables y^2, z^2 denote the attributes of the relation R^2, and the variables x^3, z^3 denote the attributes of the relation R^3. Thus an instance (R^1, R^2, R^3) satisfies (6) iff (3) is true[1]. If this atom is added to first order logic, we still remain within existential second order logic. Can we axiomatize this atom in the same sense as (4) and (5) have been axiomatised?

My contention is that the surprising connection Abramsky's paper establishes between the mathematics of quantum physics and database theory is an indication that there may be a general multiverse logic underlying phenomena not only in physics and computer science, but also in social choice, biology and philosophy. Such a logic does not exist yet, but I believe that it will exist.

Pietro Galliani made the interesting observation that (6) added to first order logic gives exactly inclusion logic and hence on finite models fixed point logic.

BIBLIOGRAPHY

[1] Pietro Galliani. Inclusion and exclusion dependencies in team semantics: On some logics of imperfect information. *Annals of Pure and Applied Logic*, 163(1):68 – 84, 2012.
[2] Pietro Galliani and Lauri Hella. Inclusion logic and fixed point logic. CSL 2013 to appear, arXiv:1304.4267 [cs.LO].
[3] Erich Grädel and Jouko Väänänen. Dependence and independence. *Studia Logica*, 101(2):233–236, 2013.
[4] Juha Kontinen and Jouko Väänänen. On definability in dependence logic. *J. Log. Lang. Inf.*, 18(3):317–332, 2009.
[5] Jouko Väänänen. Multiverse set theory and absolutely undecidable propositions. In Juliette Kennedy, editor, *Interpreting Gödel*. Cambridge University Press. to appear.

---

[1]The new atom resembles but is different from the so called *join dependency* atom $\star(xy, yz, xz)$ which would say of *one* relation that it is the join of its projections to xy, xy and xz, respectively. Atoms of the type $\star(xy, yz, xz)$ can be expressed in independence logic.

[6] Jouko Väänänen. *Dependence logic*, volume 70 of *London Mathematical Society Student Texts*. Cambridge University Press, Cambridge, 2007.

Response to ten Cate and Väänänen

SAMSON ABRAMSKY

I shall respond briefly to the remarks by the two commentators on my paper.

1 Balder ten Cate

Balder ten Cate's comments firstly give a nice summary of classical relational database theory, and he confirms the striking and surprising nature of the connections with foundations of quantum mechanics which are exposed in my paper. He then goes on to describe some of the recent developments in database theory, such as data exchange and data integration, which concern the relationships between database schemas, rather than properties of individual schemas and their instances. He explains how the question of consistency of a collection of schemas with respect to constraints is equivalent to the existence of a legal global database, and can be seen as a generalisation of the existence of a universal relation instance, which as explained in my paper is the database equivalent to the existence of a local hidden variable explanation of given experimental data. He suggests that the connections discussed in my paper might open up a new perspective on this problem.

> Even if a data integration setting is inconsistent, it may still be realizable by a quantum mechanical process. Could there be meaningful (and effectively computable) notions of quantum-consistency and of quantum-certain answers in this context?

This is an intriguing idea, which I find hard to evaluate. One could consider quantum states as "resources" which are able to generate a family of finite tables by interpreting attributes as measurements, following the dictionary described in my paper. This makes mathematical sense: could it be interesting from a database point of view? Could Bell inequalities play a natural rôle here, in the "logical" format described in [2]? These might be interesting questions to pursue, which could extend and deepen the correspondence described in my paper.

2 Jouko Väänänen

Jouko Väänänen suggests that the surprising connection described in my paper, and which it is suggested there point towards a general logic of contextuality, can be taken considerably further.

My contention is that the surprising connection Abramsky's paper establishes between the mathematics of quantum physics and database theory is an indication that there may be a general multiverse logic underlying phenomena not only in physics and computer science, but also in social choice, biology and philosophy. Such a logic does not exist yet, but I believe that it will exist.

The general form of logic he suggests takes as its starting point the recent developments in logics of dependence and independence, in which he has played the leading rôle. He suggests that a further extension of this paradigm is needed, to a "multi-relation" version. He suggests a new atom, whose semantics are the existence of a relation gluing some given ones, and asks whether it can be axiomatised, in the same fashion that dependence and independence atoms have been. We make a few comments:

- There seems an interesting parallel between Väänänen's suggestion of a multi-relation version of independence logic, and ten Cate's suggestion of extending the QM-database correspondence to database notions involving multiple schemas. Perhaps a multiverse logic would naturally correspond to such database notions, in the same way that independence atoms correspond to various classical notions of database dependency?

- It would be interesting to gain a deeper insight into the relationship between logical approaches to capturing contextuality phenomena, as proposed by Väänänen, and the category- and sheaf-theoretic approach which I and my collaborators have been developing. A useful intermediate step might be to obtain a more algebraic and abstract form of logics of dependence, which is less dependent on concrete set-theoretic semantics. One may also note the recent proposal by Tobias Fritz and Cecilia Flori (unpublished) of an algebraic notion of "gleave", a presheaf equipped with a gluing operation, directly motivated by [1], and of a similar nature to the atom proposed by Väänänen.

Overall, Väänänen's vision of a logic of contextuality of very wide scope is one which I find appealing. The challenge will be to combine genuine breadth and depth. I suspect that this will require the combination and synthesis of several different technical languages, currently seen as rather distinct.

BIBLIOGRAPHY

[1] S. Abramsky and A. Brandenburger. The sheaf-theoretic structure of non-locality and contextuality. *New Journal of Physics*, 13(2011):113036, 2011.
[2] S. Abramsky and L. Hardy. Logical Bell Inequalities. *Physical Review A*, 85:062114, 2012.

Qualitative Extensive Games with Short Sight: A More Realistic Model

CHANJUAN LIU, FENRONG LIU AND KAILE SU

ABSTRACT. The notion of *short sight*, proposed by Grossi and Turrini, weakens the assumption that every player is able to perceive the entire game structure. This paper outreaches the work of Grossi and Turrini towards a more realistic modeling. First, we propose two alternative solution concepts in games with short sight besides the one they have presented. We then give several interesting types of sight functions, and analyze the connection between games with short sight and $\alpha - \beta$ pruning, and design an algorithm for strategy choosing in the actual process of playing such games. Moreover, we study the dynamics of players' sight and show the corresponding preference change caused by sight change.

1 Introduction

Game theory studies decision-making of players in competitive situations to achieve optimal goals. In traditional game theory, there are two important aspects of extensive games, known as *quantified preferences over outcomes* and *assumption of common knowledge*, calling for potential improvement. A game of quantified preferences assumes that all outcomes are linearly ordered, usually determined by a utility function. However, as claimed by Slade [13], not all outcomes can be e-valuated purely by utility functions in certain games. In reality, to obtain the exact payoff of an outcome is often difficult or computationally expensive, and a player is often forced to make a decision in presence of imprecise or incomplete information, in which situation two available options to a player may be incomparable [10]. Therefore, it is useful to study the notion of preference by qualitative binary relations, especially when there does not exist a linear order on all options in hand.

Another assumption in most of the existing works is that the entire structure of a game is common knowledge to all players. Again, this assumption is sometimes too strong. In a game like chess, the actual game space is exponential in the size of the game configuration, and may have a computation path too long to be effectively handled by most existing computers. Sub-optimal solutions are often desirable by considering only limited information or bounded steps foreseeable by a player that has relatively small amount of computation resources.

Several attempts have been made to achieve a closer match with reality. In recent years, Halpern et al. studied games with unawareness [6, 7], where players may have no access to the whole game tree when they make decisions due to their unawareness of other player's strategies. Grossi and Turrini pushed further beyond Halpern's work by proposing the concept of *games with short sight* [5], in which players can only see part of the game tree. Nevertheless, Grossi and Turrini only offered a general frame for sight functions. They did not study the relationship between different classes of sight functions and players, nor did they explore dynamics of sight functions.

In this paper, we propose a more realistic model for extensive games, which represents players' preference qualitatively instead of using numeric payoffs. We outreach the initial model of extensive games with short sight in several aspects, including characterizing types of players in terms of their *sight functions*. Interestingly, we connect the α-β pruning, a well-known and realistic search algorithm for two-player zero-sum games [3], with *sight filtration*. Furthermore, in order to compute an equilibrium path of qualitative game playing, we introduce a modified backward induction algorithm. Besides, we investigate sight dynamics with respect to different types of players, and show that changes of players' sight would lead to the changes of their preferences.

This paper is organized as follows: The next section introduces extensive games and subgame perfect equilibrium. Then we study the model of games with short sight. After that, we discuss sight dynamics and preference change. Finally, we concludes the paper with further research issues.

2 Extensive Games with perfect information

In this section, we recall the definition of extensive games with perfect information and subgame perfect equilibrium in these games.

DEFINITION 1. (Extensive game) An extensive game is a tuple $G=(N,V,A,t, \Sigma_i, \succeq_i)$, where (V, A) is a tree with V a set of nodes or vertices including a root v_0, and $A \subseteq V^2$ a set of arcs. N is a non-empty set of the players, and $\succeq_i$ is a preference relation over V^2 for each player i. For any two nodes v and v', if $(v, v') \in A$, we call v' a *successor* of v, thus A is also regarded as the successor relation. Leaves are the nodes that have no successors, denoted by Z. t is turn function assigning a member of N to each non-terminal node. Σ_i is a non-empty set of strategies. A strategy of player i is a function $\sigma_i : \{v \in V \backslash Z |\ t(v) = i\} \to V$ which assigns a successor of v to each non-terminal node when it is i's turn to move.

For a node v we write $l(v)$ for the distance from the root v_0 to v. As usual, $\sigma = (\sigma_i)_{i \in N}$ represents a strategy profile which is a combination of strategies from all players and Σ represents the set of all strategy profiles. For any $M \subseteq N$,

σ_{-M} denotes the collection of strategies in σ excluding those for players in M. We define an outcome function $O : \Sigma \to Z$ assigning leaf nodes to strategy profiles, i.e., $O(\sigma)$ is the outcome if the strategy profile σ is followed by all players starting from the root v_0. $O(\sigma_{-M})$ is the set of outcomes players in M can enforce provided that the other players strictly follow the strategy profile σ.

The preference relation we define is different from the conventional ones, e.g., [12]: In the literature preference is often a linear order over leaves, while in this paper it is defined as a partial order over nodes in V. We assume that players may not be able to precisely determine entire computation paths leading to leave nodes, and allow them to make estimations or even conjecture a preference between non-terminal nodes. We use $\succ$ to denote the strict part of $\succeq$.

Given a game G and a node $v \in V$, a subgame $G|_v$ is a tuple $(N|_v, V|_v, A|_v, t|_v, \Sigma_i|_v, \succeq_i|_v)$ which is G restricted to the nodes reachable from v, following the conventional definitions [12]. We write $O|_v(\sigma|_v)$ for the outcome of the subgame $G|_v$ following the strategy profile σ. As the preference relation $\succeq_i$ is total on V, we sometimes write $\succeq_i$ for $\succeq_i|_v$ if it is clear from the context.

DEFINITION 2. (Subgame perfect equilibrium) Take a finite extensive game G. A strategy profile σ^* is a subgame perfect equilibrium (SPE) if for every player i, non-terminal node v for which $t(v) = i$, there is no σ_i such that $O|_v(\sigma_i, \sigma^*_{-i}|_v) \succ_i O|_v(\sigma^*_i|_v, \sigma^*_{-i}|_v)$, where σ_i is any strategy available to i in the subgame $G|_v$ of G that follows node v.

3 Short Sight in Extensive Games

In traditional game theory, players are usually assumed to have common knowledge of the whole game structure, which is unrealistic especially in large games such as chess and checkers. In this section, we introduce games with short sight in which players' sight is limited, and explore some crucial aspects of this kind of games, such as the different types of restrictions imposed on sight functions, and a solution algorithm for computing SPE.

3.1 Qualitative Games with Short Sight

We first propose a definition of sight function adapted from [5].

DEFINITION 3. (Sight function). Let $G = (N, V, A, t, \Sigma_i, \succeq_i)$ be an *extensive game*. A short sight function for G is a function $s : N \times V \setminus Z \to 2^V \setminus \emptyset$, associating to each player a nonempty set of nodes that the player can see from a node in the game. We sometimes write s_i for the *curried* sight function of type $V \setminus Z \to 2^V \setminus \emptyset$ with $i \in N$. A sight function has the following properties.

1. For each player i and node v, $s_i(v) \subseteq V|_v \setminus \emptyset$ and $|s_i(v)| < \omega$, i.e. the sight of i at v consists of a finite nonempty set of nodes extending v.

2. $v' \in s_i(v)$ implies that $v'' \in s_i(v)$ for every $v'' \lhd v'$ with $v'' \in V|_v$, i.e. players' sight is closed under predecessors, where relation $\lhd$ is the transitive closure of the successor relation A.

Intuitively, the function s associates any choice point with vertices that each player can see.

DEFINITION 4. (Qualitative game with short sight). A qualitative game with short sight (Qgss) is a tuple $S = (G, s)$ where G is a finite extensive game and s a sight function.

3.2 Sight Filtration and Solution Concepts of Qgss

Each game with short sight yields a family of finite extensive games, one for each non-terminal node $v \in V \setminus Z$:

DEFINITION 5. (Sight-filtrated extensive game) Take S as a Qgss given by (G, s) with $G = (N, V, A, t, \Sigma_i, \succeq_i)$. Given any non-terminal node v, a tuple $S\lceil_v$ is a finite extensive game by sight-filtration:
$S\lceil_v = (N\lceil_v, V\lceil_v, A\lceil_v, t\lceil_v, \Sigma_i\lceil_v, \succeq_i\lceil_v)$ where

- $N\lceil_v = N$;

- $V\lceil_v = s_{t(v)}(v)$, which is the set of nodes within the sight of $t(v)$ from node v. The terminal nodes in $V\lceil_v$ are the nodes in $V\lceil_v$ of maximal distance, denoted by $Z\lceil_v$;

- $A\lceil_v = A \cap (V\lceil_v)^2$;

- $t\lceil_v = V\lceil_v \setminus Z\lceil_v \to N$ so that $t\lceil_v(v') = t(v, v')$;

- $\Sigma_i\lceil_v$ is the set of strategies for each player available at v and restricted to $s(v)$. It consists of elements $\sigma_i\lceil_v$ such that $\sigma_i\lceil_v(v') = \sigma_i(v, v')$ for each $v' \in V\lceil_v$ with $t\lceil_v(v') = i$;

- $\succeq_i\lceil_v = \succeq_i \cap (V\lceil_v)^2$.

Accordingly, we define the outcome function $O\lceil_v$: $\Sigma\lceil_v \to Z\lceil_v$ assigning leaf nodes of $S\lceil_v$ to strategy profiles.

The following lemma shows the relation between sight-filtrated extensive games and games with short sight.

LEMMA 6. *Each sight-filtrated extensive game $S\lceil_v$ is a special qualitative game with short sight, in which players' sight at every decision point $u \in V\lceil_v$ are all the nodes in the subgame of $S\lceil_v$ that follows u.*

We first define two solution concepts for Qgss: sight-compatible best response and Nash equilibrium, which are adapted from traditional definitions [4, 11]. A sight-compatible best response for player i (Nash equilibrium) in Qgss S is consistent with a best response for player i (Nash equilibrium) in each sight-filtrated extensive game $S\lceil_v$.

DEFINITION 7. (Sight-compatible best response).

Let $S = (G, s)$ be a Qgss and $S\lceil_v$ be the sight-filtrated extensive game at v. A strategy profile σ^* is a sight-compatible best response for i if for every nonterminal node v, there is no $\sigma_i\lceil_v$ such that $O\lceil_v (\sigma_i\lceil_v, \sigma_{-i}^*\lceil_v) \succ_i$ $\lceil_v O\lceil_v(\sigma_i^*\lceil_v, \sigma_{-i}^*\lceil_v)$, where $\sigma_i\lceil_v$ is any strategy available to i. σ^* is a *sight-compatible Nash equilibrium*(SCNE) of S if it is a sight-compatible best response for every player $i \in N$.

However, these solutions ignore the sequential structure of the game. Now we will introduce an algorithm solving Qgss, obtaining a refined solution for extensive games.

A common method for determining subgame perfect equilibria of a finite extensive game is the backward induction (BI) algorithm [1], which assumes that players know the final outcomes of the whole game. As we have already mentioned, short-sighted players may not see the whole game tree, due to the limitation on computing power, memory, tactics, etc. We will propose an algorithm that computes a sequence of nodes determined by the optimal strategies of all players considering restricted sight.

The main idea lies in the following analysis. Given a Qgss S, at each decision point v, player $t(v)$ is facing a sight-filtrated extensive game $S\lceil_v$. What he can achieve the best is to find a successor node of v maximizing his own profit within his current sight, which is the subgame perfect equilibrium of $S\lceil_v$. The players play in turns, choosing a best successor node at each point, until reaching a terminal node of S. Thus the crucial task for solving games with short-sighted players is in searching for the SPE of $S\lceil_v$ at each intermediate node v.

The following algorithm QBI (qualitative backward induction) is the one that meets our requirements. While the BI algorithm mainly deals with games with numeric payoffs, we make our algorithm applicable to the qualitative preference model. For convenience, we define a virtual node with lowest preference, denoted by VLP. For each node v and player i in the game, $v \succeq_i$ VLP. In addition, the function *dominated*(G, v) in this algorithm means that v is a dominated node. In other words, v has a sibling node from which all reachable endpoints are preferred by the player to all reachable endpoints from v itself [2]. *Prune*(G, v) is an operation which prunes the subtree starting at v, i.e., it cuts off node v and all the nodes extending v in G. This function allows us to make a pretreatment by cutting off those branches that cannot lead to the best outcomes for $t(v)$ before searching G.

The theorem below show the correctness of the algorithm QBI.

Algorithm 1: QBI

---

1 QBI(G, v)

   Input: A finite extensive game G, node v

   Output: A node sequence $T(v)$ starting at v, which is determined by an SPE of G

2 **begin**

3 | **if** $v \in Z$ **then** return v;

4 | $current\_best\_outcome \leftarrow$ VLP ;

5 | **for** $v_{successor} \leftarrow leftmost\_branch$ **to** $rightmost\_branch$ **do**

6 | | **if** $dominated(G, v_{successor})$ **then**

7 | | $Prune(G, v_{successor})$; continue;

8 | | $new\_outcome \leftarrow$ final node of QBI($G, v_{successor}$);

9 | | **if** $new\_outcome \succeq_{t(v)} current\_best\_outcome$ **then**

10 | | | $current\_best\_outcome \leftarrow new\_outcome$;

11 | | | $\sigma_{t(v)}(v) \leftarrow v_{successor}$;

12 | $T(v) \leftarrow (v, \text{QBI}(\sigma_{t(v)}(v)))$;

13 | **return** $T(v)$;

14 **end**

THEOREM 8. *The path returned by running QBI over the sight-filtrated extensive game $S\lceil_v$ of a qualitative game with short sight S corresponds to a SPE of $S\lceil_v$.*

Proof. Suppose it does not hold, i.e., strategy profile $\sigma*$ determining the path $T(v)$ returned by QBI(G, v) is not a SPE of $G = S\lceil_v$. Then there exists player i, strategy σ_i and node v, such that $O|_v(\sigma_i^*|_v, \sigma_{-i}^*|_v) \prec_i O|_v(\sigma_i, \sigma_{-i}|_v)$. By line 9-11 of QBI, we know $\sigma_{t(v)}(v) \neq \sigma_{t(v)}^*(v)$. Thus, $T(v)$ should not be determined by σ^*. Contradiction. ∎

Based on the new algorithm, we define subgame perfect equilibrium of qualitative games with short sight as follows.

DEFINITION 9. (Sight-compatible subgame perfect equilibrium). Let $S = (G, s)$ be a Qgss and $S\lceil_v$ be the sight-filtrated extensive game at v. A strategy profile σ^* is a sight-compatible SPE of S if for every nonterminal node v, there exists a strategy profile $\sigma\lceil_v$ that is a subgame perfect equilibrium of $S\lceil_v$ and $\sigma_{t(v)}\lceil_v(v) = \sigma_{t(v)}^*(v)$

.

3.3 Types of Players

The definition of sight functions varies in different games. This can be characterized in terms of the number of steps that a player can see from the current position, or in terms of different sets of nodes that satisfy a certain property. In order to develop a better understanding on sight functions, we define the following two general types of players.

DEFINITION 10. (Types of players with short sight).

- *k-step-type* players ($k \in \mathbb{N}$) are those who only see the nodes that are within k steps ahead of the current point.

- *property-based-type* (or *p-based-type*) players, whose sight are the nodes satisfying certain properties.

We let the sight of a *p-based-type* player be all the nodes satisfying certain properties. This type is useful in games where views of players are restricted. For example, a player can only see positions reachable by straight lines from his current position in a maze. The set of nodes that can be seen by a player may change as the game evolves.

EXAMPLE 11. Consider a (single-player) maze game show in Figure 1. The actual layout of the game is on the left hand side of the figure, where A is the *entry* position and E is the *exit*. The standing points of the player are labeled by letters from a to q. On the right hand side is the corresponding game tree, showing all the possible (either successful or unsuccessful) paths a player can walk through. At each decision point, players have at most three options: turning left, turning right or going straight. Going backward is not allowed. Suppose player i is a *p-based* player and the property Pr is specified as "in a straight line with his current position". For example, $s(A) = \{A, a, f, h\}$ and $s(b) = \{b, o, c, k\}$ represent his sight at the starting point A, and the turning point b respectively.

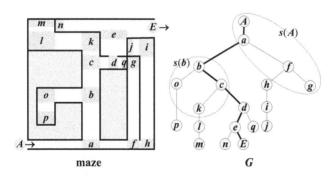

Figure 1. Sight of *p-based* player in a *maze* game

In real game scenarios, for each player i, it is his sight at the points in his turn, rather than the sight at other points, that affects his current strategy. What a player can see in other nodes may of little concern during his current decision making.

The proposition below describes the connection between player types and the sight function characterizations.

PROPOSITION 12. *For a player i in a qualitative game with short sight $S = (G, s)$*

- *i is a k-step-type player iff at any node v with $t(v) = i$, $s_i(v) = V_k \cap V|v$, where $V_k = \{v_k| \, l(v_k) \leq l(v) + k\}$.*

- *i is a p-based-type player iff at any node v with $t(v) = i$, $s_i(v) = Pr(v) \cap V|v$, where $Pr(v)$ is the set of nodes satisfying the given property Pr with respect to the current position v.*

One may regard a k-step player as a special type of p-based player. In Example 11, the property $Pr(v)$ is the set of nodes that can be connected to v by a straight line.

3.4 α-β Pruning

In the community of artificial intelligence, α-β pruning [8, 9] is a well-known search algorithm used for machine playing of two-player zero-sum games. As a matter of fact, this algorithm can be seen as a special game with short sight. The following proposition draws a connection between sight-filtration and α-β pruning.

PROPOSITION 13. *Take $G=(N, V, A, t, \Sigma, \succeq_i)$ as a two-player zero-sum extensive game. And let G' be the game after α-β pruning on G at the starting point v_0 given by $(N', V', A', t', \Sigma', \succeq_i')$. Then, there exists a game with short sight $S = (N, V, A, t, \Sigma, \succeq_i, s)$, for which $S\lceil_{v_0}= G'$, where $S\lceil_{v_0}$ is the sight-filtrated extensive game of S at v_0.*

In practice, researchers often limit the algorithm to a finite depth k, since only very simple games can have their entire game tree searched in a short time. Interestingly, this corresponds to the sight of one special kind of player: k-step-player, as formulated in the following proposition.

PROPOSITION 14. *Let $G = (N, V, A, t, \Sigma, \succeq_i)$ be a two-player zero-sum extensive game. And for any node v, take $G^v = (N^v, V^v, A^v, t^v, \Sigma^v, \succeq_i^v)$ as the subgame starting from v when the depth of the search tree of G are restricted to k, i.e., for any node v' in V^v, $l(v') \leq l(v) + k$. Then there exists a game with short sight $S = (N, V, A, t, \Sigma, \succeq_i, s)$, in which $S\lceil_v= G^v$ for any node v. That is, each v is mapped to a game for the k-step-type player $i = t(v)$.*

4 Sight dynamics

It is natural that players' sight may alter in different settings. We focus on characterizing sight dynamics in this section.

Generally, there are two kinds of changes for players' sight.

DEFINITION 15. (Operations for sight change) Let S be a game with short sight given by $((N, V, A, t, \Sigma_i, \succeq_i), s)$. Two basic dynamic operations for sight change in S are defined as follows:

- $s^\uparrow$: Expansion of sight. More formally, $s^\uparrow(v) = s(v) \cup V_{add}$, where $V_{add} \subseteq \bigcup_{w \in W} V|w$ and $W \subseteq s(v)$, i.e., add some of the vertices extending v to the original sight $s(v)$.

- $s^\downarrow$: Contraction of sight. More formally, $s^\uparrow(v) = s(v) \backslash (V_{sub} \cup \bigcup_{w \in V_{sub}} s(v)|w)$, where $V_{sub} \subseteq s(v)$, i.e., subtract some vertices V_{sub} which were in the original sight $s(v)$ and those extending V_{sub}.

The above definition only shows the general case of sight change. In order to characterize sight changes more finely, we now study sight dynamics in much detail with respect to players' types.

DEFINITION 16. Expansion of sights with respect to the types of the players are:

(i) For *k-step-type* players, the result of sight expansion is they can see the nodes that are within $k + c$ steps from the current point, where c is a nonzero natural number.

(ii) For *property-based-type* players, suppose he can see the nodes with property ψ at v, then $s^\uparrow(v)$ would be nodes satisfying $\psi \vee \psi'$, where ψ' is another property.

Similarly, contraction of sight could also be studied in detail in the light of players' types.

DEFINITION 17. Expansion of sights with respect to the types of the players are:

(i) For $k - step - type$ players, the result of sight contraction of these players are they can see the nodes that are within $k - c$ steps from the current point, where c is a nonzero natural number.

(ii) For $property - based - type$ players, suppose he can see the nodes with property ψ at v, then $s^\downarrow(v)$ would be nodes satisfying ψ', where $\psi' \rightarrow \psi$.

Corresponding preference change

The change of players sight has an direct impact on players' preferences. And further, preference change caused by sight change will affect players' choices when they make decisions.

EXAMPLE 18. Consider the following scenario in chess shown by Figure 2 with current state v_0 and $t(v_0) = white$. Suppose if $white$ chooses the branch v_1v_2, he can capture one $black$ piece, but in the next round, he will lose two pieces when he get to v_3 from v_2. While if $white$ chooses the branch $v_1'v_2'$, he will lose one piece, but can capture two $black$ pieces by going from v_2' to v_3'. Now first assume the sight of $white$ is $s(v_0) = \{v_0, v_1, v_1', v_2, v_2'\}$, then obviously her preference between the two terminal nodes v_2, v_2' is $v_2 \succeq v_2'$, and naturally, she would choose the branch v_1v_2. However, if player $t(v_0)$ can see one more steps forward (i.e., $s'(v_0) = \{v_0, v_1, v_1', v_2, v_2', v_3, v_3'\}$), then her preference would be $v_3' \succeq v_3$, and she would choose the branch $v_1'v_2'v_3'$ instead.

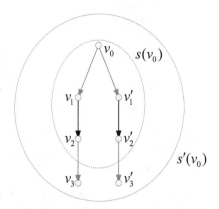

Figure 2. A chess scenario for sight change

The following theorem shows the correspondence between sight change and the preference change caused by it.

THEOREM 19. *(Representation theorem)*

Let S be a game with short sight given by $((N, V, A, t, \Sigma_i, \succeq_i), s)$. The for any node v, the extensive game yields by S at v is $S\lceil_v = (N\lceil_v, V\lceil_v, A\lceil_v, t\lceil_v, \Sigma_i\lceil_v, \succeq_i\lceil_v)$.

1. *when sight $s(v)$ at v expanded to $s^\uparrow$, the preference of $t(v)$ in $S\lceil_v$ would change from $\succeq_i\lceil_v = \succeq_i \cap (V\lceil_v)^2$ to $\succeq_i^\uparrow\lceil_v = \succeq_i \cap (V^\uparrow\lceil_v)^2$.*

2. *when sight $s(v)$ at v contracted to $s^\downarrow$, the preference of $t(v)$ in $S\lceil_v$ would change from $\succeq_i\lceil_v = \succeq_i \cap (V\lceil_v)^2$ to $\succeq_i^\downarrow\lceil_v = \succeq_i \cap (V^\downarrow\lceil_v)^2$.*

Proof. When player $t(v)$'s sight is $s(v)$, according to Definition 5, we have that her preference $\succeq_i\lceil_v$ in sight-filtrated extensive game $S\lceil_v$ is $\succeq_i\lceil_v = \succeq_i \cap (V\lceil_v)^2$. If her sight $s(v)$ is expanded to $s^\uparrow(v)$, then the sight-filtrated extensive game would be

$S^{\uparrow}\lceil_v = (N\lceil_v, V^{\uparrow}\lceil_v, A^{\uparrow}\lceil_v, t^{\uparrow}\lceil_v, \Sigma_i^{\uparrow}\lceil_v, \succeq_i^{\uparrow}\lceil_v)$, where $V^{\uparrow}\lceil_v = s^{\uparrow}(v)$. Then again by Definition 5, $\succeq_i^{\uparrow}\lceil_v = \succeq_i \cap (V^{\uparrow}\lceil_v)^2$. The case for sight contraction can be proved similarly. ∎

5 Conclusion and Future Work

In summary, towards a more realistic model for extensive games, we first introduced the model of qualitative games with short sight and investigated the three solution concepts in these games. Then we studied short sight in several respects, characterizing different types of players by their sight functions, connecting α-β pruning with sight filtration, and obtaining a backward induction algorithm for computing an equilibrium path of qualitative game playing. Furthermore, we studied sight dynamics with respect to types of players. And we showed that changes of players' sight would lead to the alteration of players' preferences.

As for future work, we are interested in investigating a dynamic logic for our model of games with short sight, formally characterizing interesting new features of games due to short sight.

BIBLIOGRAPHY

[1] Robert J. Aumann. Backward induction and common knowledge of rationality. *Games and Economic Behavior*, 8(1):6–19, 1995.
[2] Johan Van Benthem and Amélie Gheerbrant. Game solution, epistemic dynamics and fixed-point logics. *Fundam. Inform.*, 100(1-4):19–41, 2010.
[3] D. Edwards and T. Hart. The alpha-beta heuristic. Technical Report 30, MIT, 1963.
[4] Drew Fudenberg and Jean Tirole. *Game Theory*. MIT Press, 1991.
[5] Davide Grossi and Paolo Turrini. Short sight in extensive games. In *AAMAS*, pages 805–812, 2012.
[6] Joseph Y. Halpern and Leandro Chaves Rêgo. Extensive games with possibly unaware players. In *AAMAS*, pages 744–751, 2006.
[7] Joseph Y. Halpern and Leandro Chaves Rêgo. Extensive games with possibly unaware players. *CoRR*, abs/0704.2014, 2007.
[8] Donald E. Knuth and Ronald W. Moore. An analysis of alpha-beta pruning. *Artificial Intelligence*, 6(4):293–326, 1975.
[9] Yew Jin Lim and Wee Sun Lee. Properties of forward pruning in game-tree search. In *proceedings of the 21st national conference on Artificial intelligence - Volume 2*, AAAI'06, pages 1020–1025. AAAI Press, 2006.
[10] Fenrong Liu. *Reasoning About Preference Dynamics*. Springer, 2011.
[11] John F. Nash. Equilibrium points in n-person games. *Proceedings of the National Academy of Sciences of the United States of America*, 36(1):48–49, 1950.
[12] Martin J Osborne. *An Introduction to Game Theory*, volume 2. Oxford University Press, 2004.
[13] Stephen Slade. Qualitative decision theory. In *In Proceedings of DARPA Workshop on Case-Based Reasoning*, pages 339–350. Morgan Kaufmann, 1991.

Comments on Liu, Liu and Su

Thomas Ågotnes

1 Summary and Main Contributions

The paper models and analyses interaction between rational agents, taking the standard model of an extensive game as a starting point and relaxing what the authors refer to as two commonly made idealising assumptions. First, that agents' preferences over outcomes can be linearly ordered and represented using utilities. Second, that the entire structure of the game is common knowledge to all agents. The former assumption is relaxed by representing preferences usuing partial rather than linear orders, the latter by modeling agents with *short sight* who can only see a certain part of the game in each node. In particular, the paper defines new solution concepts for games with short sight as well as a backward induction algorithm for computing equilibria, discuss different types of short sight, and briefly discuss the relationship to α-β pruning and the dynamics of sight change. The framework is closely based on earlier work by Grossi and Turrini [1], extending it in several ways.

On the one hand, there has been great interest in formalising and analysing the theoretical foundations behind the reasoning and interaction of resource bounded agents in recent years, in different communities including game theory and logic. I certainly agree with the authors that the "short sigh" model is an interesting approach to that problem. On the other hand, there is a long tradition in artificial intelligence (AI), going back more than 60 years, on game playing with limited resources, employing very similar approaches. The latter type of research is sometimes more pragmatic and/or applied, and in my view more theoretical work is important in order to understand the foundations of interaction betwen resource-based agents. Thus, I think this line of theoretical work is important. I find the paper very interesting and its contributions valuable. The paper certainly sheds more light on the idea of short sight, and the new solution concepts make sense.

2 Comments and Questions

I first have some questions about the framework.

1. The relationship to the framework in [1] is not entirely clear to me. The authors say (in the abstract) that the model in the current paper is "more

realistic". What is meant by this? It is also said that "In this paper, we propose a more realistic model for extensive games, which represent players'
preference qualitatively instead of using numeric payoff" (this is related to
the first motivating assumption mentioned in the summary above). However, qualitative preferences rather than numeric payoff are used in standard
game theory text books such as [2] and indeed also in [1].

2. I am a little confused about whether the games considered in the paper are
required to be finite or not. Definition 1 first says that it defines the notion
of "extensive game" (no qualification about finiteness), then says "a finite
extensive game is..", but then neither requires the state space nor horizons to
be finite. However, the definition of the O function on the next page assumes
finite horizons. But, then again, in definition 3 it is assumed that the state
space can be infinite.

3. The preference relations are defined over both non-terminal and terminal
nodes. Are there no restrictions on them, for example on preferences between a node and its immediate successor? One can easily think of non-
intuitive situations that can be modeled otherwise. Consider, for example, a
single agent, one non-terminal node and one terminal node. Does it make
sense that the agent prefers the non-terminal node over the terminal node?
Or the other way around? (On a more minor note I am a little confused that
the preference relation is first defined to be a partial order, but then, in the
next paragraph, is assumed to be total. However, the latter assumption does
not seem to be used anywhere so this is probably not important).

4. I wonder to what extent Theorem 1 is really about sight-filtrated games. To
me Algorithm 1 seems to be much more general, taking an arbitrary game
as input and outputting a path. Theorem 1 says that the algorithm computes
a SPE of a sight-filtrated extensive game; can it be generalised to say that
it computes the SPE of any game? In any case, as the algorithm itself is
general and takes a game and produces a path, it would be interesting to see
concrete examples of how it differs from other algorithms. Of course, one
difference is that it allows partially ordered preferences, but it is not clear to
be that this difference is significant.

5. One of the two main motivations for the paper is to generalise earlier work
to allow for partially ordered preferences rather than requiring linear orders
(as, e.g., induced by utilities). It would be interesting to hear more explicitly
how that relaxed assumption makes a difference for the analysis.

Then, some questions about related work.

6. As mentioned, many of the ideas used in the paper are well known in the artificial intelligence (AI) community. Games like chess and checkers have been benchmark problem in this field since its beginning, and the key problem the current paper tries to deal with, i.e., that players, neither human or machine, can evaluate the whole game tree in practice, has been known equally long. In fact, already in 1950 Shannon [3] proposed a solution to this problem similar to the one studied in the current paper: first, define preferences not only over terminal nodes but also over internal nodes by using a heuristic *evaluation function*, and, second, use a *cutoff test* to determine when to stop going deeper in the game tree and use the evaluation function. This is now a standard technique in AI game playing [4, 5], and has been studied extensively. Indeed, it was one of a combination of techniques used by IBM's *Deep Blue* chess program to defeat world champion Garry Kasparov in 1996. To continue with chess as an example, non-terminal nodes are often evaluated by computing the material value of the pieces on the board, but the cutoff test is in fact more problematic, having to define *quiescent* positions in which, e.g., favorable captures cannot be made. This is exactly the point the authors make in Example 2, but is very well known in AI and has been the topic of extensive research.

   It is also well known that the combination of minimax and evaluations of non-terminal nodes has some inherent weaknesses [6, 7].

   The authors of the current paper has based their research on the recent framework by Grossi and Turrini, but it is not clear to me how the latter relates to the extensive existing body of work on game playing with limited "sight" in AI. In particular, how do the results in the current paper apply to existing research on imperfect real-time decisions in AI game playing?

7. One of the main motivations of the paper is that it is unrealistic that players have common knowledge of the structure of the game. Such situations have been studied more generally in *epistemic game theory* [8]. What is the relationship between the models and results in the current paper (and in the Grossi and Turrini paper) and those of epistemic game theory? Can the former be framed in the latter? There seem to be some hidden assumptions about knowledge; for example, what is it assumed that one agent knows about another agent's sight function? The latter is of course crucial for decision making.

8. In [9] a particular solution concept, viz. winning strategies, under the assumption that agents have bounded resources, are studied in a logical setting. A particular case is that a strategy can only discern between n different situations, or, said another way, that an agent can only make n different

choices. Intuitively there seem to be a relationship to agents with different types of sight functions. Is there?

BIBLIOGRAPHY

[1] Grossi, D., Turrini, P.: Short sight in extensive games. In: *Proceedings of the 11th International Conference on Autonomous Agents and Multiagent Systems* - Volume 2. AAMAS '12, Richland, SC, International Foundation for Autonomous Agents and Multiagent Systems (2012) 805–812

[2] Osborne, M.J.: *An Introduction to Game Theory.* Oxford University Press: Oxford, England (2004)

[3] Shannon, C.: Programming a computer for playing chess. *Philosophical Magazine* **41**(4) (1950) 256–275

[4] Russell, S., Norvig, P.: *Artificial Intelligence: A Modern Approach,*. Prentice-Hall, Englewood Cliffs, NJ, ISBN 0-13-103805-2, 912 pp., 1995 (1995)

[5] Pearl, J.: *Heuristics: Intelligent Search Strategies for Computer Problem Solving.* Addison-Wesley (1984)

[6] Beal, D.F.: An analysis of minimax. In Clarke, M.R.B., ed.: *Advances in Computer Chess* 2, Edinburgh University Press (1980) 103–109

[7] Nau, D.S.: Pathologies on game trees revisited, and an alternative to minimaxing. *AIJ* **21**(1–2) (1983) 221–244

[8] Perea, A.: *Epistemic game theory: Reasoning and choice.* Cambridge University Press (2012)

[9] Ågotnes, T., Walther, D.: A logic of strategic ability under bounded memory. *Journal of Logic, Language and Information* **18**(1) (2009) 55–77

Comments on Liu, Liu, and Su

PAOLO TURRINI

I would first of all like to thank the authors for embarking upon such a challenging enterprise, by which I am increasingly fascinated: the study of rationality in *concrete game play*, where factors such as thinking time, capacity to think ahead, capacity of evaluating uncertain positions, play a big role, certainly a bigger role than the one they play in standard game-theoretic models.

> *Usually economic models do not spell out the procedures by which decisions of the economic units are made; here, we are interested in models in which procedural aspects of decision making are explicitly included.*[p.1]

Ariel Rubinstein couldn't put it more clearly in his *Modelling Bounded Rationality*, where, further on, he observes something extremely relevant for our purposes:

> *one of the most interesting modeling issues in the study of games with boundedly rational players is that of modeling players who have limited ability to understand the extensive game they face. The game of chess is a good example for illustrating some of our concerns (...) Modeling chesslike behavior in games is still a puzzle. I am not aware of any analytical model that deals satisfactorily with this issue.* [p.131-132]

On the right level of analysis The paper borrows many notions from the the work of [1]. The authors' intention is however to push that model framework, towards more realistic models of *chesslike* behaviour, with the following modifications:

- the criteria that players use when evaluating what they can see induce a partial order, not a total preorder as in [1].

- there are various *dynamic* aspects, such as sight change (Definition 9), and differences in players types', i.e. k-level sight, property-based sight functions (Definition 8) etc.

It seems to me that a more exact way of stating what it means to be build a realistic model of games with short sight should be done along Rubinstein's advice: making the procedural aspects of players' reasoning *explicit in the models*. The question that first comes to mind is then whether the extra features studied in the paper can help capturing players' reasoning in games with short sight.

I am not particularly convinced of the use of adopting partial order inducing evaluation criteria, besides the obvious added value which lies in their generality, to give further insight on the procedural aspects of game play with short sight.

The algorithm used in [1] to show the existence of sight-compatible subgame perfect equilibria relies on Kuhn's theorem, which states that every finite extensive game of perfect information has a subgame perfect equilibrium. I see that one could adapt the definition of subgame perfect equilibrium to games where players' preference relations are partial orders over the set of terminal nodes. For instance saying that in every subgame the SPE strategy is *not* dominated by any other strategy starting from that subgame (rather than saying that it dominates all other strategies starting from that subgame). However this new definition yields the same equilibrium strategies of the standard definition applied to the same game with the only difference that incomparable outcomes for a player are treated as indifferent for that player, i.e. the same outcome of a special total preorder. Intuitively, when we take a decision in favour of an option only on the grounds of some given evaluation criteria, we have already implicitly compared it to the other available ones.

I do think on the other hand that the operations of sight and preference change described in the paper, in particular when coupled with property-based preference relations [2] or with an order on a set of properties of the model at hand (P-sequences) [3], can effectively be used to capture the logic of adversarial search, i.e., the way computer games (and possibly humans as well) explore, evaluate and decide between available moves when the game tree is too large to be fully explored.

I will try to elaborate on this point.

Realizing I was wrong: using sight-preference dynamics to restore evaluation coherence Theorem 2 shows how preference change can be induced by sight change. The idea is simple: a sight restriction (or sight filtration, to use the terminology in the paper) is coupled with a preference order over the sight's terminal nodes. Modifying the sight induces a new preference relation, because the terminal nodes change. I believe the approach is rather powerful and could be used in more complex scenarios. One could for instance model a process of re-evaluation of intermediate positions once a player realizes to have made a mistake in evaluating some nodes (or some moves). An example of *evaluative incoherence* is sketched in Figure 1 and could be solved in several ways. In games like chess, when one

could safely assume that the other players are *opponents*, maxmin type of update
works perfectly, even in less obvious cases. The idea is sketched in Figure 2.

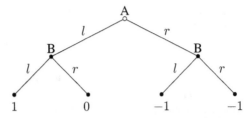

Figure 1: Evaluative incoherence. Player A evaluates the foreseeable consequences of l, in this case (l, l) (l, r), (r, l) and (r, r) and assigns them a value
(notice, I am using numbers only to ease the comparison, everything I am going to
say can be done within the qualitative account of preferences). After having done
this, no reasonable player can keep preferring r to l, if he ever did. Clearly, such
evaluation should be updated to accordingly to one where l is better than r.

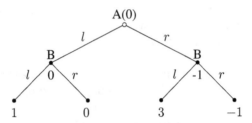

Figure 2: Maxmin value. Player A evaluates the foreseeable consequences of l,
in this case (l, l) (l, r), (r, l) and (r, r) and assigns them a value. Only, now she
knows to be playing *against* player B. Thereby, B nodes can be given a MIN
value, A nodes a MAX value. Notice, the overall evaluation displays an internal
coherence.

Modelling adversarial search   I have always been struck by how adversarial
search algorithms [6] work, coupling the α-β pruning method to the maxmin evaluation. I give an example in Figure 3. A player is in the process of choosing among
some actions of which some consequences can be foreseen, and he needs to decide
which one to take. He starts with an action, say a, and exploring and evaluating its
consequences up to a certain depth (by the way, it is not always easy to understand
when to stop searching). Having done this, he can assign a MIN value to a, say 2,
which is the payoff he can guarantee to himself by playing a. Now α-β pruning

enters the picture. Suppose that in the process of exploring a second action b, A comes across a set of outcomes, following b, which van be assigned a MIN value 1. The reasoning is this: if such a set cannot be avoided, then the further exploration of the consequences of action b is pointless.

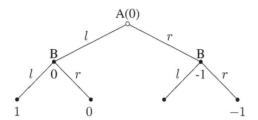

Figure 3: The logic of adversarial search. Player A evaluates the foreseeable consequences of l and r, in this case (l, l) and (l, r), and (r, r) (she cannot see or does not explore (r, l)). Even so, a MIN value can safely be assigned as, notice, there is no need to explore and evaluate (r, l).

The paper sketches a description of α-β search, but it seems to me that still some steps need to be taken to model the kind of reasoning described above. Yet, the building blocks for doing so are already there: sight change and preference dynamics. And doing so will amount to bringing forth the research on bounded rationality, constructing a powerful model of concrete game play which even Rubinstein should be aware of.

BIBLIOGRAPHY

[1] Davide Grossi and Paolo Turrini, *Short Sight in Extensive Games*, AAMAS, 2012.
[2] Franz Dietrisch and Christian List, *Where do preferences come from?*, International Journal of Game Theory, 2012.
[3] Fenrong Liu, *Changing for the better: preference dynamics and agent diversity*, ILLC Dissertation Series, 2008.
[4] Martin Osborne and Ariel Rubinstein, *A course in game theory*, 1994.
[5] Ariel Rubinstein, *Modelling Bounded Rationality*, 1998.
[6] Stuart Russel and Peter Norvig, *Artificial Intelligence: A Modern Approach*, 2012.

Response to Turrini and Ågotnes

CHANJUAN LIU, FENRONG LIU AND KAILE SU

We would like to express our appreciation to Paolo Turrini and Thomas Ågotnes for their very helpful comments. Many of their observations are guidelines for our future research, here in this note we attempt to clarify a few issues and answer a small selection from their questions.

1 From idealized to realistic: representation

Turrini starts his comments with an observation made by Rubinstin [11] that procedural aspects of decision making should be explicitly included in current game models. This is exactly what motivated our research in the first place. We found the game models employed in classical game theory too *simple* in the sense that many interesting and important aspects of real decision making and game playing are neglected. In fact, after we read the work by Grossi and Turrini [5], we were inspired by the notion of *short sight* and *prioritized preferences* and the way they intergrate those notions with game theory. Indeed, as in earlier work of one of us, preferences are based on a limited set of reasons, while it is also true players cannot see the whole structure of a game tree all the time.

The goal of our paper is to move towards even more realistic scenarios, adding three further aspects to the model proposed in [5]. On the basis of short sight, we investigate how we can use that notion to define different types of players. We also generalize the assumption of linearly ordered priorities into partial orders, taking more realistic incomparable priorities into account. Finally, to model more complex game scenarios, we also study dynamic changes of preference and sight, a common phenomenon in game playing.

Related to these ambitions, Ågotnes correctly points out that resource-bounded agents have been studied in AI for a long time. In fact, resources are something we had in mind when studying short-sighted players. Why do some players see more in a game tree, while others see less? This may well be a matter of differences in resources that are available to players. For instance, if some agent has a larger memory space and more calculations are allowed, she might see more of the game tree. We are aware of some earlier research in this respect, for instance [1], [3] and [2]. There is clearly a connection between those works and our perspective. Similarly, reasoning patterns of non-ideal real agents have attracted logicians' attention in recent years (see, for instance, [13], [8], and [9]). The general issue in much of this work, as we see it, has been to find the right angle or abstraction level for modeling differences between agents. We plan to study those issues in the future, drawing on the existing literature.

Similar concerns have led us to generalizing linearly ordered priorities to partial orders. Partial order allows us to model situations in which two priorities occur that are hard to compare. Classical game theory usually adopts utility functions to represent preference, making linearity a built-in assumption. Our paper employs qualitative representation of preference as, essentially, a binary ordering on outcomes of a game, making partial orders a natural mathematical format. Right now, as the reviewers note, the results we have obtained with partial orders do not look surprising, or dramatically different from the linear case. However, we do think this generalization is wise and significant, at least for realistic modeling. For instance, some natural events of preference change can change linear orders into partial ones (think of conflicting authorities concerning an action under consideration), showing that linearly may just be too fragile a constraint on preference in a dynamic setting. We hope to develop these themes in future research, and bring more people to appreciate partial orders in this area.

2   From idealized to realistic: computation

In addition to semantic modeling concerns, on the computational side, too, we have obtained some results toward more realistic settings. To compute an equilibrium path of qualitative game playing, we introduced a modified backward induction algorithm, called qualitative backward induction (QBI) . QBI is part of the algorithm for finding solutions in games with short sight: it should be run at every decision point. The main difference with the traditional backward induction algorithm [4] is that QBI is applicable to qualitative preference models (having a virtual lowest preference). Also, we make a pretreatment before searching by means of the operation $dominated(G, v)$. (See also [12] on the logic of qualitative versions of backward induction.)

Ågotnes mentions connections to AI-style algorithms in game play. In fact, we did provide one sample of just this: connecting sight filtration with α-β-pruning ([6]), a well-known realistic search algorithm for two-player zero-sum games. Much more can be done, and this is what we will pursue in future work, tapping into the long-standing computational experience in model checking techniques and automated deduction by one of our group.

3   Adding knowledge

Ågotnes mentions potential connections between our work and epistemic game theory. We, too, have come to realize that both short-sight games and epistemic game analysis involve players' knowledge of the game. In fact, the epistemic dimension is something we would like to explore explicitly – though, in our view, one should first make the following distinction. Epistemic game theory (cf. the recent book [10]) is about games with imperfect information, where players may lack knowledge of where they are, or at least, about games where players lack information about the future. But short sight captures a different situation, in which players know where they are but may be too short-sighted to see the whole structure in the future. This subtlety is one of several things that needs to be resolved before we can frame games with short sight in epistemic game theory. But we do think there is a genuine potential for such a connection.

Returning to our work, there are several further places in which knowledge plays a role. For instance, once we have types of players in the model, there is an issue of whether one agent knows the other agent's types. This would call for explicit modeling of players' knowledge about others. We think that this can be done using epistemic operators in familiar ways, and one particular avenue that we are currently exploring is this. In [7], we have added a logical calculus to the analysis in the present paper, introducing explicit modalities for short sight and preference. It is easy to extend that language with epistemic modalities, and then we can immediately express many interesting situations involving knowledge.

4 Varia

Finally, we make some remarks concerning some smaller points raised by the commentators.

Turrini's comments concerning dynamic operations and types of players are very instructive. His examples show the power of these notions in restoring evaluation coherence and capturing adversarial search. The situation shown in his Figure 1 is exactly what we try to illustrate with Example 2 in our paper. Figure 3 shows the application in modeling adversarial search from a new perspective. Exploring potential applications like this is precisely what we intend to do.

Ågotnes points out the often substantial difference between finite and infinite games, and wonders what happens to our results in this transition. In our paper, we were not thinking of extensive games with short sight as necessarily finite, and all our results apply to infinite games.

Ågotnes notes that preference relations can be between all nodes in a game tree, a player may even prefer non-terminal nodes to terminal ones, or the other way around. Theoretically, this is indeed possible. But our understanding is that, in real game play when players make decisions, they usually choose among the children of the current node. That is, players only compare siblings, rather than non-terminal and terminal nodes.

BIBLIOGRAPHY

[1] Thomas Ågotnes and Dirk Walther. A logic of strategic ability under bounded memory. *Journal of Logic, Language and Information*, 18(1):55–77, 2009.

[2] Natasha Alechina, Mark Jago, and Brian Logan. Modal logics for communicating rule-based agents. In A. Perini G. Brewka, S. Coradeschi and P. Traverso, editors, *Proceedings of the 17th European Conference on Artificial Intelligence (ECAI 2006)*. IOS Press, 2006.

[3] Natasha Alechina, Brian Logan, and Mark Whitsey. A complete and decidable logic for resource-bounded agents. In *AAMAS 2004*, pages 606–613, 2004.

[4] Robert J. Aumann. Backward induction and common knowledge of rationality. *Games and Economic Behavior*, 8(1):6–19, 1995.

[5] Davide Grossi and Paolo Turrini. Short sight in extensive games. In *AAMAS*, pages 805–812, 2012.

[6] Donald E. Knuth and Ronald W. Moore. An analysis of alpha-beta pruning. *Artificial Intelligence*, 6(4):293–326, 1975.

[7] Chanjuan Liu, Fenrong Liu, and Kaile Su. A logic for extensive games with short sight. In Davide Grossi, Huaxin Huang, and Olivier Roy, editors, *Proceedings of LORI-IV*. Springer, 2013.

[8] Fenrong Liu. Diversity of agents and their interaction. *Journal of Logic, Language and Information*, 18(1):23–53, 2009.

[9] Fenrong Liu and Yanjing Wang. Reasoning about agent types and the hardest logic puzzle ever. *Minds and Machines*, 23(1):123–161, 2013.

[10] Andrés Perea. *Epistemic game theory*. Cambridge University Press, 2012.

[11] Ariel Rubinstein. *Modelling Bounded Rationality*. MIT press, 1998.

[12] Johan van Benthem and Amélie Gheerbrant. Game solution, epistemic dynamics and fixed-point logics. *Fundam. Inform.*, 100(1-4):19–41, 2010.

[13] Johan van Benthem and Fenrong Liu. Diversity of logical agents in games. *Philosophia Scientiae*, 8:163–178, 2004.

Strategies in Games: A Perspective from Logic and Computation[1]

R. RAMANUJAM

Strategies in extensive form games are often compared with plans, and then similarities and differences discussed. We suggest that it is useful to compare them with programs, which fits well in a logic and computation perspective.

1 Structure in strategies

Consider a two-player win/lose extensive form game of perfect information. The notion of strategy is central to game theory and is defined to be a map from player nodes to edges incident at that node. In this sense, a strategy for player i is a complete specification of choices at nodes at which player i exercises choice. We can apply the backward induction procedure and compute a winning strategy for one of the players. If the outcomes of the game were given by payoffs to players (rather than binary win/lose outcomes), we can compute optimal strategies for players as well. When plays yield distinct outcomes, the solution of the game has nice properties.

Such a superquick summary suffices if one considers extensive form games as a step en route to outcome based analysis of games in normal form. The hurry is then justified since "all relevant factors about players' information" can be addressed in the normal form representation, especially when we work with expected utility and study equilibria that predict "stable rational choice". When stability is studied with respect to deviations from such predictions, it seems reasonable that deviations refer only to the implication of choices on outcomes. Deviations based on other considerations are then declared extraneous.

Why then should there be any difference when this story is read from a *computational* perspective ? If there is, does it matter ? What does a computational perspective consist of ? What relation does it have to logic ? These are, in our opinion, important questions to address. What we do here is to offer only a few brushstrokes in answer, and point to one sketch in [2].

Consider a game in which a player has two choices, say 0 and 1 and fix an ordering of the nodes of the game tree. Now consider a strategy that picks, at any

---

[1]This summary is based on [2].

player node n, the move 1 if the nth Turing machine halts on empty input, and 0 otherwise (with respect to a fixed enumeration of Turing machines). In the absence of any reason to eliminate this strategy, such as dominance or admissibility, it might actually accomplish something. Yet this is a patently absurd though precise specification, since we assume that the strategy space has some underlying reasoning in which the game structure figures prominently, and is not an arbitrary function space. Thus a slogan: *strategies have logical structure.*

This suggests that strategies might be weaker than (total) functions that specify action in fine detail, and logic is better at abstractions in general. But then we need to be wary of very loose specifications. We think of strategies as providing advice to a player, following which the player can be sure of doing her "best". But consider an advice: "consider all possibilities, don't ignore any detail, however trivial it may seem; give your 100% best". Earnest it is, but rather useless for the actor. "Making the first million is hard, after that it's all very easy; so do all your planning for making a million", is wise, but little help either. Therefore we say: *a strategy should constitute an effective procedure.*

Implementability by finite resources then becomes an important aspect of strategizing and this further dictates strategy structure. We are then directed to study the resources expended by a player while strategizing. These can be in the form of informational resources such as observations during play, belief formation and revision, communication etc. These could be computational resources such as memory, and reasoning depth. In any case, making such resource dependence explicit is part of what we term a 'computational perspective'.

Once we see strategies as logically specified and computationally implemented, a natural corollary to work not with a single game but a *class of games*. A single tree of size 123 can of course be specified by a formula in some logic, but such a formula is likely to be highly uninformative. Similarly an algorithm can indeed be developed to work on that single tree, utilizing all its distinctive features, but such an algorithm is likely to yield little computational insight. But logic and algorithms are much better at uniformizing a class of situations, and offer valuable tools and techniques for doing so.

At this juncture, we should point that when the game tree is sufficiently large, it is much easier to reason about it as an infinite (or potentially infinite) tree. This is true, for instance, in reasoning about chess. Such an argument has been applied by game theorists in the context of finite vs infinite horizons in the context of finitely repeated games when the number of repetitions is large. Reasoning that is not sensitive to constants and that does not accord a special status to 'end' nodes can be generic but important for strategization. Once again we find that uniform reasoning for a class of games makes sense.

What then, is strategizing for a class of games ? Such a notion seems absurd, until we note that heuristics are precisely of this kind. *Tit for tat, copy cat, go with the winner,* etc are heuristics rather than strategies, applicable to a wide class of games. We spoke above of strategies being thought of as partial functions (or even as relations, extending the argument a little), and heuristics then belong comfortably in this class.

The idea of strategies being not quite complete plans but more, despite being relational in nature, has been pointed out by game theorists repeatedly ([3]). Consider a game where player 1 has a choice of a and b at the root node. The game ends with the move a giving player 1 a payoff higher than any that would be obtained in the subtree that b leads to. A rational player would surely choose a, and hence the choice would constitute a coherent, complete *plan*. Yet the notion of behavioural strategy demands a specification of choices further down in the b-subtree as well. This is necessary, because the rationality of player 2's choices at a node n in that subtree can be evaluated only by consideration of what 1 might do in future at n, and indeed, if 1 deviates from the rational choice at the root node, 2 would be called upon to exercise such reasoning. Thus the notion of strategy involves not only planning but assertion of players' beliefs about other players. These are expectations and contingency plans, but importantly, require some *flexibility* on the part of the player *to adapt to situations*.

However, if the notion of strategy is to reflect such flexibility, we need to pause when we speak of a player's "choice of strategy". The image of selecting from a set that such a phrase evokes needs to be replaced by one of carrying a whole bag of partial strategies along, making provisional choices, replacing one and picking another during course of play. Indeed, the idea of **strategy composition** seems to reflect this image more naturally than strategy selection. Indeed, this is akin to program composition and local reasoning. In particular, having access to a library of programs that achieve small local objectives and composing them for the situation at hand is not only a good way of programming but also good for strategizing 'on the fly'.

Thus, if strategies are to possess logical and computational structure, then so also should games: rather than being seen as a description of physical events that occur during course of play, we need to see games as classes of situations as perceived by players, incorporating all elements that are considered relevant for strategizing, but also abstracting away details that strategizing ignores.

In summary, we suggest that strategies be considered as partial but implementable plans that can be composed, and that strategies possess a logical structure that ensures certain properties: *observability, flexibility, compositionality, perceptivity.*

2   The proposal

We propose a syntax of **strategy specifications** as follows [1].

---

[1]The technical development is only for two-player games for ease of presentation; the treatment generalizes easily and uniformly to multi-player games, but with considerable technical mess.

Let $P^i = \{p_0^i, p_1^i, \ldots\}$ be a countable set of proposition symbols where $\tau_i \in P_i$, for $i \in \{1, 2\}$. Let $P = P^1 \cup P^2 \cup \{leaf\}$. τ_1 and τ_2 are intended to specify, at a game position, which player's turn it is to move. *leaf* specifies whether the position is a terminal node.

Further, the logic is parametrized by the finite alphabet set $\Sigma = \{a_1, a_2, \ldots, a_m\}$ of players' moves and we only consider game arenas over Σ.

Let $Strat^i(P^i)$, for $i = 1, 2$ be the set of strategy specifications given by the following syntax:

$$Strat^i(P^i) := [\psi \mapsto a_k]^i \mid \sigma_1 + \sigma_2 \mid \sigma_1 \cdot \sigma_2 \mid \pi \Rightarrow \sigma$$

where $\pi \in Strat^{\bar{i}}(P^1 \cap P^2)$, $\psi \in Past(P^i)$ and $a_k \in \Sigma$.

There are very few operators for composing strategies, since our aim is to illustrate the rationale outlined above (rather than propose a definitive expressively complete set). The basic strategy is of the form $[\psi \mapsto a_k]^i$, which reflects the *observability* criterion. Player i makes observations as play progresses, and if ψ is true in this observational history, she chooses to play a_k. Note that this is not a property of game positions but an invariant on the game tree: at every game history that satisfies ψ and it's player i's turn to move, the choice is a_k. The specification is partial, since it is silent on what to do when ψ is false.

The operator $\sigma_1 + \sigma_2$ reflects the *flexibility* criterion, since it stands for choice, and hence allows nondeterminism in strategy specification. The operator $\sigma_1 \cdot \sigma_2$ is a conjunction, and is the basic unit of *compositionality*: typically $[\psi_1 \mapsto a_1]_i \cdot [\psi_2 \mapsto a_2]_i$ specifies the use of a heuristic that combines actions to be taken in different situations. $\pi \Rightarrow \sigma$ is the principal operator that embodies *perceptivity*: if the opponent behaviour conforms to the invaraint specification π, respond with σ. This reflects the considerations above that differentiate strategies and plans, including deviation from rational behaviour.

Note the absence of *complementation* as an operator. For one thing, in terms of strategic reasoning, the negation of a strategy offers many interpretations, most of which seem unnatural. For another, given that a strategy specification corresponds to a set of trees, complementation on tree languages is technically challenging, and the criterion of strategy as effective procedure would limit the kind of negation one can work with. All this is not to say that we cannot easily define a complementation operator, but that we are wary of doing so.

We show how every strategy specification can be implemented as a tree automaton with output (or more generally, a tree to word transducer). Such realizability is important for developing a compositional theory of strategies. The algebraic structure of these operators and their interactions is yet to be studied, and constitutes an important research direction.

The syntax of observations ψ is constrained to be a simple modal logic of past tense, and thus there are no explicit constructors relating to players' beliefs. However, strategy specifications being interpreted as invariants on the tree, and the ability to combine observations on alternate histories, does confer an expressivity that incorporates some epistemic attitudes.

In [2], we study a simple modal logic whose characteristic modality is $\sigma \leadsto_i \beta$ where σ is a strategy specification and β is an outcome specification, with the intended meaning: playing σ ensures β for player i. The technical results relate to decidability of satisfiability and model checking and a complete axiomatization of validity, but these are more assurances that the formalism of strategy specifications can be embedded in canonical modal reasoning than results on reasoning about games *per se*.

In [1] we extend the structure of strategy specifications further, with process-like operators such as $\sigma_1 \frown \sigma_2$ for applying the strategy σ_1 and nondeterministically switching to σ_2. Combined with the ability to test for conditions, this extends strategy specifications considerably. (This enables us to study a strategic stability notion: when the player does not need to switch between strategies.)

In summary, we advocate a compositional structure for strategies, much like that used in programming, and back the advocacy by a technical proposal with semantics as tree languages and theorems on automaton construction for strategies and complete axiomatization. The expressiveness and algebraic structure of such strategization awaits further exploration.

BIBLIOGRAPHY

[1] S. Paul, R. Ramanujam, and S. Simon. Stability under strategy switching. In *Proceedings of the 5th Conference on Computability in Europe, CiE 2009*, volume 5635 of *LNCS*, pages 389–398. Springer, 2009.

[2] R. Ramanujam and S. Simon. A logical structure for strategies. In *Logic and the Foundations of Game and Decision Theory (LOFT 7)*, volume 3 of *Texts in Logic and Games*, pages 183–208. Amsterdam University Press, 2008. An initial version appeared in the Proceedings of the 7th Conference on Logic and the Foundations of Game and Decision Theory (LOFT06) under the title *Axioms for Composite Strategies*, pages 189-198.

[3] Ariel Rubinstein. Comments on the interpretation of game theory. *Econometrica*, 59:909–924, 1991.

Comments on Ramanujam and Simon

OLIVIER ROY

Defeasible plans, action negation and resources sensitivity - A comment on Ram Ramanujam's computational perspective strategies

Since von Neumann and Morgenstern we have been used to think big when it comes to strategies. The classical definition of the objects of choice in extensive games is that they specify an action to take for every possible situation a player has to move. And if this is not complex enough, the usual addendum is that strategies even tell one what to do at nodes the strategies themselves exclude.

It's a common place observation that these fully-specified entities are far remote from what we usually call a strategy. Ramanujam and Simon want us to take this observation seriously. What they propose is no less than a new look at the *theory*[1] of games, where the traditional notion of a strategy is generalized to partial and, more importantly, property-based plans of action.

There is of course more to choose from in this richer universe. Are there principles that may help us focus on some interesting or maybe useful sub-family of such plans? Ramanujam and Simon's answer is "yes", or at least this is what I take to be the gist of their *computational perspective* on strategy. For them plans should be simple enough to be accessed, combined and applied by human or computational agents with limited resources.

In this comment I will to focus on two conceptual issues that are raised by Ramanujam and Simon's paper. One important motivation behind their proposal, I think, is to do game theory with a notion of strategy that is closer to the one that are used by human and computational agents. From that point of view I will argue that defeasiblity is important. But, as I will point out, Ram and Sunil's framework need to extended to capture this phenomenon. The second point I want to make rest on a combination of two points made in the paper: on the one hand the view that strategies "give advice" on what to do, and on the other hand the fact that

---

[1] I emphasize "theory" here because the idea of re-doing game theory with simpler object of choices has been around in cognitive science and behavioral game theory for a while. I have in mind here the "fast and frugal" heuristic approaches to decision making, or "dual process theories". Indeed, Ramanujam and Simon do mention at several places that one of their aim is to have a notion of strategy that is general enough to capture what we would intuitively call heuristics or, one might add, rules of thumbs.

"action negation" is bracketed in that paper. I will argue that, precisely because of the former view, a treatment of action negation seems to be of high importance.

Defeasibility of plans. Many of the partial plans we make are defeasible. Take the following:

If your think you won't make the deadline, ask politely for an extension.   (1)

This is surely an evergreen in academia. But very few of us would apply it as is. If I know that a colleague of mine has already asked for an extension, I might rather decide not to clutter the organizers' mailbox with extension requests, and just wait for the extension to be announced on the conference website. In other word, it seems perfectly fine to hold the following, *in combination with* (1):

IF (your think you won't make the deadline

       AND you know someone else already asked for an extension)

          THEN just wait for the official announcement in the website.   (2)

In short, plans usually admits exceptions. They are defeasible.

It seems difficult to naturally capture this kind of non-monotonicity in Ramanujam and Simon's approach. Let ϕ be "you think you won't make the deadline", ψ be "you know someone else already asked for an extension", a_1 be "ask politely for an extension" and a_2 be "wait for the official announcement on the website." In Ramanujam and Simon's notation (1) and (2) corresponds respectively to:

$$[\varphi \Rightarrow a_1]$$ (3)

$$[\varphi \wedge \psi \Rightarrow a_2]$$ (4)

The strategy σ consisting of the combination of both would then be:

$$[\varphi \Rightarrow a_1] \bullet [\varphi \wedge \psi \Rightarrow a_2]$$ (5)

But assuming that a_1 and a_2 are mutually exclusive, in the current proposal this strategy would would prescribe the impossible action $\emptyset$ in all situations where $\phi \wedge \psi$ are true.

There are various ways one might think of enriching Ramanujam and Simon's proposal in order to capture this kind of non-monotonicity. One conservative extension would be to add some designated atomic propositions N, true in all normal situations. In exceptional cases, i.e. where $\psi \to \neg N$, the actual rendering of (5) above would be:

$$[\varphi \wedge N \Rightarrow a_1] \bullet [\varphi \wedge \psi \Rightarrow a_2]$$ (4')

A less *ad hoc* approach might be to use tools from non-monotonic logic or belief revision to interpret $[\varphi \Rightarrow a_1]$ terms. One way to do this for instance would be to equip oneself with a normality ordering of the states in the game arena or it's unfolding, and view $[\varphi \Rightarrow a_1]$ as the strategy that says to play a_1 in all the most normal φ-states. This seems like a natural extension of the current framework. And the conditional character of the basic constructs for strategies in Ram and Sunil's framework invites such a non-monotonic reading anyways.

Action negation. Ramanujam and Simon make the interesting observation that strategies can be seen as advice on how to play. In philosophers' terminology, strategies can be given a normative reading, telling the players what they *should* play at different choice nodes. Going back to the basic construct, a term like (4) would then be read:

IF your think you won't make the deadline

THEN you should ask politely for an extension.   (6)

This sort of advice is very often negative. Take the following:

IF your think you won't make the deadline

THEN you should NOT say you'll submit on time.   (7)

More generally, some widely-used notions of rationality in game or decision theory have this character. Strict dominance, for instance, enjoins *not* to play strategies that would make one strictly worse off in all possible circumstances. In short, these sort of advice-giving strategies make negative prescriptions. They tell the player not to play a given action.

Action negation is a notoriously difficult topics in philosophy of action, and Ramanujam and Simon rightly observe that adding them would it non-trivially complexify the calculus of strategies that they propose. So on the one hand their decision (not to say "strategy") not to venture on that topic seems like a wise decision.

On the other hand, given the omnipresence of negative prescriptions and the importance of *assessing* the rationality of certain strategies in games, it seems urgent to extend the current framework with negated strategy terms. What would that look like? Maybe a way to keep things under control would be to allow negation to occur *within* basic strategy constructs,

$$[\varphi \Rightarrow \neg a_1] \qquad (8)$$

interpreted as the strategy that prescribes to play any action but a_1 in case φ is true. One could then define negation of basic terms using the following, and not allowing negation of any more complex strategy constructs.

$$\neg[\varphi \Rightarrow a_1] \equiv_{df} [\varphi \Rightarrow \neg a_1] \tag{9}$$

How much more complex would be the resulting calculus? I do not know. But this seems like a natural way to re-introduce action negation, and one that, to go back to the preceding point, some people have actually argued is in fact the right way to understand negation of counterfactual conditionals.

Comments on Ramanujam and Simon

YDE VENEMA

Introduction

In this short note I gather some comments on Ramanujam's work [2], which itself can be seen as a summary of two earlier papers [3, 1]. This work aims for a formal, logic-based account of extensive games with perfect information, that allows for explicit, object-level reasoning about *strategies*. The author draws a parallel between strategies and programs or processes, with the purpose of taking logical and computational aspects of strategies into account.

Ramanujam's main point is that strategies have a logical *structure*, and that it makes sense to take this structure into account when analyzing strategies. Concretely, he inductively defines a formal algebraic language of *syntax specifications*, of which the operators reflect some of the criteria that he imposes on the structure of strategies, such as observability, flexibility, compositionality, and perceptiveness. On the basis of this he develops a logic that is somewhat reminiscent of PDL (propositional dynamic logic) in the sense that the syntax has two sorts, one for formulas and one for strategy specifications. Finally, for the resulting logic the authors of [3] provide an axiomatization for which they prove completeness.

I found the ideas conveyed in this paper quite stimulating. Being a relative outsider in the area of game theory in general and of reasoning about strategies in particular, some of my comments will be somewhat superficial; next to these, I raise some more concrete technical questions as well.

On the structure of strategies

One of the most interesting questions that the paper raises, is the following:

Question 1. What is a/the natural structure of strategies?

Clearly this question depends on context and will not have a unique answer. Ramanujam discusses some interesting properties of strategy structure, and builds up his language of strategy specifications on the basis of these. He clearly indicates that in this work his aim is not to propose a definitive expressively complete set of operators for composing strategies, but there is one aspect of strategy structure that I would certainly be interested to see developed in some more detail, namely, how a *global* strategy is composed of *local* strategies. Here 'local' has both a spatial and a temporal dimension: the point is that while playing a complex game, agents may temporarily focus on subgoals that relate to part of the overall game, and their global strategy consists of a plan on how to put these local strategies together.

Take the game of chess as an example. A local, aggressive strategy could be to build up an attack by exercising a well-defined amount of pressure on a certain corner of the board; such a strategy would only make sense if it is paired with another local strategy, now defensive, that makes sure that while this gathering of forces is in swing, the opponent does not gain too much control over another part of the board. Observe that these two strategies correspond to respectively a *liveness* and a *safety* condition.

As a second example, an *infinite game* can often be seen as an infinite sequence of finite-length games; a strategy of a player could consist of guaranteeing that she establishes some well-defined task in each round of the game. Concretely, in a two-player infinite parity game, this could mean that the maximal priority reached during this round of the game has the right parity.

These examples support Ramanujam's claim that strategies resemble programs, since they are composed using the same collection of constructors, such as parallel and sequential composition. This would suggest an approach that is even closer to that of propositional dynamic logic or game logic. Here strategies (or strategy specifications) are inductively built up from tests and atomic strategies (i.e., actions), using some repertoire of strategy constructors, including *sequential composition* and *iteration*, so that composite strategies can be defined such as 'repeat strategy σ until property ϕ has been established'.

As an aside, I would like to raise the question in what sense the structure of strategies can directly be derived from that of games. One reason for suggesting this is that the strategy of a player in an $n + 1$-player game can itself be seen as an n-player game.

Language

Clearly, mirroring the question about the right structure of strategies one could ask what is a/the natural language for reasoning about strategies. And as usual, answering this question will involve a trade-off between expressiveness and computational feasibility. Leaving these issues aside for now, here I would like to raise two questions concerning the language Ramanujam and Simon introduce in [3] — let us denote their system by R.

First a rather specific question. A key construct in this language is the connective $\leadsto_i$, with $\sigma \leadsto_i \phi$ meaning that agent i, playing a strategy (conforming to specification) σ, ensures an outcome of formula β. The latter sentence has a 'safety' interpretation, in the sense that every continuation of the play in which player i sticks to σ, the formula β will continue to hold. In the line of my chess example above, I wonder whether one should not add a 'liveness' version of this operator as well, stating that every continuation of the play in which i sticks to σ, eventually a position is reached where β is true. As far as I can see, such a formula would properly increase the expressive power of the language, and would certainly be a useful instrument in analyzing the power of players to achieve certain outcomes.

A more technical question that I would like to raise is the following:

Question 2. Where in the landscape of modal logics is the system R of Ramanujam and Simon situated?

Note that if we find the proper location of a logic in this wider landscape, many properties of the logic may follow already from general principles.

In the case of the logic R, my conjecture would be that it can be embedded in the alternation free fragment of the modal μ-calculus over the modal signature with basic modalities $\{\langle a \rangle, \langle \bar{a} \rangle \mid a \text{ an atomic action}\}$. To verify this, one would first need to find out whether the modalities $\langle - \rangle$ and $\leadsto_i$ can be defined as certain least fixpoint connectives; this is obvious for $\langle - \rangle$, but would involve more work for the other connective. And second, one needs to check whether the formula $(\sigma)_i : c$ is expressible in the language without this construct; but observe that the axioms A6(a-c) look very much like rewrite rules to this effect.

Logic

Finally, I would like to bring up two issues regarding the completeness result for the deductive system that Ramanujam and Simon develop in [3]. Both of these were prompted by the remark of Ramanujam [2] that "these [results] are more assurances that the formalism of strategy specifications can be embedded in canonical modal reasoning than results on reasoning about games *per se*".

First of all, this remark begs the question about the purpose of completeness theorems, and of deductive systems in general, in this branch of applied logic.

And second, if we are interested in the embedding of the completeness result in 'canonical modal reasoning", then the following problem presents itself naturally:

Question 3. Can the completeness result in [3] be seen as an instance of a general completeness theorem?

In basic modal logic, Sahlqvist's theorem provides such a general completeness result. Unfortunately, all proofs of this result heavily rely the compactness of the logics, a property that is not satisfied by the kind of logic we are dealing with here. In the context of extensions of basic modal logics with fixpoint connectives, some results have been obtained in [4, 5], but these results do not seem to be immediately applicable.

BIBLIOGRAPHY

[1] S. Paul, R. Ramanujam, and S. Simon. Stability under strategy switching. In *Proceedings of the 5th Conference on Computability in Europe (CiE 2009)*, volume 5635 of LNCS, pages 389–398.

Springer, 2009.

[2] R. Ramanujam. Strategies in games: a perspective from logic and computation, 2013.

[3] R. Ramanujam and S. Simon. A logical structure for strategies. In *Logic and the Foundations of Game and Decision Theory (LOFT 7)*, volume 3 of Texts in Logic and Games, pages 183–208. Amsterdam University Press, 2008.

[4] L. Santocanale and Y. Venema. Completeness for at modal fixpoint logics. *Annals of Pure and Applied Logic*, 162:55–82, 2010.

[5] L. Schröder and Y. Venema. Flat coalgebraic fixed point logics. In *Proceedings of 21st International Conference on Concurrency Theory (CONCUR 2010)*, volume 6269 of LNCS, pages 524–538. Springer, 2010.

Response to Roy and Venema

R. RAMANUJAM

Both the readers have raised important questions that go right to the core of the 'Logical structure for strategies' project. I also see them as an acceptance of the basic premise, the questions being on the architecture of the edifice that needs to be built as yet. I thank them for the insightful remarks.

1 Yde Venema's comments

There is one aspect of strategy structure that I would certainly be interested to see developed in some more detail, namely, how a *global* strategy is composed of *local* strategies. ... As an aside, I would like to raise the question in what sense the structure of strategies can directly be derived from that of games.

Indeed yes, there is much to be done here. Propositional game logic uses a structure of game composition, and what we have in our paper is one of strategy composition, both using the flavour of propositional dynamic logic. Naturally these are related and interdependent notions: strategy structure reflects subgame structure and the latter imposes constraints on strategization. Indeed, in [1], we argue that we need to consider game-strategy pairs: after playing σ_1 in game g_1, playing σ_2 in game g_2 ensures an outcome α. But this is a limited answer; a more satisfactory one would be a formal relationship between game logic and strategy logic. The map from game logic to strategy logic would also answer the question about deriving strategy structure from that of games. We have worked on this, but do not have a satisfactory result as yet.

I wonder whether one should not add a 'liveness' version of this operator as well, stating that in every continuation of the play in which i sticks to σ eventually a position is reached where ϕ is true.

This is a very valuable suggestion. In some sense, the operator does allow a limited form of liveness: we can see ϕ as an invariant maintained by the strategy σ no matter what the opponent achieves. We could separately assert that the invariant implies a certain eventuality. This is in line with program assertions, where we specify loop invariants as safety properties and separately ensure termination or other such liveness properties. But a combined operator would be worth exploring as well.

Where in the landscape of modal logics is this system situated?

Indeed yes, the logic can be embedded in the alternation-free fragment of the μ-calculus as suggested. In fact this question also ties in with the formal relationship between game logic and strategy logic. Referring to the logic of game - strategy pairs in [1], we can show that propositional dynamic logic (PDL) can be embedded in it ([2]), but that is about all we can assert. We also have some conjectures relating logics of strategy composition to monadic second order logics on trees, but no definite results. Delineating precisely the expressiveness of structural reasoning about strategies is important, but as yet unclear.

Can the completeness result here be seen as an instance of a general completeness theorem?

Firstly I must explain the remark that led to this question. Our remark on the completeness theorem being more a demonstration that "the formalism of strategy specifications can be embedded in canonical modal reasoning than results on reasoning about games *per se*" should have been better contextualised. The remark pertains to the weakness of axiomatizing validities in the light of reasoning in games described earlier. In a logic that can describe game trees and strategy structure, one would expect inference of the form $\Gamma \vdash \alpha$ where Γ is a theory describing the game and α an assertion on a strategy ensuring an outcome. In such an inference system, the inference rules would (presumably) be guided by strategy composition, reflecting the pattern of reasoning that a player follows in the course of game play. That the Hilbert-style system presented is far from this objective is the gist of the remark, and developing a proof system of this two-sided kind would be worthwhile.

Now to the question. I certainly believe that this result is an instance of a general completeness theorem, but as pointed out by Venema, in logics without compactness, such characterizations are hard to obtain. Indeed, the references he has given only add to the conviction that modal fixpoint logics are perhaps the most natural logical formalism for game logics and strategy logics, but then perhaps it is also necessary to seriously limit the power of fixpoint quantification as appropriate for strategic reasoning.

In summary, Venema asks for a theoretical justification for the mix of operators for strategy composition, one that delineates the scope of the logic in terms of expressiveness and completeness, perhaps in model theoretic terms. This is certainly important. We have taken the line that the choice of operators be guided by 'practical' considerations, that of building a library of useful strategy specifications that can be composed online, but any such mix needs to be evaluated against the expressiveness considerations suggested by Venema.

2   Olivier Roy's comments

Defeasibility: Much of our planning tends to be defeasible, and hence it is natural to suggest that strategizing in games should reflect this. The example chosen by Olivier to illustrate this point is illuminating. Here we have two tests, ϕ and ψ that are not exclusive, but lead to advice a_1 and a_2 that are exclusive. Olivier formulates the scenario as $(\phi \mapsto a_1) \cdot ((\phi \wedge \psi) \mapsto a_2)$ and then, rightly, points out that this leads to a null advice. However, one can argue that the correct formulation here would be:

$$((\phi \wedge \neg\psi) \mapsto a_1) \cdot ((\phi \wedge \psi) \mapsto a_2)$$

since we would (politely) seek extension only if we don't know that someone has already asked for one. But such a fix begs the spirit of the suggestion. Plans should admit exceptions, and listing the negation of every possible exception alongwith a test is both silly and infeasible.

Much better would be a default style reading of $\phi \mapsto a$, with the intended meaning: as a rule, when ϕ holds play a, providing for exceptions separately. The suggestion that this be implemented by an ordering on states is welcome but buys new problems, since not only is what is considered an exception subjective and player-dependent, but so also is the ability to recognize exceptions and act on them. However, any theory of play that wishes to seriously engage with online reasoning by players does need to admit non-monotonicity of the kind pointed out by Olivier, and the current proposal allows only ad hoc mechanisms that are unsatisfactory.

A remark here: we have suggested that strategy structure can be seen as akin to that of programs. In the same vein, it is worth considering exception handling in programming languages: they provide primitives for raising exceptions and for handling them, without drastically altering program semantics. Perhaps there is a similar soution for logical treatment as well.

Action negation: This is an extremely interesting topic for research, both from a technical viewpoint and from that of the underlying philosophy. In the summary, I had remarked that "this is not to say that we cannot easily define a complementation operator, but that we are wary of doing so". The wariness stems from actual attempts that we made earlier, each one leading to some technical marshland or the other.

As Olivier suggests, one can consider a restricted operator such as $\phi \mapsto \neg a$ to mean that the test for ϕ entails playing any action but a. When the set of possible moves is finite this can be coded up using disjunction over the remaining actions. According to the semantics, when the test succeeds, if any of the remaining moves is enabled, one would be chosen. The trouble is when the only available move is a. The spirit of negation would mean that the strategy is not viable, and incoporating such a notion systematically into the semantics is possible, but not easy. Our attempts of this kind proved to be messy so we let it go, hoping that more competent attempts later by others would lead to better solutions.

There is a philosophical implication as well. As Olivier rightly remarks, the normative reading of strategies comes with many uses of free negation. But then this also suggests a reading of $\phi \mapsto \neg a$ to mean that playing a when ϕ holds would lead to something 'bad' for the player (like in the case of strictly dominated strategies). This might be seen as a more natural form of negation for games, but again, going down this road leads to a technical morass as well.

The project of strategy composition built on property-based plans of action needs to look at many such issues to have any chance of providing logical foundations for game theory. Plan negation and non-monotonic reasoning are surely central features for such a development.

BIBLIOGRAPHY

[1] R. Ramanujam and S. Simon. Dynamic logic on games with structured strategies. In *Proceedings of the 11th International Conference on Principles of Knowledge Representation and Reasoning (KR-08)*, pages 49–58. AAAI Press, 2008.

[2] Sunil Simon. *A logical study of strategies in games*. PhD thesis, IMSc, Chennai, 2010.

Refuting Random 3CNF Formulas in Propositional Logic[1]

SEBASTIAN MÜLLER AND IDDO TZAMERET

ABSTRACT. Random 3CNF formulas constitute an important distribution for measuring the average-case behavior of propositional proof systems. Lower bounds for random 3CNF refutations in many propositional proof systems are known. Most notable are the exponential-size resolution refutation lower bounds for random 3CNF formulas with $\Omega(n^{1.5-\varepsilon})$ clauses (Chvátal and Szemerédi [13], Ben-Sasson and Wigderson [9]). On the other hand, the only known non-trivial upper bound on the size of random 3CNF refutations in a non-abstract propositional proof system is for resolution with $\Omega(n^2/\log n)$ clauses, shown by Beame et al. [5]. In this paper we show that already standard propositional proof systems, within the hierarchy of Frege proofs, admit short refutations for random 3CNF formulas, for sufficiently large clause-to-variable ratio. Specifically, we demonstrate polynomial-size propositional refutations whose lines are TC^0 formulas (i.e., TC^0-Frege proofs) for random 3CNF formulas with n variables and $\Omega(n^{1.4})$ clauses. The idea is based on demonstrating efficient propositional correctness proofs of the random 3CNF unsatisfiability witnesses given by Feige, Kim and Ofek [19]. Since the soundness of these witnesses is verified using spectral techniques, we develop an appropriate way to reason about eigenvectors in propositional systems. To carry out the full argument we work inside weak formal systems of arithmetic and use a general translation scheme to propositional proofs.

1 Introduction

This work deals with the average complexity of propositional proofs. Our aim is to show that standard propositional proof systems, within the hierarchy of Frege proofs, admit short random 3CNF refutations for a sufficiently large clause-to-variable ratio, and also can outperform resolution for random 3CNF formulas in this ratio. Specifically, we show that most 3CNF formulas with n variables and at least $cn^{1.4}$ clauses, for a sufficiently large constant c, have polynomial-size in n propositional refutations whose proof-lines are constant depth circuits with threshold gates (namely, TC^0-Frege proofs). This is in contrast to resolution (that

---

[1]This summary is based on the paper "Short Propositional Refutations for Dense Random 3CNF Formulas" appearing in proceedings of LICS 2012.

can be viewed as depth-1 Frege) for which it is known that most 3CNF formulas with at most $n^{1.5-\epsilon}$ clauses (for $0 < \epsilon < \frac{1}{2}$) do not admit sub-exponential refutations [13, 9]. The main technical contribution of this paper is a propositional characterization of the random 3CNF unsatisfiability witnesses given by Feige at al. [19]. In particular we show how to carry out certain spectral arguments inside weak propositional proof systems such as TC^0-Frege. The latter should hopefully be useful in further propositional formalizations of spectral arguments. This also places a stream of recent results on efficient refutation algorithms using spectral arguments—beginning in the work of Goerdt and Krivelevich [23] and culminating in Feige et al. [19]—within the framework of propositional proof complexity. Loosely speaking, we show that all these refutation algorithms and witnesses, considered from the perspective of propositional proof complexity, are not stronger than TC^0-Frege.

Background in proof complexity. Propositional proof complexity is the systematic study of the efficiency of proof systems establishing propositional tautologies (or dually, refuting unsatisfiable formulas). *Abstractly* one can view a propositional proof system as a deterministic polynomial-time algorithm A that receives a string π ("the proof") and a propositional formula Φ such that there exists a π with $A(\pi, \Phi) = 1$ iff Φ is a tautology. Such an A is called an *abstract proof system* or a *Cook-Reckhow proof system* due to [17]. Nevertheless, most research in proof complexity is dedicated to more concrete or structured models, in which proofs are sequences of lines, and each line is derived from previous lines by "local" and sound rules. Perhaps the most studied family of propositional proof systems are those coming from propositional logic, under the name Frege systems, and their fragments (and extensions). In this setting, proofs are written as sequences of Boolean formulas (proof-lines) where each line is either an axiom or was derived from previous lines by means of simple sound derivation rules. The *complexity* of a proof is just the number of symbols it contains, that is, the total size of formulas in it. And different proof systems are compared via the concept of *polynomial simulation*: a proof system P polynomially-simulates another proof system Q if there is a polynomial-time computable function f that maps Q-proofs to P-proofs of the same tautologies. The definition of Frege systems is sufficiently robust, in the sense that different formalizations can polynomially-simulate each other [38]. It is common to consider fragments (or extensions) of Frege proof systems induced by restricting the proof-lines to contain presumably weaker (or stronger) circuit classes than Boolean formulas. This stratification of Frege proof systems is thus analogous to that of Boolean circuit classes: Frege proofs consist of Boolean formulas (i.e., $\mathbf{NC}^1$) as proof-lines, TC^0-Frege (also known as Threshold Logic) consists of TC^0 proof-lines, Bounded Depth Frege has $\mathbf{AC}^0$ proof-lines, depth-d Frege has circuits of depth-d proof-lines, etc. In this framework, the resolution system can be viewed as *depth*-1 *Frege*. Similarly, one usually considers extensions of the Frege system such as $\mathbf{NC}^i$-Frege, for $i > 1$, and $\mathbf{P/poly}$-Frege (the latter is polynomially equivalent to the known Extended Frege system, as shown by Jeřábek [29]). Restrictions (and extensions) of Frege proof systems form a hierarchy with respect to polynomial-simulations, though it is open whether the hierarchy is proper. It thus constitutes one of the main goals of proof complexity to understand the above hierarchy of Frege systems, and to separate different propositional proof systems, that is, to show that one proof system does not polynomially simulate another proof system. These questions also relate in a certain sense to the hierarchy of Boolean circuits (from $\mathbf{AC}^0$, through, $\mathbf{AC}^0[p]$, $\mathbf{TC}^0$, $\mathbf{NC}^1$, and so forth; see [15]). Many separations between propositional proof systems (not just in the Frege hierarchy) are known. In the case of Frege proofs there are already known separations between certain fragments of it (e.g., separation of depth-d Frege from depth $d + 1$ Frege was shown by Krajíček [30]). It is also known that $\boldsymbol{TC}^0$-Frege is strictly stronger than both resolution and bounded depth Frege proof systems, since, e.g., $\boldsymbol{TC}^0$-Frege admits polynomial-size proofs of the propositional pigeonhole principle, while resolution and bounded depth Frege do not (see [26] for the resolution lower bound, [1] for the bounded depth Frege lower bound and [16] for the corresponding $\boldsymbol{TC}^0$-Frege upper bound). *Average-case proof complexity via the random 3CNF model.* Much as in algorithmic research, it is important to know the average-case complexity of propositional proof systems, and not just their worst-case behavior. To this end one usually considers the model of random 3CNF formulas, where m clauses with three literals each, out of all possible $2^3 \cdot \binom{n}{3}$ clauses with n variables, are chosen independently, with repetitions (however, other possible distributions have also been considered in the literature). When m is greater than cn for some sufficiently large c (say, $c = 5$), it is known that with high probability a random 3CNF is unsatisfiable. (As m gets larger the task of refuting the 3CNF becomes easier since we have more constraints to use.) In average-case analysis of proofs we investigate whether such unsatisfiable random 3CNFs also have short (polynomial-size) refutations in a given proof system. The importance of average-case analysis of proof systems is that it gives us a better understanding of the complexity of a system than merely the worst-case analysis. For example, if we separate two proof systems in the average case—i.e., show that for almost all 3CNF one proof system admits polynomial-size refutations, while the other system does not—we establish a stronger separation. Until now only weak proof systems like resolution and Res(k) (for $k \leq \sqrt{\log n / \log \log n}$; the latter system introduced in [32] is an extension of resolution that operates with kDNF formulas) and polynomial calculus (and an extension of it) were analyzed in the random 3CNF model; for these systems exponential lower bounds are known for random 3CNFs (with varying number of clauses) [13, 5, 9, 4, 39, 2, 8, 3, 22]. For random 3CNFs with n variables and $n^{1.5-\epsilon}$ ($0 < \epsilon < \frac{1}{2}$) clauses it is known that there are no sub-exponential size resolution refutations [9]. For many proof systems, like cutting planes (CP) and bounded depth Frege ($\mathbf{AC}^0$-Frege), it is a major open problem to prove random 3CNF lower bounds (even for number of clauses near the threshold of unsatisfiability, e.g., random 3CNFs with n variables and $5n$ clauses). The results mentioned above only concerned lower bounds. On the other hand, to the best of our knowledge, the only known non-trivial polynomial-size *upper bound* on random kCNFs refutations in any non-abstract propositional proof system is for resolution. This is a result of Beame et al. [5], and it applies for fairly large number of clauses (specifically, $\Omega(n^{k-1}/\log n)$). *Efficient refutation algorithms.* A different kind of results on refuting random kCNFs were investigated in Goerdt and Krivelevich [23] and subsequent works by Goerdt and Lanka [24], Friedman, Goerdt and Krivelevich [21], Feige and Ofek [20] and Feige [18]. Here, one studies efficient refutation *algorithms* for kCNFs. Specifically, an *efficient refutation algorithm* receives a kCNF (above the unsatisfiability threshold) and outputs either "unsatisfiable" or "don't know"; if the algorithm answers "unsatisfiable" then the kCNF is required to be indeed unsatisfiable; also, the algorithm should output "unsatisfiable" with high probability (which by definition, is also the correct answer). Such refutation algorithms can be viewed as *abstract* proof systems having short proofs on the average-case: $A(\Phi)$ is a deterministic polytime machine whose input is only kCNFs (we can think of the proposed proof π input as being always the empty string). On input Φ the machine A runs the refutation algorithm and answers 1 iff the refutation algorithm answers "unsatisfiable"; otherwise, A can decide, e.g. by brute-force search, whether Φ is unsatisfiable or not. (In a similar manner, if the original efficient refutation algorithm is *non-deterministic* then we also get an abstract proof system for kCNFs; now the proof π that A receives is the description of an accepting run of the refutation algorithm.) Goerdt and Krivelevich [23] initiated the use of *spectral methods* to devise efficient algorithms for refuting kCNFs. The idea is that a kCNF with n variables can be associated with a graph on n vertices (or directly with a certain matrix). It is possible to show that certain properties of the associated graph witness the unsatisfiability of the original kCNF. One then uses a spectral method to give evidence for the desired graph property, and hence to witness the unsatisfiability of the original kCNF. Now, if we consider a random kCNF then the associated graph essentially becomes random too, and so one may show that the appropriate property witnessing the unsatisfiability of the kCNF occurs with high probability in the graph. The best (with respect to number of clauses) refutation algorithms devised in this way work for 3CNFs with at least $\Omega(n^{1.5})$ clauses [20]. Continuing this line of research, Feige, Kim and Ofek [19] considered efficient *non-deterministic* refutation algorithms (in other words, efficient *witnesses* for unsatisfiability of 3CNFs). They established the currently best (with respect to the number of clauses) efficient, alas non-deterministic, refutation procedure: they showed that with probability converging to 1 a random 3CNF with n variables and at least $cn^{1.4}$ clauses has a polynomial-size witness, for sufficiently big constant c. The result in the current paper shows that all the above refutation algorithms, viewed as abstract proof systems, *are not stronger (on average) than TC^0-Frege*. The short TC^0-Frege refutations will be based on the witnesses from [19], and so the refutations hold for the same clause-to-variable ratio as in that paper.

Our result. The main result of this paper is a polynomial-size upper bound on random 3CNF formulas refutations in a proof system operating with constant-depth threshold circuits (known as Threshold Logic or TC^0-Frege; see e.g., [10] for a definition). Since Frege and Extended Frege proof systems polynomially simulate TC^0-Frege proofs, the upper bound holds for these proof systems as well. (The actual formulation of TC^0-Frege is not important since different formulations, given in [12, 34, 10, 37, 16], polynomially simulate each other.)

THEOREM 1. *With probability $1-o(1)$ a random 3CNF formula with n variables and $cn^{1.4}$ clauses (for a sufficiently large constant c) has polynomial-size TC^0-Frege refutations.*

Beame et al. [5] and Ben-Sasson and Wigderson [9] showed that with probability $1 - o(1)$ resolution does not admit sub-exponential refutations for random 3CNF formulas when the number of clauses is at most $n^{1.5-\epsilon}$, for any constant $0 < \epsilon < 1/2$. Therefore, Theorem 1 shows that TC^0-Frege has an exponential speed-up over resolution for random 3CNFs with at least $cn^{1.4}$ clauses (when the number of clauses does not exceed $n^{1.5-\epsilon}$, for $0 < \epsilon < 1/2$). The main result contributes to our understanding (and possibly to the development of) refutation algorithms, by giving an explicit logical characterization of the Feige et al. [19] witnesses. This places a stream of recent results on refutation algorithms using spectral methods, beginning in Goerdt and Krivelevich [23], in the propositional proof complexity setting (showing essentially that these algorithms can be carried out already in TC^0-Frege).

Overview of the argument. Here we outline the proof of the main theorem (Theorem 1). We need to construct certain TC^0-Frege proofs. For this purpose we use the theory VTC^0 introduced in [37] (cf. [16]): any proof of a Σ_0^B formulas in VTC^0 can be translated into polynomial-size TC^0-Frege proofs. Our construction consists of the following steps:

I. Formalize the following statement as a first-order formula:

$$\forall \text{ assignment } A \ \Big(\mathbf{C} \text{ is a 3CNF and } w \text{ is its FKO unsatisfiability witness} \longrightarrow \atop \text{exists a clause } C_i \text{ in } \mathbf{C} \text{ such that } C_i(A) = 0 \Big), \qquad (1)$$

where an *FKO witness* is a suitable formalization of the unsatisfiability witness defined by Feige, Kim and Ofek [19]. The corresponding predicate is called *the FKO predicate*.

II. Prove formula (1) in the theory VTC^0.

III. Translate the proof in Step **II** into a family of propositional TC^0-Frege proofs (of the family of propositional translations of (1)). By a known result from [16], this will be a polynomial-size propositional proof (in the size of **C**). The translation of (1) will consist of a family of propositional formulas of the form:

$$[\![\text{C is a 3CNF and } w \text{ is its FKO unsatisfiability witness}]\!] \longrightarrow$$
$$[\![\text{exists a clause } C_i \text{ in } \textbf{C} \text{ such that } C_i(A) = 0]\!], \tag{2}$$

where $[\![\cdot]\!]$ denotes the mapping from first-order formulas to families of propositional formulas. By the nature of the propositional translation (second-sort) variables in the original first-order formula translate into a collection of propositional variables. Thus, (2) will consist of propositional variables derived from the variables in (1).

IV. For the next step we first notice the following two facts:

(i) Assume that $\underline{\textbf{C}}$ is a random 3CNF with n variables and $cn^{1.4}$ clauses (for a sufficiently large constant c). By [19], with high probability there exists an FKO unsatisfiability witness $\underline{w}$ for $\underline{\textbf{C}}$. Both $\underline{w}$ and $\underline{\textbf{C}}$ can be encoded as finite sets of numbers, as required by the predicate for 3CNF and the FKO predicate in (1). Let us identify $\underline{w}$ and $\underline{\textbf{C}}$ with their encodings. Then, assuming (1) was formalized correctly, assigning $\underline{w}$ and $\underline{\textbf{C}}$ to (1) satisfies the *premise* of the implication in (1).

(ii) Now, by the definition of the translation from first-order formulas to propositional formulas, if an object α satisfies the predicate $P(X)$ (i.e., $P(\alpha)$ is true in the standard model), then there is a propositional assignment of $0, 1$ values that satisfies the propositional translation of $P(X)$. Thus, by Item (i) above, there exists an $0, 1$ assignment ζ that satisfies the premise of (2) (i.e., the propositional translation of the premise of the implication in (1)).

In the current step we show that after assigning ζ to the conclusion of (2) (i.e., to the propositional translation of the conclusion in (1)) one obtains precisely $\neg\underline{\textbf{C}}$ (formally, a renaming of $\neg\underline{\textbf{C}}$, where $\neg\underline{\textbf{C}}$ is the 3DNF obtained by negating $\underline{\textbf{C}}$ and using the de Morgan laws). **V.** Take the propositional proof obtained in (**III**), and apply the assignment ζ to it. The proof then becomes a polynomial-size TC^0-Frege proof of a formula $\phi \to \neg\underline{\textbf{C}}$, where ϕ is a propositional sentence (without variables) logically equivalent to TRUE (because ζ satisfies it, by (**IV**)). From this, one can easily obtain a polynomial-size TC^0-Frege refutation of $\underline{\textbf{C}}$ (or equivalently, a proof of $\neg\underline{\textbf{C}}$).

The bulk of our work lies in (**I**) and especially in (**II**). We need to formalize the necessary properties used in proving the correctness of the FKO witnesses and show that the correctness argument can be carried out in the weak theory. There are two main obstacles in this process. The first obstacle is that the correctness of the witness originally is proved using spectral methods, which assumes that eigenvalues and eigenvectors are over the *reals*; whereas the reals are not defined in our weak theory. The second obstacle is that one needs to prove the correctness of the witness, and in particular the part related to the spectral method, *constructively* (formally in our case, inside VTC^0). Specifically, linear algebra is not known to be (computationally) in TC^0, and (proof-complexity-wise) it is conjectured that TC^0-Frege do not admit short proofs of the statements of linear algebra, such as statements relating to inverse matrices and the determinant properties (see [40, 28]). The first obstacle is solved using rational approximations of sufficient accuracy, and showing how to carry out the proof in the theory with such approximations. The second obstacle is solved basically by constructing the argument in a way that exploits non-determinism (i.e., in a way that enables supplying additional witnesses for the properties needed to prove the correctness of the original witness; e.g, all eigenvectors and all eigenvalues of the appropriate matrices in the original witness).

Acknowledgements   Sebastian Müller is supported by the Marie Curie Initial Training Network in Mathematical Logic - MALOA - From MAthematical LOgic to Applications, PITN-GA-2009-238381. Iddo Tzameret is supported in part by the National Basic Research Program of China Grant 2011CBA00300, 2011CBA00301, the National Natural Science Foundation of P. R. China; Grant 61033001, 61061130540, 61073174.

BIBLIOGRAPHY

[1] Miklós Ajtai. The complexity of the pigeonhole principle. In *Proc. 29th Ann. IEEE Symp. Found. Comp.*, 346–355, 1988.

[2] Michael Alekhnovich. Lower bounds for k-DNF resolution on random 3-CNFs. In *Proceedings of the Annual ACM Symposium on the Theory of Computing*, 251–256, 2005.

[3] Michael Alekhnovich and Alexander A. Razborov. Lower bounds for polynomial calculus: non-binomial case. In *Proc. 42nd Ann. IEEE Symp. Found. Comp.*, 190–199, 2001.

[4] A. Atserias, Maria Luisa Bonet, and J. Esteban. Lower bounds for the weak pigeonhole principle and random formulas beyond resolution. *Inform. and Comp.*, 176:152–136, 2002.

[5] Paul Beame, Richard Karp, Toniann Pitassi, and Michael Saks. The efficiency of resolution and Davis-Putnam procedures. *SIAM J. Comput.*, 31(4):1048–1075, 2002.

[6] Eli Ben-Sasson. *Expansion in Proof Complexity*. PhD thesis, Hebrew University, Jerusalem, Israel, September 2001.

[7] Eli Ben-Sasson and Yonatan Bilu. A gap in average proof complexity. *Elec. Colloq. Comput. Complexity (ECCC)*, 2002. TR02-003.

[8] Eli Ben-Sasson and Russell Impagliazzo. Random CNF's are hard for the polynomial calculus. *Comput. Complexity*, pages 1–19, 2010.

[9] Eli Ben-Sasson and Avi Wigderson. Short proofs are narrow—resolution made simple. *J. ACM*, 48(2):149–169, 2001.

[10] Maria Luisa Bonet, Toniann Pitassi, and Ran Raz. On interpolation and automatization for Frege systems. *SIAM J. Comput.*, 29(6):1939–1967, 2000.

[11] Samuel R. Buss. *Bounded Arithmetic*, volume 3 of *Studies in Proof Theory*. Bibliopolis, 1986.

[12] Samuel R. Buss and Peter Clote. Cutting planes, connectivity, and threshold logic. *Arch. Math. Logic*, 35(1):33–62, 1996.

[13] Vašek Chvátal and Endre Szemerédi. Many hard examples for resolution. *J. Assoc. Comput. Mach.*, 35(4):759–768, 1988.

[14] Matthew Clegg, Jeffery Edmonds, and Russell Impagliazzo. Using the Groebner basis algorithm to find proofs of unsatisfiability. In *Proceedings of the 28th Annual ACM Symposium on the Theory of Computing* pages 174–183, 1996.

[15] Stephen Cook. *Theories for Complexity Classes and Their Propositional Translations*, pages 175–227. Complexity of computations and proofs, Jan Krajíček, ed. Quaderni di Matematica, 2005.

[16] Stephen Cook and Phuong Nguyen. *Logical Foundations of Proof Complexity*. ASL Perspectives in Logic. Cambridge, 2010.

[17] Stephen A. Cook and Robert A. Reckhow. The relative efficiency of propositional proof systems. *J. Symbolic Logic*, 44(1):36–50, 1979.

[18] Uriel Feige. Refuting smoothed 3CNF formulas. In *Proc. 48th Ann. IEEE Symp. Found. Comp.*, 407–417, 2007.

[19] Uriel Feige, Jeong Han Kim, and Eran Ofek. Witnesses for non-satisfiability of dense random 3CNF formulas. In *Proc. 47th Ann. IEEE Symp. Found. Comp.*, 2006.

[20] Uriel Feige and Eran Ofek. Easily refutable subformulas of large random 3CNF formulas. *Theory of Computing*, 3(1):25–43, 2007.

[21] Joel Friedman, Andreas Goerdt, and Michael Krivelevich. Recognizing more unsatisfiable random k-SAT instances efficiently. *SIAM J. Comput.*, 35(2):408–430, 2005.

[22] Nicola Galesi and Massimo Lauria. Optimality of size-degree trade-offs for polynomial calculus. *ACM Trans. Comput. Log.*, 12(1), 2011.

[23] A. Goerdt and M. Krivelevich. Efficient recognition of random unsatisfiable k-SAT instances by spectral methods. In *Annual Symposium on Theoretical Aspects of Computer Science*, pages 294–304, 2001.

[24] Andreas Goerdt and André Lanka. Recognizing more random unsatisfiable 3-SAT instances efficiently. *Electronic Notes in Discrete Mathematics*, 16:21–46, 2003.

[25] P. Hajek and P. Pudlak. *Metamathematics of First-order Arithmetic*. Perspectives in Mathematical Logic. Springer-Verlag, 1993.

[26] Armin Haken. The intractability of resolution. *Theoret. Comput. Sci.*, 39(2-3):297–308, 1985.

[27] Roger A. Horn and Charles R. Johnson. *Matrix Analysis*. Cambridge University Press, 1985.

[28] Pavel Hrubeš and Iddo Tzameret. Short proofs for the determinant identities. In *44th ACM Symp. on Theory of Computing*, 2012.

[29] Emil Jeřábek. Dual weak pigeonhole principle, Boolean complexity, and derandomization. *Ann. Pure Appl. Logic*, 129(1-3):1–37, 2004.

[30] Jan Krajíček. Lower bounds to the size of constant-depth propositional proofs. *J. Symbolic Logic*, 59(1):73–86, 1994.

[31] Jan Krajíček. *Bounded arithmetic, propositional logic, and complexity theory*, volume 60 of *Encyclopedia of Mathematics and its Applications*. Cambridge University Press, 1995.

[32] Jan Krajíček. On the weak pigeonhole principle. *Fund. Math.*, 170(1-2):123–140, 2001.

[33] Jan Krajíček. A proof complexity generator. *Proc. 13th Intern. Congress of Logic, Methodology and Philosophy of Science, 2007*, Stud. Logic and the Found. Math. 2009.

[34] Alexis Maciel and Toniann Pitassi. On $\mathrm{ACC}^0[p^k]$ Frege proofs. In *Proceedings of the Annual ACM Symposium on the Theory of Computing 1997*, pages 720–729, 1999.

[35] Sebastian Müller and Iddo Tzameret. Short Propositional Refutations for Dense Random 3CNF Formulas *Elec. Colloq. Comput. Complexity (ECCC)*, 2011. TR11-006.

[37] Phuong Nguyen and Stephen A. Cook. Theories for TC^0 and other small complexity classes. *Log. Meth.in Comp. Sci.*, 2(1), 2006.

[38] Robert Reckhow. *On the lengths of proofs in the propositional calculus*. PhD thesis, University of Toronto, 1976.

[39] Nathan Segerlind, Sam Buss, and Russell Impagliazzo. A switching lemma for small restrictions and lower bounds for k-DNF resolution. *SIAM J. Comput.*, 33(5):1171–1200, 2004.

[40] Michael Soltys and Stephen Cook. The proof complexity of linear algebra. *Ann. Pure Appl. Logic*, 130(1-3):277–323, 2004.

Comments on Müller and Tzameret

HANS DE NIVELLE

1 Introduction

The general topic of the paper is the study of average length of proofs of propositional formulas in various calculi. Quite a few lower bounds are known, but not so many upper bounds. The current paper improves this situation by giving a polynomial average upper bound for 3CNF formulas with clause to variable ratio of $\Omega(n^{1.4}/n)$ when using a Frege proof system. It should be noted that these formulas are above the satisfiability threshold, which implies that they are nearly always unsatisfiable.

The paper is well-written. I liked reading the paper. There are no problems with English grammar, and technical notions are well-explained. The accompanying seven page introduction is excellent. Not being an expert in the area of the paper, I was still able to understand it, which is a fact that I attribute to the authors.

On the practical side, I would wish to see more connection to (or at least awareness of) modern developments in SAT-solving. Between 1995 and 2005, there has been a truly spectacular progress in practical algorithms and calculi for SAT-solving. Propositional logic is now routinely used in planning, verification of Boolean circuits, finite model searching, and theorem proving with data structures. It is a pity that the authors ignore this, and that the only reference related to practical SAT-solving is [2], which is from 1962. It would be good to include more modern references, see e.g. [1] for an overview.

I liked the system TC^0-Frege that is given on page 11 of the paper, and also system VTC^0 and the formalizations that are made in it.

2 General Questions

- Is there a nice way to express XOR (or $\leftrightarrow$) using the threshold operator of TC^0-Frege? (I mean XOR with arity > 2, like in $x_1 \oplus x_2 \oplus \cdots \oplus x_n$)? XOR occurs quite a lot in practice, e.g. in addition circuits, and standard satisfiability methods seem to have no good way of handling it.

- I understand that TC^0-Frege is good at concisely expressing pigeon hole principles and proving them. Are there other classes of formulas that are naturally expressed and proven in TC^0, or in other Frege systems?

Do there exist formulas that do not contain the threshold operators $\text{Th}_i(A_1, \ldots, A_n)$, not counting instances with $i = n$ or $i = 1$, for which TC^0-Frege has shorter proofs than resolution?

Or asked in a different way: Is it necessary to use the added threshold operators in the goal in order to be able to enjoy the advantage of short proofs of TC^0-Frege, or can it also be shorter on formulas containing only standard operators?

- I understand that the method of finding short proofs is based on the translation of Feige-Kim-Ofek unsatisfiability witnesses into the system VTC^0, and proving in this calculus that the existence of FKO witness implies unsatisfiability of the original formula.

  In a second stage, the VTC^0 proof is translated into propositional logic, and it is shown that there is a short proof in TC^0-Frege for the translated formula.

  Can the second translation be used as a general method for proving formulas in VTC^0 by translation to propositional logic? Would it be effective? I ask this, because such approaches are being used in practice.

- What could people working in practical theorem proving learn from your ressearch area? Are there calculi that they should look at, or additional axioms that they should use?

  Current-generation theorem provers are based on a combination of backtrack search and resolution. (See [1], Chapter 3,4,5) This is a description of the standard DPLL algorithm with learning:

  1. If all variables have an assignment, then report the satisfying interpretation.

     Otherwise, guess a truth value for an unassigned variable.

  2. Apply unit resolution as much as possible.

  3. If step 2 results in a conflict, then derive a new clause by tracing back the unit resolutions to the last guess. This guess has now become refutable by unit resolution, which means that its opposite is now derivable by unit resolution. Proceed to step 2.

     If step 2 did not result in a conflict, then go back to step 1.

  Since this method constructs resolution proofs, it is clear that its efficiency is bounded by the efficiency of resolution. Are there any other reasoning principles, or axioms that could be included in this calculus to make it more efficient?

- Could you comment on how meaningful average case complexity is as a tool for predicting proof size for propositional logic? The predictive content of the average depends very much on standard deviation, which I think is rather high for propositional logic. Are there attempts to estimate standard deviation or other indicators for the probability distribution?

- Can you explain where the number 1.4 comes from? Is it $\sqrt{2}$, or does it have different origine?

3 Review-Style Remarks

I know this report is not a review, but there are still some review-like questions and comments that I would like to pose:

On page 4 of the accompanying introduction, near the bottom of the page, you write 'w is its FKO unsatisfiability witness.' Does C have only one witness, or can it have more? If it can have more than one, the sentence should be reformulated as 'w is an FKO unsatisfiability witness of C'

On page 5 of the accompanying introduction, 2nd line: What is in the family of propositional TC^0-Frege proofs? Are the members obtained from different assignments A? Please explain from which parameter the family originates.

There is a general problem with layout of relativized quantifications in the paper: In a quantification of form $\forall x \leq X \exists y \leq Y F(x, y)$, some separation must be put between X and and $\exists$, and also between Y and F.

BIBLIOGRAPHY

[1] Armin Biere, Marijn Heule, and Hans van Maaren en Toby Walsh. *Handbook of Satisfiability*. IOS Press, 2009.
[2] Martin Davis, George Logemann, and Donald Loveland. A machine program for theorem proving. *Communications of the ACM*, 5(7):394–397, 1962.

Comments on Müller and Tzameret

XINSHUN ZHAO

The paper under comment investigates proof complexity of TC^0-Frege system. The following is proved: For random 3CNF formulas with n variables and $cn^{1.4}$ clauses, with high probability they are unsatisfiable and have polynomial-size TC^0-Frege refutations.

This result, together with a former result saying that almost all 3CNF formulas with clause-variable ratio $cn^{1.5-\epsilon}$ have no polynomial-size resolution refutations, implies that resolution cannot p-simulate TC^0-Frege system. However, that resolution $<_p$ Frege system was first shown by Cook and Coullard in 1987.

I suggest to add somewhere a proposition like: the Satisfiability problem for 3CNF formulas with clause $O(n^d)$ remains NP-complete for all (or some) $d > 0$ (e.g. $d = 1.4$). This proposition will show the more significance of the result of the paper. (I believe, to study proof complexity of formulas for which SAT becomes in PTIME is of less importance).

In page 5, the authors discuss efficient refutation algorithms which are incomplete algorithm but with high probability output correct answer. It is well-known that there are polynomial-time (incomplete) algorithms which solve the satisfiability problem with high probability (i.e., it outputs "satisfiable" or "unknown", if answers "satisfiable" the formula is really satisfiable). However, for UNSAT, according to my intuitive feeling I guess there unlikely exists a polynomial-time algorithm which with high probability outputs correct answers (i.e., it outputs "unsatisfiable" or "unkown", and if it outputs "unsatisfiable" then the input formula is unsatisfiable). Even for subclasses for which UNSAT remains co-NP, such an algorithm might not exist. Based on this observation, I think it is important and interesting to look for classes of unsatisfiable formulas almost all of which have short refutations. That is, the result of this paper is interesting. However, in line 12 of page 15, the sentence "$A(\Phi)$ is a deterministic polytime machine $\cdots\cdots$" is confusing for me. This is because existence of poly-size refutation does not mean there is a poly-time algorithm computing such a refutation.

As my understanding, the main aim of proof complexity is to show there is no polynomial proof system for propositional logic. If succeed we will show NP$\neq$co-NP which is believed to be true. I believe that all the existing proof system have been proved non polynomial. The average proof complexity is relatively new for me. But I still believe that it is more interesting to have results like: for some class of formulas, almost all of unsatisfiable formulas in this class have no polynomial-size proof with respect to a fixed proof system.I want to ask the following question: does such kind of result imply some conjecture on complexity classes based on some random computational model. For example, Suppose SAT problem for 3CNF formulas with clause-variable ratio $O(n^{1.5})$ remains NP complete, does the result of Beame, Karp et al imply NP=RP? I guess the answer is likely no. Then what is the main motivation to study average proof complexity?

Finally, a little bit confusing: many times it says "a predicate or a formula holds" (my understanding: holds in N_2), e.g. Theorem 7.2, Corollary 7.3, but sometimes say "VTC^0 proves a formula", e.g. in Section 5. Or what is the exact role played by VTC^0?

Response to de Nivelle and Zhao

IDDO TZAMERET

1 Discussion

Q: What could people working in practical theorem proving learn from your research area? Are there calculi that they should look at, or additional axioms that they should use? Current-generation theorem provers are based on a combination of back- track search and resolution. (See [1], Chapter 3, 4, 5)

This is a description of the standard DPLL algorithm with learning:

1. If all variables have an assignment, then report the satisfying interpretation. Otherwise, guess a truth value for an unassigned variable.

2. Apply unit resolution as much as possible.

3. If step 2 results in a conflict, then derive a new clause by tracing back the unit resolutions to the last guess. This guess has now become refutable by unit resolution, which means that its opposite is now derivable by unit resolution. Proceed to step 2.

   If step 2 did not result in a conflict, then go back to step 1.

Since this method constructs resolution proofs, it is clear that its efficiency is bounded by the efficiency of resolution. Are there any other reasoning principles, or axioms that could be included in this calculus to make it more efficient?

A: It is not entirely unreasonable to view proof complexity as the theoretical foundation of efficient theorem proving (though, perhaps this should not be over-stressed, as proving concrete lower bounds on propositional proof size, with relation to the NP vs. $coNP$, is at the center of proof complexity). There are tight and well known connections between the resolution refutation system (and its refinements) and practical SAT-solving (via DPLL-based algorithms). However, as far as I know, not much is known to connect "strong" proof systems with the practice of efficient theorem proving. The informal qualification *strong* here means proof systems that no (unconditional) lower bound is known for them.

As for our result, let me provide two relevant points, the first explains what apparently we cannot deduce from our result about practical SAT-solving, and the second what we possibly can:

(i) Our result constitutes an upper bound on (a certain kind of) average-case tautologies for Threshold Logic. Threshold Logic, is quite a *strong* proof system relative to resolution (for instance, no super-polynomial lower bound on the size of Threshold Logic proofs is currently known, while many such lower bounds are known for resolution). The fact that the proof system is strong means also that to formulate a proof-search algorithm based on Threshold Logic (in a similar spirit that DPLL is a proof-search algorithm for tree-like resolution), would probably be highly non-trivial. Note that even to formulate practical SAT-solvers that follow (namely, search for) *weaker* propositional proof systems such as cutting planes, or weak extensions of resolution, like k-DNF resolution (for k a constant), is a apparently a difficult open problem in the SAT solving community.

(ii) The *proof-search question* is an active stream of research in proof complexity. In this respect, the central concept dealt with in the literature is that of *automatizability* of proof systems [3]: a propositional proof system P is *automatizable* if there is a polynomial p and an algorithm, that given a tautology T runs in time $p(s)$ and outputs a P-proof of T, where s is the minimal size P-proof of T (if the input T is not a tautology we do not require anything from the algorithm). For Threshold Logic (i.e., TC^0-Frege), it is already known from Bonet et al. [3] that this proof system does *not* admit an efficient proof-search algorithm, namely Threshold Logic is known not to be automatizable, based on certain assumptions from computational complexity. However, a recent subsequent result [8], shows that to obtain an efficient *deterministic refutation algorithm* for random 3CNF formulas (with the same clause density as our result), techniques from proof complexity can be useful. In particular, [8] shows that assuming we can prove the existence of short refutations of a certain combinatorial principle (formally, the *3XOR principle*) in a weak enough proof system, then we get an efficient deterministic algorithm that can verify the unsatisfiability of most 3CNF formulas (with n variables and $\Omega(n^{1.4})$ clauses). The paper [8] further shows a weak proof system that does admit short refutations of the 3XOR principle (though this proof system is not weak enough to obtain the desired deterministic efficient refutation algorithm). For more recent work on proof-search algorithms and its relation to SAT-solving, see for example [2].

Q: Is there a nice way to express XOR using the threshold operator of TC^0-Frege? (I mean XOR with arity > 2, like in $x_1 \oplus x_2 \oplus \ldots \oplus x_n$)? XOR occurs quite a lot in practice, e.g. in addition circuits, and standard satisfiability methods seem to have no good way of handling it.

A: Threshold gates can express easily the predicate $x_1 + \ldots + x_n = c$ for any natural number $0 \le c \le n$, as follows: $\mathrm{Th}_2(\mathrm{Th}_c(x_1, \ldots, x_n), \neg \mathrm{Th}_{c+1}(x_1, \ldots,$

x_n)). Thus, the XOR operator can be expressed as: $\bigwedge_{c=1,3,5,...} (x_1+...+x_n = c)$.

Q: Do there exist formulas that do not contain the threshold gates $\mathrm{Th}_i(A_1, ..., A_n)$, excluding instances with $i = n$ or $i = 1$, for which $\boldsymbol{TC^0}$-Frege has shorter proofs than resolution?

Or asked in a different way: Is it necessary to use the added threshold operators in the goal in order to be able to enjoy the advantage of short proofs of $\boldsymbol{TC^0}$-Frege, or can it also be shorter on formulas containing only standard operators?

A: The answers for the first question is yes, and for the second question is no. Every family of CNF formulas that have polynomial-size $\boldsymbol{TC^0}$-Frege refutations, but no polynomial-size resolution (or constant-depth Frege) refutations, constitutes such an example. Take for example the Pigeonhole principle tautologies that have polynomial-size $\boldsymbol{TC^0}$-Frege refutations but no polynomial-size resolution refutations [7]. The pigeonhole principle tautologies do not use threshold gates (excluding the threshold gates $\mathrm{Th}_i(x_1, ..., x_n)$ with $i = n$ or $i = 1$). Nevertheless, it should be noted that, the way to efficiently prove the pigeonhole principle is by reducing the original formulation (that does not use threshold gates) to a formulation that does use such gates.

Q: I understand that the method of finding short proofs is based on the translation of Feige-Kim-Ofek unsatisfiability witnesses into the system $\boldsymbol{VTC^0}$, and proving in this calculus that the existence of FKO witness implies unsatisfiability of the original formula.

In a second stage, the $\boldsymbol{VTC^0}$ proof is translated into propositional logic, and it is shown that there is a short proof in $\boldsymbol{TC^0}$-Frege for the translated formula.

Can the second stage, that of translating $\boldsymbol{VTC^0}$ proofs into propositional proofs be used as a general method for proving formulas in $\boldsymbol{VTC^0}$ by translation to propositional logic? Would it be effective? I ask this, because such approaches are being used in practice.

A: This sounds like an interesting idea to pursue. However, I see a simple obstacle that might hinder this approach. I will attempt to point some of these obstacles in what follows.

Let us restrict our discussion to two-sorted theories of arithmetic (as in [4]), and recall the correspondence between first-order (two sorted) theories and propositional logic. Let Σ_0^B be a formula $F(X)$ in the language of arithmetic, with X being a free string variables and where we assume for the sake of simplicity that there are no other free string or number variables in F. Assume that $F(X)$ has no string quantifiers and all number quantifiers in $F(X)$ are bounded. Then one can associate to $F(X)$ a *family* of polynomial-size (in n) propositional formulas $\{F_n\}_{n=1}^{\infty}$, such that this family consists of only tautologies iff $F(X)$ is a true formula (that is, $F(X)$ holds in the standard (two-sorted) model of arithmetic). By [4], if $\boldsymbol{VTC^0}$ proves $F(X)$ then there is a polynomial $p(n)$ such that for any n, there is a Threshold Logic proof of F_n whose size is at most $p(n)$. This means that if there *exists* a proof of $F(X)$ in $\boldsymbol{VTC}^0$ then the corresponding family $\{F_n\}_{n=1}^{\infty}$ of propositional tautologies have *polynomial-size* proofs in Threshold Logic.

Therefore, in order to establish that $F(X)$ is provable in the theory via the approach suggested in the question, we need to find an *infinite* family of proofs. This is obviously impossible to do. (We may try to use the same approach to find whether $F(n)$, for a constant n, is provable in the theory. But then $F(n)$ is a formula with no string variables, and so this case is not interesting in itself.)

Q: Could you comment on how meaningful average case complexity is as a tool for predicting proof size for propositional logic? The predictive content of the average depends very much on standard deviation, which I think is rather high for propositional logic. Are there attempts to estimate standard deviation or other indicators for the probability distribution?

A: If I understand correctly the question, you ask whether demonstrating short refutations for random 3CNF formulas (with some clause density) gives good evidence that most unsatisfiable propositional formulas have short refutation in a proof system.

This is a matter of debate, and is a question of an empirical nature (namely, it depends on what kind of CNF formulas are used as input in practice). A common answer is "no", because random 3CNF upper bounds (for a certain clause density) show only that most 3CNF formulas (in the prescribed clause density) admit short refutations, but it gives no indication on different formulas. It also assumes that the 3CNF formulas are distributed according to a certain (uniform) distribution, which might not be the case for industrial SAT-solving. On the other hand, with respect to the resolution refutation system, random CNF formulas were found useful in the practical development of SAT-solvers (as reported, e.g., in [1] Chapter 8; see also [6]).

Q: Can you explain where the number 1.4 comes from? Is it $\sqrt{2}$, or does it have different origin?

A: The number 1.4 comes entirely from Feige, Kim and Ofek combinatorial analysis. They managed to show that their short unsatisfiability witness exists with high probability for 3CNF formulas with n variables and $c \cdot n^{1.4}$ clauses, for c a sufficiently large constant. As far as I know, this is not known to be tight, that is, it might be possible that the same result also holds for $c \cdot n^b$ clauses, where $b < 1.4$.

Acknowledgements

Iddo Tzameret is supported in part by the National Basic Research Program of China Grant 2011CBA00300, 2011CBA00301, the National Natural Science Foundation of P. R. China; Grant 61033001, 61061130540, 61073174.

BIBLIOGRAPHY

[1] Armin Biere, Marijn Heule, Hans van Maaren and Toby Walsh (Eds.). *Handbook of Satisfiability*. Frontiers in Artificial Intelligence and Applications, IOS Press 2009.

[2] Albert Atserias. *The Proof-Search Problem between Bounded-Width Resolution and Bounded-Degree Semi-Algebraic Proofs*. in the Proceedings of 16th International Conference on Theory and Applications of Satisfiability Testing (SAT), Lecture Notes in Computer Science vol. 7962, pp. 1-17, Springer-Verlag, 2013.

[3] Maria Luisa Bonet, Toniann Pitassi, and Ran Raz. On interpolation and automatization for Frege systems. *SIAM J. Comput.*, 29(6):1939–1967, 2000.

[4] Stephen Cook and Phuong Nguyen. *Logical Foundations of Proof Complexity*. ASL Perspectives in Logic. Cambridge, 2010.

[5] Uriel Feige, Jeong Han Kim, and Eran Ofek. Witnesses for non-satisfiability of dense random 3CNF formulas. In *Proc. 47th Ann. IEEE Symp. Found. Comp.*, 2006.

[6] C. P. Gomes, B. Selman, N. Crato, and H. Kautz. *Heavy-tailed phenomena in satisfiability and constraint satisfaction problems*. J. Automat. Reason., 24 (1-2):67-100, 2000.

[7] Armin Haken. The intractability of resolution. *Theoret. Comput. Sci.*, 39(2-3):297–308, 1985.

[8] Iddo Tzameret. Sparser random 3CNF refutation algorithms and feasible interpolation. *Manuscript*, 2013.

Part III

Linguistics has long been a natural companion to logic, ever since Aristotle expounded the basic principles of grammar alongside with those of logic. These connections picked up depth and width in the 1950s and 1960s with the work of Montague, and led to the current area of logic-based grammars and logical semantics of natural language. But natural language is only one arena where human cognition plays out, and despite some initial barriers of mistrust or misunderstanding, contacts between logic and empirical cognitive science, too, have been coming to the fore in recent decades.

Gerhard Jäger (University of Tübingen) discusses the recent interface of game theory and natural language semantics in 'Interpretation games with variable costs.' Successful communication can be cast in terms of game-theoretic equilibria, in a tradition going back to Lewis' signaling games for communication, but applied by now to a wide variety of pragmatic phenomena. Jäger discusses a recent successful version of this approach in terms of the game-theoretic solution algorithm of iterated best response, and points out that it makes wrong predictions in cases where the cost of messages becomes a factor that needs to be taken into account. The result of the analysis is more delicate models for game-theoretic pragmatics. *Robert van Rooij* (University of Amsterdam) points out some ways in which game theory can deliver even more notions of relevance for natural language than Jäger shows, but also draws attention to the failure of game-theoretic approaches as they stand to account for the simple fact that signaling is useful and effective. *Yungcheng Zhou* (Tsinghua University) contrasts the game-theoretic approach to natural language communication with that of Grice and Stalnaker, and raises some questions about relative merits.

Congjun Yao and *Chongli Zou* (Chinese Academy of Social Sciences) show in 'Hybrid categorial type logics and the formal treatment of Chinese' how various constructions in the grammar of the Chinese language require sophisticated adaptations of current type-logical grammars. Rejecting an indiscriminate use of additional modalities, they show how various resource-conscious base logics can be employed together, once we give a mechanism for passing information between them, and choose suitable new categories maintaining descriptive flexibility. *Arno Bastenhof* (University of Utrecht) relates these proposals to the rich recent literature on categorial type logics and grammars, and points at alternative solutions, including assigning different lambda terms, that might be considered for Chinese. *Sylvain Pogodalla* (INRIA, Nancy) inquires about the more systematic logical properties of the systems proposed by Yao and Zou, and also asks for a more pregnant description of those grammatical peculiarities of Chinese that should serve as benchmarks for systems of logical grammar.

'An ontology-based approach to metaphor cognitive computation' by *Huaxin Huang, Xiaoxi Huang, Beishui Liao*, and *Cihua Xu* (Zhejiang University, Hangzhou) proposes a new approach to the ubiquitous cognitive phenomenon of under-standing metaphors. The approach combines the conceptual metaphor theory of Lakoff and Johnson and the conceptual blending theory by Fauconnier and Turner, and uses an approach via formal ontologies and description logics to build them into computational models of metaphorical reasoning. *Emiliano Lorini* (CNRS and IRIT, Toulouse) points out how metaphors function in direct contact with agents' knowledge and belief states, and asks how the model would accommodate these, thus relating to how meanings arise in a setting of theory of mind that also allows for the crucial role of emotion. *John Woods* (University of British Columbia, Vancouver) and *Peter Bruza* (Queensland University of Technology, Brisbane) discuss how slippery the empirical facts of metaphors can be that a theory needs to be measured against. Moreover, they raise concerns about how the proposed description logic approach would deal with the behavior of prototypes, and the subtle compositional effects of complex metaphorical expressions.

Dag Westerståhl (Stockholm University) looks at a perennial topic in the history of logic and language in 'Negation and quantification. A new look at the square of opposition'. He confronts this famous diagram of quantifiers, internal, and external negations with modern generalized quantifier theory, and shows how it occurs all across natural language, including many observations on specific categories of determiners. This approach enables us to see what is special and what is general about the original square, and its role in the natural logic of language users. *Fengkui Ju* (Beijing Normal University) asks what roles squares play exactly in understanding negation and quantification, and also draws attention to the famous problem of the 'missing corner', the lack across languages of one single word for the expression "not all". *Lawrence Moss* (Indiana University, Bloomington) points out another mathematical realization of the square of opposition, suggesting a different message about definability, and asks what such variants mean. He also wonders whether any structure like the square has occurred in other logic traditions, Indian or Chinese.

In 'Vague classes and a resolution of the Bald Man Paradox', *Beihai Zhou* (Peking University) and *Liying Zhang* (Central University of Finance and Economics, Beijing) propose a new solution to the classical Sorites Paradox, where vague predicates and apparently plausible logical principles generate the contradiction that everyone (or, no one) is bald. Unlike common approaches in terms of many truth values, their method involves the notion of an agent-dependent vague class, whose behavior is governed by a set of principles formulated entirely in first-order logic. *Frank Veltman* (University of Amsterdam) recalls the long history of the Sorites, and the crucial role in its logical diagnosis of the principle of equivalence of observationally indistinguishable objects. He worries whether the authors' solution is not too far from ordinary linguistic practice: perhaps, rules for using vague predicates need only work 'normally'. *Wen-fang Wang* (Yang-Ming University, Taipei) relates the methodology of the proposal to existing approaches to vagueness, and wonders about its perhaps limited descriptive scope. Are there always paradigmatic cases as the vague class account seems to assume?

Noah Goodman and *Dan Lassiter* (Stanford University) pursue an ambitious project in 'Probabilistic semantics and pragmatics: Uncertainty in language and thought', at the interface of cognitive science, philosophy, probability theory, and logic. They propose a rich probabilistic model that can account for many concrete features for natural language understanding through a mechanism of belief update in communication, that supports joint inferences of speakers and hearers. To deal with important aspects such as compositionality, a system of stochastic lambda calculus is used as a formal underpinning. *Shushan Cai* (Tsinghua University) places this approach in a historical context where semantics and pragmatics need to be analyzed together, in an empirical cognitive setting, and points out the relevance of de Saussure's famous theory of signs. *Thomas Icard* (Stanford University) inquires into the role of logic in the authors' enterprise: where do we find the fast simple inferences of natural logic, what about common knowledge in communication, and more generally, how does logic emerge at all on a probabilistic substratum?

Cristina Bicchieri (University of Pennsylvania, Philadelphia), *Azi Lev-On* (Ariel University), and *Alex Chavez* (University of Michigan, Ann Arbor) report on experiments investigating the functioning of trust between agents in 'The medium or the message? Communication relevance and richness in trust games'. Looking at the actual empirical role of content (relevant vs. irrelevant communication) and media-richness (face-to-face vs. computer-mediated communication), they find many interesting things, including trust being positively influenced by relevance, but also expectations predicting reciprocity between subjects better than trust itself does. *Guo Meiyun* (Southwest University, Chongqing) and *Jianying Cui* (Sun Yat-sen University, Guangzhou) ask what minimal notion of rationality might underlie face-to-face and computer-mediated communication, and point out possible links with classical work in game theory, as well as current studies of trust in dynamic logics of agency and games. *Jakub Szymanik* (University of Amsterdam) wonders whether the experimental findings might have a simpler explanation than the one suggested by the authors, makes a suggestion to this effect, and also asks for an explicit cognitive model of the agents engaged in the trust experiment.

Interpretation Games with Variable Costs[1]

GERHARD JÄGER

1 Interpretation games

Game theory has proved a versatile tool to model the Grice-style reasoning processes enabling interlocutors to figure out *what is meant* from *what is said* and contextual information. In this position statement, I will give a brief overview of the *Iterated Best Response* model of game-theoretic pragmatics, a particular incarnateion of this general program that has emerged within the past five years ([1, 2, 3, 4, 5, 6, 7]). In this brief communication, I will rely mainly on the version of the model as devoloped by Michael Franke ([1, 2]) and propose a modification regarding the treatment of message costs.

The contextual information is modeled as an *interpretation game*, a variant of signaling games in the sense of [9]. It consists of a set of types T, a finite set of of messages M, a set of actions A, a prior probability distribution $\mathbf{p}$ over T (with $\mathbf{p} \in \Delta^+(T)$, i.e. $\mathbf{p}$ assigns a strictly positive probability to each type), a cost function $\mathbf{c} \in M \mapsto [0, \infty)$, two utility functions $u_s, u_r \in T \times A \mapsto \mathbb{R}$, and an interpretation function $\| \cdot \| \in M \mapsto POW(T)$.

A history of a game is a sequence $\langle t, m, a \rangle \in T \times M \times A$. The payoff of sender and receiver are defined over histories as:

$$
\begin{aligned}
u_s(t, m, a) &= u_s(t, a) - \mathbf{c}(m) \\
u_r(t, m, a) &= u_r(t, a)
\end{aligned}
$$

A (behavioral) sender strategy is a stochastic function σ from types to messages, i.e. a function from types to probability distributions over messages. Likewise, a receiver strategy is a function ρ from messages to probability distributions over actions.

In this paper we will only consider games where $T = A$ and where both u_s and u_r are Kronecker delta functions, i.e.

$$
u_{s/r}(t, a) = \begin{cases} 1 & \text{if } t = a \\ 0 & \text{else} \end{cases}
$$

---

[1]This summary is based on the chapter "Game theory in semantics and pragmatics" submitted to *Semantics. An International Handbook of Natural Language Meaning.*

Let us consider a standard example, the emergence of a scalar implicature triggered by the determiner *some*.

(1)   a. Joe ate some of the cookies. (m_{some})

        b. Joe ate all of the cookies. (m_{all})

If it is understood by both interlocutors that the listener is interested in how many cookies Joe ate, answer (1a) triggers the implicature that he did not eat all of the cookies.

The corresponding interpretation game can concisely be represented as in Table 1: The first two rows represent the two types $t_{\exists\neg\forall}$ (where Joe ate some but not all

	m_{some}	m_{all}	**p**
$t_{\exists\neg\forall}$	1	0	$1/2$
$t_\forall$	1	1	$1/2$
c	0	0	

Table 1. Some-all game

of the cookies) and $t_\forall$ (where he ate all of them). The first two columns represent the messages ((1a) and (b) respectively). The Boolean entries in the upper left 2×2 sub-table gives the interpretation function. The last row gives the cost function **c** and the last column the prior probabilities **p**.

2   The IBR sequence

Iterated Best Response (IBR) is a protocol to select a strategy profile for such a game via iterated back-and-forth reasoning. The default strategy σ_0 of an honest sender maps each type t to a uniform distribution over all messages m with $t \in \|m\|$. Likewise, the default strategy ρ_0 of a trusting receiver maps each message m to a uniform distribution over all actions a with $a \in \|m\|$.

A rational sender who believes that the listener plays ρ_0 will use a strategy σ_1 that maximizes her expected utility given that belief. The same holds *mutatis mutandis* for a sophisticated receiver who beliefs that the sender plays σ_0. Such players are called *level-1*. A rational player believing that their partner is a level-1 player will play a best response to σ_1/ρ_1, making them level-2 players, etc.

The following definitions formalize this basic idea:

DEFINITION 1 (Best response).

$$BR_s(t, \rho) = \arg\max_{m'} \rho(t|m') - \mathbf{c}(m')$$

$$BR_r(m, \sigma) = \begin{cases} \arg\max_a \sigma(m|a)\mathbf{p}(a) & \text{if } \max_a \sigma(m|a) > 0 \\ \|m\| & \text{else} \end{cases}$$

DEFINITION 2 (IBR sequence).

$$\sigma_0(m|t) = \begin{cases} 1/|\{m'|t\in\|m'\|\}| & \text{if } t \in \|m\| \\ 0 & \text{else} \end{cases}$$

$$\rho_0(a|m) = \begin{cases} 1/|\{a'|a'\in\|m\|\}| & \text{if } a \in \|m\| \\ 0 & \text{else} \end{cases}$$

$$\sigma_{k+1}(m|t) = \begin{cases} 1/|BR_s(t,\rho_k)| & \text{if } m \in BR_s(t, \rho_k) \\ 0 & \text{else} \end{cases}$$

$$\rho_{k+1}(a|m) = \begin{cases} 1/|BR_r(m,\sigma_k)| & \text{if } a \in BR_r(m, \sigma_k) \\ 0 & \text{else} \end{cases}$$

The IBR sequence for the some-all game converges quickly:

$$\sigma_0 = \begin{array}{c} \\ t_{\exists\neg\forall} \\ t_\forall \end{array}\begin{array}{cc} m_{\text{some}} & m_{\text{all}} \\ \left(\begin{array}{cc} 1 & 0 \\ 1/2 & 1/2 \end{array}\right) \end{array} \qquad \rho_0 = \begin{array}{c} \\ m_{\text{some}} \\ m_{\text{all}} \end{array}\begin{array}{cc} t_{\exists\neg\forall} & t_\forall \\ \left(\begin{array}{cc} 1/2 & 1/2 \\ 0 & 1 \end{array}\right) \end{array}$$

$$\sigma_1 = \begin{array}{c} \\ t_{\exists\neg\forall} \\ t_\forall \end{array}\begin{array}{cc} m_{\text{some}} & m_{\text{all}} \\ \left(\begin{array}{cc} 1 & 0 \\ 0 & 1 \end{array}\right) \end{array} \qquad \rho_1 = \begin{array}{c} \\ m_{\text{some}} \\ m_{\text{all}} \end{array}\begin{array}{cc} t_{\exists\neg\forall} & t_\forall \\ \left(\begin{array}{cc} 1 & 0 \\ 0 & 1 \end{array}\right) \end{array}$$

$$\sigma_2 = \sigma_1 \qquad\qquad\qquad\qquad \rho_2 = \rho_1$$

In the fixed point (σ_1, ρ_1), m_{some} induces a scalar implicature, i.e. it is associated exclusively with the type $t_{\exists\neg\forall}$ by both players.

3 Problematic examples

While this model captures a sizeable number of pragmatic inferences correctly, there are a few examples where the IBR sequence does not converge to the intuitively correct equilibrium. Two such games are shown in Table 2.

The *some-but-not-all* game (see [1]) is an extension of the some-all game where a third message is admitted:

(2) Joe ate some but not all of the cookies. (m_{sbna})

$$
\begin{array}{c|ccc|c}
 & m_{\text{some}} & m_{\text{all}} & m_{\text{sbna}} & \mathbf{p} \\
\hline
t_{\exists\neg\forall} & 1 & 0 & 1 & 1/2 \\
t_{\forall} & 1 & 1 & 0 & 1/2 \\
\hline
\mathbf{c} & 0 & 0 & 0.1 &
\end{array}
\qquad
\begin{array}{c|ccc|c}
 & m_{\text{open}} & m_{\text{open-h}} & m_{\text{open-a}} & \mathbf{p} \\
\hline
t_{h} & 1 & 1 & 0 & 2/3 \\
t_{a} & 1 & 0 & 1 & 1/3 \\
\hline
\mathbf{c} & 0 & 0.1 & 0.1 &
\end{array}
$$

Table 2. Some-but-not-all game (left) and I-implicature game (right)

This message is only true in $t_{\exists\neg\forall}$, and it is more complex than the other two messages. This is captured by assigning it a small cost of 0.1. This leads to the IBR sequence:

$$
\sigma_0 = \begin{array}{c|ccc}
 & m_{\text{some}} & m_{\text{all}} & m_{\text{sbna}} \\
t_{\exists\neg\forall} & 1/2 & 0 & 1/2 \\
t_{\forall} & 1/2 & 1/2 & 0
\end{array}
\qquad
\rho_0 = \begin{array}{c|cc}
 & t_{\exists\neg\forall} & t_{\forall} \\
m_{\text{some}} & 1/2 & 1/2 \\
m_{\text{all}} & 0 & 1 \\
m_{\text{sbna}} & 1 & 0
\end{array}
$$

$$
\sigma_1 = \begin{array}{c|ccc}
 & m_{\text{some}} & m_{\text{all}} & m_{\text{sbna}} \\
t_{\exists\neg\forall} & 0 & 0 & 1 \\
t_{\forall} & 0 & 1 & 0
\end{array}
\qquad
\rho_1 = \begin{array}{c|cc}
 & t_{\exists\neg\forall} & t_{\forall} \\
m_{\text{some}} & 1/2 & 1/2 \\
m_{\text{all}} & 0 & 1 \\
m_{\text{sbna}} & 1 & 0
\end{array}
$$

$$
\sigma_2 = \sigma_1 \qquad\qquad\qquad \rho_2 = \rho_1
$$

In the fixed point, message m_{some} is never used, and it is interpreted according to its literal meaning. This outcome is odd because we would expect the scalar implicature not to be affected by the additional message.

The second game shown in Table 2 is inspired by the following example:

(3)   a. John opened the door. (m_{open})

          b. John opened the door with the handle. ($m_{\text{open-h}}$)

          c. John opened the door with an axe. ($m_{\text{open-a}}$)

While the literal meaning of (3a) is true both in a scenario where John opened the door with the handle (type t_h) and in a scenario where he uses an axe (t_a), we would normally interpret it in the sense of (3b). This is based on the facts that doors are typically opened rather with handles than with axes, and that the unspecific message (3a) is shorter than the more specific (3b). This type of inference is called *I-implicature* by [8].

The basic features of this example are captured by the assumptions that t_h has a higher prior probability than t_a and that m_{open} is less costly than $m_{\text{open-h}}$ or $m_{\text{open-a}}$.

The IBR sequence is

$$
\sigma_0 = \begin{array}{cc} & \begin{array}{ccc} m_{\text{open}} & m_{\text{open-h}} & m_{\text{open-a}} \end{array} \\ \begin{array}{c} t_h \\ t_a \end{array} & \left(\begin{array}{ccc} 1/2 & 1/2 & 0 \\ 1/2 & 0 & 1/2 \end{array} \right) \end{array}
\qquad
\rho_0 = \begin{array}{cc} & \begin{array}{cc} t_h & t_a \end{array} \\ \begin{array}{c} m_{\text{open}} \\ m_{\text{open-h}} \\ m_{\text{open-a}} \end{array} & \left(\begin{array}{cc} 1/2 & 1/2 \\ 1 & 0 \\ 0 & 1 \end{array} \right) \end{array}
$$

$$
\sigma_1 = \begin{array}{cc} & \begin{array}{ccc} m_{\text{open}} & m_{\text{open-h}} & m_{\text{open-a}} \end{array} \\ \begin{array}{c} t_h \\ t_a \end{array} & \left(\begin{array}{ccc} 0 & 1 & 0 \\ 0 & 0 & 1 \end{array} \right) \end{array}
\qquad
\rho_1 = \begin{array}{cc} & \begin{array}{cc} t_h & t_a \end{array} \\ \begin{array}{c} m_{\text{open}} \\ m_{\text{open-h}} \\ m_{\text{open-a}} \end{array} & \left(\begin{array}{cc} 1 & 0 \\ 1 & 0 \\ 0 & 1 \end{array} \right) \end{array}
$$

$$
\sigma_2 = \begin{array}{cc} & \begin{array}{ccc} m_{\text{open}} & m_{\text{open-h}} & m_{\text{open-a}} \end{array} \\ \begin{array}{c} t_h \\ t_a \end{array} & \left(\begin{array}{ccc} 1 & 0 & 0 \\ 0 & 0 & 1 \end{array} \right) \end{array}
\qquad
\rho_2 = \begin{array}{cc} & \begin{array}{cc} t_h & t_a \end{array} \\ \begin{array}{c} m_{\text{open}} \\ m_{\text{open-h}} \\ m_{\text{open-a}} \end{array} & \left(\begin{array}{cc} 1/2 & 1/2 \\ 1 & 0 \\ 0 & 1 \end{array} \right) \end{array}
$$

$$
\sigma_3 = \sigma_1 \qquad\qquad\qquad \rho_3 = \rho_1
$$

This sequence does not converge to a fixed point but keeps alternating between (σ_1, ρ_1) and (σ_2, ρ_2). The intuitively correct strategy profile would be (σ_2, ρ_1) though.

4 Variable costs

The reason for these wrong prediction is similar in both cases. When designing her best response to ρ_0 in the some-but-not-all game in type $t_{\exists\neg\forall}$, the sender has the choice between (a) a cheap message with induces a sub-optimal posterior probability for the correct type and (b) a costly message that does uniquely identify the correct type. As long as costs are nominal, the rational choice is the more specific costly message. If the sender would choose the cheap message with some small probability, the IBR sequence would converge to the correct fixed point. The same applies to the best response to ρ_0 in the I-implicature game.

So far we assumed that message costs are common knowledge. The mentioned problems can be remedied if the receiver does not know the sender's costs for sure in advance. To formalize this intuition, we generalize the definition of intepretation games given above in the following way: instead of a vector $\mathbf{c}$ of costs, the specification of a game contains a probability density function $\mathcal{C}$ over $\mathbf{c} \in [0, \infty)^{|M|}$, i.e. over cost vectors $\mathbf{c}$.

The cost density $\mathcal{C}$ is common knowledge between the players while the sender knows the exact cost vector. So technically speaking, t and $\mathbf{c}$ represent the sender's type, and $\mathbf{p}, \mathcal{C}$ jointly define the receiver's prior probability distribution over sender types. As t affects the receiver's payoff while $\mathbf{c}$ does not, it is conceptually justified to keep these two aspects of the game structure separate though.

Regarding the some-but-not-all game, let us assume that $\mathcal{C}$ has the following properties:

$$\int_0^\infty \mathcal{C}(0,0,x)dx = 1$$

$$\int_{1/2}^\infty \mathcal{C}(0,0,x)dx = \alpha > 0$$

In words, we assume that m_{some} and m_{all} are costless with probability 1, and that there is a positive probability α that $\mathbf{c}(m_{\text{sbna}}) \geq 1/2$.

With these assumptions, the IBR sequence now comes out as

$$\sigma_0 = \begin{array}{c} t_{\exists\neg\forall} \\ t_\forall \end{array} \begin{pmatrix} m_{\text{some}} & m_{\text{all}} & m_{\text{sbna}} \\ 1/2 & 0 & 1/2 \\ 1/2 & 1/2 & 0 \end{pmatrix} \qquad \rho_0 = \begin{array}{c} m_{\text{some}} \\ m_{\text{all}} \\ m_{\text{sbna}} \end{array} \begin{pmatrix} t_{\exists\neg\forall} & t_\forall \\ 1/2 & 1/2 \\ 0 & 1 \\ 1 & 0 \end{pmatrix}$$

$$\sigma_1 = \begin{array}{c} t_{\exists\neg\forall} \\ t_\forall \end{array} \begin{pmatrix} m_{\text{some}} & m_{\text{all}} & m_{\text{sbna}} \\ \alpha & 0 & 1-\alpha \\ 0 & 1 & 0 \end{pmatrix} \qquad \rho_1 = \begin{array}{c} m_{\text{some}} \\ m_{\text{all}} \\ m_{\text{sbna}} \end{array} \begin{pmatrix} t_{\exists\neg\forall} & t_\forall \\ 1/2 & 1/2 \\ 0 & 1 \\ 1 & 0 \end{pmatrix}$$

$$\sigma_2 = \begin{array}{c} t_{\exists\neg\forall} \\ t_\forall \end{array} \begin{pmatrix} m_{\text{some}} & m_{\text{all}} & m_{\text{sbna}} \\ \alpha & 0 & 1-\alpha \\ 0 & 1 & 0 \end{pmatrix} \qquad \rho_2 = \begin{array}{c} m_{\text{some}} \\ m_{\text{all}} \\ m_{\text{sbna}} \end{array} \begin{pmatrix} t_{\exists\neg\forall} & t_\forall \\ 1 & 0 \\ 0 & 1 \\ 1 & 0 \end{pmatrix}$$

$$\sigma_3 = \begin{array}{c} t_{\exists\neg\forall} \\ t_\forall \end{array} \begin{pmatrix} m_{\text{some}} & m_{\text{all}} & m_{\text{sbna}} \\ 1 & 0 & 0 \\ 0 & 1 & 0 \end{pmatrix} \qquad \rho_3 = \begin{array}{c} m_{\text{some}} \\ m_{\text{all}} \\ m_{\text{sbna}} \end{array} \begin{pmatrix} t_{\exists\neg\forall} & t_\forall \\ 1 & 0 \\ 0 & 1 \\ 1 & 0 \end{pmatrix}$$

$$\sigma_4 = \sigma_3 \qquad\qquad\qquad\qquad \rho_4 = \rho_3$$

For the I-implicature game to come out correctly, we demand that

$$\int_0^\infty \int_0^\infty \mathcal{C}(0,x,y)dxdy = 1$$

$$\int_0^\infty \int_{1/2}^\infty \mathcal{C}(0,x,y)dxdy = \alpha > 0$$

$$\int_{1/2}^\infty \int_0^\infty \mathcal{C}(0,x,y)dxdy = \beta$$

$$\alpha\mathbf{p}(t_a) > \beta\mathbf{p}(t_h)$$

This means that $\mathbf{c}(m_{\text{open}}) = 0$ with certainty, the marginal probability that $\mathbf{c}(m_{\text{open}-h}) > 1/2$ is a positive value α, and α is at least half as high as (or, more generally, by at least a factor $\mathbf{p}(t_h)/\mathbf{p}(t_a)$ higher than) the marginal probability that $\mathbf{c}(m_{\text{open}-a}) > 1/2$.

Under these assumptions, we arrive at the IBR sequence

$$\sigma_0 = \begin{array}{c} \\ t_h \\ t_a \end{array}\begin{array}{c} m_{\text{open}} \\ \begin{pmatrix} 1/2 \\ 1/2 \end{pmatrix} \end{array}\begin{array}{c} m_{\text{open}-h} \\ 1/2 \\ 0 \end{array}\begin{array}{c} m_{\text{open}-a} \\ 0 \\ 1/2 \end{array} \qquad \rho_0 = \begin{array}{c} \\ m_{\text{open}} \\ m_{\text{open}-h} \\ m_{\text{open}-a} \end{array}\begin{array}{cc} t_h & t_a \\ \begin{pmatrix} 1/2 & 1/2 \\ 1 & 0 \\ 0 & 1 \end{pmatrix} \end{array}$$

$$\sigma_1 = \begin{array}{c} \\ t_h \\ t_a \end{array}\begin{array}{c} m_{\text{open}} \\ \begin{pmatrix} \alpha \\ \beta \end{pmatrix} \end{array}\begin{array}{c} m_{\text{open}-h} \\ 1-\alpha \\ 0 \end{array}\begin{array}{c} m_{\text{open}-a} \\ 0 \\ 1-\beta \end{array} \qquad \rho_1 = \begin{array}{c} \\ m_{\text{open}} \\ m_{\text{open}-h} \\ m_{\text{open}-a} \end{array}\begin{array}{cc} t_h & t_a \\ \begin{pmatrix} 1 & 0 \\ 1 & 0 \\ 0 & 1 \end{pmatrix} \end{array}$$

$$\sigma_2 = \begin{array}{c} \\ t_h \\ t_a \end{array}\begin{array}{c} m_{\text{open}} \\ \begin{pmatrix} 1 \\ 0 \end{pmatrix} \end{array}\begin{array}{c} m_{\text{open}-h} \\ 0 \\ 0 \end{array}\begin{array}{c} m_{\text{open}-a} \\ 0 \\ 1 \end{array} \qquad \rho_2 = \begin{array}{c} \\ m_{\text{open}} \\ m_{\text{open}-h} \\ m_{\text{open}-a} \end{array}\begin{array}{cc} t_h & t_a \\ \begin{pmatrix} 1 & 0 \\ 1 & 0 \\ 0 & 1 \end{pmatrix} \end{array}$$

$$\sigma_3 = \sigma_2 \qquad\qquad\qquad\qquad \rho_3 = \rho_2$$

5   Summary of main claims

1. As spelled out in more detail in the work cited above, the IBR model of game theoretic pragmatics provides a comprehensive formalization of the neo-Gricean program. It correctly captures a wide range of pragmatic inferences. These include

   - scalar implicatures,
   - free choice readings,

- ignorance implicatures,
- Horn's division of pragmatic labor,
- the pragmatics of measure phrases, and
- putative embedded scalar implicatures.

2. In its standard formulation, the IBR model systematically leads to wrong predictions in scenarios where a cheap non-specific message competes with a more specific costly message. The common assumption of nominal costs implies that a rational sender will always opt for the costly messages in such a case. Assuming non-nominal costs would solve the problem but appears to be an *ad hoc* decision. In this brief paper I argued that the problem can be overcome if we give up the assumption that message costs are common knowledge. As long as the receiver cannot exclude with certainty that costs are non-nominal, IBR converges to the intuitively correct fixed point in the problematic examples.

BIBLIOGRAPHY

[1] Franke, M. (2009). *Signal to Act: Game Theory in Pragmatics*. Ph.D. thesis, University of Amsterdam.

[2] Franke, M. (2011). Quantity implicatures, exhaustive interpretation, and rational conversation. *Semantics and Pragmatics*, **4**(1):1–82.

[3] Franke, M. and G. Jäger (2012). Bidirectional optimization from reasoning and learning in games. *Journal of Logic, Language and Information*, **21**(1):117–139.

[4] Jäger, G. (2008). Applications of game theory in linguistics. *Language and Linguistics Compass*, **2/3**:408–421.

[5] Jäger, G. (2011). Game-theoretical pragmatics. In J. van Benthem and A. ter Meulen, eds., *Handbook of Logic and Language*, pp. 467–491. Elsevier, 2nd edition.

[6] Jäger, G. (2012). Game theory in semantics and pragmatics. In C. Maienborn, P. Portner, and K. von Heusinger, eds., *Semantics. An International Handbook of Natural Language Meaning*, volume 3, pp. 2487–2516. de Gruyter, Berlin.

[7] Jäger, G. and C. Ebert (2009). Pragmatic rationalizability. In A. Riester and T. Solstad, eds., *Proceedings of Sinn und Bedeutung 13*, number 5 in SinSpeC. Working Papers of the SFB 732, pp. 1–15. University of Stuttgart.

[8] Levinson, S. C. (2000). *Presumptive Meanings*. MIT Press.

[9] Lewis, D. (1969). *Convention*. Harvard University Press, Cambridge, Mass.

Comments on Jäger

ROBERT VAN ROOIJ

1 The use of game theory in linguistic pragmatics

The enterprise of Gricean pragmatics can be summarised as exploring the inferences (beyond the semantic content of an utterance) licensed by the assumption that a speaker intends to contribute cooperatively to some shared conversational end, and the extent to which these inferences match those which real hearers and speakers respectively draw and intend their listeners to draw.

There is an inherent self-referential loop in this description: a speaker must have expectations about a hearer's interpretative response (the additional inferences he might be capable of imagining or willing to draw), while the hearer must tailor his response to his notion of these speaker expectations, and so on into deeper nestings of belief operators. Classical game theory studies just such situations of interlocked multi-agent belief and action, and I fully agree with Jäger that game theory is the obvious tool to use to account for pragmatic reasoning. Following Franke (2009), Jäger's paper does a great job to show that one's initial thought that game theory might be used in linguistic pragmatics were fully justified, most spectacularly, perhaps, by proving that his IBR model of pragmatic reasoning can account for so-called embedded implicatures in a 'global' way, where many linguistic pragmaticists have claimed that this is impossible. This, I believe, already shows that linguists should take these game theoretical tools serioulsy.

However, I believe that (a) even standard game theory (with its standard assumptions) has, in fact, more to offer than what Jäger actually uses to account for pragmatic inferences, but also that (b) game theory (with its standard assumptions) fails miserably as a theory of genuine strategic communication to account for many other examples that count, intuitively, as rhetorically smart uses of language.

Recall that Gricean pragmatics is deeply rooted in the Cooperative Principle: the notion that a (normal) conversation has a common purpose or direction, shared by all participants. But natural examples of speaker-hearer conflict are not hard to find, from lying and keeping secrets in a court of law through to avoiding a sensitive topic with dinner guests or feigning interest in a boring story badly told. Cases where speaker and hearer interest diverge —partially or wholly— should be incorporated into the pragmatic agenda, for they can be studied no less systematically than their pure cooperative cousins; in other words, I would like to make the case for a broader notion of pragmatics: This notion of pragmatics should *at least* also cover cases where one can intuitively also make pragmatic inferences in cases of *partial* conflict: preferences that diverge enough to allow for interesting strategic reasoning, but not so widely that standard game theory predicts that no communication is possible at all. But (b), there will also cases where standard game theory predicts wrongly

Game theory has more to offer   More concretely, suppose two people are arguing about the conflict between Israel and Palestine; one, claiming that the blame lies with Palestine, says:

(1)   Most Israelis voted for peace.

The argumentative setting clearly distinguishes this example from more standard cases of scalar implicature. If each conversational participant attempts to convince the other of his own position by whatever means possible (including giving partial, misleading or even incorrect information), then Grice's cooperativity principle simply seems not to hold. In this particular case, the inference that not all Israelis voted for peace seems to be one the speaker *does not want* the hearer to make (indeed, she would prefer that he does *not* make this inference, as it weakens her argument), and that it therefore is not a conversational implicature of this utterance. Still we believe that it seems intuitively just as reasonable an inference as a standard cooperative scalar implicature that the speaker wants to convey and foreground; but if it is arguably not an inference derivable from Gricean cooperativity. It can be derived as an inference straightforwardly however, by making use of standard game theoretical tools (more in particular, using so-called persuasion games)

Communicate more than predicted   Though Jäger's paper proves that game theory can offer new insights to account for strategic communication, it also has its *limitations*: it predicts that in many cases genuine communication is impossible where, in fact, we see it all the time.

We normally (try to) use language just to *influence* the *behavior* of others in a purely rational way to our own benefit. Standard game theory assumes that this is common knowledge. But by doing so, standard game theory predicts that any purely manipulative use of language is impossible. The reason is, basically, that if the adressee is also rational (as is assumed), she will see through any attempt of manipulation of the speaker and will not take the new information at face value. But if the speaker is rational as well (as is assumed in standard game theory), she will see trough this in turn, and will realize that it doesn't make sense to try to use language manipulatively. Thus, communication is predicted to be possible only if, and as far as, the preferences are aligned. But this immediately gives rise to a

problem. It seems that agents – human or animal – also send messages to each other, even if the preferences are less harmonically aligned. Why would they do that? In particular, how could it be that natural language is used for communication even in these unfavorable circumstances? I strongly believe that if we really want game-theoretical pragmatics going beyond Gricean pragmatics, we have to answer this question. But to do so, we have to give up some of the most central, but also most idealistic, assumptions of game theory.[1]

- In standard game theory it is assumed that players model the game in the same way: it is common knowledge what game is played. But this seems like a highly idealized assumption. Is it not the case that players might model the game differently, or at least view others as modeling it as differently? Feinberg (2008), and recent work of de Jager and Franke, shows that if this possibility is taken into account, a new rationale for communication can show up.

- In his model of pragmatic reasoning, Jäger makes use of (iterated) elimination of strategies that violate the canons of rationality, i.e. that are strongly dominated. This procedure crucially depends, however, on a very strong epistemic assumption: *common knowledge of rationality*; not only must every agent be ideally rational, everybody must also know of each other that they are rational, and they must know that they know it, and so on ad infinitum. Unfortunately, the assumption of common knowledge of rationality is highly unrealistic. But by giving it up, it gives us a new opportunity to explain deceptive and manipulative language use.

  Indeed, it can be argued that wherever we do see attempted deceit in real life we are sure to find at least a belief of the deceiver (whether justified or not) that the agent to be deceived has some sort of limited reasoning power that makes the deception at least conceivably successful. Some agents are more sophisticated than others, and think further ahead. Jäger's IBR model makes use of such agents, but doesn't use its potential to explain manipulative communication in unfavorable circumstances.

- It can hardly be doubted that our decisionmaking violates normative prescriptions endorsed by game theorists. Especially due to the work of Kah-

---

[1] Attemps have been made within standard game theory as well to account for such cases. *Reputation* effects of lying in repeated games have been proposed to explain reliable communication. But experiments show that communication takes place even in one-shot games. To account for these cases, it is standardly assumed that reliable communication is possible, if we assume that signals can be too *costly* to fake. But assuming that messages of natural languages can be costly seems, to a large extent, counterintuitive.

neman and Tversky, it is now widely accepted among psychologists that the idea that we choose by maximizing expected utility is inaccurate. What is interesting for us is that due to the fact that people don't behave as rationally as the standard normative theory prescribes, it becomes possible for smart communicators to *manipulate* them: to convince them to do something that goes against their own interest. Perhaps the most common way to influence a decision maker making use of the fact that he or she does not choose in the prescribed way is by *framing*.

By necessity, a decisionmaker interprets her decision problem in a particular way. A different interpretation of the same problem may sometimes lead to a different decision. But this shows that manipulating the way information is presented can influence decision making. Framing effects are predicted by Kahneman and Tversky's Prospect Theory: a theory that implements the idea that our behavior is only boundedly rational. But if correct, it is this kind of theory that should be taken into account in any serious analysis of persuasive language use, whether this is called pragmatics or rhetoric.

2   Some additional comments/questions

- One of the most spectacular things shown in Jäger's paper is his successful treatment of so-called embedded implicatures. Indeed, I fully agree that this is a very nice feature of the IBR-model. Following Franke (2009), Jäger counts the so-called free-choice inferences as one of them. But there is some worry here: whereas standard implicatures can always be cancelled, it is not at all sure that free-choice inferences can. In fact, Ede Zimmermann has argued a few years ago that the inference is, in fact, a semantic inference.

- (Much) more generally, Jäger's use of game theory to account for pragmatic inferences assumes that there exists a strict distinction between semantics and pragmatics, and that semantics gives rise to a proposition expressed. But how natural are these assumptions? Wouldn't it be natural to use game theory also to partly determine what is said? More radically, perhaps that is all that is needed. And indeed, Grice (in his account of speaker's meaning) used a very similar type of self-referential reasoning to ground meanings in intentions. Davidson even concluded that conventional semantic meaning is only a myth. What would be left of the IBR model of pragmatic reasoning if semantics is assumed to be less autonomous, or not even existing?

Comments on Jäger

YUNCHENG ZHOU

"Game theory in semantics and pragmatics" is an introduction to a game-theoretic approach to a linguistic theory, with a focus on the Iterated Best Response Model (IBR). It is claimed that IBR is used to model neo-Gricean pragmatics according to which the derivation of what is meant from what is said is supposed to be based on fundamental principles of rational engagement among interlocutors. A general game-theoretic linguistic theory is first introduced to pave the way to the unfolding of IBR model. The article ends with application of IBR to some analyses of specific problems in semantics and pragmatics. In this comment, I will first give a general account of what a game-theoretic linguistic theory is able to do and indicate the problems it has. Then, I will use Stalnaker's possible-world apparatus to model Grice's reasoning process in order to clarify what Grice's conception of meaning is. Finally, I will point out a misgiving.

Game theory offers a vivid picture of the generative process of decision-making among the agents who engage with each other. An act of decision-making in a game is a response to other agents' acts or potential acts. The goal is to win the game: to maximize the gain and minimize the loss. Nash equilibrium tells us that there is a point in a game at which all agents involved could have the most gains and the least loss. This is achieved through iterative maximization of expected utilities. When a Nash equilibrium is arrived at, all acts are systematically constrained to a certain extent to which a flourishing state emerges in a game.

Game theory could be used to model the process of communication. In the process of communication, the interlocutors try their best to have themselves understood. In response to a linguistic act, an interlocutor maximizes its cognitive effects. The process of maximization is iteratively going on until a Nash equilibrium is reached. This is a point where interlocutors arrive at a common understanding regarding what they are talking about. An analysis is then made to get clear what is required in order for the communication to reach a Nash equilibrium. The desiderata in question is claimed to point to a way to understanding the utterance of a sentence in the communication. So what we can get from a game theoretic linguistic theory is a formal analysis of a conversational process, which might provide a way to understanding linguistic acts. But it has little to do with working out the meaning conveyed by the linguistics acts. In particular, it does not seem to conform to Grice's reasoning process of speaker meaning.

According to Grice, in issuing a linguistic act, the speaker intends to produce a certain belief in the audience, intends that the linguistic act be recognized as intending to produce a certain belief in the audience, and that the recognition enable the audience to have the belief produced. Suppose a speaker uttered a sentence in (1).

(1) Joe ate some of the cookies.

In uttering (1), the speaker intended to induce in the audience a belief that Joe ate some of the cookies. In order to attain that goal, the speaker needed to intend that her utterance be recognized as intending to induce that belief and this recognition enables the audience to have a belief that Joe ate some of the cookies.

What is the belief brought about by the utterance of (1)? Is it that Joe ate some of the cookies or that Joe ate some but not all of the cookies? This has to do with the effect on the context the utterance of (1) brings about. Accordingly, the utterance of (1) is supposed to enable the audience to exclude those possible worlds in the context which are incompatible with the utterance of (1). What does it mean to say that a possible world is incompatible with the utterance of (1)? In order to illustrate the problem, let me use Stalnaker's possible-world apparatus to model a propositional concept. ([3])

$$A.$$

	W_1	W_2
W_1	T	T
W_2	T	F

In A, W_1 is a possible world where Joe ate some of the cookies and W_2 is a possible world where Joe ate all of the cookies. The horizontal rows indicate two different propositions that are expressed with the utterance of (1). The vertical columns indicate the two possible truth valuations. Notice that the proposition expressed in W_1 is different from that expressed in W_2. The proposition in W_1 could be made true in the possible world W_1 as well as in W_2, while the proposition in W_2 can only be made true in W_1. Now the question is which proposition is expressed by the utterance of (1). According to Stalnaker, neither proposition is the one expressed by the utterance of (1). The proposition uttered by the utterance of (1) is the diagonal proposition which is true in W_1 but false in W_2. This shows that the meaning conveyed by the utterance of (1) does not only depend on semantic fact but also on contextual fact. So W_2 is the possible world that is incompatible with the utterance of (1). W_2 is to be excluded so as the utterance of (1) has the effect it has. That is to say, the belief produced is that Joe ate some but not all of the cookies.

In order to produce this belief, the speaker also intends that the utterance of (1) be recognized as working this way. In other words, it is presupposed that the linguistic act has the effect on the context it has. In the Stalnaker's model, this is realized through an operation: project the diagonal proposition onto the horizontal. With this operation, A transforms into B.

$B.$

	W_1	W_2
W_1	T	F
W_2	T	F

The diagonal proposition in B is a proposition that is true in W_1 and false in W_2, the same as the diagonal proposition of A. This shows that when the utterance of (1) is recognized as intending to produce the required belief what we get is exactly the belief that the utterance of (1) is intended to produce. This completes the calculation of Grice's reasoning process.

Two points are crucial at this juncture. One is that Grice's reasoning process does not involve *de facto* competing interactions among interlocutors. What is essentially needed is a complex of speaker intentions. The other is that the working out of speaker meaning is revealed through objective characterization of distinguishing possible states of affairs rather than through calculation of subjective probabilities.

As to the second point, let me go a little bit into the details of the example. Suppose the expected utilities of two possible readings of (1) are 1 and 0.5 respectively. What do we say as to the second reading, viz., Joe ate all of the cookies? Does the speaker think that the proposition is true or not? This difficult situation is removed through the process of maximization that occurs after the hearer's calculation. The problem is that there exists a gap between what the speaker intends and what the hearer ascribes to the speaker's intention. At the end of the day, a linguistic act is only meaningful depending on iterative ascription of speaker intention rather than speaker intention *per se*.

BIBLIOGRAPHY

[1] Grice, Paul. 1989. *Studies in the Way of Words*. Harvard University Press.
[2] Jäger, Gerhard. 2011. Game Theory in Semantics and Pragmatics,
     http://www.sfs.uni-tuebingen.de/ gjaeger/publications.
[3] Stalnaker, Robert. 1978. *Assertion, Context and Content*, pp.78-95, Oxford University Press.

Response to van Rooij and Zhou

GERHARD JÄGER

1   Reply to Robert van Rooij

In his thoughtful comments, Robert van Rooij brings up a series of important issues regarding the application of game theoretic tools to problems in semantics and pragmatics. My target paper *Game Theory in Semantics and Pragmatics* elaborates one particular implementation of the general game theoretic program, and I fully agree that there are many routes worth exploring that were not taken in this paper.

Let me briefly reply to the comments in turn:

Game theory has more to offer   The target article exclusively deals with scenarios where the speaker has a genuine interest to honestly supply information to the receiver. The interests of speaker and listener are assumed to diverge only insofar as the speaker also tries to minimize signaling costs, which may provide an incentive to be less informative than the listener would prefer. It goes without saying that in many conversational settings, the interests of the interlocutors are less in alignment. This includes debates, as van Rooij correctly points out.

The model that I develop in the target article has several aspects. On the one hand, there is the game, i.e. the space of types, messages and actions, as well as the prior probabilities and the utility function. The solution concept, Iterated Best Response, is not tied to the game construction as such and can be applied to other classes of games as well. In the example at hand, we could imagine a game where the actions of the listener are not isomorphic to the types but are rather different opinions that are being reached as the result of the conversation, such as *Israel is to blame* or *The Palestines are to blame*, and the interlocutors may have diverging preferences regarding the action been taken. If we assume that the speaker is not allowed to provide false information, we have to adjust the utility function accordingly.

For the sake of the argument, we can model this scenario by the following game, which slightly extends the format of standard signaling games:

- Nature chooses one of the two propositions ϕ_1 (*Israel is to blame.*) or ϕ_2 (*The Palestines are to blame.*) with probability q and $1-q$ respectively. Neither player observes the outcome of this move. The value of q is known to the receiver but not to the sender. The sender assumes a uniform probability density for the receiver's prior over q.

- Nature chooses one of two possible worlds w_1 (*Most but not all Israelis voted for peace.*) and w_2 (*All Israelis voted for peace.*) and shows the outcome to the sender. In ϕ_1, the probabilities of w_1, w_2 are $0.8, 0.2$, and in ϕ_2 they are $0.2, 0.8$ respectively.

- The sender transmits one of the two signals MOST and ALL to the receiver.

- The receiver chooses one of two actions a_1 and a_2.

MOST is true in both worlds, while ALL is only true in w_2.

The sender always prefers action a_2 over a_1. Furthermore she is supposed to say the truth. Accordingly, I assume that she gains a payoff of 1 if a_2 is chosen and 0 otherwise. Additionally, she has to pay a penalty of 10 if she utters a false message.

The receiver's task is to identify the value of ϕ, i.e. it equals 1 if he chooses a_i in ϕ_i, and 0 otherwise.

This game gives rise to the IBR below. Note that the entries in the receiver's sequence do not contain the behavioral strategy of the receiver at this reasoning level per se, but the sender's expectation of the receiver's behavior.

σ_0	MOST	ALL
w_1	1	0
w_2	.5	.5

ρ_0	a_1	a_2
MOST	0.5	0.5
ALL	0.2	0.8

σ_1	MOST	ALL
w_1	1	0
w_2	0	1

ρ_1	a_1	a_2
MOST	0.6	0.4
ALL	0.2	0.8

σ_2	MOST	ALL
w_1	1	0
w_2	0	1

ρ_2	a_1	a_2
MOST	0.8	0.2
ALL	0.2	0.8

(σ_2, ρ_2) form a fixed point. Here a scalar implicature emerges. The message MOST is only used in w_1 by the sender, and the receiver's posterior probability of w_1 given MOST is 1. (This in turn increases his posterior probability of ϕ_1 to $\frac{0.8q}{0.8q+0.2(1-q)}$. Therefore a_1 will be his optimal choice provided $q \in (0.2, 1]$, and this has a posterior probability of 0.8 for the sender. The other entries in the ρ-sequence are computed by analogous applications of Bayes rule.)

The intuitive reason for this effect is quite similar to the parallel example in the target article. In σ_1, the sender prefers ALL over MOST in w_2 because it would yield a higher expected payoff. In w_1 the message ALL would be sub-optimal though. Therefore MOST is only used in w_1, which is, in turn, anticipated by the receiver.

Communicate more than predicted   As van Rooij correctly points out, the IBR model predicts that any attempt of deception on the side of the sender would be anticipated by the receiver. Therefore information transmission is predicted to ensue only if both parties benefit from it. This is in stark contrast to the widespread use of lies, deceptions, malevolent manipulation etc. The IBR model affords a straightforward modifications that can easily account for these effects. First, the assumption that interlocutors always play the game according to the fixed point strategy is an idealization that can be given up without compromising the model as such. In fact, there are several recent applications of IBR (or a very similar variant thereof) to experimental data ([1]) indicating that participants in behavioral experiments have a rather limited reasoning depth. This can easily be captured by the model if the fixed point assumption is replaced by a probability distribution over inference depths. In other words, and contrary to van Rooij's claim, the IBR model does not require common knowledge of rationality. Quite the opposite: the move away from Nash equilibrium concepts towards a hierarchical model that IBR replaces the classical assumption of perfect rationality by a more bounded version of rationality.

The second possible modifications concerns the epistemic assumptions that are part of the game construction. As illustrated by the example above, IBR is compatible with more complex communication games where, for instance, the sender has only partial knowledge of the receiver's epistemic state. There are many further options along these lines, including non-identical prior assumptions of the players regarding the structure of the game and each others types.

Likewise, the notion of a *best* response within the IBR model can be replaced by other, psychologically more realistic decision protocols. In fact, the already mentioned [1] does a step in this direction by assuming a stochastic choice rule rather than categorical utility maximization. Applying prospect theory in this context to account for framing effects is an option worth exploring.

Some additional comments/questions

- Van Rooij expresses doubts regarding the analysis of free-choice inferences as quantity implicatures. An in-depth discussion of this issue would go beyond the scope of this exchange of commentaries here. Suffice it to say that it seems to me that free-choice inference can actually be canceled, as illustrated by

  (1) You may take an apple of a pear

  (2) ... but I don't know which.

In isolation, (1) suggests that the addressee has a free choice between apples and pears. If the sentence is continued as in (2), this inference is canceled.

- Van Rooij raises the issue whether the strict separation between literal meaning and pragmatically conveyed information is really viable. This is a very profound issue that cannot be dealt with here in any depth, of course. An extension of the IBR model to ambiguity/underspecification/vagueness is certainly an attractive route to explore. One could imagine a model where the sender's type does not just encompass factual information about the world but also truth conditions for messages with an underspecified semantics. In such a model, the receiver's task is not just to figure out which possible world he is in but also, in a sense, which game the sender is playing.

  I am somewhat skeptical though whether a theory of language interpretation according to which "conventional semantic meaning is only a myth" (a point of view van Rooij ascribes to Davidson) is really viable. After all, languages have to be learned and the form-meaning correlation is to a large degree arbitrary. I am not fully convinced that these basic facts can be accounted for without the assumption of conventionalized meaning.

2   Reply to Yungcheng Zhou

In his commentary, Zhou contrasts the model of communication that was developed in the target article to Paul Grice's ([2]) theory of non-natural meaning. I am grateful for bringing up this connection. In fact, Grice's notion of non-natural meaning provides, I think, a good justification for the asummption that interlocutors use the fixed-point strategy profile of the IBR sequence.

Grice assumes that the following three conditions are necessary and sufficient for a speaker Sally to mean$_{nn}$ that the listener Robin performs action a by sending the message s (rephrased in the terminology of signaling games):

- By sending s, Sally intends to make Robin perform a.

- By sending s, Sally intends Robin to recognize her intention to make him perform a.

- By sending s, Sally intends that Robin's recognition of her intention to make him perform a induces him to perform a.

Suppose Sally is in type t, and she believes Robin to play according to ρ. In IBR parlance, this amounts to saying that Sally is at least in σ_1, i.e. she plays a best response to some belief about Robin. In this scenario, there is a set of actions $A_\rho = \{a | \exists s.\rho(a|s) > 0\}$ that Sally believes she can induce by sending some message. Assuming that Sally plays a best response to ρ amounts to saying that she chooses her message in such a way that her utility is maximized, i.e. she induces the game history (t, s, a) rather than an alternative (t, s', a'). If this history is unique in maximizing Sally's expected payoff, this amounts to saying that she expects s to induce a, and she prefers this history over all alternative histories. This seems a plausible rendering of the notion *By sending s, Sally intends Robin to perform a.*

If there are ties in utilities, we would have to replace the notions *the message s* and *the action a* by *one of the messages* $s_1, \ldots, s_n$ and *one of the actions* $a_1, \ldots, a_n$. In the sequel, I will assume that there are no ties; a reformulation in terms of sets is straightforward though.

If Sally furthermore believes Robin to be rational and to believe that she is rational, she expects s to induce in Robin the posterior belief that she is in a type t' where she prefers (t', s, a) over all alternative histories (t', s', a'). In other words, she expects Robin to recognize her intention to make him do a. Within the IBR sequence, this holds if Sally is at least in σ_3.

If, finally, ρ is part of a fixed point, i.e. a Nash equilibrium, action a is rational for Robin according to his posterior beliefs upon observing s if and only if Sally is in a state where she intends Robin to perform a. Therefore if observing s induces him to believe that Sally intends him to do a, he can conclude that a is the rational thing to do for him, and he will in fact perform a.

The equivalence between *a is the rational thing for Robin to do* and *Sally intends Robin to do a* does not always hold though if ρ is not part of a fixed point, no matter how high up in the IBR sequence it occurs.

To summarize, if a signaling game is played in a Nash equilibrium, the sender always means$_{nn}$ to induce the action that she in fact induces. A formal proof of this fact would require a full formalization in terms of epistemic game theory. This is an attractive project for a later occasion.

Zhou furthermore hints at an intriguing connection between implicature computation and Stalnaker's ([3]) possible-world apparatus. To do justice to this proposal, a more thorough discussion would be required, however. Therefore I will refrain from long elaborations here. It seems to me that the critical point here is the issue how propositional concepts are assigned to utterances. According to Zhou, the sentence *Joe ate some of the cookies* is interpreted as the propositional concept in the left panel of Table 1. To fully spell out this approach, a principled account is

	w_1	w_2
w_1	T	T
w_2	T	F

	w_1	w_2
w_1	T	F
w_2	T	T

Table 1. Possible propositional concepts

needed to exclude an interpretation of the sentence in question as the propositional concept in the right panel.

BIBLIOGRAPHY

[1] Judith Degen, Michael Franke, and Gerhard Jäger. Cost-based pragmatic inference about referential expressions. In Markus Knauff, Michael Pauen, Natalie Sebanz, and Ipke Wachsmuth, editors, *Proceedings of CogSci 2013*, pages 376–381, Austin, TX, 2013. Cognitive Science Society.

[2] Herbert Paul Grice. Meaning. *Philosophical Review*, 66:377–388, 1957.

[3] Robert Stalnaker. Assertion. In Peter Cole, editor, *Pragmatics*, number 9 in Semantics & Pragmatics, pages 315–32. Academic Press, New York, 1978.

Hybrid Categorial Type Logics and the Formal Treatment of Chinese

CONGJUN YAO AND CHONGLI ZOU

ABSTRACT. There are four versions of the Lambek calculus in terms of Association and Permutation rules. It is only of restricted value to employ any single one of them in analyzing of natural languages, thus a multimodal categorial system which has different analyzing modes is required for this purpose. One way is to introduce structural modalities which allow controlled involvement of structural rules. But extensive use of structural modalities tends to result in very complex analyses. This fact tends to favor the selection of stronger system for the base level, a move which is associated with loss of possibly useful resource sensitivity. Another method is to create hybrid categorial type logics by combining multiple substructural logics, restricting the structural rules, and adding rules of inclusion. A weaker logic employing the latter method is used to assign new categories, so the sensitivity to linguistic resources is kept in the system, and the stronger logic of the hybrid system maintains the descriptive flexibility for the information of linguistic objects. Because of these advantages, it can adequately explain various natural language phenomena, including non-constituent coordination, extraction, and Chinese topicalised sentences.

1 The Origin of Hybrid Categorial Type Logics

We give four modalities: n (non-associative, non- permutative), a (associative, non-permutative), c (non-associative, permutative), b (associative, permutative). o_i is the product connective corresponding to any modality $i \in \{n, a, c, b\}$, while $\overset{i}{\leftarrow}$ and $\overset{i}{\rightarrow}$ are the entailment connectives.

(1) $\qquad A \Rightarrow A \quad \text{(id)} \qquad\qquad \dfrac{\Phi \Rightarrow B \quad \Gamma[B] \Rightarrow A}{\Gamma[\Phi] \Rightarrow A} \ [\text{cut}]$

(2) $\qquad \dfrac{(\Gamma, B)^{\circ_i} \Rightarrow A}{\Gamma \Rightarrow A \overset{i}{\rightarrow} B} \ [\overset{i}{\rightarrow} R] \qquad\qquad \dfrac{\Phi \Rightarrow C \quad \Gamma[B] \Rightarrow A}{\Gamma[(\Phi, C \overset{i}{\rightarrow} B)^{\circ_i}] \Rightarrow A} \ [\overset{i}{\rightarrow} L]$

$$\frac{(\Gamma, B)^{\circ_i} \Rightarrow A}{\Gamma \Rightarrow A \overset{i}{\leftarrow} B} \ [\overset{i}{\leftarrow} R] \qquad\qquad \frac{\Phi \Rightarrow C \quad \Gamma[B] \Rightarrow A}{\Gamma[(B \overset{i}{\leftarrow} C, \Phi)^{\circ_i}] \Rightarrow A} \ [\overset{i}{\leftarrow} L]$$

$$\frac{\Gamma \Rightarrow A \quad \Phi \Rightarrow B}{(\Gamma, \Phi)^{\circ_i} \Rightarrow A \circ_i B} \ [\circ_i R] \qquad\qquad \frac{\Gamma[(B, C)^{\circ_i}] \Rightarrow A}{\Gamma[B \circ_i C] \Rightarrow A} \ [\circ_i L]$$

(3) $$\frac{\Gamma[(B, (C, D)^{\circ_i})^{\circ_i}] \Rightarrow A}{\Gamma[(B, C)^{\circ_i}, D)^{\circ_i}] \Rightarrow A} \ [A] \qquad\qquad \frac{\Gamma[(B, C)^{\circ_i}] \Rightarrow A}{\Gamma[(C, B)^{\circ_i}] \Rightarrow A} \ [P]$$

The system which contains the rule (1) and (2) is the non-associative Lambek Calculus which is called NL with $\{\circ_n, \overset{n}{\leftarrow}, \overset{n}{\rightarrow}\}$ as its set of connectives; The addition of [A] to NL will result in the associative Lambek Calculus which is called L with $\{\circ_a, \overset{a}{\leftarrow}, \overset{a}{\rightarrow}\}$ as its set of connectives; The addition of [P] to NL will result in the permutative Lambek Calculus which is called NLP with $\{\circ_c, \overset{c}{\leftarrow}, \overset{c}{\rightarrow}\}$ as its set of connectives; The addition of [A] and [P] to NL will result in the permutative and associative Lambek Calculus which is called LP with $\{\circ_b, \overset{b}{\leftarrow}, \overset{b}{\rightarrow}\}$ as its set of connectives. Each substructure offers an analyzing mode. However, the following symbols are usually used to represent the connectives employed in these four substructures.

associative	permutative	connectives	corresponding system
−	−	$\odot, \varnothing, \reflectbox{$\varnothing$}$	NL
−	+	$\ominus, \ominus{-}, {-}\ominus$	NLP
+	−	$\bullet, \backslash, /$	L
+	+	$\otimes, {-}\!\circ, \circ\!{-}$	LP

Most of the categorial systems consist in a single substructure, that is, they have only one analyzing mode. However, the comprehensive analysis of natural languages needs more than one representing mode. For instance, strict tree structure (NL) is better in analyzing one kind of linguistic phenomena (say, some bounding phenomena) while the more flexible structure (say, system L) is better in analyzing another kind of linguistic phenomena. The treatment of varieties found in the cross-linguistic studies enhances the need of multi-mode of analysis. Thus a system which contains different modes of analysis is necessary for it will permits us to choose a mode which encode a structure relevant only to a certain kind of linguistic phenomenon to analyze them.

For this purpose, some scholars used modal operators which are also called structural modalities. Structural modalities are unary operators that allow for controlled involvement of structural rules. This method firstly needs to select a specific logic as basis logic, which set the default characteristics of resource sensitivity. The system may contain a modified structural rule which is only applicable to those categories marked with the known modalities , so as to allow controlled access to higher substructural levels. For instance, a permutative modality Δ allows controlled involvement of permutative rule, thus the system containing the following rules:

(4)
$$\frac{\Delta\Gamma \Rightarrow A}{\Delta\Gamma \Rightarrow \Delta A} \; [\Delta R] \qquad\qquad \frac{\Gamma[B] \Rightarrow A}{\Gamma[\Delta B] \Rightarrow A} \; [\Delta L]$$

$$\frac{\Gamma[\Delta B, C] \Rightarrow A}{\Gamma[C, \Delta B] \Rightarrow \Delta A} \; [\Delta P]$$

The restricted permutative rule allows the permutation in the formula of ΔX, that is, the linearity of hypothesis in the formula of ΔX is weakened. The left rule permits the free deletion of Δ-. Thus, a modality can be used to represent the extraction. We can add this modality and its rules to calculus L. Since L is subject to the linearity, np\s and s/np are used to represent respectively the sentences which is missing a np in the left side or in the right side. However, a type of s/(Δnp) corresponds to the sentence in which there is no np in a position, therefore, it is applicable to the analysis of extraction, which means the position of extraction may occurs on the non- peripheral part of the sentence. The following is an instance of derivation of *"who Kim sent away"*.

(5)

$$
\frac{\begin{array}{c}
\frac{\begin{array}{c}
\frac{np \Rightarrow np \qquad s \Rightarrow s}{(np, np\backslash s)^\bullet \Rightarrow s} \; [\backslash L]
\end{array}}{\begin{array}{c}
pp \Rightarrow pp \qquad (np, ((np\backslash s)/pp, pp)^\bullet)^\bullet \Rightarrow s
\end{array}} \; [/L] \\
\frac{np \Rightarrow np \qquad (np, ((np\backslash s)/pp, pp)^\bullet)^\bullet \Rightarrow s}{(np, ((((np\backslash s)/pp)/np, np)^\bullet, pp)^\bullet)^\bullet \Rightarrow s} \; [/L] \\
\frac{}{(np, ((((np\backslash s)/pp)/np, \Delta np)^\bullet, pp)^\bullet)^\bullet \Rightarrow s} \; [\Delta L] \\
\frac{}{(np, (((np\backslash s)/pp/np, (\Delta np, pp)^\bullet)^\bullet)^\bullet \Rightarrow s} \; [A] \\
\frac{}{(np, (((np\backslash s)/pp/np, (pp, \Delta np)^\bullet)^\bullet)^\bullet \Rightarrow ss} \; [\Delta P] \\
\frac{}{(np, ((((np\backslash s)/pp)/np, pp)^\bullet, \Delta np)^\bullet)^\bullet \Rightarrow s} \; [A] \\
\frac{}{(np, ((((np\backslash s)/pp)/np, pp)^\bullet)^\bullet, \Delta np)^\bullet \Rightarrow s} \; [A] \\
(np, ((((np\backslash s)/pp)/np, pp)^\bullet)^\bullet \Rightarrow s/\Delta np
\end{array}}{(rel/(s/\Delta np), (np, (((np\backslash s)/pp)/np, pp)^\bullet)^\bullet)^\bullet \Rightarrow rel} \quad \frac{rel \Rightarrow rel}{} \; [/L]
$$

It is also possible to introduce the other structural rules by adoption of controlled involvement of other structural modalities. However, the involvement of structural modalities will result in various problems, the-oretical, computational, as well as practical. For instance, where they are used extensively , unduly complicated accounts tend to result. The complexity of syntactic analyses tends to encourage the selection of the highest workable level possible for the default level of the logic. However, this selection will result in the loss of possibly useful resource sensitivity. The discarding of resource sensitivity may not always be in the long term interests of developing adequate linguistic accounts.

The other way to realize the goal is to merger the substructures into a single multi-modal system, that is, the connectives which are used in the different substructures are co-existent in a single composite logic. Thus the multi-modal system will allow the changeable categories which represent the transition between the different structures (an operator in one substructure can be rewritten into a corresponding operator in another substructure). This method eliminates the requirement of structural modalities, and thus avoids the problems caused by structural modalities. Such a method is call hybrid method, and such a logic is called hybrid categorical type logic.

2   Construction of Hybrid Categorial Type Logics

In this section, firstly we will demonstrate the ways in which more than one substructure can be merged into a hybrid categorical type logic. Supposed that we have a system which contains the connectives used in those different substructures, and in which the logic rule (1) and (2) can be used freely, but the structural rules (3) can be used restrictively, as illustrated in the following:

(6)
$$\frac{\Gamma[(B,(C,D)^\circ)^\circ] \Rightarrow A}{\Gamma[(B,C)^\circ,D)^\circ] \Rightarrow A} \; [A] \quad (\circ \in \{\bullet, \otimes\})$$

$$\frac{\Gamma[(B,C)^\circ] \Rightarrow A}{\Gamma[(C,B)^\circ] \Rightarrow A} \; [P] \quad (\circ \in \{\ominus, \otimes\})$$

By means of the rule (1), (2) and (6), we obtain a system in which the different substructures are coexistent but there is no correlation between these substructures. Actually, it is the rule (7) (Inclusive Rule) that incurs the correlation. (Note: a relation $<$ which is on the product operators is employed in the affiliate condition of the application of the rule (7). Here, for any product $\circ'$ and $\circ$, $\circ' < \circ$ iff $< \circ', \circ > \in \{< \odot, \bullet >, < \odot, \ominus >, < \odot, \otimes >, < \bullet, \otimes >, < \ominus, \otimes >\}$. This rule supports $A \circ_i B \Rightarrow A \circ_i B(\circ_j < \circ_i)$(as illustrated by (8), which reveals the correlation of product connectives in the different substructures. The employment of this rule will also enable us to obtain the correlation of entailment connectives in different substructures:

$A \overset{j}{\leftarrow} B \Rightarrow A \overset{i}{\leftarrow} B(\circ_j < \circ_i)$. More generally, for any two connectives which are of the relation ¡, the conversions similar to (8) and (9) can be obtained. The rule which further correlates the different substructures is the interaction rule (10).

(7)
$$\frac{\Gamma[(B,C)^{\circ_j}] \Rightarrow A}{\Gamma[(B,C)^{\circ_i}] \Rightarrow A} \ [<] \quad (\circ_j < \circ_i)$$

(8)
$$\frac{\dfrac{B \Rightarrow B \qquad A \Rightarrow A}{(A,B)^{\circ_j} \Rightarrow A \circ_j B} \ [\circ_j R]}{\dfrac{(A,B)^{\circ_i} \Rightarrow A \circ_j B}{A \circ_i B \Rightarrow A \circ_j B} \ [\circ_i L]} \ [<] \quad (\circ_j < \circ_i)$$

(9)
$$\frac{\dfrac{B \Rightarrow B \qquad A \Rightarrow A}{(AB,B)^{\circ_j} \Rightarrow A} \ [\overset{j}{\leftarrow} L]}{\dfrac{(A \overset{j}{\leftarrow} B, B)^{\circ_i} \Rightarrow A}{A \overset{j}{\leftarrow} B \Rightarrow A \overset{i}{\leftarrow} B} \ [\overset{i}{\leftarrow} R]} \ [<] \quad (\circ_j < \circ_i)$$

(10)
$$\frac{\Gamma[(B,(C,D)^{\circ_i})^{\circ_j}] \Rightarrow A}{\Gamma[(B,C)^{\circ_j},D)^{\circ_i}] \Rightarrow A} \ a \quad (\circ_j < \circ_i)$$

$$\frac{\Gamma[(B,C)^{\circ_i},D)^{\circ_j}] \Rightarrow A}{\Gamma[(B,(C,D)^{\circ_j})^{\circ_i}] \Rightarrow A} \ b \quad (\circ_j < \circ_i)$$

$$\frac{\Gamma[(B,C)^{\circ_i},D)^{\circ_j}] \Rightarrow A}{\Gamma[(B,D)^{\circ_j},C)^{\circ_i}] \Rightarrow A} \ c \quad (\circ_j < \circ_i)$$

There are some other theorems we will take into consideration. note that although the composition (11a) is not allowed within a non-associative level, the same type may be composed if the result is in an associative level, as in (11b), proven in (12). (the derivational steps marked [< *] corresponds to the multiple uses of [<]). Similarly, two counter directional arguments cannot be reordered purely within a non-associative system, as in (11c). However, if the result type is in an associative level, the order of them can be changed, as in (11d), proven in (13).

(11)
$$
\begin{array}{llll}
a & * & (A\varnothing B, B\varnothing C)^{\circ} \Rightarrow A\varnothing C & b \quad (A\varnothing B, B\varnothing C)^{\bullet} \Rightarrow A/C \\
c & * & (B\varnothing A)\varnothing C) \Rightarrow B\varnothing (A\varnothing C) & d \quad (B\varnothing A)\varnothing C) \Rightarrow B\backslash(A/C)
\end{array}
$$

(12)

$$\dfrac{\dfrac{\dfrac{C \Rightarrow C \qquad B \Rightarrow B}{(B\varnothing C, C)^{\circ} \Rightarrow B}\ [\varnothing L] \qquad A \Rightarrow A}{(A\varnothing B, (B\varnothing C, C)^{\circ})^{\circ} \Rightarrow A}\ [\varnothing L]}{\dfrac{(A\varnothing B, (B\varnothing C, C)^{\bullet})^{\bullet} \Rightarrow A}{\dfrac{((A\varnothing B, B\varnothing C)^{\bullet}, C))^{\bullet} \Rightarrow A}{(A\varnothing B, B\varnothing C)^{\bullet} \Rightarrow A/C}\ [A]}}\ [<*]}{}\ [/R]$$

(13)

$$\dfrac{\dfrac{\dfrac{C \Rightarrow C \qquad \dfrac{B \Rightarrow B \qquad A \Rightarrow A}{(B, B\varotimes A)^{\circ} \Rightarrow A}\ [\varotimes L]}{(B, ((B\varotimes A)\varnothing C, C)^{\circ})^{\circ} \Rightarrow A}\ [\varnothing L]}{\dfrac{(B, ((B\varotimes A)\varnothing C, C)^{\bullet})^{\bullet} \Rightarrow A}{\dfrac{((B, ((B\varotimes A)\varnothing C)^{\bullet}, C)^{\bullet} \Rightarrow A}{\dfrac{(B, ((B\varotimes A)\varnothing C)^{\bullet} \Rightarrow A/C}{(B\varotimes A)\varnothing C \Rightarrow B\backslash(A/C)}\ [/R]}\ [/R]}}\ [A]}{}\ [<*]}{}$$

So far, we have all components of a hybrid categorial type logic. However, in some cases, it is unnecessary to include all these four substructures. Supposing that we combine two substructures L and LP to form a hybrid categorial type logic, the set of connectives of this logic is $\{\bullet, \backslash, /, \otimes, \multimap, \circ\!-\}$, and its rules include (1), (2), (14), (15), and (16).

(14)

$$\dfrac{\Gamma[(B, (C, D)^{\circ})^{\circ}] \Rightarrow A}{\Gamma[(B, C)^{\circ}, D)^{\circ}] \Rightarrow A}\ [A]\quad (\circ \in \{\bullet, \otimes\}) \qquad\qquad \dfrac{\Gamma[(B, C)^{\otimes}] \Rightarrow A}{\Gamma[(C, B)^{\otimes}] \Rightarrow A}\ [P]$$

(15)

$$\dfrac{\Gamma[(B, C)^{\bullet}] \Rightarrow A}{\Gamma[(B, C)^{\otimes}] \Rightarrow A}\ [<]$$

(16)

$$\dfrac{\dfrac{\Gamma[(B, (C, D)^{\otimes})^{\bullet}] \Rightarrow A}{\Gamma[(B, C)^{\bullet}, D)^{\otimes}] \Rightarrow A}\ a}{\dfrac{\Gamma[(B, C)^{\otimes}, D)^{\bullet}] \Rightarrow A}{\Gamma[(B, D)^{\bullet}, C)^{\otimes}] \Rightarrow A}\ c} \qquad\qquad \dfrac{\Gamma[(B, C)^{\otimes}, D)^{\bullet}] \Rightarrow A}{\Gamma[(B, (C, D)^{\bullet})^{\otimes}] \Rightarrow A}\ b$$

Alhough associative Lambek Calculus allows composition, directionally mixed compositions such as (17a) are not allowed. Nevertheless, such types may be composed if the result of composition is in the permutative linear substructure, as in (17b), proven in (18).

(17)   a $*(A/B, C\backslash B)^{\bullet} \Rightarrow C\backslash A$ b $(A/B, C\backslash B)^{\otimes} \Rightarrow A \circ\!- C$

(18)

$$\dfrac{\dfrac{\dfrac{\dfrac{\dfrac{\dfrac{\dfrac{C \Rightarrow C \quad B \Rightarrow B}{(C, C \backslash B)^{\bullet} \Rightarrow B} \; [\backslash L] \quad A \Rightarrow A}{(A/B, (C, C \backslash B)^{\bullet})^{\bullet} \Rightarrow A} \; [/L]}{(A/B, (C, C \backslash B)^{\otimes})^{\otimes} \Rightarrow A} \; [< *]}{(A/B, (C \backslash B, C)^{\otimes})^{\otimes} \Rightarrow A} \; [P]}{((A/B, C \backslash B)^{\otimes}, C)^{\otimes} \Rightarrow A} \; [A]}{((A/B, C \backslash B)^{\otimes} \Rightarrow A \multimap C} \; [\multimapinv R]}$$

As it is, the use of $(,)^{\otimes}$ allows for the derivation of various 'unwanted' sequents as well, For example,

$(A/B, C \backslash B)^{\otimes} \Rightarrow C \multimap A, (C \backslash B, A/B)^{\otimes} \Rightarrow A \multimapinv C,$ and $(C \backslash B, A/B)^{\otimes} \Rightarrow C \multimap A$

As it is understood, conforming to standard categorical practice, that word order is to be read off from the ordering of hypotheses. To this end, we should redefine the language recognized by a hybrid categorical grammar, taking into account only those derivations whose conclusion $\Gamma \Rightarrow s$ excludes the use of $(,)^{\otimes}$ in Γ. For example, (18) may now be adapted as follows:

(19)

$$\dfrac{\dfrac{\dfrac{\dfrac{\dfrac{\dfrac{\dfrac{C \Rightarrow C \quad B \Rightarrow B}{(C, C \backslash B)^{\bullet} \Rightarrow B} \; [\backslash L] \quad A \Rightarrow A}{(A/B, (C, C \backslash B)^{\bullet})^{\bullet} \Rightarrow A} \; [/L]}{((A/B, C)^{\bullet}, C \backslash B)^{\bullet} \Rightarrow A} \; [A]}{((A/B, C)^{\otimes}, C \backslash B)^{\bullet} \Rightarrow A} \; [< *]}{((C, A/B)^{\otimes}, C \backslash B)^{\bullet} \Rightarrow A} \; [P]}{(C, (A/B, C \backslash B)^{\bullet})^{\otimes} \Rightarrow A} \; 16b}{(A/B, C \backslash B)^{\bullet} \Rightarrow C \multimap A} \; \multimap R$$

This conclusion has two features. On the one hand, it excludes the use of $(,)^{\otimes}$; On the other hand, it is identical to the forward crossed composition rule ($> B_X$) in CCG.

This instance implies that linear implication in a hybrid system might allow treatment of the non-peripheral extraction, e.g. with relative pronouns bearing type rel/(s $\multimapinv$ np), eliminating the need for a permutative structural modality.

3 Application of Hybrid Categorial Type Logic

Firstly, let's take a hybrid system which contains only NL and L. The set of connectives of this logic is $\{\{\bullet, \backslash, /, \odot, \varnothing, \wr\}$, and its rules include (1), (ref1.2), (20), (21), and (22).

(20)
$$\frac{\Gamma[(B,(C,D)^\circ)^\circ] \Rightarrow A}{\Gamma[(B,C)^\circ,D)^\circ] \Rightarrow A} [A] \quad (\circ \in \{\bullet\})$$

(21)
$$\frac{\Gamma[(B,C)^\circ] \Rightarrow A}{\Gamma[(B,C)^\bullet] \Rightarrow A} [<]$$

(22)
$$\frac{\dfrac{\Gamma[(B,(C,D)^\bullet)^\circ] \Rightarrow A}{\Gamma[(B,C)^\circ,D)^\bullet] \Rightarrow A}\, a}{\Gamma[(B,C)^\bullet,D)^\circ] \Rightarrow A}\, c \qquad \frac{\Gamma[(B,C)^\bullet,D)^\circ] \Rightarrow A}{\Gamma[(B,(C,D)^\circ)^\bullet] \Rightarrow A}\, b$$

The non-associative structure NL to the hybrid system can be used as the major structure in analysis of lexicon, which constitutes some advantages in analysis of language. For instance, if we subcategorize the complimentary constituents which are of a non-associative functor type (say of the form $A \varnothing B$ or $B \varnothing A$), we can be sure that the adopted complimentary constituents is the projection of head words in some lexical unit, not the composite constituents. On the other hand, there is no obstacle to the free use associative rules in compositional operation if necessary, for we have $X \varnothing Y \Rightarrow X/Y$. The treatment of some conjunction of non-constituents depends on association. If it is legitimate to coordinate subject and verb, the sentence, such as(i)Mary spoke and Susan whispered, to Bill (in which each conjunct can be analyzed as s/pp), can be explainable. However, it is expelled in a pure non-associative system, say, NL. In a hybrid system, the coordination is assumed to occur in the associative substructure, then even they are assigned a non-associative category, it is also explainable. For instance, the conjuncts in (i) can be deduced as s/pp, as illustrated in (23).

(23)
$$\frac{\dfrac{pp \Rightarrow pp \quad \dfrac{np \Rightarrow np \quad s \Rightarrow s}{(np, np \varnothing s)^\circ \Rightarrow s}[\varnothing L]}{\dfrac{(np, ((np \varnothing s) \varnothing pp, pp)^\circ)^\circ \Rightarrow s}{\dfrac{(np, ((np \varnothing s) \varnothing pp, pp)^\bullet)^\bullet \Rightarrow s}{\dfrac{(np, ((np \varnothing s) \varnothing pp^\bullet), pp)^\bullet \Rightarrow s}{np, \quad (np \varnothing s) \varnothing pp)^\bullet \quad \Rightarrow s/pp}[A]}[< *]}[\varnothing L]}}{}$$

np, $(np \varnothing s) \varnothing pp)^\bullet$ $\Rightarrow s/pp$
Mary   spoke
Susan   shispered

Similarly, we can analyze the coordination of non-constituents in Chinese with the hybrid system, and NL can be used to as the major structure. For instance, "John xihuan er Smith bu xihuan pijiu" (John likes but Smith doesn't like beer), it is deducible that the type of the conjuncts are s/np, because $(np, (np \diagdown s) \varnothing np)^{\bullet} \Rightarrow s/np$ is theorem, as illustrated by (24).

(24)

$$
\cfrac{
\cfrac{
np \Rightarrow np \qquad
\cfrac{
\cfrac{np \Rightarrow np \qquad s \Rightarrow s}{(np, np \diagdown s)^{\circ} \Rightarrow s} [\diagdown L]
}{(np, ((np \diagdown s) \varnothing np, np)^{\circ})^{\circ} \Rightarrow s} [\varnothing L]
}{
\cfrac{
\cfrac{(np, ((np \diagdown s) \varnothing np, np)^{\bullet})^{\bullet} \Rightarrow s}{(np, ((np \diagdown s) \varnothing np^{\bullet}), np)^{\bullet} \Rightarrow s} [< *]
}{np, \quad (np \diagdown s) \varnothing np)^{\bullet} \qquad \Rightarrow s/np} [A]
}
}{} [/R]
$$

np,	Xihuan
John	likes
John	buxihuan
Smith	doesn't like
Smith	

(24) involves the coordination of subject and transitive verb, which constitutes a non-standard analysis, the type of the coordination is s/np.

In the sentence "John jinchang er Smith henshao he pijiu" (John often but Smith seldom drinks beer), it is deduced that the type of conjuncts are $s/(np \diagdown s)$, for $(np, (np \diagdown s) \varnothing (np \diagdown s))^{\bullet} \Rightarrow s/(np \diagdown s)$ is theorem, as illustrated by (25) which involves the coordination of subject and adverbial.

(25)

$$
\cfrac{
np \diagdown s \Rightarrow np \diagdown s \qquad
\cfrac{np \Rightarrow np \qquad s \Rightarrow s}{(np, np \diagdown s)^{\circ} \Rightarrow s} [\diagdown L]
}{
\cfrac{(np, ((np \diagdown s) \varnothing (np \diagdown s), np \diagdown s)^{\circ})^{\circ} \Rightarrow s}{
\cfrac{(np, ((np \diagdown s) \varnothing (np \diagdown s), np \diagdown s)^{\bullet})^{\bullet} \Rightarrow s}{
\cfrac{(np, ((np \diagdown s) \varnothing (np \diagdown s)^{\bullet}), np \diagdown s)^{\bullet} \Rightarrow s}{np, \quad (np \diagdown s) \varnothing (np \diagdown s))^{\bullet} \qquad \Rightarrow s/(np \diagdown s)} [A]
} [< *]
} [\varnothing L]
} [/R]
$$

np,	jing chang
John	often
John	hen shao
Smith	seldem
Smith	

In the sentence "John heshui henkuai er pijiu henman" (John dink water quickly but beer slowly), it is deduced that the type of conjuncts are $((np \diagdown s) \varnothing np)/(np \diagdown s)$, for $(np, (np \diagdown s) \diagdown (np \diagdown s))^{\bullet} \Rightarrow ((np \diagdown s) \varnothing np) \diagdown (np \diagdown s)$ is theorem, as illustrated by (26) which involves the coordination of object and adverbial.

$$
\begin{array}{c}
\dfrac{\quad np \Rightarrow np \qquad\qquad s \Rightarrow s \quad}{(np, np\text{\tiny ⊗}s)^{\circ} \Rightarrow s} \ [\text{⊗}L]
\end{array}
$$

(26)

$$
\cfrac{\cfrac{\cfrac{\cfrac{\cfrac{\cfrac{\cfrac{np \Rightarrow np \qquad \cfrac{np\text{⊗}s \Rightarrow np\text{⊗}s \qquad \cfrac{np \Rightarrow np \qquad s \Rightarrow s}{(np, np\text{⊗}s)^{\circ} \Rightarrow s}\ [\text{⊗}L]}{(np,(np\text{⊗}s,(np\text{⊗}s)\text{⊗}(np\text{⊗}s))^{\circ})^{\circ} \Rightarrow s}\ [\varnothing L]}{(np,(((np\text{⊗}s)\varnothing np, np)^{\circ},(np\text{⊗}s)\text{⊗}(np\text{⊗}s))^{\circ})^{\circ} \Rightarrow s}\ [\varnothing L]}{((((np\text{⊗}s)\varnothing np, np)^{\circ},(np\text{⊗}s)\text{⊗}(np\text{⊗}s))^{\circ} \Rightarrow np\text{⊗}s}\ [\text{⊗}R]}{((((np\text{⊗}s)\varnothing np, np)^{\bullet},(np\text{⊗}s)\text{⊗}(np\text{⊗}s))^{\bullet} \Rightarrow np\text{⊗}s}\ [< *]}{(((np\text{⊗}s)\varnothing np,(np,(np\text{⊗}s)\text{⊗}(np\text{⊗}s))^{\bullet})^{\bullet} \Rightarrow np\text{⊗}s}\ [A]}{np,\quad (np\text{⊗}s)\text{⊗}(np\text{⊗}s))^{\bullet} \quad \Rightarrow ((np\text{⊗}s)\varnothing np)\backslash(np\text{⊗}s)}\ [\backslash R]
$$

np,	(np⊗s)⊗(np⊗s))•	⇒ ((np⊗s)∅np)\(np⊗s)
John	hen kuai	
John	very quickly	
Smith	hen man	
Smith	very slowly	

If we want to give the type of the conjunct in "Mama gei liao never yikuai tang er erzi yikuai binggan" (The mother gave her daughter a piece of candy and her son a piece of cookies), it is better to see what kind of category can be deduced from np, as in (27).

(27)

$$
\cfrac{\cfrac{np \Rightarrow np \qquad s \Rightarrow s}{(s\varnothing np, np)^{\circ} \Rightarrow s}\ [\varnothing L]}{np \Rightarrow (s\varnothing np)\text{⊗}s}\ [\text{⊗}R]
\qquad\qquad
\cfrac{\cfrac{np \Rightarrow np \qquad s\varnothing np \Rightarrow s\varnothing np}{((s\varnothing np)\varnothing np, np)^{\circ} \Rightarrow s\varnothing np}\ [\varnothing L]}{np \Rightarrow ((s\varnothing np)\varnothing np)\text{⊗}(s\varnothing np)}\ [\text{⊗}R]
$$

Because $np \Rightarrow (s\varnothing np)\text{⊗}s$ and $np \Rightarrow ((s\varnothing np)\varnothing np)\text{⊗}(s\varnothing np)$ are theorems, noun phrases can be assigned the category of $(s\varnothing np)\text{⊗}s$ and $((s\varnothing np)\varnothing np)\text{⊗}(s\varnothing np)$. Supposing that the direct object is assigned the former while the indirect object is assigned the latter, we can deduce that the type of conjuncts are $(s\varnothing np)\varnothing np\backslash s$, because $(((s\varnothing np)\varnothing np)\text{⊗}(s\varnothing np),(s\varnothing np)\text{⊗}s)^{\bullet} \Rightarrow (s\varnothing np)\varnothing np/s$ is theorem, as illustrated by (28) which involves the coordination of direct object and indirect object.

(28)

$$\cfrac{\cfrac{\cfrac{\cfrac{\cfrac{(s\varnothing np)\varnothing np \Rightarrow (s\varnothing np)\varnothing np \qquad s\varnothing np \Rightarrow s\varnothing np}{((s\varnothing np)\varnothing np, ((s\varnothing np)\varnothing np)\diagdown(s\varnothing np)) \Rightarrow s\diagdown np}[\diagdown L] \qquad s \Rightarrow s}{(((s\varnothing np)\varnothing np, ((s\varnothing np)\varnothing np)\diagdown(s\varnothing np))^{\circ}, (s\varnothing np)\diagdown s)^{\circ} \Rightarrow s}[\diagdown L]}{(((s\varnothing np)\varnothing np, ((s\varnothing np)\varnothing np)\diagdown(s\varnothing np))^{\bullet}, (s\varnothing np)\diagdown s)^{\bullet} \Rightarrow s}[<*]}{((s\varnothing np)\varnothing np, (((s\varnothing np)\varnothing np)\diagdown(s\varnothing np), (s\varnothing np)\diagdown s)^{\bullet})^{\bullet} \Rightarrow s]}[A]}{\begin{array}{lll} (((s\varnothing np)\varnothing np)\diagdown(s\varnothing np)), & (s\varnothing np)\diagdown s)^{\bullet} & \Rightarrow (s\varnothing np)\varnothing np\backslash s \\ \text{nv er} & \text{yi kuai tang} & \\ \text{daughter} & \text{a piece of candy} & \\ \text{er zi} & \text{yi kuai bing gan} & \\ \text{Smith} & \text{a piece on cookie} & \end{array}}$$

T_{top} (rule of topicalization), proposed by Steedman ([1, 20]), is a syntactic rule based on functional composition. It is formalized as follows:

$$T_{top} : X \rightarrow S_T/(S/X)$$

T_{top} is capable of assigning appropriate categories to topicalized materials, which, in turn, leads to right syntactic concatenation for simple topical sentences. For example, this Hybrid system can be used to analyze the Mandarin Chinese sentence with the topicalized object. However, in Mandarin Chinese, except object extracted topicalized sentences, topicalized phrases can also come from other syntactic locations as well as co-refer with overt pronouns in situ, for instance, indirect object extraction (*Nage qigai wo gei-le wukuai qian* 'That baggar$_i$ I gave t_i 5 bucks'), subject modifier extraction (*Nage qigai wo gei-le wukuai qian* 'This clothes$_i$ fabric (of) t_i is good'), subject modifier extraction (*Nimen ban wo zui xihuan Yangfan* 'Among your class$_i$ I like Yangfan t_i the most'), head of subject extraction (*Haizi tamen-de tinghua* 'Kids$_i$ their t_i are behaved'), head of object extraction (*Xiangyan wo xihuan chou hunan-de* 'Cigarettes$_i$ I like smoking t_i from Hunan'), oblique extraction (*Ruanjian wo shi menwaihan* 'Software$_i$ I am a laymen of t_i'), object extraction with in-situ pronoun (*XiaoLongnv Yangguo ai ta* 'Xiaolongnv$_i$' Yangguo loves her$_i$), Subject extracted with in-situ pronoun (*Leifeng ta jiang yongchuibuxiu* 'Leifeng$_i$ he$_i$ is immortal'), indirect object with in-situ pronoun (*Nage qigai wo gei ta wukuai qian* 'That baggar$_i$ I give him$_i$ 5 bucks'). If we unanimously assign all these extracted strings nominal category $S_T/(s/np)$, right results do not necessarily follow. Though Steedman's formalization is very enlightening, we find two things that are not completely satisfactory. One is, it adds one more syntactic rule, which contradicts CCG's upholding cross-linguistic invariant principle. Second, there are more than one types of typicalized sentences in Mandarin Chinese, which is beyond T_{top}'s reach. Since CCG keeps cross-linguistic variance in lexicon rather than grammar, we would like to find a way out by assigning syntactic categories in lexicon. We will disregard Steedman's T_{top} and assign different categories needed for syntactic concatenation to topicalized materials directly:

1. Object extraction without in-situ pronoun: $s_T/(s/np)$;

2. Subject, object and indirect object extraction with in-situ pronoun: s_T/s;

3. Double bject extraction without in-situ pronoun: $s_T/(s/np)$;

4. Modifier extraction within complex subject: $s_T \varnothing ((np \varnothing np)\backslash s), (s_T/((np/np) \multimap s)$ or $s_T/((np/np)\backslash s)$;

5. Modifier extraction within complex object: $s_T/(s/np)$;

6. Head of complex object extraction: $s_T \varnothing (s/np), (s_T/(s \multimap np)$ or $s_T/(s/np)$

7. Oblique extraction: $s_T \varnothing/(np \otimes s) \otimes (np \otimes s))$.

Now we can derive these topicalized sentences within hybrid Categorial type logical systems. For economy, we will only elaborate with some representative instances. Firstly, we will derive some topicalized sentences within our NL+L system.

(29) is the derivation of sentence with the topicalized complex subject modifier.

$$
(29) \quad
\cfrac{
\cfrac{
\cfrac{
\cfrac{
\cfrac{
\cfrac{
np \Rightarrow np \qquad s \Rightarrow s
}{
np \Rightarrow np \qquad (np, np \otimes s)^{\circ} \Rightarrow s
}\,[\otimes L]
}{
((np \varnothing np, np)^{\circ}, np \otimes s)^{\circ} \Rightarrow s
}\,[\varnothing L]
}{
((np \varnothing np, np)^{\bullet}, np \otimes s)^{\circ} \Rightarrow s
}\,[<]
}{
(np \varnothing np, (np, np \otimes s)^{\circ})^{\bullet} \Rightarrow s
}\,[22b]
}{
(np, np \otimes s)^{\circ} \Rightarrow (np \varnothing np)\backslash s
}\,[\backslash R] \qquad s_T \Rightarrow s_T
}{
(s_T/((np \varnothing np)\backslash s)), \quad (np, \quad np \otimes s)^{\circ})^{\circ} \Rightarrow s_T
}\,[\varnothing L]
$$

$(s_T/((np\varnothing np)\backslash s)),$	$(np,$	$np\otimes s)^{\circ})^{\circ}$
Zhe-jian yifu	buliao	bu-cuo
This-CL clothes	fabric	not-bad

As to this clothes, the fabric (it uses) is pretty good.

(30) is the derivation of sentence with topicalized complex head word of object.

(30)

$$
\cfrac{
np \Rightarrow np \quad
\cfrac{
np \Rightarrow np \quad
\cfrac{
\cfrac{
\cfrac{np \Rightarrow np \quad s \Rightarrow s}{(np, np \diagdown s)^{\circ} \Rightarrow s}\,[\diagdown L]
}{(np, ((np \diagdown s)\varnothing np, np)^{\circ})^{\circ} \Rightarrow s}\,[\varnothing L]
}{\begin{array}{c}(np, ((np \diagdown s)\varnothing np, (np \varnothing np, np)^{\circ})^{\circ})^{\circ} \Rightarrow s \\ \hline (np, ((np \diagdown s)\varnothing np, (np \varnothing np, np)^{\bullet})^{\circ})^{\circ} \Rightarrow s \\ \hline (np, (((np \diagdown s)\varnothing np, np \varnothing np)^{\circ}), np)^{\bullet})^{\circ} \Rightarrow s \\ \hline ((np, ((np \diagdown s)\varnothing np, np \varnothing np)^{\circ}))^{\circ}, np)^{\bullet} \Rightarrow s \\ \hline ((np, ((np \diagdown s)\varnothing np,)^{\circ}), np \varnothing np)^{\circ}, \Rightarrow s/np\end{array}}{}
}{}
}{}
$$

[∅L] [<] [(22a)] [(22a)] [/R]

$$s_T \Rightarrow s_T \qquad [/L]$$

$(s_T \varnothing (s/np),$	$((np,$	$(np \diagdown s)\varnothing np)^{\circ},$	$np \varnothing np)^{\circ})^{\circ}$	$\Rightarrow s_T$
Jiu	wo	xihuan he	guizhou de	
Alcohol	I	like drink	Guizhou nominalizer	

In terms of alcohol, I like the ones from Guizhou.

(31) is the derivation of sentence with topicalized Oblique consititient.

(31)

$$
\cfrac{
np \diagdown s \Rightarrow np \diagdown s \quad
\cfrac{
\cfrac{
\cfrac{
\cfrac{np \Rightarrow np \quad s \Rightarrow s}{(np, np \diagdown s)^{\circ} \Rightarrow s}\,[\diagdown L]
}{(np, (np \diagdown s, (np \diagdown s)\diagdown (np \diagdown s))^{\circ})^{\circ} \Rightarrow s}\,[\diagdown L]
}{(np, (np \diagdown s, (np \diagdown s)\diagdown (np \diagdown s))^{\bullet})^{\circ} \Rightarrow s}\,[<]
}{((np, np \diagdown s)^{\circ}, (np \diagdown s)\diagdown (np \diagdown s))^{\bullet} \Rightarrow s}\,[(22a)]
}{(np, np \diagdown s)^{\circ} \Rightarrow s/(np \diagdown np)\diagdown (np \diagdown s)}\,[/R]
}{}
$$

$$s_T \Rightarrow s_T \qquad [\varnothing L]$$

$(s_T \varnothing (s_T/((np \diagdown s)\diagdown (np \diagdown s)))),$	$(np,$	$np \diagdown s)^{\circ})^{\circ}$	$\Rightarrow s$
ruanjian		wo	shimenwaihan
software		I	being a layman

As for software, I am a layman.

Next, let's take a hybrid system which contains only L and LP. Under the rules proposed in section 2 for this hybrid system, a sentence which lacks a np somewhere can be deduced by means of $s \circ\!\!- np$, as illustrated in (32).

(32)

$$\cfrac{\cfrac{\cfrac{\cfrac{\cfrac{\cfrac{pp \Rightarrow pp \quad \cfrac{np \Rightarrow np \quad s \Rightarrow s}{(np, np\backslash s)^{\bullet} \Rightarrow s}[\backslash L]}{(np, ((np\backslash s)/pp, pp)^{\bullet})^{\bullet} \Rightarrow s}[/L]}{(np, ((((np\backslash s)/pp)/np, np)^{\bullet}, pp)^{\bullet})^{\bullet} \Rightarrow s}[/L]}{(np, ((((np\backslash s)/pp)/np, np)^{\circledcirc}, pp)^{\bullet})^{\bullet} \Rightarrow s}[<]}{(np, ((((np\backslash s)/pp)/np, pp)^{\bullet}, pp)^{\circledcirc})^{\bullet} \Rightarrow s}[16c]}{(np, (((np\backslash s)/pp)/np, pp)^{\bullet})^{\bullet}, np)^{\circledcirc} \Rightarrow s}[16a]}{(np, (((np\backslash s)/pp)/np, pp)^{\bullet})^{\bullet} \Rightarrow s \circ\!\!-\, np}[\circ\!\!-\, R]$$

$(rel/(s \circ\!\!-\, np),$	$(np,$	$(((np\backslash s)/pp)/np, \quad pp)^{\bullet})^{\bullet})^{\bullet},$	$\Rightarrow rel$
who	Kim	sent	away

(with $rel \Rightarrow rel$ over $[/L]$)

In this way, we explain the non-peripheral extraction in the hybrid system containing L and LP. However, if "who" is assigned with the category of $rel/(s/np)$ or $rel/(s\varnothing np)$, then non-peripheral extraction is not explainable.

Now, we will derive some topicalized sentences within our L+LP system

(33) is the derivation of sentence with the topicalized Object without in-situ pronoun.

(33)

$$\cfrac{\cfrac{\cfrac{\cfrac{\cfrac{\cfrac{(np\backslash s)\backslash(np\backslash s) \Rightarrow (np\backslash s)\backslash(np\backslash s) \quad \cfrac{np \Rightarrow np \quad s \Rightarrow s}{(np, np\backslash s)^{\bullet} \Rightarrow s}[\backslash L]}{(np, ((np\backslash s)/((np\backslash s)\backslash(np\backslash s)), (np\backslash s)\backslash(np\backslash s))^{\bullet})^{\bullet} \Rightarrow s}[/L]}{(np, ((((np\backslash s)/((np\backslash s)\backslash(np\backslash s)))/np, np)^{\bullet}, (np\backslash s)\backslash(np\backslash s))^{\bullet})^{\bullet} \Rightarrow s}[/L]}{(np, (((np\backslash s)/((np\backslash s)\backslash(np\backslash s)))/np, (np, (np\backslash s)\backslash(np\backslash s))^{\bullet})^{\bullet})^{\bullet} \Rightarrow s}[A]}{(np, (((np\backslash s)/((np\backslash s)\backslash(np\backslash s)))/np, (np, (np\backslash s)\backslash(np\backslash s))^{\circledast})^{\bullet})^{\bullet} \Rightarrow s}[<]}{(np, (((np\backslash s)/((np\backslash s)\backslash(np\backslash s)))/np, ((np\backslash s)\backslash(np\backslash s), np)^{\circledast})^{\bullet})^{\bullet} \Rightarrow s}[P]}{(np, (((np\backslash s)/((np\backslash s)\backslash(np\backslash s)))/np, (np\backslash s)\backslash(np\backslash s))^{\bullet}, np)^{\circledast})^{\bullet} \Rightarrow s}[(16a)]$$

$$\cfrac{(np, (((np\backslash s)/((np\backslash s)\backslash(np\backslash s)))/np, (np\backslash s)\backslash(np\backslash s))^{\bullet}, np)^{\circledast} \Rightarrow s}{(np, (((np\backslash s)/((np\backslash s)\backslash(np\backslash s)))/np, (np\backslash s)\backslash(np\backslash s))^{\bullet})^{\bullet} \Rightarrow s \circ\!\!-\, np}[\circ\!\!-\, R]$$

$(s_T/(s \circ\!\!-\, np),$	$(np,$	$(((np\backslash s)/((np\backslash s)\backslash(np\backslash s)))/np,$	$(np\backslash s)\backslash(np\backslash s))^{\bullet})^{\bullet})^{\bullet},$	$\Rightarrow s_T$
Na-ge pingguo	Zhangsan	chi	le	
That-CL apple	Zhangsan	eat	Perf	

(with $s_T \Rightarrow s_T$)

As to that apple, Zhangsan has eaten (it).

(34) is the derivation of sentence with the topicalized complex subject modifier.

(34)

$$
\cfrac{
  \cfrac{
    np \Rightarrow np \qquad
    \cfrac{
      \cfrac{np \Rightarrow np \qquad s \Rightarrow s}{(np, np\backslash s)^\bullet \Rightarrow s}\ [\backslash L]
    }{}
  }{
    \cfrac{
      \cfrac{
        \cfrac{
          \cfrac{((np/np, np)^\bullet, np\backslash s)^\bullet \Rightarrow s}{(np/np, (np, np\backslash s)^\bullet)^\bullet \Rightarrow s}\ [A]
        }{(np/np, (np, np\backslash s)^\bullet)^\otimes \Rightarrow s}\ [<]
      }{(np, np\backslash s)^\bullet \Rightarrow (np/np) \multimap s}\ [\multimap R]
    }{}\ [/L]
  }\qquad s_T \Rightarrow s_T
}{(s_T/((np/np) \multimap s), \quad (np, \quad np\backslash s)^\bullet)^\bullet \Rightarrow s_T}\ [/L]
$$

$(s_T/((np/np) \multimap s),$	$(np,$	$np\backslash s)^\bullet)^\bullet$
Zhe-jian yifu	buliao	bu-cuo
This-CL clothes	fabric	not-bad

As to this clothes, the fabric (it uses) is pretty good.

If we associate the type $S_T/((np/np)\backslash s)$ to Zhe-jian yifu instead of $(S_T/((np/np) \multimap s)$, we can analyze *Zhe-jian yifu buliao bu-cuo* only in L as the following derivation shows:

(35)

$$
\cfrac{
  \cfrac{
    np \Rightarrow np \qquad
    \cfrac{np \Rightarrow np \qquad s \Rightarrow s}{(np, np\backslash s)^\bullet \Rightarrow s}\ [\backslash L]
  }{
    \cfrac{
      \cfrac{
        \cfrac{((np/np, np)^\bullet, np\backslash s)^\bullet \Rightarrow s}{(np/np, (np, np\backslash s)^\bullet)^\bullet \Rightarrow s}\ [A]
      }{(np, np\backslash s)^\bullet \Rightarrow (np/np)\backslash s}\ [\backslash R]
    }{}\ [/L]
  }\qquad s_T \Rightarrow s_T
}{(s_T/((np/np)\backslash s), \quad (np, \quad np\backslash s)^\bullet)^\bullet \Rightarrow s_T}\ [/L]
$$

$(s_T/((np/np)\backslash s),$	$(np,$	$np\backslash s)^\bullet)^\bullet$
Zhe-jian yifu	buliao	bu-cuo
This-CL clothes	fabric	not-bad

As to this clothes, the fabric (it uses) is pretty good.

(36) is the derivation of sentence with the topicalized complex object modifier.

(36)

$$
\cfrac{
\cfrac{
\cfrac{
\cfrac{
\cfrac{
\cfrac{
\cfrac{
\cfrac{
\cfrac{
np \Rightarrow np \qquad
\cfrac{np \Rightarrow np \qquad s \Rightarrow s}{(np, np\backslash s)^{\bullet} \Rightarrow s}\,[\backslash L]
}{(np, ((np\backslash s)/np, np)^{\bullet})^{\bullet} \Rightarrow s}\,[/L]
}{}
}{}
}{}
}{}
}{}
}{}
}{}
$$

(st/((np/np) ⊸ np),   ((np,   (((np\s)/np)•,   np)•)•,   ⇒ sT
Ni-men-ban de   Wo   shouxuan   Yangfan
You-plural-class nominalizer I   first-pick   Yangfan

Among all students in your class, Yangfan is my first choice.

np ⇒ np   s ⇒ s
(np, np\s)• ⇒ s   [\L]
np ⇒ np   (np, np\s)• ⇒ s   [/L]
(np, ((np\s)/np, np)•)• ⇒ s   [/L]
np ⇒ np   ((np, ((np\s)/np)•, np)• ⇒ s [A]
((np, ((np\s)/np)•, (np/np, np)•)• ⇒ s   [/L]
((np, ((np\s)/np)•, (np/np, np)⊗)• ⇒ s   [<]
((np, ((np\s)/np)•, (np, np/np)⊗)• ⇒ s   [P]
(((np, (np\s)/np)•, np)•, np/np)⊗ ⇒ s   [A]
(np/np, ((np, (np\s)/np)•, np)•)⊗ ⇒ s   [P]
((np, (np\s)/np)/np)•, np)• ⇒ np/np ⊸ s   [⊸ R]   sT ⇒ sT

(37) is the derivation of sentence with topicalized complex head word of subject.

(37)

$$
\cfrac{
\cfrac{
\cfrac{
\cfrac{
\cfrac{
\cfrac{
np \Rightarrow np \qquad
\cfrac{np \Rightarrow np \qquad s \Rightarrow s}{(np, np\backslash s)^{\bullet} \Rightarrow s}\,[\backslash L]
}{((np/np, np)^{\bullet}, np\backslash s)^{\bullet} \Rightarrow s}\,[/L]
}{(np/np, (np, np\backslash s)^{\bullet})^{\bullet} \Rightarrow s}\,[A]
}{(np/np, (np, np\backslash s)^{\otimes})^{\otimes} \Rightarrow s}\,[<*]
}{(np/np, (np\backslash s, np)^{\otimes})^{\otimes} \Rightarrow s}\,[P]
}{(np/np, (np\backslash s, np)^{\otimes}) \Rightarrow s \multimap np}\,[\multimap R]
\qquad s_T \Rightarrow s_T
$$

(sT/((np ⊸ np)\s),   (np/np, np\s)⊗)•   ⇒ sT   [/L]
Haizi   ta-men-de tinghua

As to kids, it is theirs that are obedient.

(38) is the derivation of sentence with topicalized complex head word of object.

(38)

$$
\cfrac{
\cfrac{
np \Rightarrow np \quad
\cfrac{
np \Rightarrow np \quad
\cfrac{
np \Rightarrow np \quad
\cfrac{
\cfrac{
\cfrac{
np \Rightarrow np \quad \cfrac{s \Rightarrow s}{}
}{(np, np\backslash s)^{\bullet} \Rightarrow s} [\backslash L]
}{(np, ((np\backslash s)/np, np)^{\bullet})^{\bullet} \Rightarrow s} [/L]
}{(np, ((np\backslash s)/np, (np/np, np)^{\bullet})^{\bullet})^{\bullet} \Rightarrow s} [\varnothing L]
}{(((np, (np\backslash s)/np)^{\bullet}, np/np)^{\bullet}, np)^{\bullet} \Rightarrow s} [A*]
}{(((np, (np\backslash s)/np)^{\bullet}, np/np)^{\bullet}, np)^{\otimes} \Rightarrow s} [<]
}{((np, ((np\backslash s)/np,)^{\bullet}), np/np)^{\bullet}, \Rightarrow s \circ\!\!- np} [\circ\!\!- R]
\quad s_T \Rightarrow s_T
}{\Rightarrow s_T} [/L]
$$

$(s_T/(s \circ\!\!- np)$,	$((np,$	$(np\backslash s)/np)^{\bullet},$	$np/np)^{\bullet})^{\bullet}$	
Jiu	wo	xihuan he	guizhou de	
Alcohol	I	like drink	Guizhou nominalizer	

In terms of alcohol, I like the ones from Guizhou.

If we associate the type $s_T/(s/np)$ to *jiu* instead of $s_T/(s \circ\!\!- np)$, we can analyze *Jiu wo xihuan he guizhou de* only in L as the following derivation shows:

(39)

$$
\cfrac{
\cfrac{
np \Rightarrow np \quad
\cfrac{
\cfrac{
\cfrac{
\cfrac{
\cfrac{
np \Rightarrow np \quad \cfrac{s \Rightarrow s}{}
}{(np, np\backslash s)^{\bullet} \Rightarrow s} [\backslash L]
\quad np \Rightarrow np
}{(np, ((np\backslash s)/np, np)^{\bullet})^{\bullet} \Rightarrow s} [/L]
}{(np, ((np\backslash s)/np, (np/np, np)^{\bullet})^{\bullet})^{\bullet} \Rightarrow s} [\varnothing L]
}{(((np, (np\backslash s)/np)^{\bullet}, np/np)^{\bullet}, np)^{\bullet} \Rightarrow s} [A*]
}{((np, ((np\backslash s)/np,)^{\bullet}), np/np)^{\bullet}, \Rightarrow s/np} [/R]
}{}
\quad s_T \Rightarrow s_T
}{\Rightarrow s_T} [/L]
$$

$(s_T/(s/np)$,	$((np,$	$(np\backslash s)/np)^{\bullet},$	$np/np)^{\bullet})^{\bullet}$	
Jiu	wo	xihuan he	guizhou de	
Alcohol	I	like drink	Guizhou nominalizer	

In terms of alcohol, I like the ones from Guizhou.

Another instance that can be analyzed by our hybrid system is given in (40):

(40)

$$
\cfrac{
\cfrac{
\cfrac{
\cfrac{
\cfrac{
\cfrac{
\cfrac{np \Rightarrow np \qquad s \Rightarrow s}{(np, np\backslash s)^{\bullet} \Rightarrow s}[\backslash L]
}{(np, ((np\backslash s)\backslash(np\backslash s)\backslash(np\backslash s))^{\bullet})^{\bullet} \Rightarrow s} \quad \cfrac{np\backslash s \Rightarrow np\backslash s}{}[\backslash L]
}{(np, (((np\backslash s)/np, np)^{\bullet}, (np\backslash s)\backslash(np\backslash s))^{\bullet})^{\bullet} \Rightarrow s} \quad np \Rightarrow np \;[\backslash L]
}{(np, (((np\backslash s)/np, np)^{\otimes}, (np\backslash s)\backslash(np\backslash s))^{\bullet})^{\bullet} \Rightarrow s}[<]
}{(np, (((np\backslash s)/np, (np\backslash s)\backslash(np\backslash s))^{\bullet}, np)^{\otimes})^{\bullet} \Rightarrow s}[(16c)]
}{(np, \quad ((np\backslash s)/np, \quad (np\backslash s)\backslash(np\backslash s))^{\bullet})^{\bullet}, \quad np)^{\otimes} \Rightarrow s}[(16a)]
}{}
$$

Zhangsan	chi	le	fan
Zhangsan	eat	Perf	meal

Zhangsan finished his meal already.

The antecedent of conclusion contain a $(,)^{\otimes}$, so we can go on to do it, as the followings:

(41)

$$
\cfrac{
\cfrac{
\cfrac{
\cfrac{
\cfrac{
\cfrac{
\cfrac{np \Rightarrow np \qquad s \Rightarrow s}{(np, np\backslash s)^{\bullet} \Rightarrow s}[\backslash L]
}{(np, ((np\backslash s)\backslash(np\backslash s)\backslash(np\backslash s))^{\bullet})^{\bullet} \Rightarrow s} \quad \cfrac{np\backslash s \Rightarrow np\backslash s}{}[\backslash L]
}{(np, (((np\backslash s)/np, np)^{\bullet}, (np\backslash s)\backslash(np\backslash s))^{\bullet})^{\bullet} \Rightarrow s} \quad np \Rightarrow np \;[\backslash L]
}{(np, (((np\backslash s)/np, np)^{\otimes}, (np\backslash s)\backslash(np\backslash s))^{\bullet})^{\bullet} \Rightarrow s}[<]
}{(np, (((np\backslash s)/np, (np\backslash s)\backslash(np\backslash s))^{\bullet}, np)^{\otimes})^{\bullet} \Rightarrow s}[(16c)]
}{(np, ((np\backslash s)/np, (np\backslash s)\backslash(np\backslash s))^{\bullet})^{\bullet}, np)^{\otimes} \Rightarrow s}[(16a)]
}{(np, \quad (np, \quad ((np\backslash s)/np, (np\backslash s)\backslash(np\backslash s))^{\bullet})^{\bullet})^{\otimes}, \quad \Rightarrow s}[/L]
$$

fam	Zhangsan	chi	le
meal	Zhangsan	eat	Perf

Zhangsan finished his meal already.

This sentence is also right in Chinese, but we usually view it as sentence with the topicalized Object without in-situ pronoun. But (40) and (41) violate our assuming: the derivational conclusion $\Gamma \Rightarrow s$ excludes the use of $(,)^{\otimes}$ in Γ.

In (40) and (41), transitive verb and auxiliary verb are adjoined , which makes it impossible to enter in LP if [<] is not used.

The analysis given above confirms that a hybrid method supports the general characteristics of language model. That is, the entailment connectives in the weakest logic available are used to encode syntactic information into lexical units, and the stronger logic in the hybrid system allows the (functional) analysis of the minor linguistic phenomena (which is a more flexible analysis). Of course, it is unlikely for the instances given above to cover all possibilities of encoding linguistic information into lexical units. Actually, the application of hybrid categorial type logic includes the analysis of bounded phenomena of pronoun and reflexive pronoun, and also of quantification.

BIBLIOGRAPHY

[1] Ades, Anthony E. and Steedman, Mark J: On the order of words. Linguistics and Philosophy, 4 (4):517 - 558 (1982)

[2] Barry, G., Hepple, M., Leslie,N. and Morrill, G.: Proof figures and structural opetator for categorial grammar. In: Proc. of EACL-5, Berlin (1991).

[3] Dosen, K.: Modal translations in substructural logics. In: Report, pp.10-90, Universitat Konstanz (1990)

[4] Hepple, M.: A general framework for hybrid substructural categorial logics. In: IRCS Report, pp.94-14 (1993)

[5] Hepple, M.: Mixing Modes of Linguistic Description in Categorial Grammar. In: Proc. of EACL-7 (1995)

[6] Hepple, M.: Hybrid Categorial Logic. In: Bull of the IGPL, Vol.3 No. 2,3, pp.344-355 (1995)

[7] Hepple, M.: A Dependency-based Approach to Bounded & Unbounded Movement. In: Becker, T. and Krieger,H (ed.) Proc. of the Fifth Meeting on Mathematics of Language(MOL-5), DFKI-D-97-02 (1997)

[8] Lambek, J.: The mathematics of sentence structure. American Mthematical Monthly 65, 154-170(1958)

[9] Lambek, J.: On the calculus of syntactic types. In Jakobson, R. (ed.), Mathematical Society, Structure of Language and its Mathematical Aspects, Proc. of the Symposia in Applied Mathematics XII, American (1961)

[10] Moortgat, M.: Categorial investigations: logical and linguistic aspects of the lambek calculus. Foris. Dordrecht (1988)

[11] Moortgat, M: Generalized quantification and discontinuous type constructors. To appear in Bunt, H. & van Horck, A. (eds) Discontinuous constituency. De Gruyter, Berlin (1991)

[12] Moortgat, M. and Morrill , G: Heads and phrases: Type calculus for dependency and constituency. To appear in: Journal of Language, Logic and Information (1991)

[13] Moortgat, M. and Oehrle, R: Logical parameters and linguistic variation. Lecture notes on categorical grammar. Fifth European Summer School IN Logic, Language and Information. Lisbon (1993)

[14] Moortgat, M. and Oehrle, R.: Adjacency, dependency and order.In: Proc. of the 9th Amsterdam Colloquium (1994)

[15] Morrill , G.: Rules and derivations: binding phenomena and coordination in categorial logic. DYANA Deliverable R1.2.D, ESPRIT Basic Research Action BR3175 (1990)

[16] Morrill , G.: Discontinuity and pied-piping in categorial grammar. Research Report LSI-93-18-R. Department de Llenguatges I Sistemes Informatics, Universitat Politecnica de Catalunya (1990)

[17] Morrill , G., Leslie, N., Hepple, M. and Barry, G.: Categorial deductions and structural operations. In Barry, G. and Morrill, G. (Eds), Studies in Categorial Grammar. Edinburgh Working Papers in Cognitive Science, Volume 5. Centre for Cognitive Science, University of Edinburg (1990)

[18] Morrill, G. and Solias, M.T.: Tuples, discontinuity, and gapping in categorial grammar. In: Proc. of the 6th Conferences of the European Chapter of the Association for Computational Linguistics(EACL-6), Utrecht(1993)

[19] Oehrle, R . and Zhang, S.: Lambek calculus and preposing of embedded subjecte. Chicago Linguistic Society 25 (1989)

Congjun Yao and Chongli Zou

[20] Steedman, Mark J.: Dependency and coordination in the grammar of Dutch and English. Language,61:523-56.(1985)
[21] van Benthem, J.: The semantics of variety in categorical grammar. In Report 83-29, Department of Mathematics, Simon Fraser University (1983). Also in W. Buszkowski, W. Marciszewki and J. van Benthem(eds), Categorial Grammar, Vol.25, Linguistic and Literary Studies in Eastern Europe, John Benjamins, Amsterdam/Philadelphia (1988)
[22] Versmissen, K.: Categorial Grammar, modalities, and algebraic semantics. In: Proc. of EACL-6, Utrecht (1992)
[23] Wansing, W.: Formulas-as-typess for a hierarchy of sublogics of Intuitionistic Propositional Logic. Ms. Institut fur Philosophie, Freie Universitat Berlin (1990)

Comments on Yao and Zou

ARNO BASTENHOF

This paper provides a type-logical analysis of topicalization and non-constituent coordination in Mandarin Chinese, while at the same time making a case for Hepple's proposal to combine the Lambek calculus **L** together with its variations **NL**, **NLP** and **LP** into a single "hybrid" logic.[1] While the paper succeeds in its discussion and analysis of the linguistic data, I have found that in certain respects its execution leaves room for further improvement. For example, despite the inclusion of a substantial bibliography at the end, none of the resources found therein were referenced (nor alluded to) in the main text, while conversely the works of Ades and Steedman cited in §3 do not appear in it. Moreover, while the authors have provided a helpful exposition on hybrid categorial grammar to make their paper self-contained, they have refrained from discussing its treatment of word order, differing notably from standard categorial practice. This latter point particularly causes confusion when confronted with the later linguistic analyses, seemingly projecting word order on sequents containing structural connectives that satisfy commutativity. Below, I shall address this issue in more detail, while suggesting various alternative solutions besides Hepple's by drawing both from some of the bibliographic entries already found in this paper but which were left undiscussed, as well as from certain articles that were excluded which reflect more recent developments. I will conclude with a brief evaluation of the authors' arguments against the use of structural modalities, again referring to more recent developments which I think may shed a different light on the issue.

To start with, consider the conclusion of (20), reproduced here:

$$(\text{ rel}/(s \multimap np) , (\text{ np } , (((np/s)/pp)/np , \text{ pp })^{\otimes})^{\otimes})^{\bullet} \Rightarrow \text{rel}$$
$$\text{who} \qquad \text{Kim} \qquad \text{sent} \qquad \text{away}$$

As it is, the use of $(,)^{\otimes}$ allows for the derivation of various 'unwanted' sequents as well, assuming it is understood, conforming to standard categorial practice, that word order is to be read off from the ordering of hypotheses. For example,

$$(\text{ rel}/(s \multimap np) , (\text{ pp } , (\text{ np } , ((np/s)/pp)/np)^{\otimes})^{\otimes})^{\bullet} \Rightarrow \text{rel}$$
$$\text{who} \qquad \text{away} \quad \text{Kim} \qquad \text{sent}$$

---

[1] I will occasionally speak of *multimodal* categorial grammar, in reference to terminology used by the closely related proposals found in [6] and [7].

The same problem affects most of the analyses of the Chinese data as well. Hepple ([3], §6) already made note of this fact, and went on to describe a procedure for obtaining word order from the derivational semantics of a proof. This rather crucial technicality, however, seems absent in the exposition on hybrid categorial grammar found in the authors' discussion. I would like to point to two alternative solutions, one tying in to the multimodal tradition, the other to Hepple's use of the semantics of a proof. The first takes inspiration from Moortgat and Oehrle's '93 and '94 papers ([6], [7]), already referred to in the bibliography. Theirs is a proposal very similar to Hepple's, though differing in a number of respects. Besides inverting $<$ in their take on rule (7) (a discrepancy I will henceforth ignore), they aditionally propose a number of 'interaction rules'. Adapted to Hepple's notation, two particular instances would read as follows:

$$\frac{\Gamma[(\Delta_1,(\Delta_2,\Delta_3)^{\otimes})^{\bullet}] \Rightarrow A}{\Gamma[((\Delta_1,\Delta_2)^{\bullet},\Delta_3)^{\otimes}] \Rightarrow A} \ (a) \qquad \frac{\Gamma[((\Delta_1,\Delta_3)^{\otimes},\Delta_2)^{\bullet}] \Rightarrow A}{\Gamma[((\Delta_1,\Delta_2)^{\bullet},\Delta_3)^{\otimes}] \Rightarrow A} \ (b)$$

Interestingly, the authors' (10) describes a similar rule, though differing in its nesting of $(,)^{\otimes}$ and $(,)^{\bullet}$ and having been left unmotivated because of its absence in the subsequent analyses. With (a) and (b), however, it seems possible to go back to using the ordering of hypotheses for determining word order. To this end, we should redefine the language recognized by a hybrid categorial grammar, taking into account only those derivations whose conclusion $\Gamma \Rightarrow s$ excludes the use of $(,)^{\otimes}$ in Γ.[2] For example, (20) may now be adapted as follows:

$$\frac{\dfrac{\dfrac{\dfrac{\dfrac{(\ldots \text{etc} \ldots)}{(\text{np},((((\text{np}/s)/\text{pp})/\text{np},\text{np})^{\bullet},\text{pp})^{\bullet})^{\bullet} \Rightarrow s}}{(\text{np},((((\text{np}/s)/\text{pp})/\text{np},\text{np})^{\otimes},\text{pp})^{\bullet})^{\bullet} \Rightarrow s}<}{(\text{np},((((\text{np}/s)/\text{pp})/\text{np},\text{pp})^{\bullet},\text{np})^{\otimes})^{\bullet} \Rightarrow s}(b)}{\dfrac{((\text{np},((((\text{np}/s)/\text{pp})/\text{np},\text{pp})^{\bullet})^{\bullet},\text{np})^{\otimes} \Rightarrow s}{(\text{np},((((\text{np}/s)/\text{pp})/\text{np},\text{pp})^{\bullet})^{\bullet} \Rightarrow s\multimap\text{np}}\ \multimap R}(a) \qquad \text{rel} \Rightarrow \text{rel}}{(\text{rel}/(s\multimap\text{np})),(\text{np},((((\text{np}/s)/\text{pp})/\text{np},\text{pp})^{\bullet})^{\bullet})^{\bullet} \Rightarrow \text{rel}}\ /L$$

A different approach may be found in the more recent resurgence of interest in Curry's distinction between tecto- and phenogrammar ([1]), as witnessed by the rise of abstract categorial grammars ([2]) and λ-grammars ([9]), though predated by similar efforts by Oehrle (e.g., [10]). Roughly, similar to Hepple's proposals for his hybrid logic, word order is determined using a Montagovian semantics, the difference being that it is now stipulated inside the lexical component rather than

---

[2]While seemingly an ad-hoc restriction, we note that in the Lambek-Grishin calculus ([5]) as well something similar happens. While several additional structural connectives are employed besides the usual counterpart for multiplicative conjunction, only the latter is used in practice for combining lexical materials.

being read off from its derivational counterpart. Thus, in the simplest execution of these ideas, a word is now assigned two λ-terms, the one, as usual, serving as an intermediate representation of its denotation, while the other (succumbing to Chomskyan terminology) roughly determines its moment of 'spell-out' relative to its arguments. Given the authors' discussion in §1 on the "unduly complicated accounts" resulting from the use of structural modalities when compared to Hepple's hybrid logic, it may be of interest to note that the alternative approach presently sketched allows for an even further simplification of one's categorial apparatus. Particularly, the need for the structurally more fine-grained **(N)L** and **NLP** dissapears, and one can make do with **LP** only. On the other hand, many of the technical machinery used in multimodal analyses can still be retained by transferring them to the semantic component, as shown by the aforecited [9].

To conclude with, I wish to return briefly to the authors' favourable judgement, already briefly alluded to above, of Hepple's methods when compared to the use of modal operators. While several of their objections against the latter were already voiced by Hepple as well, more recent developments go a long way towards alleviating the potential pitfalls raised by the use of structural modalities. For example, Vermaat ([11]) made only very limited use of modal operators in describing a small number of structural rules (resulting in not too complicated accounts), while furthermore arguing extensively for their cross-linguistic applicability. Moreover, Moot ([8]) offered a favourable analysis of both their computational expressivity and complexity, although it should be noted that his analogous results for the Lambek-Grishin calculus were later partially refuted by Melissen ([4]). Interestingly enough, the authors' type-logical analyses of Mandarin seem easily adaptable to Vermaat's setting, thus providing further evidence for the latter's cross-linguistic interest.

BIBLIOGRAPHY

[1] Haskell B. Curry. Some Logical Aspects of Grammatical Structure. In Roman O. Jakobson, editor, *Structure of Language and its Mathematical Aspects, volume 12 of Symposia on Applied Mathematics*, pages 56–68. American Mathematical Society, 1961.

[2] Philippe de Groote. Towards abstract categorial grammars. In *Proceedings of the 39th Annual Meeting on Association for Computational Linguistics*, ACL '01, pages 252–259. Association for Computational Linguistics, 2001.

[3] Mark Hepple. Mixing modes of linguistic description in categorial grammar. In *Proceedings of the seventh conference on European chapter of the Association for Computational Linguistics*, EACL '95, pages 127–132. Morgan Kaufmann Publishers Inc., 1995.

[4] Matthijs Melissen. The generative capacity of the Lambek-Grishin calculus: a new lower bound. In *Proceedings of the 14th international conference on Formal grammar*, FG'09, pages 118–132. Springer-Verlag, 2011.

[5] Michael Moortgat. Symmetric categorial grammar: residuation and galois connections. *CoRR*, abs/1008.0170, 2010.

[6] Michael Moortgat and Richard Oehrle. Logical parameters and linguistic variation. Lecture notes on categorial grammar. In *Fifth European Summer School in Logic, Language and Information, Lisbon*, 1993.

[7] Michael Moortgat and Richard Oehrle. Adjacency, dependency and order. In *Proceedings of the Ninth Amsterdam Colloquium*, 1994.

[8] Richard Moot. Lambek Grammars, Tree Adjoining Grammars and Hyperedge Replacement Grammars. In *Proceedings of the Ninth International Workshop on Tree Adjoining Grammars and Related Formalisms (TAG+ 9)*, pages 1–8, 2008.

[9] Reinhard Muskens. Separating Syntax and Combinatorics in Categorial Grammar. *Research on Language and Computation*, 5(3):267–285, 2007.

[10] Richard T. Oehrle. Term-labeled categorial type systems. *Linguistics & Philosophy*, 17(6):633–678, 1994.

[11] Willemijn Vermaat. *The Logic of Variation*. LOT Dissertation Series 121. Netherlands Graduate School of Linguistics / Landelijke (LOT), 2005.

Comments to Yao and Zou

Sylvain Pogodalla

1 Introduction

The paper "Hybrid Categorial Type Logics and the Formal Treatment of Chinese" presents the modeling of several linguistic phenomena occurring in Chinese. It relies on the formal framework of Hybrid Categorial Logics (HCL). This framework takes benefit from the ability to parametrize resource-sensitive logical systems with structural properties (such as associativity and commutativity), resulting in four systems:

NL: non-associative and non-commutative Lambek calculus [5];

NLP: non-associative and commutative Lambek calculus;

L: associative and non-commutative Lambek calculus [4];

LP: associative and commutative Lambek calculus, also known as (the intuitionistic fragment of) Linear Logic [1].

Each of the binary connectives of these systems can be indexed by a *composition mode i* to allow for the formulas[1] $\mathcal{F}$ where $\mathcal{A}$ is a set of atomic types:

$$\mathcal{F} ::= \mathcal{A} \mid \mathcal{F} \backslash_i \mathcal{F} \mid \mathcal{F} /_i \mathcal{F} \mid \mathcal{F} \bullet_i \mathcal{F}$$

The key point of hybrid systems is to offer a way to move from one deduction system to another with *inclusion* and *interaction postulates* [6] such as (1) or (2).

(1)
$$\frac{\Gamma[(\Delta_1, \Delta_2)^{\circ_i}] \Rightarrow A}{\Gamma[(\Delta_2, \Delta_1)^{\circ_i}] \Rightarrow A}$$

(2)
$$\frac{\Gamma[((\Delta_1, \Delta_2)^{\circ_i}, \Delta_3)^{\circ_j}] \Rightarrow A}{\Gamma[(\Delta_1, (\Delta_2, \Delta_3)^{\circ_j})^{\circ_i}] \Rightarrow A}$$

---

[1] And the corresponding family of the structural connectives $(\cdot, \cdot)^{\circ_i}$ that describes the structured antecedent in the sequent formulation of the calculus.

which allow for interaction between structures and formulas lying in different systems.

The authors emphasize that their modeling consider multimodality restricted to *binary* connectives, and that they favor not considering *unary* connectives as MultiModal Categorial Grammar (MMCG) [6] usually does. They show how this hybrid system allows them to give interesting accounts of several phenomena in Chinese. They focus in particular on various ways to express topicalization and conjunction of non constituents.

The modeling they propose brings to my attention different questions. Some of them, discussed in Section 2, relate to the formal system itself and to the motivations of using a multimodal system with only binary connectives rather than unary ones. They also relate to the design chosen for the interaction postulates. Another set of questions that Section 3 points out relate to the modeling capacity of the approach, and to the possible over-generation of the system.

2   On Formal Properties of Hybrid Categorial Grammars

2.1   Unary and Binary Operations

The paper puts a strong emphasis on the advantages of systems that use multimodality and structural postulates for binary connectives over those using unary connectives: "extensive use of structural modalities tends to result on very complex analyses" (in the abstract), "where [structural modalities] are used extensively, unduly complicated accounts tend to result" (end of the first section). However, we would like to have a better characterization of the nature of the involved complexity. Does it relate to the number of postulates? To the design of lexical categories and grammar engineering? To some computational issues?

It would be of a great value that the authors discuss the formal properties of the framework of hybrid logics with respect to MMCG. In particular, it is worth mentioning that there exists embedding theorems [2, 3, 6] between the different logics by mean of $\Box$ and $\Diamond$ unary connectives. These theorems use translations $(\cdot)^{\sharp}$ from formulas of a strong logic $\mathcal{L}_1$ into formulas of some weaker logic $\mathcal{L}_1$[2]. They state that

$$\mathcal{L}_1 \vdash A \Rightarrow B \text{ iff } \mathcal{L}_0 \Diamond + \mathcal{R}_{\Diamond} \vdash A^{\sharp} \Rightarrow B^{\sharp}$$

where $\mathcal{L}_0 \Diamond$ is $\mathcal{L}_0$ augmented with the unary connectives and $\mathcal{R}_{\Diamond}$ the relevant structural postulates. Would it be possible that any lexical item whose category requires some constraint relaxation and is thus expressed in a strong logic is modeled in the weaker logic **L** using the unary connectives. For instance, the topicalization of noun phrases of example (24) is modeled with $s_T/(s \multimap np)$[3] using **LP**. Would

---

[2]Different kinks of translations are considered in the above mentioned papers, but it seems to me it is the ones that are relevant here.

[3]We follow the usual way to express the right implication of **L** with / and the implication of **LP** with $\multimap$ rather than with indexed connectives.

it be possible to model this phenomenon using the unary connectives, possibly $s_T/(\Box s/np)$ and the adequate translation of the structured antecedent?

2.2 Postulates

There also are some questions regarding the structural postulates and the inter-action between the different composition modes. The paper should make them precise. The inclusion postulates are clearly stated. It is also clear which modes allow for permutation or associativity or combination of both. However, it should be explicitly stated that the associativity rule is bidirectional, as the examples show. That is both

$$\frac{\Gamma[((\Delta_1, \Delta_2)^\circ, \Delta_3)^\circ] \Rightarrow A}{\Gamma[(\Delta_1, (\Delta_2, \Delta_3)^\circ)^\circ] \Rightarrow A} \; [A]$$

and

$$\frac{\Gamma[(\Delta_1, (\Delta_2, \Delta_3)^\circ)^\circ] \Rightarrow A}{\Gamma[((\Delta_1, \Delta_2)^\circ, \Delta_3)^\circ] \Rightarrow A} \; [A]$$

stand, what is usually written

$$\frac{\Gamma[((\Delta_1, \Delta_2)^\circ, \Delta_3)^\circ] \Rightarrow A}{\Gamma[(\Delta_1, (\Delta_2, \Delta_3)^\circ)^\circ] \Rightarrow A} \; [A]$$

More importantly, the examples the paper provides don't make any use of the interaction postulate (rule (10) of the paper). Does it mean that it is not useful for the given examples but are useful for other phenomena? Does it mean the authors prefer not to consider this postulate? It would be interesting to comment on the scope of this postulate, both from a cross-linguistic perspective and from a Chinese language perspective. Also note that the interaction postulate slightly differs from the one given in [2]:

$$\frac{\Gamma[(\Delta_1, (\Delta_2, \Delta_3)^{\circ_i})^{\circ_j}] \Rightarrow A}{\Gamma[((\Delta_1, \Delta_2)^{\circ_i}, \Delta_3)^{\circ_j}] \Rightarrow A} \; \text{here vs.} \quad \frac{\Gamma[(\Delta_1, (\Delta_2, \Delta_3)^{\circ_j})^{\circ_i}] \Rightarrow A}{\Gamma[((\Delta_1, \Delta_2)^{\circ_i}, \Delta_3)^{\circ_j}] \Rightarrow A} \; \text{in [2]}$$

3   Linguistic Modeling with Hybrid Categorial Grammars

The end of the paper is devoted to providing analysis of linguistic phenomena in Chinese. It focuses on topicalization and conjunction of non constituents.

3.1   Specificities of Chinese

While the modeling that are proposed here are convincing, it would help the reader to have a better exposition of the specificities of Chinese. Regarding topicalization, as far as I can see, only example (24) "topicalized complex head word of subject" really needs inclusion postulates.

For instance in example (22), we can associate the type $s_T/((np/np)\backslash s)$ to *Zhe-jian* instead of $s_T/((np/np) \multimap s)$ and still analyze *Zhe-jian yifu buliao bu-cuo* as the following derivation shows:

$$\cfrac{\cfrac{\cfrac{np \Rightarrow np \quad \cfrac{np \Rightarrow np \quad s \Rightarrow s}{(np, np\backslash s)^\bullet \Rightarrow s}\,[\backslash\text{L}]}{((np/np, np)^\bullet, np\backslash s)^\bullet \Rightarrow s}\,[/\text{L}]}{\cfrac{(np/np, (np, np\backslash s)^\bullet)^\bullet \Rightarrow s}{(np, np\backslash s)^\bullet \Rightarrow (np/np)\backslash s}\,[\backslash\text{R}]}\,[\text{A}] \qquad s_t \Rightarrow s_T}{\underset{\text{Zhe-jian yifu}}{(s_T/((np/np)\backslash s),} \quad \underset{\text{buliao}}{(np,} \quad \underset{\text{bu-cuo}}{np\backslash s)^\bullet)^\bullet} \; \Rightarrow s_T}\,[/\text{L}]$$

In order to require the commutative implication, the example should illustrate an extraction that is not peripheral, as in (4) where the cleft word relates to the object rather than to the subject, contrasting with (3). Note that with the type $s_T/((np/np) \multimap s)$ for *zhe-ge* would enforce the grammaticality of (4)[4] as the derivation (5) shows.[5] Such contrasts would help the reader without any knowledge of Chinese to grasp the insights of the modeling proposed here.

(3)   *zhe-ge ban shang*   *xuesheng*   *xihuan*   *zhuxi*
        this class in   students   like   chairman
        $s_T/((np/np) \multimap s)$ np $(np\backslash s)/np$ np

'As to this class, the (its) students like the chairman'

(4)   *zhe-ge ban shang*   *zhuxi*   *xihuan*   *xuesheng*
        this class in   chairman   likes   students
        $s_T/((np/np) \multimap s)$ np $(np\backslash s)/np$ np

'As to this class, the chairman likes its students'

(5)

$$\cfrac{\cfrac{\cfrac{\cfrac{\cfrac{\cfrac{\cfrac{\cfrac{\cfrac{np \Rightarrow np \quad s \Rightarrow s}{(np, np\backslash s)^\bullet \Rightarrow s}\,[\backslash\text{L}] \quad np \Rightarrow np}{(np, ((np\backslash s)/np, np)^\bullet)^\bullet \Rightarrow s}\,[/\text{L}]}{((np, (np\backslash s)/np)^\bullet, np)^\bullet \Rightarrow s}\,[\text{A}]}{((np, (np\backslash s)/np)^\bullet, (np/np, np)^\bullet)^\bullet \Rightarrow s}\,[/\text{L}]}{((np, (np\backslash s)/np)^\bullet, (np/np, np)^\otimes)^\bullet \Rightarrow s}\,[<]}{((np, (np\backslash s)/np)^\bullet, (np, np/np)^\otimes)^\bullet \Rightarrow s}\,[\text{P}]}{(((np, (np\backslash s)/np)^\bullet, np)^\bullet, np/np)^\otimes \Rightarrow s}\,[(2)]}{(np/np, ((np, (np\backslash s)/np)^\bullet, np)^\bullet)^\otimes \Rightarrow s}\,[\text{P}]}{((np, (np\backslash s)/np)^\bullet, np)^\bullet \Rightarrow (np/np) \multimap s}\,[\multimap\text{R}] \quad s_t \Rightarrow s_T}{\underset{\text{zhe-ge ban shang}}{(s_T/((np/np) \multimap s),} \quad \underset{\text{zhuxi}}{((np,} \quad \underset{\text{xihuan}}{(np\backslash s)/np)^\bullet,} \quad \underset{\text{xuesheng}}{np)^\bullet)^\bullet} \; \Rightarrow s_T}\,[/\text{L}]$$

---

[4] To the best of my knowledge, this reading seems at least hard to get.

[5] Note that the derivation uses interaction postulate (2) . It's possible that other examples don't. I let it to Chinese speakers!

A similar comment apply to example (23). The following derivation:

$$
\dfrac{
\dfrac{
np\backslash s \Rightarrow np\backslash s \quad
\dfrac{np \Rightarrow np \quad s \Rightarrow s}{(np, np\backslash s)^\bullet \Rightarrow s}\ [\backslash L]
}{
\dfrac{
\dfrac{
\dfrac{(np, (np\backslash s, (np\backslash s)\backslash(np\backslash s))^\bullet)^\bullet \Rightarrow s}{((np, np\backslash s)^\bullet, (np\backslash s)\backslash(np\backslash s))^\bullet \Rightarrow s}\ [A]
}{(np, np\backslash s)^\bullet \Rightarrow s/((np\backslash s)\backslash(np\backslash s))}\ [/R]
}{(s_T/(s/((np\backslash s)\backslash(np\backslash s))), (np, np\backslash s)^\bullet)^\bullet \Rightarrow s_T}
}\ [\backslash L]
\quad s_T \Rightarrow s_T
}{}\ [/L]
$$

shows that (6) is grammatical with this typing.[6]

(6) *Ni-men-ban*   *we*   *shouxuan Yangfan*
     You-PL-class   I   first-pick Yangfan
     $s_T/(s/((np\backslash s)\backslash(np\backslash s)))$ np $(np\backslash s)$

And (7) illustrates a similar remark for example (25)

(7) *Jiu*   *wo*   *xihuan he*   *guizhou de*
     Alcool   I   like drink   Guizhou NOMI
     $s_T/(s/np)$ np $(np\backslash s)/np$ np/np
     $(s_T/(s/np), ((np, (np\backslash s)/np)^\bullet, np/np)^\bullet)^\bullet \Rightarrow s_T$

None of these examples makes use of an inclusion or interaction postulate. So it is really important to provide relevant examples focusing on the requirement Chinese makes on peripheral or non-peripheral extraction. The examples for conjunction don't make clear either what the specificities for Chinese are. If there are none, related works to conjunction for other languages should be mentioned.

Finally, one could improve the understanding of the reader in providing glosses with a better alignment. It also seems that the examples include expressions that are provided with a single type, such as *Zhe-jian* : $s_T/((np/np) \multimap s)$ whereas it is itself a combination of words. It would be worth providing the very lexical types so that we can better understand the combination even if the derivations make use of partially evaluated expression. It amounts to ask: could $(s_T/((np/np) \multimap s))/n$ be a suitable type for *zhe* and n a suitable type for *jian*?

3.2   Parsing, Proof-Search, and Over-generation

My final comment has to do with choosing which sequent to prove. Indeed, as soon as several composition modes are available, one needs to specify the ones that are used in the structured antecedent of the sequent to prove. This choice is very important since choosing one composition mode or the other could prevent

---

[6]I don't quite understand why the authors call this case "object modifier" since its rather a verb phrase modifier that is hypothesized in the derivation.

to use relevant structural postulates in the derivations and make them fail. So what are the strategies? Do we have to try all the possible composition modes? And what happens in case several structures would allow us to prove the sequent?

It is indeed striking that examples (22) and (25) only make use of the **L** composition mode, while (23) and (24) mix **L** and **LP** composition modes, and (26) uses only **LP** composition mode. The latter is very surprising because it implies that the grammar will generate all the permutations: if

$$(np \quad , \quad ((np\backslash s)/np \quad , \quad ((np\backslash s)\backslash(np\backslash s) \quad , \quad np)^{\otimes})^{\otimes})^{\otimes} \quad \Rightarrow s$$
$$\text{Zhangsan} \quad \text{chi} \qquad\qquad \text{le} \qquad\qquad\qquad \text{fan}$$

is derivable, so are *Zhangsan chi fan le*, *Zhangsan le chi fan*, *Zhangsan le fan chi*, *Zhangsan fan chi le*, *Zhangsan fan le chi*, *chi Zhangsan le fan*, *chi Zhangsan fan le* etc. I really wonder if all of them are acceptable.

4   Conclusion

This paper provides an interesting use of hybrid systems to model Chinese. Defining more precisely the specificity the system should handle would improve the relevance of the approach. As such systems require fine-tuning to avoid under- and over-generation, automatic systems such as Grail [7] could be used.

BIBLIOGRAPHY

[1] Jean-Yves Girard. Linear logic. *Theoretical Computer Science*, 50:1–102, 1987.

[2] Mark Hepple. Hybrid categorial logics. *Bulletin of the IGPL*, 2(2,3):343–355, 1995.

[3] Natasha Kurtonina and Michael Moortgat. Structural control. In Patrick Blackburn and Maarteen de Rijke, editors, *Specifying Syntactic Structures*, Studies in Logic, Language and Information, chapter 4, pages 75–113. CSLI Publications and FoLLI, 1997.

[4] Joachim Lambek. The mathematics of sentence structure. *American Mathematical Monthly*, 65(3):154–170, 1958.

[5] Joachim Lambek. On the calculus of syntactic types. In R. Jacobsen, editor, *Structure of Language and its Mathematical Aspects*, Proceedings of Symposia in Applied Mathematics, XII. American Mathematical Society, 1961.

[6] Michael Moortgat. Categorial type logics. In Johan van Benthem and Alice ter Meulen, editors, *Handbook of Logic and Language*, pages 93–177. Elsevier Science Publishers, Amsterdam, 1996.

[7] Richard Moot. Grail: an automated proof assistant for categorial grammar logics. In R.C. Backhouse, editor, *Proceedings of Calculemus/User Interfaces for Theorem Provers*, pages 120–129, 1998. Grail 3 is available at http://www.labri.fr/perso/moot/grail3.html.

Response to Bastenhof and Pogodalla

Congjun Yao and Chongli Zou

First we want to express our thanks to Sylvain Pogodalla and Arno Bastenhof for their precious and insightful comments. We agree that there is still room for improvement on our paper, and we will do so, and perfect it at a later stage.

1 Replies to both Arno Bastenhof and Sylvain Pogodalla

Both Bastenhof and Pogodalla pointed out two main problems. One is that our paper hardly makes theoretical contributions to Hybrid Categorical Type Logics, because we failed to take into account its new developments. The other is our failing to consider the over-productivity of the system constructed in our paper when applied to Chinese. For these two issues, our reply is as follows.

Actually, we began to study Hybrid Categorical Type Logics only last year, and the literature that we have read is far from sufficient. It is obvious that many new developments in the field have not been touched, not to speak of their application in the formal treatment of Chinese.

As far as we know, only few scholars have been involved in research in this field, and so our purpose was to undertake a first pilot study on the application of Hybrid Categorical Type Logics in analyzing Chinese structures. Therefore it is inevitable that there is little innovation on the theory of this framework and little in-depth analysis in our paper.

We did not consider over-generation at first but at the present time, we have perfected our old derivations by using interaction rules. In this way we can take into account only those derivations whose conclusion $\Gamma \Rightarrow s$ excludes the use of $(,)^{\otimes}$ in Γ, and eliminate the derivation of various "unwanted" sequents.

2 Reply to Sylvain Pogodalla

There are also some other problems pointed out by Sylvain Pogodalla also pointed out some further problems, to which we respond now.

Firstly, Prof. Pogodalla said:"As far as I can see, only example (24) 'topicalized complex head word of subject' really needs inclusion postulates" and "None of these examples makes use of an inclusion or interaction postulate".

We think that these doubts are based on assigning categories that are different from the ones we had in mind. In particular, we think that topicalized constituents can be assigned different categories. For example, the modifier extraction within complex subject can be assigned categories $s_T \emptyset((np \emptyset np) \backslash s), (s_T/((np/np) \multimap s)$ or $s_T/((np/np) \backslash s)$ respectively.

Secondly, because "zhe-ge ban shang zhuxi xihuan xuesheng" is ungrammatical in Chinese, we think the following proof given by Professor Pogodalla is not valid

(4) *zhe-ge ban shang*   *zhuxi*   *xihuan*   *xuesheng*
      this class in   chairman   likes   students
      $(s_T/((np/np) \multimap s)$ np $(np \backslash s)/np$ np
      'As to this class, the chairman likes students'

Finally, since "Ni-men-ban de wo shouxuan yangfan" is a sentence with a topicalized complex object modifier, we think the following analysis given by Prof. Pogodalla is not acceptable

(6) *Ni-men-ban*   *wo*   *shoucuan*   *Yangfan*
      You-PL-class   I   first-pick   Yangfan
      $(s_T/(s/((np \backslash s)\ (np \backslash s)))$ np $(np \backslash s)$

3 Adjustments

With the help of these comments we have improved our paper in a number of ways, both as to coverage and as to theoretical design.

We have changed the order of the derivations in the old paper. We now first take a hybrid system which contains only NL and L, and the non-associative structure NL of the hybrid system is used as the major structure in analysis of the lexicon. Then we give the derivation of sentences with Chinese non-constituent conjunction, topicalized complex subject modifier, topicalized complex head word of the object, and topicalized oblique constituent. Within L+LP, we give derivations of sentences with non-peripheral extraction, topicalized object without in-situ pronoun, topicalized complex subject modifier, topicalized complex object modifier, topicalized complex head word of subject, topicalized complex head word of object, and sentences with transitive verb and auxiliary verb adjoined.

Moroever, on the theoretecial system design side, we have now introduced the following new interaction rules:

(10) $\dfrac{\Gamma[(B,(C,D)^{\circ_i})^{\circ_j}] \Rightarrow A}{\Gamma[(B,C)^{\circ_i},D)^{\circ_j}] \Rightarrow A} \ a \quad (\circ_j < \circ_i) \qquad \dfrac{\Gamma[(B,C)^{\circ_i},D)^{\circ_j}] \Rightarrow A}{\Gamma[(B,(C,D)^{\circ_j})^{\circ_i}] \Rightarrow A} \ b \quad (\circ_j < \circ_i)$

$\dfrac{\Gamma[(B,C)^{\circ_i},D)^{\circ_j}] \Rightarrow A}{\Gamma[(B,D)^{\circ_j},C)^{\circ_i}] \Rightarrow A} \ c \quad (\circ_j < \circ_i)$

Using these rules, we perfect our old derivations, so that we can take into account only those derivations whose conclusion $\Gamma \Rightarrow s$ excludes the use of $(,)^{\otimes}$, and thereby eliminate the derivation of various 'unwanted' sequents.

An Ontology-Based Approach to Metaphor Cognitive Computation. A Summary[1]

XIAOXI HUANG, HUAXIN HUANG, BEISHUI LIAO AND CIHUA XU

Language understanding is one of the most important characteristics for human beings. As a pervasive phenomenon in natural language, metaphor is not only an essential thinking approach, but also an ingredient in human conceptual system. Many of our ways of thinking and experiences are virtually represented metaphorically. With the development of the cognitive research on metaphor, it is urgent to formulate a computational model for metaphor understanding based on the cognitive mechanism, especially with the view to promoting natural language understanding. Many works have been done in pragmatics and cognitive linguistics, especially the discussions on metaphor understanding process in pragmatics and metaphor mapping representation in cognitive linguistics. In this paper, a theoretical framework for metaphor understanding based on the embodied mechanism of concept inquiry is proposed. Based on this framework, ontology is introduced as the knowledge representation method in metaphor understanding, and metaphor mapping is formulated as ontology mapping. In line with the conceptual blending theory, a revised conceptual blending framework is presented by adding a lexical ontology and context as the fifth mental space, and a metaphor mapping algorithm is proposed.

Metaphor understanding involves many factors. Besides the transference of literal meaning and utterance meaning for general language understanding, there are factors of acquisition of embodied concept knowledge (embodied cognition), taxonomy of conceptual knowledge (categorization theory), constraints on conceptual mapping (Image Schemas, basic relations, context), subjective factor of agent, and the analogical association mechanism in conceptual mappings. In light of these analyses, we propose a general framework to describe the process of metaphor understanding. The main components of this model include:

- **Multi-domain Knowledge Base** includes the linguistic knowledge base for parsing and literal meaning extraction, the basic conceptual knowledge base

---

[1]This summary is based on [8].

for embodied concepts and the conceptual knowledge base about conceptual domain.

- **Parsing and Literal Meaning Extraction** provides the function to parse and extract the literal meaning of metaphorical sentence.

- **Associative Mechanism**. Association is a psychological process, which means bringing ideas or concepts together in memory or imagination. From the view of cross-domain mapping in metaphor, the associative mechanism is actually a bisociation [1, 11].

- **Reasoning Mechanism**. Reasoning is constructed on the premise of knowledge bases. It is a constraint on associative mechanism, and is used to select the mapping results.

- **Epistemic State** denotes agent's mental state, which considers the subjective factors of agent in metaphor understanding [10].

Therefore, the process of metaphor understanding can be described as follows: Firstly, in a given context, through analyzing the linguistic form and literal meaning of the metaphorical utterance, obtain its metaphorical elements and their corresponding conceptual domain; Secondly, under the current epistemic status, with the context and conceptual knowledge base, generate analogical mappings between concepts through association mechanism and select similarities between the source domain and the target domain [5, 12]; Finally, understand the metaphorical meaning through the reasoning mechanism. The reasoning results of metaphor understanding can be extended to the embodied conceptual knowledge base and conceptual knowledge base convinces that knowledge in the process of metaphor cognition is accumulative.

Metaphor Knowledge Representation

Knowledge plays a crucial role in metaphor understanding, and metaphor understanding involves cross-domain mappings, so, it is important to design a multi-domain knowledge base. We adopt ontology[6] as the tool to organize concepts, hereafter, we will introduce conceptual blending to describe metaphor mapping based on ontology. We define conceptual primitives to represent knowledge for metaphor understanding, which can be divided into two types: nodes and links. *Conceptual Constituent, Relation Name, Theta Role Name, Individual Name* are nodes, and *Taxonomy of Concepts and Relations, Part-Whole Relation, Instance, Equivalence, Metaphorical Relation* are links. From the view of graph theory, node is vertex of graph, and link is the edge between vertexes.

The concept and formula can be defined with the syntax and semantics of Description Logic as follows. Let C_1 and C_2 be concept constituents, a and b are individual names, R is a relation name, T_1 and T_2 are theta role names, then:

- C_1 and C_2 are concepts;

- $C_1 \sqcap C_2, \neg C_1$ and $\exists R.C_1$ are concepts;

- $T_1 : C_1$ is concept.

- $C_1(a)$, $R(a,b)$, $C_1 \equiv C_2$, $C_1 @ C_2$, $C_1 \sqsubseteq C_2$ and $C_1 \sim C_2$ are all atomic formulas;

- the boolean composition of atomic formulas is formula, e.g., if α and β are formulas, then $\alpha \wedge \beta$ and $\neg \alpha$ are formulas.

The semantic model for conceptual primitives is a tuple $< \triangle^{\mathcal{I}}, \cdot^{\mathcal{I}} >$, where $\triangle^{\mathcal{I}}$ is a non-empty set (the domain of the interpretation) and $\cdot^I$ is an interpretation function, which assigns to every atomic concept C a set $C^{\mathcal{I}} \subseteq \triangle^{\mathcal{I}}$, to every relation R a binary relation $R^{\mathcal{I}} \subseteq \triangle^{\mathcal{I}} \times \triangle^{\mathcal{I}}$, to every theta role T an element $T^{\mathcal{I}} \in N_T(N_T \subseteq \triangle^{\mathcal{I}})$, and to every individual a an element $a^{\mathcal{I}} \in \triangle^{\mathcal{I}}$. Then, the semantic interpretation function can be extended to complex concept description and formulas by the classical inductive definitions. A formula φ is true under $\mathcal{I}$, denoted as $\mathcal{I} \models \varphi$, and we call $\mathcal{I}$ a model of φ. If there is a model for φ, we say φ is satisfiable. For a concept C, if there is a model $\mathcal{I}$, such that $C^{\mathcal{I}} \neq \emptyset$, then we say C is satisfiable.

Given a domain space $< \mathcal{D}, \mathcal{W} >$, where $\mathcal{D}$ is a domain and $\mathcal{W}$ is a set of maximal states of affairs of such domain(also called possible worlds), an *ontological model* is defined as a tuple $\mathcal{O} =< \mathcal{C}, \mathcal{H}_{\mathcal{C}}, \mathcal{R}, \mathcal{H}_{\mathcal{R}}, \mathcal{T}, \mathcal{A}, \mathcal{D} >$, where, $\mathcal{C}$ is set of concepts, $\mathcal{H}_{\mathcal{C}} \subseteq \mathcal{C} \times \mathcal{C}$ denotes the inclusion between concepts; $\mathcal{R} \subseteq \mathcal{C} \times \mathcal{C}$ is the set of properties, denoting the binary relations between concepts; $\mathcal{H}_{\mathcal{R}} \subseteq \mathcal{R} \times \mathcal{R}$ denotes the inclusion between properties; $\mathcal{T} \subseteq \mathcal{C} \times \mathcal{TH}$ denotes the theta role property, where $\mathcal{TH} \subseteq \mathcal{D}$ is the set of theta roles; $\mathcal{A}$ is the axiom set provided by the description language; $\mathcal{D}$ is the domain.

Concepts are represented as words in natural language. In order to get the appropriate concept one word represent, we introduce an auxiliary ontology called L*exical Ontology*, denoted as a tuple $\Omega = (L, LR, M, \eta, SR)$,where L is a set of words, called the vocabulary of Ω; LR is the set of lexical relations, and is the description of relations between words, such as antonym, participle, derivation, and so on; M is the set of meanings(or senses) of Ω, every $m \in M$ is a meaning represented by a word, and every meaning has a corresponding concept in ontology; $\eta : L \rightarrow 2^M$ is a function. For any word $l \in L$, $\eta(l) = \{m_1, m_2, ..., m_k\}$, denotes that l has k meanings; SR is the set of relations between concepts, such as hyponymy, part-whole and so on, the same as in ontology.

Metaphor Mapping

As we know, metaphor involves the mapping between two conceptual domains.

On the basis of Ontological Model and Lexical Ontology, we will discuss the formalization of metaphor from two levels: linguistic level and conceptual level. On the linguistic level, the formalization of metaphor is to tag metaphor roles on the parsing result of a metaphor utterance, which was discussed in [14] and [9]. On the conceptual level, we introduce mental spaces as source domain and target domain, and the representation of mental spaces as ontological model.

In the conceptual blending theory, the mental spaces are represented as frames [4]. Here, we will use ontology to describe mental spaces, and design different ontological models for specific mental spaces according to their functions in blending.

Generic space is described as a general ontology, providing background knowledge for conceptual mapping in the blending. The generic space ontology is defined as GO. The main functions of GO include:(a) providing world knowledge; (b) providing support for mappings between input spaces, such as similarity computation, semantic relations, and so on. The generic space ontology is constructed on the basis of HowNet [3] and Extended-HowNet[2].

Input spaces represent the knowledge of source concept and target concept, called source ontology IO_S and target ontology IO_T. According to conceptual blending theory [4], input space is an online structure which is generated according to the ingredient of metaphor during an agent understanding a metaphor, so, there are two steps to form an input space. Firstly, obtain corresponding metaphorical words in tagged sentence resulted from metaphor identification and tagging module, which was implemented in [9] and [7]. Secondly, for the words tagged as source of metaphor, we can get the corresponding concepts through Text-to-Onto approach based on lexical ontology and Generic Ontology, and then create the source ontology IO_S. Analogously, the target ontology IO_T is created from the words tagged as target of metaphor.

Therefore, the four-space model of conceptual integration network can be extended to a five-space network . In this revised network model, lexical ontology and context information (i.e. the added space) are constraints on input spaces

As we know, ontological model can be regarded as a graph, then, conceptual blending can be converted to looking for isomorphism, match and mapping relations between graphs. In conceptual blending theory [4], there are three cognitive operators, i.e. *Composition, Completion* and *Elaboration*. Before running these three operators, elements in two input spaces are projected into the blended space.

We define a mapping operator Φ to represent the corresponding relation between input spaces, namely $\Phi : O_1 \times O_2 \rightarrow \Pi$, where O_1 and O_2 are mental spaces denoted by ontology, and Π is a set of all the candidate element mappings between O_1 and O_2. To generate as many candidate mappings as possible, referring to the *Triangulation Rule* and *Squaring rule* in Sapper [13], the mapping operator consists of three steps corresponding to the cognitive operators in conceptual blending respectively: 1) generate all possible candidate mappings through *Triangulation Rule* and *Squaring rule*; 2) evaluate the candidates; 3) project these elements into blended space selectively.the details of the algorithm **GenerateCandidates** for mapping operator Φ can refer to [8]. The mapping operator Φ takes on the Composition operator in conceptual blending.

After the candidate mappings are generated, according to the optimality principles and governing principles for compression in conceptual blending theory, we present a quantification equation to evaluate a candidate mapping, called Integrated Degree. Suppose $m = (nodeT, nodeS)$ is a set of candidate mappings, then, the integrated degree of m is defined as $I(m) = \frac{(m.size-1)\times 2}{nodeT.rSize+nodeS.rSize}$, where $rSize$ denotes the number of edges in connection with the node, and size is the number of candidate mappings which are raised from $nodeT$ and $nodeS$. The candidate mappings with integrated degree over some threshold are projected into the blended space. An example shows the above procedure can be found in [8] in details.

BIBLIOGRAPHY

[1] Margaret A. Boden. *The Creative Mind: Myths and Mechanisms*. Routledge, London, 2003.
[2] Keh-Jiann Chen, Huang Shu-Ling, Shih Yueh-Yin, and Chen Yi-Jun. Extended-hownet: A representational framework for concepts. In *OntoLex 2005- Ontologies and Lexical Resources IJCNLP-05 Workshop*, Jeju Island, South Korea, 2005.
[3] Zhen Dong Dong and Qiang Dong. *Hownet And the Computation of Meaning*. World Scientific Publishing Company, Singapore, 2006.
[4] Gilles Fauconnier and Mark Turner. *The Way We Think: Conceptual Blending and The Mind's Hidden Complexities*. Basic Books, 2002.
[5] D. Gentner. Structure-mapping: A theoretical framework for analogy. *Cognitive Science*, 7:155–170, 1983.
[6] T. Gruber. A translation approach to portable ontology specification. *Knowledge Acquisition*, 5:199–220, 1993.
[7] Xiao Xi Huang. *Research on Some Key Issues of Metaphor Computation*. PhD thesis, Zhejiang University, Hangzhou, China (in Chinese), 2009.
[8] Xiao Xi Huang, Hua Xin Huang, Bei Shui Liao, and Ci Hua Xu. An ontology-based approach to metaphor cognitive computation. *Minds and Machines*, 23(1):105–121, 2013.
[9] Xiao Xi Huang, Hua Xin Huang, Ci Hua Xu, Weiying Chen, and Rong Bo Wang. A novel pattern matching method for chinese metaphor identification and classification. In *Proceedings of the 2011 International Conference on Artificial Intelligence and Computational Intelligence (AICI'11)*, pages 104–114, Taiyuan, China, 2011. Springer.
[10] Xiao Xi Huang and Chang Le Zhou. A logical approach for metaphor understanding. In *Proceedings of 2005 IEEE International Conference on Natural Language Processing and Knowledge Engineering*, pages 268–271, Wuhan, China, 2005. IEEE Computer Society.
[11] Francisco Camara Pereira. *Creativity and artificial intelligence : a conceptual blending approach*. Mouton de Gruyter, Berlin, 2007.
[12] Eric Charles Steinhart. *The Logic of Metaphor: Analogous Parts of Possible Worlds*. Kluwer Academic Publishers, Dordrecht, 2001.
[13] Tony Veale. *Metaphor, Memory and Meaning: Symbolic and Connectionist Issues in Metaphor Interpretation*. PhD thesis, Trinity College, Dublin, 1995.
[14] Yun Yang, Chang Le Zhou, Xiao Jun Ding, Jia Wei Chen, and Xiao Dong Shi. Metaphor recognition: Chmeta, a pattern-based system. *Computational Intelligence*, 25(4):265–301, 2009. 10.1111/j.1467-8640.2009.00349.x.

Comments on Huang, Huang, Liao and Xu

EMILIANO LORINI

The authors propose a computational approach to metaphor understanding based on some existing cognitive theories of metaphorical reasoning such as conceptual metaphor theory by Lakoff and Johnson and conceptual blending theory by Fauconnier and Turner. The approach is interesting as it builds a connection between the area of computational models of metaphorical reasoning and metaphor understanding and the area of formal ontologies, of which the most representive example is description logic.

There are a number of issues that I would like the authors to address in their work. I think that these clarifications are fundamental in order to clarify the connections between the theory of metaphorical reasoning and metaphor understanding proposed by the authors and some important research topics in cognitive science and artificial intelligence. Let me discuss these aspects in more detail.

Metaphors and beliefs The authors emphasize the importance of an agent's epistemic state in understanding metaphors. For example, the agent's interpretation of the sentence "market is sea" highly depends on the agent's conceptual knowledge about "markets" and "seas". An agent a might consider "sea" a dangerous place, whereas a different agent b might consider it a safe place. These two different intepretations of the concept "sea" may lead to different understandings of the metaphorical expression. In particular, agent a may understand the metaphor "market is sea" as "market is full of danger", while agent b may understand it as "market is a safe place".

Thus, the following observation is in order: since metaphor understanding is knowledge-dependent, a formal model of metaphor understanding should be able to represent the epistemic and doxastic mental states of agents.

Indeed, as it is well-known in cognitive science, an agent's conceptual knowledge can be represented in terms of frames or scripts [6, 8] which are nothing but sets of fundational beliefs representing an agent's unproblematic interpretation of the context of the situation where her action and perception are situated. For example, an agent's conceptual knowledge about (the concept of) "restaurant" is made of beliefs such as "restaurants have waiters", "restaurants have tables", "restaurants have chairs", etc. Thus, when an agent believes that he has entered in a restaurant, that agent can can then reasonably expect to see some waiters, some tables, some chairs, etc.

The computational model of metaphor understanding presented by the authors has nothing to say about this fundamental relationship between an agent's beliefs representing her conceptual knowledge and the agent's capability to understand metaphors. My suggestion for the authors is to explain how their formal model of metaphor understanding can be extended in order to incorporate this kind of beliefs. This clarification could be of interest for people working at the intersection between cognitive science and formal epistemology, as it will build a connection between the cognitive theory of metaphorical reasoning and existing logical theories of epistemic states (e.g., knowledge and belief) [2] and their dynamics [9].

Metaphors and theory of mind Another issue that I would like the authors to clarify is the crucial role of theory of mind (ToM) in metaphor understanding. Theory of mind [3] refers to the cognitive capacity (i) to attribute mental states to self and to others, and (ii) to use these attributions to explain and predict the actions of self and others. Agents express metaphors during conservation and dialogue. From this perspective, metaphor production and metaphor understanding are, by definition, social activities that involve a speaker (i.e., the agent who express the metaphor) and a hearer (i.e., the agent who has to understand it). Therefore, an agent's capacity to understand metaphors relies on her more general capability to ascribe beliefs to the speaker and to understand the speaker's communicative intention [4]. I wonder whether the computational model of metaphor understanding proposed by the authors can be extended in order to incorporate a minimal level of mindreading that can help the hearer to understand the *meaning* of the metaphor uttered by the speaker.

Metaphors and emotions A third point that it would be interesting to discuss is the relationship between the computational approach to metaphorical reasoning and metaphor understanding proposed by the authors and existing theory of emotions. As emphasized by psychologists [7], language that humans use to talk about emotions is full of metaphorical expressions. The main reason is that emotions have an elusive, transient quality that is difficult to describe using literal expressions. Examples of emotional metaphors are for example "I was beside myself with grief when I failed the exam" or "she is absolutely cheesed off with her job". Given this fundamental connection between emotions and metaphors, I would suggest the authors to explain how their model of metaphor understanding can be combined with existing formal and computational models of emotions developed in the area of artificial intelligence (see, e.g., [5, 1]) in order to endow agents with the capability to understand metaphorical emotion expressions on the basis of a comprehensive formal theory of emotions.

BIBLIOGRAPHY

[1] M. Dastani and J.-J. Ch. Meyer. Agents with emotions. *International Journal of Intelligent Systems*, 25(7):636–654, 2010.

[2] R. Fagin, J. Halpern, Y. Moses, and M. Vardi. *Reasoning about Knowledge*. MIT Press, Cambridge, 1995.

[3] A. I. Goldman. Theory of mind. In E. Margolis, R. Samuels, and S. Stich, editors, *Oxford Handbook of Philosophy and Cognitive Science*. Oxford University Press, 2012.

[4] H. P. Grice. Utterer's meaning and intentions. *Philosophical Review*, 78:147–177, 1969.

[5] E. Lorini and F. Schwarzentruber. A logic for reasoning about counterfactual emotions. *Artificial Intelligence*, 175(3-4):814–847, 2011.

[6] M. Minsky. A framework for representing knowledge. In R. J. Brachman and H. J. Levesque, editors, *Readings in Knowledge Representation*, pages 245–262. Kaufmann, Los Altos, CA, 1985.

[7] A. Ortony and L. Fainsilber. The role of metaphors in descriptions of emotions. In Y. Wilks, editor, *Theoretical Issues in Natural Language Processing*, pages 178–182. Erlbaum, Hillsdale, NJ, 1989.

[8] R. C. Schank and R. Abelson. *Scripts, Plans, Goals, and Understanding*. Earlbaum Assoc., Hillsdale , NJ, 1977.

[9] H. van Ditmarsch, W. van der Hoek, and B. Kooi. *Dynamic Epistemic Logic*, volume 337 of *Synthese Library*. Springer, 2007.

Comments on Huang, Huang, Liao and Xu

JOHN WOODS AND PETER D. BRUZA

"An ontology-based approach to metaphor cognitive computation" is an engaging contribution to the computational modelling of metaphor. The paper is fashioned within a conceptual blending framework, enriched in selective ways. "Based on this framework", say its authors, "ontology is introduced as the knowledge representation method in metaphor understanding, and metaphor mapping is formulated as ontology mapping." For ease of reference let's call this "the ontology mapping model" or, for short, the "OM model". The OM model is fashioned from well-known theoretical and conceptual elements, supported by a growing literature. The ensuing synthesis is a good example of a prominent way of modelling concepts instantiated in complex patterns of real-life human behaviour.

Our comments here will be divided into two sections. In section 1 we will discuss the relationship between a formal mechanism such as OM and the data the machinery is intended to model. Section 2 shifts attention to the formal mechanism itself, and briefly considers its mathematical and operational adequacy.

1   Data

It is not uncommon that a theorist's models will be highly idealized and its theorems radically distant from behavioural facts on the ground. Sometimes ideal models are chosen to facilitate the mathematical expressibility of its provisions. Sometimes simplication is the rationale. In the first instance, mathematical expressibility trumps the desideratum of empirical engagement. In the second instance, the rationale is to start small – evocative of the idea that it is best not to bite off more than we can chew, and that big journeys are best achieved in modest, well-managed steps. Naturally enough, the established modelling methodologies give rise to models of varying distance from the ground. This raises an obvious question: How accurate is it to say of a distant model that it is a formal representation of behavioural facts on the ground? An obvious answer is that it depends on the model's distance from them. We ourselves aren't so sure.

It would be wrong to leave the impression that any of this is news to the formal modelling community, still less that its concern for the facts of real-life is sham. Of all the ways in which a formal modeller might show his respect for the ground-data – to take a simplified example, – suppose that the data the ground D exhibit properties $P_1, ..., P_n$ and that a theory M that formally represents them gives lexical or syntactico-semantic recognition to only some proper subset of the P_i. It is sometimes conceded that when this happens the intended formalization of D is marred – even discredited – by over-simplification. One method of repair is to enlarge the representational capacities of M by extending it to M^+. Early examples of this are the early agent-centred temporal systems of logic and AI.

It is not to be denied that most of the going time-and-action models to date are as distant from the facts of human behaviour on the ground as logics that make no room for agents or actions or times. But it can reasonably be argued that, if otherwise properly made, M^+ could be a better way of modelling D than M is. Here is why. Let D^+ be the subset of D whose formalization is sought by M^+. We might suppose that empirical investigation discloses the presence in D^+ of properties $P_1, ..., P_j$ and various patterns of interrelation between and among them. If the empirical investigator wanted a *theory* of such interconnections, a standard way of providing it would be to conceptualize the P_i and its patterns of connection in ways that confer upon them the sought-for theoretical significance – explanatory coherence say. We now see that the notion of *structure* has entered our story in a load-bearing way. These data themselves are structured, and their conceptualizations in turn make for conceptual structure. The present point can now be simply stated. If it is intrinsic to formal methods to handle at best a proper subset of properties on the ground, it is best to pick that subset of them that best reveals structure on the ground. Why? Because it is a bettable strategy to think that structure present in conceptualizations of those data will have an as yet unexamined presence in the data not covered by M^+. Then the job for the formal modeller is to try to effect patterns of representability between the formal elements of M^+ and the elements of D^+ in ways that match structure to structure. And now the basic idea is clear. You won't get those structures right unless you admit the right kinds of elements in the right numbers. So if it is the structure of human thinking that you're after you won't get the structure you want unless reasoning-agents and reasoning-actions have a formal presence in your M^+. The moral: Try in M^+ to capture those P_j that are structure-bearing.

Case by case implementation of this idea requires that the formalizer have an accurate grasp of what the P_i of his target-data actually are. So while it needn't be the case that his *theory* touches base with the ground, the same can't be said of the theorist himself. If he doesn't know what the P_i are, he lacks a fulcrum for the erection of the formal apparatus. In the present case, the target data are those discernible in the employment of metaphor by human speakers. What are its relevant properties? Real-life examples of such utterances would seem to include sentences such as

(1) Life's a barrel of laughs if you keep your pecker up.[1]
(2) With tear-soaked soul, Lucille surrendered her dignity to the fires of cruel ambition.
(3) My heart belongs to Daddy.

Perhaps this is an already troubled list. Linguists, philosophers of language, and literary scholars have long been aware of the blurriness of the distinctions between simile, analogy and metaphor. It is a trichotomy of sufficient unruliness to make it unlikely that everyone would agree that our three sentences are metaphorical in the sense marked out by the OM model for treatment. So it would not be unreasonable to ask what might further be said about this matter.

Another possible complication is an apparent ambiguity in the very word "metaphor", illustrated as follows:

(4) *Tender is the Night* is a metaphor for the psychic collapse of 1930s Europe.
(5) "In 'My heart belongs to Daddy', 'heart' is a metaphor for love."

It is easy to see that (4) and its like strain the compositional resources of the OM model (or any such other). *Tender is the Night* is not an utterance. Even if we say that in some extended sense it is a composite utterance – the fusion of the book's constituent utterances – it seems not to fit the OM apparatus. What would be the literal meaning of the compound entity that is the utterance constituted by all the book's sentences jointly? How would the OM belief-mechanisms render such entities believable? And, of course, there is presently no text that is literally about the psychic decline of 1930s to which the metaphorical text of *Tender is the Night* could be mapped in anything like a structure-preserving way.

We think the authors are entirely right in their favourable regard for the work on metaphor done by George Lakoff and colleagues ([4],[5]). As in all of Lakoff's work, we have here a deft admixture of empirical observation and conceptual analysis. Metaphor turns out to be a complex and elusive idea, a "slippery customer" as Lakoff once said. Much of this work is conjectural or promissory. Much of it is fragmentary. It is messy. How could it not be? It lies in the nature of metaphor to be messy. OM is an attempt to model such properties of metaphor as may be discerned in the least messy parts of a Lakoff-like theory. But it is not entirely clear from reading Lakoff what those properties are. It's not clear how they are structured. It is not clear here either.

The point of these remarks is to motivate an earlier question. Do we have a

---

[1]A quick word of explanation. In British English "keep your pecker up" means "keep your chin up" (which means, in turn, "look on the bright side and be focused and steadfast".) In the English spoken in other countries, such as Canada and the U.S., the phrase offers some advice about an important part of the male anatomy. That is not its meaning here.

principled command of the *data* the OM model is designed for? A related question now follows. Given that these are the data D (whatever they turn out to be), what are the model's conditions of conceptual adequacy? To what extent do we have to get the right conceptualization of the D-properties to make their structure discernible? A third question is one of what we might call the "mathematical adequacy" of the OM machinery. This is the question to which we now turn.

2 Machinery

The OM model focusses on the representation and semantics of concepts. The basis of the model is description logic (Definition 2). Its description logic foundation traces back to structured inheritance networks which were developed in the field of artificial intelligence during the 1980's. This work led to terminological logics, the forerunners of today's description logics. [2] documents this history well and also points out two deficiencies of the ontological approach. The first is the inability to deal with prototypicality which is crucial to allowing exceptions to inheritance. The second is compositionality. In the remainder of this section we will focus on it.

Description logics are cut-down versions of first order logic, the semantics of which are compositional. Semantic compositionality is expressed by the second last bulleted point of Definition 3 (Semantic interpretation). The question is whether semantic compositionality is always appropriate in relation to metaphors. Consider the example, "He is a BLACK CAT" (i.e. a "cool" coloured person). The intersective semantics proposed by Definition 3 would be rendered as the intersection of the domain of black objects and the domain of cat objects. if one assumes that the domain of CAT is restricted to elements corresponding to animals, this intersection is empty so the semantics are vacuous. Similarly the semantics of "He is a CACTUS COWBOY" are vacuous as the intersection of cactus objects and cowboy objects is empty It is well-known from Hampton's work on conceptual combinations that intersections may be over – or under – extended, a fact which challenges intersective approaches to compositional semantics. ([3]). This is related to protypicality effects that description logic can't express ([2]). Moreover, it is prototypicality that confounds compositionality. For example, the prototype of PET FISH (i.e. a goldfish) can't be seen as a composition of the prototype of PET (i.e., a furry, loveable animal) and FISH (i.e., the grey scaly thing on your dinner plate). These issues matter for the definition of semantic interpretation. It might reasonably be expected that because conceptual combinations sometimes allow for creative expressions in language, they may therefore figure quite prominently in metaphors, since these often creatively combine the senses of constituent concepts to furnish unusual interpretations which in turn underpin the force of the metaphor. Indeed, [1] argue that some conceptual combinations are classified as non-compositional because they make use of cognitive processes such as metaphor, analogy or metonymy in their interpretation. Their position highlights the connection between metaphor and semantic non-compositionality rather than semantic compositionality.

Our remarks here should not be construed as an outright rejection of the authors' approach. Concepts do underpin metaphors. However the question of how to represent concepts is still very much an open question in cognitive science. Different positions have been put forward, including the prototype view, the exemplar view, and the theory theory view? [6] contrasts these positions, asking which is best supported by the various aspects of cognition implicated in conceptual processing – e.g., learning, induction, lexical processing and conceptual understanding in children. He concludes, somewhat disappointingly, that "there is no clear, dominant winner". Despite the intervening ten or so years since Murphy made this statement, there is still no "winner". Therefore, any choice for a particular conceptual theory needs to be made mindful of its potential alignment for underpinning an account of metaphors, as well as its deficiencies.

BIBLIOGRAPHY

[1] Costello, F., & Keane, M. (2000). Efficient creativity: Constraint-guided conceptual combination. *Cognitive Science*, 24 (2), 299-349.
[2] Frixione, M., & Lieto, A. (2012). Representing concepts in formal ontologies: Compositionality vs. typicality effects. *International Journal of Logic and Logic Philosophy*, 21 (4), 391-414.
[3] Hampton, J. (1997). Conceptual combination. In K. Lamberts & D. Shank (Eds.), *Knowledge, Concepts, and Categories* (p. 133-160). Cambridge, MA: MIT Press.
[4] Lakoff, G. & Johnson, M. (1980). *Metaphors We Live By*, Chicago: The University of Chicago Press.
[5] Lakoff, G & Turner, M. (1989). *More than Cool Reason: A Field Guide to Poetic Metaphor*, Chicago: The University of Chicago Press.
[6] Murphy, G. (2002). *The Big Book of Concepts*, Cambridge, MA: MIT Press.

Response to Woods and Bruza, and Lorini

XIAOXI HUANG, HUAXIN HUANG, BEISHUI LIAO AND CIHUA XU

1 Responses to Lorini

1.1 Metaphors and Beliefs

An agent's epistemic state plays important role in metaphor understanding, as Lorini pointed out that different agent may understand a metaphor from different perspectives. The epistemic state can be regarded as a kind of context in metaphor understanding. There were some tentative researches on modeling logical system for metaphor understanding [6, 15, 14]. In [14], Steinhart introduced the possible world theory to analogous reasoning of metaphor, with the Extended Predicate Calculus (XPC), but his approach does not involve the agent's belief.

1.1.1 A tentative logic system EA for metaphor understanding

To resolve the agent's subjective factors, namely the epistemic state, including her world knowledge of the metaphorical topic, belief and intention, etc. we introduce an epistemic modal operator of epistemic logic in the logical characterization of metaphor understanding. The proposed logical system called EA (Epistemic Analogy) is follows. In EA, the epistemic modal operator U means "understand", $U_i\phi$ demonstrates that "The agent i understands ϕ". Given a set of atomic propositions $Prop$, a set of agents Agt, denoted as $Agt = \{1, 2, ..., m\}$, a set of predicates PS, a set of individual constants CS, as set of individual variables VS, a set of function FS, $FS \cup CS \cup VS$ constitutes the set of concepts. P_n and F_n denotes the n-ary predicates and n-ary functions respectively, the language $\mathcal{L}_{EA}^m$ of EA is the smallest closed set satisfying the follows:

(1) all the well-formed formulas in first-order predicate logic;

(2) if a and b are two concepts, then $a \sim b \in \mathcal{L}_{EA}^m$

(3) if $\varphi \in \mathcal{L}_{EA}^m$, then $U_i\varphi \in \mathcal{L}_{EA}^m$

where, $a \sim b$ means that a and b is indistinguishable, $U_i\varphi$ means agent i understands φ.

To define the indistinguishability between two concepts, we introduce an auxiliary function $Exchange(\varphi, a, b)$, which exchanges the elements a and b in formula φ. Then, the indistinguishability is defined as follows.

DEFINITION 1. Indistinguishability
Given a set of closed first-order formulas Γ, if there exists a closed formula φ satisfies:

- $\Gamma \vdash \varphi$, means that φ can be derived from Γ

- $\Gamma \vdash \neg Exchange(\varphi, a, b)$ or $\Gamma \vdash \neg Exchange(\varphi, b, a)$

then, we say a and b are distinguishable in Γ. Otherwise, if there is no such φ, we can say that a and b are indistinguishable in Γ, denoted as $I(\Gamma, a, b)$.

DEFINITION 2. Axiom system includes the following axioms and rules:

(PL) All the tautologies of first-order predicate logic

(AU1) $(\forall x)\varphi \rightarrow \varphi_{[x/d]}$

(AU2) $(\forall x)(\varphi \rightarrow \psi) \rightarrow (\forall x\varphi \rightarrow \forall x\psi)$

(MP) $\dfrac{\varphi, \varphi \rightarrow \psi}{\psi}$

(RM) $\dfrac{\varphi \rightarrow \psi}{U_i\varphi \rightarrow U_i\psi}$

(RG) $\dfrac{\varphi}{(\forall x)\psi}$

(DR) $\dfrac{\Gamma, I(\Gamma, a, b), \neg I(\Gamma \cup \{\neg\varphi\}, a, b)}{\varphi}$

where, axiom (PL) and modus ponens(MP) are from traditional propositional logic, while axioms (AU1), (AU2) and (RG) are axioms and rules from predicate logics. The rule (DR) is a new rule in EA. In (DR), Γ is a set of closed formulas, a and b are individual terms, φ is a first-order formula. Intuitively, the meaning of (DR) is: given Γ, a and b are indistinguishable under Γ, if a and b become distinguishable after $\neg\varphi$ is added into Γ, then , we can monotonically suppose φ is true.

An EA **model** is a tuple $< S, R, D, Agt, \pi, \tau >$ where S is the set of situations, R is the set of accessibility between situations, $R_i \subseteq S \times S$ is a serial, transitive, euclidean relation, representing the epistemic state of agent i. D is the domain, including VS, FS and PS. π is an assignment function, which assigns the truth value for every atomic proposition, assigns D^n for every n-ary predicate in every situation s. τ is also an assignment function, which interprets every function in FS (independent to situations).

A possible world w consists of a model M and a situation s, denoted as (M, s). For individual variables, the semantic is interpreted by an assignment function g, therefore, the interpretation of individual terms can be defined as follows.

$$val_{w,g}(x) = \begin{cases} g(x) & \text{if } x \in \text{VS} \\ \tau(f)(val_{w,g}(a_1), ..., val_{w,g}(a_n)) & \text{if } x \in \text{FS and } x = f(a_1, ..., a_n) \end{cases}$$

We use Ψ as the reasoning context, $M, s, g \models^{\Psi} \varphi$ means that under context Ψ, formula φ is true in possible world $w = (M, s)$ with respect to assignment function g, while $M, s, g =|^{\Psi}\varphi$ means that under context Ψ, formula φ is false in possible world $w = (M, s)$ with respect to assignment function g. Therefore, if $\varphi \notin \Psi$, we cannot infer $w, g =|^{\Psi} \varphi$ from $w, g \not\models^{\Psi} \varphi$, intuitively, this means that if a formula is not in the context, EA system cannot determine its truth value. To consider the epistemic state of agent, we introduce Ψ_s to denote the set formulas and concepts of which the agent has become aware.

DEFINITION 3. Semantic Interpretation
Given epistemic formulas φ and ψ, variable v, possible world $w = (M, s)$, the truth value of a formula is defined as follows.

- $w \models^{\Psi} p$ iff $\pi(s, p) = true$ and $p \in \Psi$ (where p is an atomic proposition)
  $w =|^{\Psi} p$ iff $\pi(s, p) = false$ and $p \in \Psi$ (where p is an atomic proposition)
  $w \models p$ iff $\pi(s, p) = true$ (where p is an atomic proposition)

- $w, g \models^{\Psi} P(t_1, ..., t_n)$ iff $(val_{w,g}(t_1), ... val_{w,g}(t_n)) \in \pi(s, p)$ and $t_1, ..., t_n \in \Psi$
  $w, g =|^{\Psi} P(t_1, ..., t_n)$ iff $(val_{w,g}(t_1), ... val_{w,g}(t_n)) \notin \pi(s, p)$ and $t_1, ..., t_n \in \Psi$
  $w, g \models P(t_1, ..., t_n)$ iff $(val_{w,g}(t_1), ... val_{w,g}(t_n)) \in \pi(s, p)$

- $w, g \models^{\Psi} \neg\varphi$ iff $w, g =|^{\Psi} \varphi$
  $w, g =|^{\Psi} \neg\varphi$ iff $w, g \models^{\Psi} \varphi$
  $w, g \models \neg\varphi$ iff $w, g \not\models \varphi$

- $w, g \models^{\Psi} \varphi \wedge \psi$, iff $w, g \models^{\Psi} \varphi$ and $w, g \models^{\Psi} \psi$
  $w, g =|^{\Psi} \varphi \wedge \psi$, iff $w, g =|^{\Psi} \varphi$ or $w, g =|^{\Psi} \psi$
  $w, g \models \varphi \wedge \psi$, iff $w, g \models \varphi$ and $w, g \models \psi$

- $w, g \models^{\Psi} (\exists x)\varphi$, iff $(\exists d \in D) : w, g_{[x/d]} \models^{\Psi} \varphi$
  $w, g =|^{\Psi} (\exists x)\varphi$, iff $(\forall d \in D) : w, g_{[x/d]} =|^{\Psi} \varphi$
  $w, g \models (\exists x)\varphi$, iff $(\exists d \in D) : w, g_{[x/d]} \models \varphi$

- $w, g \models^{\Psi} a \sim b$, iff $I(\Psi \cap \Psi_s, a, b)$, where a and b are concepts
  $w, g \dashv^{\Psi} a \sim b$, iff $\neg I(\Psi \cap \Psi_s, a, b)$, where a and b are concepts
  $w, g \models a \sim b$, iff $I(\Psi_s, a, b)$, where a and b are concepts

- $w, g \models^{\Psi} U_i \varphi$, iff $\forall t, (s, t) \in R_i, v = (M, t), v, g \models^{\Psi \cap \Psi_s} \varphi$
  $w, g \dashv^{\Psi} U_i \varphi$, iff $\exists t, (s, t) \in R_i, v = (M, t), v, g \dashv^{\Psi \cap \Psi_s} \varphi$
  $w, g \models U_i \varphi$, iff $\forall t, (s, t) \in R_i, v = (M, t), v, g \models^{\Psi_s} \varphi$

- $w \models \varphi$, iff $\forall g \in G, w, g \models \varphi$

- $\models \varphi$, iff $\forall M \forall s, w = (M, s), w \models \varphi$

THEOREM 4. *Tightening Theorem*
Given two reasoning context Ψ_1 and Ψ_2, suppose $\Psi_1 \cap \Psi_2 \neq \varnothing$, $w = (M, s)$ is a possible world, φ and ψ are formulas, if there are $w, g \models^{\Psi_1} \varphi$ and $w, g \models^{\Psi_2} \psi$, then, we have $w, g \models^{\Psi_1 \cap \Psi_2} \varphi \wedge \psi$.

THEOREM 5. *Consistency*
Suppose Γ is a consistent set of first-order closed formulas, if $I(\Gamma, a, b)$, $\neg I(\Gamma \cup \{\neg\varphi\}, a, b)$, then $\Gamma \cup \{\varphi\}$ is consistent.

Proof. Suppose $\Gamma \cup \varphi$ is inconsistent, then we have $\Gamma \vdash \neg\varphi$. From $I(\Gamma, a, b)$, we can get $I(\Gamma \cup \{\neg\varphi\}, a, b)$, this contradicts to the given condition $\neg I(\Gamma \cup \{\neg\varphi\}, a, b)$. Therefore, the assumption does not hold, namely, $\Gamma \cup \{\varphi\}$ is consistent. ∎

With the theorems and properties of EA, referring to the theory of epistemic state[8], we also propose the notion of "stable set" of epistemic formulas.

DEFINITION 6. Stable Set
Suppose Γ is a consistent set of first-order closed formulas, Σ is an operator, for any set first-order formulas S, $\Sigma(S)$ is the smallest set satisfying follows:

(1) $\Gamma \subseteq \Sigma(S)$

(2) if $\varphi \in \Sigma(S)$, and $\varphi \rightarrow \phi$, then $\phi \in \Sigma(S)$

(3) if $I(S, a, b)$, $\neg I(S \cup \{\neg\varphi\}, a, b)$, then $\varphi \in \Sigma(S)$, where a and b are both concepts

Particularly, if a set E satisfies $\Sigma(E) = E$, we say E is a stable set of Γ, and E can be regarded as agent's epistemic state. Analogous to theory of default logic[10], the stable set is an extension under Σ.

THEOREM 7. *Analogy*
Suppose S is a set of first-order closed formulas, a and b are concepts. If $\varphi \in S$, and $I(S, a, b)$, then $Exchange(S, a, b) \in \Sigma(S)$.

Proof. Set $W = S \cup \{\neg Exchange(\varphi, a, b)\}$. According to $\varphi \in S$, we have $S \vdash \varphi$, therefore $W \vdash \varphi$. With definition of W, we also have $W \vdash \neg Exchange(\varphi, a, b)$. According to Definition 1, we can get $\neg I(W, a, b)$, namely, a and b are distinguishable under W, then according to Definition 4, we have $Exchange(\varphi, a, b) \in \Sigma(S)$. ∎

Intuitively, Analogy Theorem means that: if a formula $\varphi \in S$, concepts a and b are indistinguishable at current state, then the new formula generated by $Exchange$ a and b in φ is assumed to hold.

COROLLARY 8. *If E is a stable set, $\varphi \in E$, and $I(E, a, b)$, then $Exchange$ $(\varphi, a, b) \in E$.*

Corollary 1 demonstrates that analogical reasoning is closed in EA epistemic state theory.

1.1.2 An example for application of EA
In this section, we will discuss how to understand metaphor "NP1 is NP2" using EA based on the metaphor understanding framework[3]. Metaphors with form of "NP1 is NP2" are the so-called "Denotational Metaphor" in [4]. Suppose the reasoning context is Ψ, with the lexical ontology, we can get the concepts represented by "NP1" and "NP2", denoted as C_T and C_S, and construct a formula $C_T \sim C_S$, then to understand such a metaphor is to determine the truth value of formula $C_T \sim C_S$ under Ψ in EA.

As to the example "Market is sea"[3], the target concept "MARKET" and source concept "SEA" have metaphor mapping with properties "situation", "extent" and their common superordinate concept "waters". Given a situation s, if formula $Surging \sim Danger$, then from the perspective of property "situation", we can interpret $Market \sim Sea$ through replacing the value of property "situation" of "MARKET" by the one of "SEA", and metaphor "Market is sea" can be understood as "market is full of danger". Whereas, if some agent i has a metaphorical relation "*wide expanse*" $\sim$ "*opportunity*" in her epistemic state, then we can say $U_i(Market \sim Sea)$ is true under context Ψ, namely, agent i understands the metaphor "market is sea" as "the market is greater chance" from the perspective on "extent".

1.2   Metaphors and theory of mind
As point out by the conceptual metaphor theory (CMT)[7], metaphors are cognitive structures that map aspects of source domain to target domain. In the theoretical framework proposed in the paper [3], we introduced embodied concept knowledge, which is inspired by the cognitive theory of metaphor. Revert to the question on the role of theory of mind (ToM) in metaphor understanding, raised by Lorini. Lorini explains, "from this perspective, metaphor production and metaphor under-

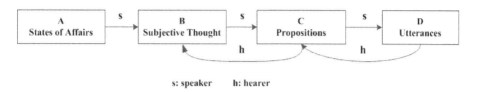

s: speaker   h: hearer

Figure 1. A simple model for language communication

standing are, by definition, social activities that involve a speaker and a hearer",
this reminds us of the pragmatic views on metaphors, such as Searle[12] and other
pragmatists. The main question Searle aims to investigate is "How do metaphorical
utterances work, that is, how is it possible for speakers to communicate to hearers
when speaking metaphorically inasmuch as they do not say what they mean?"[12]
Searle proposed eight principles for English metaphor interpretation on the view
of speech act theory. According these principles, the hearer may take three steps to
understanding a metaphor:seeing an utterance as metaphoric; have some strategies
for computing metaphoric meaning if he decides something is metaphor; have
some strategies for restricting the range of possible metaphoric meanings. This
procedure can be demonstrated by a simple model of language communication, as
in Fig. 1.

During the communication, "A" is a physically observable state of affairs in the
actual world. The speaker encodes "A" to her thoughts ("B") and form the proposition ("C") to convey her thoughts. To communicate the proposition, she must use
the language code("D") to express it. The hearer can perceive the utterance, de-
code it and try to form the proposition("C"), and then interpret it by guessing what
the speaker is most likely to convey("B"). In this procedure, there are gaps between "B" and "C", "C" and "D". The larger the gap between "B" and "C" means
the more metaphorical the speaker utters. Therefore, if we have a mechanism to
model the intention, belief and goals of the speaker in "B", and the reasoning of
multi-agent system, it is possible for us to extend the computational model with
the EA logic system mentioned above to help the hearer to guess the thought which
the speaker is to convey.

1.3   Metaphors and Emotions

Emotions lead an important role in human intelligence. For artificial intelligence, it
is a vital task to let a computer have the abilities to recognize, understand and even
to express emotions [9]. In [5], we investigated the kinds of emotional metaphors
in Chinese, and proposed characteristics of them. In recent researches of NLP,
such as opinion mining, sentiment analysis etc., researchers have realized the importance of metaphorical language [11]. From the view of formalization of e-

motions and metaphors,we think the main difficulty is to build the corresponding ontology of emotions [2]. Since emotions are mental processes of agents, in the future research, we would combine the existing formal theory of emotion into our EA system.

2   Responses to Woods & Bruza

Woods & Bruza give comments from two aspects: the relationship between formal mechanism and the data; the mathematical and operational adequacy of the formal mechanism. We response the comments in follows.

Firstly, for metaphor recognition, we have proposed a method for metaphor identification and classification in Chinese [4]. In the language level, we classify metaphorical language into two categories: referential metaphor with ten subclasses and collocational metaphor with five subclasses, and the proposed algorithm in [4] use two quantitative criterions to determine the metaphor category. In the fields of natural language processing, some approaches based on machine learning are recently proposed to discriminate an utterance into metaphorical and literal one [13].

Secondly, our model focuses on the factors of metaphor understanding, especially on knowledge representation and the metaphor mappings. The theoretic framework of metaphor understanding is based on Conceptual Metaphor Theory [7] and Conceptual Blending Theory[1]. The data for metaphor mappings are from the following sources as described in the paper [3]: upper ontology, lexical ontology and context information provided by the metaphor identification algorithm.

Lastly, metaphor is a complex phenomena in natural language, it is impossible to involve all the aspects of metaphor in a single model. Our model aims to propose a theoretic framework on the basis of state-of-the-art method of knowledge representation and our metaphor identification algorithm, and implements a tentative system in the light of cognitive theory of metaphor.

BIBLIOGRAPHY

[1] Gilles Fauconnier and Mark Turner. *The Way We Think: Conceptual Blending and The Mind's Hidden Complexities*. Basic Books, 2002.

[2] Janna Hastings, Werner Ceusters, Barry Smith, and Kevin Mulligan. The emotion ontology: enabling interdisciplinary research in the affective sciences. In *Proceedings of the 7th international and interdisciplinary conference on Modeling and using context*, CONTEXT'11, pages 119–123, Berlin, Heidelberg, 2011. Springer-Verlag.

[3] Xiao Xi Huang, Hua Xin Huang, Bei Shui Liao, and Ci Hua Xu. An ontology-based approach to metaphor cognitive computation. *Minds and Machines*, 23(1):105–121, 2013.

[4] Xiao Xi Huang, Hua Xin Huang, Ci Hua Xu, Weiying Chen, and Rong Bo Wang. A novel pattern matching method for chinese metaphor identification and classification. In *Proceedings of the 2011 International Conference on Artificial Intelligence and Computational Intelligence (AICI'11)*, pages 104–114, Taiyuan, China, 2011.

[5] Xiao Xi Huang, Yun Yang, and Chang Le Zhou. Emotional metaphors for emotion recognition in chinese text. *AFFECTIVE COMPUTING AND INTELLIGENT INTERACTION, PROCEEDINGS*, 3784:319–325, 2005.

[6] Xiao Xi Huang and Chang Le Zhou. A logical approach for metaphor understanding. In *Proceedings of 2005 IEEE International Conference on Natural Language Processing and Knowledge Engineering*, pages 268–271, Wuhan, China, 2005. IEEE Computer Society.

[7] George Lakoff and Mark Johnson. *Metaphors We Live By*. The University of Chicago Press, Chicago, 1980.

[8] J. J. Ch Meyer and W. van der Hoek. *Epistemic Logic for AI and Computer Science (Cambridge Tracts in Theoretical Computer Science)*. Cambridge University Press, 1995.

[9] Rosalin Picard. *Affective Computing*. MIT Press, Cambridge, 1997.

[10] Raymond Reiter. A logic for default reasoning. *Artificial Intelligence*, 13(1-2):81–132, 1980.

[11] Vassiliki Rentoumi, George A. Vouros, Vangelis Karkaletsis, and Amalia Moser. Investigating metaphorical language in sentiment analysis: A sense-to-sentiment perspective. *ACM Transactions on Speech and Language Processing*, 9(3):1–31, November 2012.

[12] J. R. Searle. Metaphor. In Andrew Ortony, editor, *Metaphor and Thought*, pages 83–111. Cambridge Universtiy Press, Cambridge, 2nd edition, 1993.

[13] Ekaterina Shutova and Lin Sun. Unsupervised metaphor identification using hierarchical graph factorization clustering. In *Proceedings of NAACL-HLT 2013*, pages 978–988, Atlanta, Georgia, 2013.

[14] Eric Charles Steinhart. *The Logic of Metaphor: Analogous Parts of Possible Worlds*. Kluwer Academic Publishers, Dordrecht, 2001.

[15] Wei Zhang and Changle Zhou. Study on logical description of chinese metaphor comprehension. *Journal of Chinese Information Processing*, 18(5):23–28, 2004.

Negation and Quantification. A New Look at the Square of Opposition[1]

Dag Westerståhl

1 Outline

This paper is about how the fundamental logical notions of negation and quantification interact. The subject is as old as logic itself, and appears in various forms in all ancient traditions in logic, but perhaps most explicitly and systematically in the Western tradition starting with Aristotle, which is the one I will refer to here. Many of the issues raised by Aristotle and his followers are still subject of lively debate among linguists and philosophers today.

Negation is closely tied to *opposition*: a statement and its negation are in some sense opposed to each other. There are different kinds of opposition (and philosophers love displaying all kinds of opposing pairs), but Aristotle's main distinction is between *contradictory* and *contrary* opposition or negation. The contradictory negation of *Socrates is wise* is simply *Socrates is not wise*, which is true if and only if the positive statement is false. But there are several contrary negations: *Socrates is non-wise/unwise/foolish/...*; the hallmark of contrariety is that a statement and its contrary cannot both be true.

The contrast between contrariety and contradictoriness becomes clearer, and simpler, when applied to quantified statements. Aristotle's syllogistics—the beginning of modern logic—is a systematic study of relations between the four quantifiers *all, no, some, not all*, and in particular of how they interact with negation. The contrary of *all* is *no*, whose contradictory negation is *some*. (Really it is the corresponding statements, *all A are B*, *no A are B*, etc. that have these relations, but they are easily lifted to the quantifiers themselves.) Aristotle's claims about these matters were later summarized diagrammatically and extended in the famous Square of Opposition (Fig. 1).

This square is not only an elegant way of displaying logical relations, but is intended to provide important information about the meaning of negation and quantification in natural languages. Among questions related to it that have been, and

---

[1]This summary is based on the paper 'Classical vs. Modern Squares of Opposition, and Beyond', which appeared in Beziau, Jean-Yves & Payette, Gillman eds, *The Square of Opposition : A General Framework for Cognition*, Peter Lang Publishing Group, 2012.

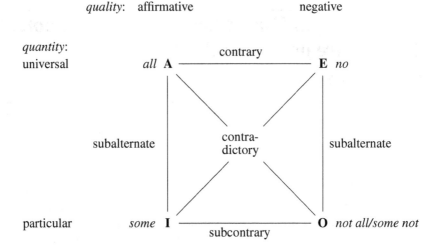

Figure 1. The Aristotelian Square

still are, intensely debated are the following (see [1] for a rich account of the history of negation):

(a) What (if anything) is less valuable/real/natural/informative about negative statements as compared to positive ones?

(b) What (if anything) makes a statement negative rather than positive?

(c) What sort of thing or state of affairs makes a negative statement *true*?

(d) Why is the **O** corner never lexically realized (in any language)?

(e) Does the **A** corner have *existential import*, as in the Aristotelian square, i.e. does *All A are B* entail (or presuppose) that there are *A*s? What about the other corners?

For example, Aristotle (presumably), most (but not all) medieval logicians, and most (but not all) modern linguists claim that *All A are B* indeed entails (or presupposes) that there are *A*s. By contrast, universal quantification for most modern logicians since Frege does *not* have existential import.

  This paper only indirectly addresses the above questions (though I will be glad to talk about them in discussion). Instead, I focus on another issue, also intimately related to the meaning of negation and quantification in natural languages, namely, what exactly are the relations along the *sides* of the square? My claim is that it is misleading to think of contrariety, subalternation, etc. here; rather, the 'horizontal' relations are one and the same: *inner negation* or *post-complement*; and the 'vertical' ones are *dual*, as in Fig. 2.

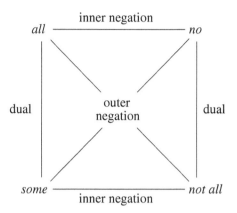

Figure 2. The modern version of Aristotle's square

The main argument is that this Modern Square—but not the Aristotelian Square—applies to all kinds of natural language quantification, not just to the Aristotelian quantifiers. Indeed, those four quantifiers instantiate a very general pattern, with examples such as *at least three, exactly six, all but five, most, many, few, more than two-thirds of the, the seven, Mary's, some students'*, ..., each one related to its negations and dual according to the modern square. Nothing similar works for the classical square. Here is an outline of the paper.

1.1   Generalized quantifiers

(Binary) *generalized quantifiers* are relations between subsets of the universe. Thus $all(A, B)$ says that A is a subset of B, $no(A, B)$ that the intersection of A and B is empty, $some(A, B)$ that it is non-empty, *at least three*(A, B) that it has at least three elements, $most(A, B)$ that it contains more elements than the difference $A - B$ does, *the seven*(A, B) that there are (exactly) seven As and that they are all B, *some student's*(A, B) that some student 'possesses' As and all the As 'possessed' by her/him are B, etc. In English and many other languages, binary quantifiers interpret determiners. In languages with so-called A quantification, they interpret adverbs, auxiliaries, or other devices for quantification.

For each such quantifier Q we can define its outer (contradictory) negation: $\neg Q(A, B)$ iff not $Q(A, B)$; its inner negation: $Q\neg(A, B)$ iff $Q(A, A - B)$; and its dual: $Q^{\mathrm{d}} = \neg(Q\neg)$. So each such Q spans a (modern) square:

$$square(Q) = \{Q, Q\neg, \neg Q, Q^{\mathrm{d}}\}$$

One verifies that $square(Q)$ has either 4 elements (the normal case) or 2 (it can happen that $Q = Q\neg$), and that each element in $square(Q)$ spans the same square. (For an overview of generalized quantifiers in language and logic, see [2].)

1.2   Classical vs. modern squares

Aristotle's square concerns just the four classical quantifiers. Trying to extend it to other quantifiers yields the following definition: A *classical square* is an arrangement of four quantifiers as traditionally ordered and with the same logical relations—contradictories, contraries, subcontraries, and subalternates—between the respective positions. Here two statements are subcontraries if they cannot both be false, and a statement is subalternate to another statement if it is implied by it.

However, the result is that, for example, the square

[**A**: *at least five*; **E**: *no*; **I**: *some*; **O**: *at most four*]

is classical. This looks very unnatural. There is no interesting sense, it seems, in which *no* is a negation of *at least five* or *at most four*. The classical Square of Opposition simply doesn't generalize in a natural way to other quantifiers.

1.3   Identifying the corners

Medieval logicians classified the corners of the square by means of the categories *quantity*: universal (*all, no*) / particular (*some, some not*) and *quality*: affirmative (*all, some*) / negative (*no, not all*). These properties are often hard to apply to other quantifiers. For example, *no fewer than five* may seem negative, and *more than four* positive. But extensionally they are the same quantifier!

There are, however, clearcut—and Aristotelian—semantic properties that can often be used to at least partly identify the corners. One is *monotonicity*; in fact, the syllogisms describe exactly the monotonicity behavior of the four Aristotelian quantifiers, and these properties apply to other quantifiers as well. Another is *symmetry*; Aristotle noted that the quantifiers in the **I** and **E** corner are symmetric (though he didn't use these terms): $Q(A, B)$ implies $Q(B, A)$. Using these properties, and the further criterion that, if possible, that the standard $square(all)$ should be a *special case* of $square(Q)$, one can in most common cases classify the corners of $square(Q)$.

1.4   Case studies

Numerical quantifiers

These are Boolean combinations of quantifiers of the form *at least n*, such as *at most six* and *exactly five*. Their squares—i.e. their patterns of negation—are interesting, in terms of how they are realized in English, and how little the classical approach says about them.

Proportional quantifiers

Typical proportional quantifiers are *most* and *at least two-thirds of the*; the other quantifiers in their squares also have natural renderings as English determiners. It is notable that—except for *most*!—the definite article is required in English. This reopens the issue of existential import (e.g. is it presupposed or entailed?), but with more clearcut examples than those in the traditional square.

Exception quantifiers

In *No student except Mary came to the opening*, the phrase *no _ except Mary* can be seen as a quantifier: *no _ except Mary*(A, B) iff $A \cap B = \{\text{Mary}\}$, with a corresponding square of opposition.

Possessive quantifiers

Possessive quantifiers like *Mary's, two students', several of most teachers'*, have interesting logical and semantic properties, also as regards their interaction with negation. (See [3] for much more on the semantics of possessives.) A main observation is that they involve *two* quantifiers, one over possessors and one over possessions (relative to some 'possessive' relation, that need not have anything to do with ownership). In *No car's tires were slashed*, $Q_1 = no$ quantifies over possessors (cars), but the quantier Q_2 over possessions (tires) is implicit; presumably in this case it is *some*. In *At least one of most teachers' pupils failed the exam*, both Q_1 (*most*) and Q_2 (*at least one*) are explicit.

This allows for several ways in which negation can apply. In addition to standard outer and inner negation, we also find what we call *middle negation*. For the sentence *All of Mary's friends didn't come to the party*, the outer negation reading is implausible (but it applies in other cases), and the inner negation reading says that none of Mary's friends showed up. This is a possible reading, but more likely is the reading that not all of them showed up (though some of them may have): middle negation. The result is that there are two kinds of squares for possessive quantifiers. For example, the 'middle' square of *Mary's* more accurately reflects how *Mary's* behaves under negation than the ordinary *square(Mary's)*. However, I show that the full (logically possible) behavior of possessives under negation can be represented in a *cube* of opposition.

2   Main claims

- It may seem—and it is usually assumed—that the only difference between the Aristotelian square of opposition and the modern *square(all)* is that *all* has existential import in the former but not the latter square. This impression is mistaken. The main difference concerns the relations along the sides of the squares. That difference is hardly visible for *square(all)*, but it becomes evident for the squares spanned by other quantifiers.

- The issue of existential import for *all* is real, and alive especially among linguists. However, it does not really concern the properties of the square, i.e. the interaction between quantification and negation (except that if *all* has existential import, its contradictory negation becomes a rather unnatural quantifier). On the other hand, the square highlights other issues about existential import, e.g. for the proportional quantifiers.

- The fact that 'modern' squares—but not traditional ones—are generated by any binary quantifier, and hence by any English determiner denotation, constitutes a main argument that they capture important facts about negation and quantification in natural languages.

- The traditional way of identifying the corners of the square in terms of quantity and quality are not sufficiently precise to be applicable to other quantifiers. In particular, 'quality' relies on identifying *negative* statements (or quantifiers), but it is very doubtful that this can be done in a coherent way. However, the criteria can be replaced by relying on precise semantic—and still very much Aristotelian!—properties of the quantifiers involved. In particular, the monotonicity behavior of quantified expression is an area where an exact study of semantics is both possible and rewarding. As to the square, a question that remains after classification is: What (if any) is the significance of the various corners of the square?

- Investigating the (modern) squares of various quantifiers is a fruitful way to study the interaction between negation and quantification in natural languages. Many questions arise that have no counterpart for the Aristotelian Square. In particular, possessives provide a rich field of application for generalized quantifier theory to natural language. Among other things, they exhibit a new form of negation, distinct from outer and inner negation, and a full description of how possessives interact with negation turns out to require a *cube* of opposition.

BIBLIOGRAPHY

[1] Laurence R. Horn. *A Natural History of Negation*. Chicago University Press, Chicago. Republished by CSLI Publications, Stanford, 2001., 1989.
[2] Stanley Peters and Dag Westerståhl. *Quantifiers in Language and Logic*. Oxford University Press, Oxford, 2006.
[3] Stanley Peters and Dag Westerståhl. The semantics of possessives. To appear in *Language*, 2013.

Comments on Westerståhl

FENGKUI JU

1 Comments

1. This paper makes such a claim: The essential characteristic of the Aristotelian square of opposition is not existential import, but the relations along the sides of it; The essential characteristic of the modern square of opposition corresponding to the Aristotelian square is also the relations along the sides of it; Therefore, the main difference between the two squares does not lie in existential import, but the relations along their sides. This claim is argued in a broad scope consisting of classical and modern quantifiers, which are generalized from the two squares. I am convinced by the claim and believe that the paper provides a novel and nice perspective to look at the two squares, which is very different from the usual viewpoints.

2. The paper presents a method to identify corners of quantified squares. Basically, the method is based on checking whether quantifiers satisfy monotonicity and symmetry. This method is very elegant and applies to many cases.

2 Questions

1. In Figure 5, the quantifier "all but at least six" is not in italics. According to the convention made in the paper, this means that the paper thinks that it does not interpret any determiners in English. What does "it does not interpret any determiners in English" mean? In my understanding, this means that the quantifier can not be expressed in English. However, as a natural language, English is universal and can express everything.

2. Exceptive quantifiers such as "no ...except ..." are considered as type $\langle 1, 1 \rangle$ quantifiers in this paper. In my opinion, it is better to treat them as type $\langle \langle 1, 1 \rangle, 1 \rangle$ quantifiers, as all sentences like "no students except foreign exchange students joined this club" involve at least three type $\langle 1 \rangle$ notions.

3. Why is there existential import with "Mary's A are B"? What about "some students' A are B" and "all students' A are B"?

4. By page 21, the sentence "not everyone's needs can be satisfied with standard products" can be truly uttered in the case that no one actually has any needs. I don't think that this is reasonable. The sentence uttered can be rewritten as the FO formula $\neg\forall x\forall y((Mx \land Ny \land Pxy) \to Sy)$. It implies that there is a person and he has a need.

5. I am confusing with the outer negation, inter negation and middle negation of possessive quantifiers. Take "Mary's pupils are bright" as an example. According to the paper, the inter negation of it is "none of Mary's pupils is bright", and the middle negation of it is "not all of Mary's pupils are bright". What is the outer negation of it?

6. The paper mentions that for the sentence "no car's tires were slashed", the quantifier "no" is over cars and "some" is over tires. I think that there is another possibility: "any" is over cars and "no" over tires. Actually, there are two readings of this sentence, although they are equivalent: (i) there is no car such that it has a tire slashed; (ii) for any car, it has no tire slashed. The two possibilities about which is over which are based on the two readings.

   In "six Mary's books are wet", the quantifier "six" is over books, the possessions. In "six students' books are wet", "six" might be over students. Is there a systematical way to tell which is over which for possessive quantifiers?

7. Like the paper says, modern quantified squares elegantly illustrate how negation interacts with quantification. However, in my opinion, this seems to be the only sense of squares; they cognitively help us to understand the interaction better and that is all. What important information can squares provide about the meaning of negation and quantification in natural languages? What would we lose if we do not use squares?

8. Why do we need to identify the corners of squares?

9. The paper claims that modern squares illustrate the interaction between negation and quantification better than classical squares. The main argument is that modern squares apply to all kinds of natural language quantification while classical ones do not. I am not convinced by this claim, although as shown in the paper, modern squares have nice properties rather than classical squares. What does "apply to quantification in natural languages" mean? How do modern squares do better than classical squares? It seems that the paper does not sufficiently answer these two questions.

10. It is an interesting fact that there is no single word having the same meaning with "not all" in any natural language. What are possible explanations of this fact?

Comments on Westerståhl

LAWRENCE S. MOSS

It is a pleasure to comment on Dag Westerståhl's article "Classical vs. modern squares of opposition, and beyond". I believe that I met Dag during the 1984–85 academic year. I was a post-doc at Stanford's Center for the Study of Language and Information, and I believe that he was an extended visitor for part of the year. We shared an interest in generalized quantifier theory: I had participated in Ed Keenan's seminars on this topic as a graduate student and even wrote a couple of papers with Ed. Dag's interest was much more serious and long-lasting. Indeed, he and Stanley Peters would in time write the most important book on this topic, [2].

My comments are about two articles: first, "Classical vs. modern squares of opposition, and beyond"; second, a summary presentation entitled "Negation and quantification: a new look at the square of opposition." I believe that the second was written especially for the "Logic Across the University" conference. Both of these are very clear, and so there is little point in my adding yet another summary. Instead, I'll offer a few questions and make a few remarks that could help some readers who are not familiar with generalized quantifier theory.

The square of opposition has been a traditional topic in logic for centuries, even though it no longer a popular topic. I would like to start by asking some questions that situate this article and this topic in larger discussions.

Question 1. Given that the conference is in China, I wonder whether ancient Chinese or Indian logicians had something like a square of opposition of their own? If so, did it resemble the ancient or modern Western square? Or was it different?

Question 2.

    a. The overall theme of the work is the reconsideration of the square of opposition, a reconsideration built on a different view of negation and quantifiers than what previously available. Is the traditional square of opposition worth studying at all?

    b. Here is a question related to the one in part (a). Parsons [1] says

> Today, logic texts divide between those based on contemporary logic and those from the Aristotelian tradition or the nineteenth

century tradition, but even many texts that teach syllogistic teach it with the forms interpreted in the modern way, so that e.g. sub-alternation is lost. So the traditional square, as traditionally interpreted, is now mostly abandoned.

Does the modern reconstruction of the square make for a topic that belongs in contemporary logic textbooks?

Question 3. This is a technical comment, leading to a question. The paper notes the square on the left below. (Actually, the square is drawn without the labels on the arrows, and without the diagonals.)

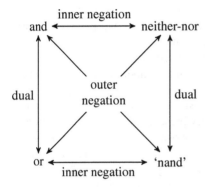
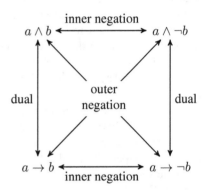

On the other hand, if we work with the truth-functional connectives in curried form, we really get the square on the right. That is, let us take the propositional functions to be of type $\mathbb{B} \to (\mathbb{B} \to \mathbb{B})$, where $\mathbb{B}$ is the set of booleans, rather than $\mathbb{B} \times \mathbb{B} \to \mathbb{B}$. This way of doing things makes the terminology of "inner" and "outer" negation a little clearer, I think. In a sense, what I have done here is to make the same move that is frequently made in generalize quantifiers. But the presentation aside, the square on the right exhibits material implication as a dual to conjunction. What do you make of this?

Question 4. Your work at the end on possessives, and also the work on numerical and proportional quantifiers in Section 4 leads me to wonder whether there is a square of cube for constructions like *more A than B are C*. What is the situation for this?

Question 5. I'd like to end on one of the speculative questions that are mentioned in the paper but not answered: Why is the O corner never realized in any language? What are your feelings on this?

BIBLIOGRAPHY

[1] T. Parsons, "The Traditional Square of Opposition", The Stanford Encyclopedia of Philosophy (Fall 2012 Edition), Edward N. Zalta (ed.),
URL = <http://plato.stanford.edu/archives/fall2012/entries/square/>.
[2] S. Peters and D. Westerståhl, Quantifiers in Language and Logic, Oxford University Press, 2006.

Response to Ju and Moss

DAG WESTERSTÅHL

I am very grateful to Fengkui Ju and to Larry Moss for their interesting (and appreciative) remarks on my paper. Space does to not permit a thorough discussion here of all their points. But I will try to say something brief about most of them, following roughly the order in which they were presented, and beginning with Fengkui's questions.

1 Expressibility

Fengkui notes (question 1) that I don't take the phrase "all but at least six" to denote an English determiner, and wonders if this means that the quantifier cannot be expressed in English, which seems strange. But there are different notions of expressibility. A specific one concerns which type $\langle 1, 1 \rangle$ generalized quantifiers directly interpret English determiners. For a clear example, which is not specific to English, take the so-called Härtig or equi-cardinality quantifier I, defined by $I(A, B) \Leftrightarrow |A| = |B|$. This quantifier has been studied by logicians, but it does not interpret any Det in English, or any similar expression in any language which serves to denote type $\langle 1, 1 \rangle$ quantifiers. The reason is that it is not *conservative* (it does not satisfy $Q(A, B) \Leftrightarrow Q(A, A \cap B)$), and conservatively seems to be a mark of all those natural language expressions.

But that doesn't mean that I cannot be expressed at all in English. In fact, a common "there" construction will do:

$I(A, B)$ iff there are as many As as Bs

I think the question of which type $\langle 1, 1 \rangle$ quantifiers are expressible by English Dets (or similar expressions in other languages) has some intrinsic interest, and I try to make these judgments in the squares drawn in the paper, since it is clear that negations and duals of Det denotations are not always themselves Det denotations. As to the specific example "all but at least six", my judgment is that this is not a well-formed English Det—in contrast with e.g. "all but at most six"!—but this may very well be a borderline case.[1]

---

[1] In this case the quantifier in question is conservative. But there are many conservative type $\langle 1, 1 \rangle$ quantifiers that are not Det denotations; for example, this one: $Q(A, B) \Leftrightarrow |A| = 10$.

2 Possessives

Possessive determiners (Fengkui's questions 3–6) interact in particularly interesting ways with negation; I discuss examples in the paper, and also in [4], where we make some points that were not really clear in the current paper, especially concerning existential import. A typical possessive Det involves two quantifiers, one (Q_1) over 'possessors' and one (Q_2) over 'possessions', as in

(1)   Some (Q_2) of most (Q_1) students' term papers got a C.

We argue at length in [4] that all possessives exhibit *possessive existential import* (PEI), which means that the quantification over 'possessions' always has existential import, *even when Q_2 itself doesn't*. So, for example,

(2)   At most two of Mary's brother's live in New York.

entails, according to our analysis, that Mary has brothers, even though *at most two* does not have existential import. Now in sentences like "Mary's/some students'/all students' A are B", Q_2 is *implicit*, but if we take it to be universal then, according to our claim, it has existential import, so these sentences entail that Mary or the relevant students 'possess' an A.

Since a possessive Det involves two quantifiers, there are several logical possible ways negation could apply, but we show that they reduce to three: outer negation (of Q_1), inner negation (of Q_2), and middle negation (= inner negation of Q_1 or, equivalently, outer negation of Q_2). But that doesn't mean that all three negations always have natural expressions in English. As Fengkui notes, inner and middle negation applied to "Mary's pupils are bright" are easily expressed, but the outer negation would be "Either Mary has no pupils, or she does but not all of them are bright". So in this case you cannot get this by inserting "not" or a similar phrase in the positive sentence. We point out that, on the other hand, "Not everyone's needs can be satisfied with standard products" has an interpretation with outer negation, according to which someone has needs that cannot be so satisfied.

Here I must thank Fengkui for noting a mistake in the paper: The first sentence after (25) is correct, but the two following sentences, starting with "Note that . . . ", should just be deleted.[2]

Fengkui also asks how we know which Dets quantify over which objects. In English, this is usually determined by form alone, but there are some subtleties. In "Six of Mary's books are wet" (the "of" is necessary!), *six* quantifies over books: this is always so for the form

Det of [NP's N]

---

[2]What I meant to say was that the *unnegated* sentence "Everyone's needs can be satisfied with standard products" can be true even if no one has any needs. But this is beside the point, and the use there of 'possessive existential import' is not what we now call PEI. Sorry!

Det denotes Q_2, the quantifier over 'possessions' (N). In "Six students' books are wet", the possessive Det has the form

[Det N]'s

in which the Det denotes Q_1, the quantifier over possessions. One type of complication is with so-called *modifying* possessives: "Six children's books are wet" actually has *two* readings: one, as above, saying that six children are such that their books are wet, and one (modifying) about *children's books*, i.e. books for children, so here *six* quantifies over (children's) books. The discussion in the paper does not concern modifying possessives. Another complication is that the partitive form [Det of NP] is also employed with *definite* (plural) NPs, which is (sometimes) distinct from the possessive case. Without going into details here (see [4], sect. 4.1), think about the *three* readings of

(3)   Two of the ten boys' books are missing.

and over which entities *two* quantifies in each case.

3   Why squares, and why corners?

(Questions 7–9) The square is a geometrical way of understanding certain facts about negation in natural languages. Aristotle didn't use it, so it certainly isn't necessary for understanding these facts. I don't think there is any deep significance of squares: it is just a common pattern, and sometimes diagrams make you see things that are not so easily visible without them. However, ever since the Middle Ages, people have tried to make much more of the squares than this. In modern times, starting in 2007 there is a series of Square Conferences, where all kinds of aspects of squares and similar geometrical pattern are explored.

As to the corners, the original classification comes from the Aristotelian square, for which their significance was cashed out in terms of the properties universal/particular and affirmative/negative. I think especially the latter distinction is inappropriate for quantifiers in general, and suggest monotonicity and to some extent symmetry instead. But is there a special significance in having, say, the monotonicity pattern $\downarrow$MON$\uparrow$? I really don't know; I just found it curious that the corners can in fact be identified in the way I suggest in most cases.

The main thing about the square, in my opinion, is that each (CONS and EXT) quantifier Q generates

$$square(Q) = \{Q, Q\neg, \neg Q, Q^d\}$$

which, as I try to illustrate, is a robust quantificational unit. The classical Aristotelian square, which partly uses different 'negations', is *not* robust: there seems to be no way to make it apply sensibly to arbitrary Det denotations.

4   The infamous O corner

(Fengkui's question 10 and Larry's question 5) Why doesn't English (or, apparently, any other language) have a single word—say, "nall"— for the quantifier in the **O** corner? And similarly for other squares: "not(A and B)" corresponds to no natural language connective (but "not(A or B)" does), and analogously for "not necessary" (which doesn't mean 'unnecessary'), "not always". etc. (Note that this is a strictly Aristotelian question: many other squares have no problem with that corner.) Surely this rather striking asymmetry has an explanation. But my feeling is that *logic* is unlikely to help here. Rather, the explanation would involve facts about pragmatics and language evolution. Horn proposes such an explanation (see e.g. [1], ch. 4.5): Very roughly, *first*, if you know exactly which A are B, and you say "Some A are B", you normally imply (the Gricean implicature of Quantity) "Not all A are B"; and vice versa. So from this pragmatic perspective, it seems unnecessary to lexicalize both the **I** and the **O** corners. *Second*, since all languages explicitly mark negation,[3] it seems more economical to lexicalize the affirmative **I** case.

5   Negation East and West

Larry's question 1 is a topic for the historian of logic. A brief but readable survey of negation in Indian and Chinese thought is [1], ch. 1.3. One issue concerns the Law of Excluded Middle and the Law of Contradiction, which underlie both classical and modern squares of opposition, but seem to be rejected in some strands of Buddhist logic. The extent of that rejection, however, appears to be debatable. A more directly relevant issue, perhaps, is the distinction between contradictory and contrary negation, which seems to be absent in the Indian tradition but was recognized in Mohist logic. Without that distinction, 'square-related' investigations could hardly get started.

6   Is the square still relevant?

(Larry's questions 2a and b) Is the traditional square of opposition still useful today? For one thing, given that it differs so little from the modern *square(all)*, comparing the two can be helpful, also for discussing the complex pragmatic and semantic issues surrounding existential import. But the most significant aspect of the traditional square is the notion of *contraries*. This notion has great interest in a linguistic study of negation, various kinds of gradable adjectives, etc., but it arises mainly at the *term* level, rather than the *proposition* level. My feeling is that the square itself is not particularly helpful, as a geometric representation, for the study of contraries.

---

[3]One could easily imagine a language which uses forms like $\langle P, a, + \rangle$ to say that a is P, and $\langle P, a, - \rangle$ to say that a is not P. But no natural language does this: the $+$ is never explicit.

What about the modern square? I think it would be pedagogically useful for understanding the interaction between quantification and negation in an introductory logic course. And in a course that presents some non-Aristotelian (generalized) quantifiers, the transfer from *square(all)* to *square(Q)* is effortless, as I tried to illustrate in the paper.

Furthermore, the square is closely tied to *syllogistics*, and I don't think a modern logic course should avoid syllogisms altogether. Rather, they should be presented from a modern perspective, belonging to a particular kind of logical system, with advantages and disadvantages. One disadvantage is of course weak expressive power, but on the other hand syllogisms encode simple *monotonicity reasoning* often used in practice. And the metalogical advantages are, as Larry has shown in a series of beautiful papers (see [3] for a start), decidability and very simple completeness proofs for the simplest systems. (Here it doesn't matter if the quantifiers in *square(all)* or in the traditional square are used; things just get a little simpler in the former case.)

7   Duals or adjoints?

Larry's question 3 starts with two distinct propositional logic squares corresponding to the Aristotelian quantified square. Both have $a \wedge b$ in the upper left corner and its contradictory negation in the diagonally opposite corner, but the first, standard, square has $a \vee b$ in the lower left corner, whereas the second has $a \to b$. Thus, the vertical relations in the second square is no longer *dual*, but rather *adjoint*, in the category-theoretic sense (that conjunction with a is (right) adjoint to implication from a is an early application of category theory to logic by [2]). I have not encountered this square before, and am not sure about its full significance. Let me just briefly comment on the relations to the quantified square.

Let us call the two squares the *dual square* and the *adjoint square*, respectively. They are both analogous, but in different ways, to the quantified square. The dual square relies on seeing universal quantification as generalized conjunction, and, dually, existential quantification as generalized disjunction. This is most clear for type $\langle 1 \rangle$ quantifiers: $\forall x Bx \leftrightarrow \wedge \{Bx : x \in M\}$ and $\exists x Bx \leftrightarrow \vee \{Bx : x \in M\}$ (M is the universe). The familiar duality between $\wedge$ and $\vee$ (de Morgan laws, etc.) transfers directly to $\forall$ and $\exists$. But the analogy extends to type $\langle 1, 1 \rangle$ quantification too: $all(A, B) \leftrightarrow \wedge \{Bx : x \in A\}$ and $some(A, B) \leftrightarrow \vee \{Bx : x \in A\}$. Similarly, $no(A, B) \leftrightarrow \wedge \{\neg Bx : x \in A\}$ and $not\ all(A, B) \leftrightarrow \vee \{\neg Bx : x \in A\}$, and we have the Aristotelian square.

In the type $\langle 1, 1 \rangle$ case, inner negation applies to the second argument. As Larry points out, this terminology becomes more natural if we view Q not as a relation but as a function, taking a set A to a function QA from sets to truth values. Now conjunction can in the same way be seen as a function, taking a truth value to a function from truth values to truth values. But then the inner negation of $a \wedge b$

becomes $a \wedge \neg b$, whose contradictory (outer) negation is equivalent to $a \rightarrow b$, and we have the adjoint square.

A slightly different way of arriving at the same result is as follows. The focus now is on the analogy between quantifiers and connectives as *binary* operators (or their curried versions). (On the earlier analogy, there was nothing binary about conjunction and disjunction; they applied to arbitrarily many formulas.) Now we can use the *Curry-Howard correspondence* to see implication as a *special case* of universal quantification: Write $all(A, B)$ as $(\forall x \in A)Bx$. Now think of Bx as a *set* (rather than a formula) for each x in the *set* A, and of $(\forall x \in A)Bx$ as the *Cartesian product* of the family of sets $\{Bx\}_{x \in A}$. So $(\forall x \in A)Bx$ consists of the functions f such that $f(x) \in Bx$ for $x \in A$. Now in the special case when Bx is the same set B for each x in A, we obtain the set $[A \rightarrow B]$ of functions from A to B. And under the Curry-Howard correspondence, $(\forall x \in A)Bx$ corresponds to universal quantification, and $[A \rightarrow B]$ to implication. Similarly, $(\exists x \in A)Bx$, the *disjoint union* of the family of sets $\{Bx\}_{x \in A}$, corresponds to existential quantification, yielding the product $A \times B$ in the special case when all the Bx are equal to B, corresponding to conjunction.

Thus, to fully maintain the analogy between the quantified square and the adjoint square, I would turn the latter upside down, so that $a \rightarrow b$ is in the upper left (**A**) corner. But this is a small detail; perhaps the analogy can be pursued much further.

8   Type $\langle 1, 1, 1 \rangle$ quantifiers

The short answer to Larry's question 4 is that if, for a type $\langle 1, 1, 1 \rangle$ quantifier Q satisfying Conservativity and Extension,[4] we define $Q\neg(A, B, C) \Leftrightarrow Q(A, B, (A \cup B) - C)$ and $Q^{d} = \neg(Q\neg)$, then

$$square(Q) = \{Q, Q\neg, \neg Q, Q^{d}\}$$

is a square of opposition perfectly analogous to the type $\langle 1, 1 \rangle$ case. In particular, for any $Q' \in \{Q, Q\neg, \neg Q, Q^{d}\}$, we have $square(Q) = square(Q')$.

BIBLIOGRAPHY

[1] Laurence R. Horn. *A Natural History of Negation*. Chicago University Press, Chicago. Republished by CSLI Publications, Stanford, 2001., 1989.
[2] F. W. Lawvere. Adjointness in foundations. *Dialectica*, 23:281–296, 1969.
[3] L. Moss. Completeness theorems for syllogistic fragments. In F. Hamm and S. Kepser, editors, *Logics for Linguistic Structures*, pages 143–173. Mouton de Gruyter, 2008.
[4] Stanley Peters and Dag Westerståhl. The semantics of possessives. To appear in *Language*, 2013.

---

[4] Here conservativity is the property that $Q(A, B, C) \Leftrightarrow Q(A, B, (A \cup B) \cap C)$ for all A, B, C.

Vague Classes and a Resolution of the Bald Paradox

BEIHAI ZHOU AND LIYING ZHANG

ABSTRACT. In this paper, we introduce the concept of a vague class, based on which we give a resolution for the bald paradox, and explain how it arises. In a formalization part, we add equality symbols $\approx_P$ and monadic predicates Θ_P to a first-order language. Based on $\approx_P$ and Θ_P, vague predicates are formalized in an extended first-order language $\mathcal{L}^*$, and then vague classes are captured in a first-order way. Capturing vague predicates (vague classes) in an extended first-order language is a new method for dealing with vagueness, different from approaches that try to capture vagueness by introducing more truth values.[1]

1 Motivation

Vagueness is a common phenomenon in natural language. The literature on vagueness flourished since the 1970s, and there are different proposals to interpret vagueness and resolve the Sorites paradox. Among these approaches, three-valued logics, fuzzy theories, super-valuationism and epistemicism are the most influential. Though influential, each approach has been questioned from a technical aspect or from its intuitive reasonability.

This paper focuses on the simplest and typical vague expressions such as bald, tries to give a concept of vague class, and based on vague classes, a resolution to the Bald Paradox is given. An adequate study of vagueness needs to answer three questions: (1) What are the nature of vague expressions? (2) What about the truth value of borderline cases? (3) Why is the conclusion of the Bald Paradox wrong, and why does the reasoning look so reasonable? This paper tries to answer these questions. Section 2 is about Questions (1) and (2), Section 3 is about (3), and Section 4 continues with the analysis in Section 2.

In this paper, "bald" and the Bald Paradox are the running example. In fact, "bald" only represents one type of vague class, but a simplest one. Vague classes as defined in this paper, based on the case of "bald", cannot cover all other types of vagueness. In order to cover more types, we need further study.

---

[1] The paradox to be discussed is known under various names in the literature, such as the Sorites, or the Paradox of the Bald Man. In this paper, for brevity, we use the term "Bald Paradox", emphasizing our focus on the functioning of the vague predicate bald.

2   Vague classes

2.1   The intuition of a vague class

In fuzzy logic, there are fuzzy sets corresponding to vague expressions. The core concept for fuzzy sets is *degree of membership*. A member does not simply belong to a set or not, there is a measures to how an object belongs to a set. For instance, for "young people", we may use age as a criterion for membership; for "bald", we may use the number of hairs. Membership can be established by 'scientific' methods.

Degree of membership depends on a measure perspective. But in our view, understanding a vague predicate in daily life usually involves a natural perspective of agents' impressions, the scientific measure perspective is something acquired.[2] So, we will introduce the concept of vague class based on impressions. Classes with unclear bounds are *vague classes*, for instance, *bald*, *young*, *fatty*, *good student*, etc.

Consider the formation of the vague class for "bald". First, there are samples of objects that we consider bald. Second, if an individual is indistinguishable from a sample with respect to being bald, then that individual is bald. Hence, a vague class has two basic elements: the *sample* and the *distinguishing power*.

For distinguishing power, we use an indistinguishability relation. Through indistinguishability surrounding the samples, some objects are gathered, forming an object class. The higher the distinguishing power, the fewer indistinguishable objects left. Classes obtained in this way come wholly from an agent's cognition. Because our distinguishing power is not stable, this kind of class is vague, it is influenced by many factors, such as context and situation, and different cognitive agents may have different distinguishing power.

Something to be highlighted is that whether an individual is bald or not depends on our overall impression. In the process of forming this impression, we did not count numbers of hairs. In other words, "bald" is not a class obtained by counting numbers of hair, it is obtained by our cognitive impression. Although the number of hairs is relevant to determining whether someone is bald or not (more than a certain amount is surely non-bald, less than a certain amount is surely bald), since adding or removing several hairs does not affect the overall impression, the number of hairs is not an exact factor for determining whether someone is bald or not.

Here, two kinds of cognitive styles are involved. One is to get an overall visual impression with 'distance', the other is a closer observation of the number of hairs. For simplicity, we call these two ways as 'macro-impression' and 'micro-count' respectively. "Bald" is a class obtained by macro-impression, not by micro-count.

---

[2]For the details of the second perspective, see 3.1.

2.2 Mathematical description of vague classes

A vague class has two basic elements: *sample* and *indistinguishability relation*.

Sample   Samples are standard individuals confirmed by an agent. Samples are always about something, This something is the sample's subject. For example, there are samples of "bald". We denote the subject of a sample as τ, and the sample set of subject τ as θ_τ.

When we say that two objects are indistinguishable, the concept of an indistinguishability relation is needed. We introduce two kinds of indistinguishability relation in this section.

Indistinguishability relation 1: $x \approx_\tau y$ (x and y are indistinguishable on τ).

This says that, in a macroscopic perspective, individuals are indistinguishable on some subject. For example, in a macroscopic perspective, x and y are indistinguishable on the subject "bald", or on the subject "tall". It is possible that x and y are indistinguishable on the subject "bald", but distinguishable on the subject "tall". Being indistinguishable on some subject does not mean being indistinguishable on all subjects.

x and y being indistinguishable on τ can also be understood as: on τ, the differences of x and y are invisible (cannot be detected), or the difference is so tiny that it can be ignored.

Indistinguishability relation 2: $x \approx_{(\tau,\varphi)} y$

In indistinguishability relation 2, a distinguishable condition or circumstance φ is added: under condition φ, on subject τ, x and y are indistinguishable.

This relation will be denoted as: $\varphi \rightarrow x \approx_\tau y$ or $x \approx_{(\tau,\varphi)} y$.

Since whether two individuals are distinguishable or not depends on an agent's cognitive ability, there are many influencing factors. For example, to determine whether a man is bald or not is influenced by viewing distance, lighting, sum of hairs, colors, etc. All these factors are denoted by φ.

Definition 1   An indistinguishable set on τ under condition φ and based on a sample x is: $\phi_{(\tau,\varphi)}(x) = \{y : x \approx_{(\tau,\varphi)} y\}$.

Definition 2   Let θ_τ be a sample set with subject τ, φ is a distinguish condition.
Vague class is: $A(\theta_\tau, \varphi) = \bigcup_{x \in \theta_\tau} \phi_{(\tau,\varphi)}(x)$
The *core* of $A(\theta_\tau, \varphi)$ is: $\kappa(A(\theta_\tau, \varphi)) = \bigcap_{x \in \theta_\tau} \phi_{(\tau,\varphi)}(x)$
The *border area* of $A(\theta_\tau, \varphi)$ is:
$$\mu(A(\theta_\tau, \pi, \varphi)) = A(\theta_\tau, \varphi) - \kappa(A(\theta_\tau, \varphi)) = \bigcup_{x \in \theta_\tau} \phi_{(\tau,\varphi)}(x) - \bigcap_{x \in \theta_\tau} \phi_{(\tau,\varphi)}(x)$$

A vague class A is a class decided by a sample set θ_τ with subject τ and distinguish condition φ, which has two parts: *core region* and *border area*.

If a distinguish condition φ and a sample set θ_τ are given, and x is an object under subject τ, we then say that x is τ under condition φ and sample set θ_τ, for short, x is τ.

Based on Definition 2, we get:

FACT 1.

x is τ, if $x \in A(\theta_\tau, \varphi)$.

x is a standard τ, if $x \in \kappa(A(\theta_\tau, \varphi))$. If x is standard then x is not vague. All samples are standard.

x is a vague τ, if $x \in \mu(A(\theta_\tau, \varphi))$.

x is a sample of τ, if $x \in \theta_\tau$.

2.3   Vague degrees

Based on vague classes, we now discuss the concept of a vague degree. There are two kinds of vague degree: vague degrees of vague classes and vague degrees of objects. We start with the first.

When $A(\theta_\tau, \varphi)$ is finite, it may have a vague degree.

Definition 3   The vague degree of a vague class $A(\theta_\tau, \varphi)$ is denoted as $\mathrm{Vd}(A(\theta_\tau, \varphi))$

$$\mathrm{Vd}(A(\theta_\tau, \varphi)) = 1 - \frac{\|\kappa(A(\theta_\tau, \varphi))\|}{\|A(\theta_\tau, \varphi)\|}$$

The greater the value of $\kappa(A(\theta_\tau, \varphi))$, the smaller the vague index of $A(\theta_\tau, \varphi)$, and the more $A(\theta_\tau, \varphi)$ approaches to a precise class.

When $\|\kappa(A(\theta_\tau, \varphi))\| = \|A(\theta_\tau, \varphi)\|$, that is to say, every individual in $\|A(\theta_\tau, \varphi)\|$ is standard, we have $\mathrm{Vd}(A(\theta_\tau, \varphi)) = 0$.

Clear classes can be defined by means of vague degrees of vague classes:

Definition 4 $A(\theta_\tau, \varphi)$ is a clear class if and only if $\mathrm{Vd}(A(\theta_\tau, \varphi)) = 0$.

Clear class are a special case of vague classes. Normal sets are clear classes.

Proposition 1 $A(\theta_\tau, \varphi)$ is a clear class if and only if $\|\mu(A(\theta_\tau, \varphi))\| = 0$. In particular, $A(\theta_\tau, \varphi)$ does not have a border area.

Thus, the reverse of a vague class is a clear class. If $A(\theta_\tau, \varphi)$ has no border area, it is a clear class.

Now, we discuss the vague degree of an object x according to a vague class $A(\theta_\tau, \varphi)$.

Positive vague degree: The *positive vague degree* says to what extent an object x is τ. This is proportional to how many samples x are indistinguishable from. Intuitively, if x is indistinguishable from all samples, then "x is τ" is not vague. (If one is indistinguishable from all samples of "bald", then the vague degree of "he is bald" is 0: one is bald for sure, which is not vague at all.)

By definition 1, we obtain:

FACT 2. Let y be a sample ($y \in \theta_\tau$). x is indistinguishable with y if and only if $x \in \phi_{(\tau, \varphi)} y$.

Definition 5 $\mathrm{vd}^+(x, A(\theta_\tau, \varphi)) = 1 - \dfrac{\|\{\phi_{(\tau,\varphi)} y : x \in \phi_{(\tau,\varphi)} y, y \in \theta_\tau\}\|}{\|\theta_\tau\|}$

$\text{vd}^+(x, A(\theta_\tau, \varphi))$ is the vague degree of 'x is τ'. If x is indistinguishable with all samples, then $\|\{\phi_{(\tau,\varphi)}y : x \in \phi_{(\tau,\varphi)}y, y \in \theta_\tau\}\| = \|\theta_\tau\|$, $\text{vd}^+(x, A(\theta_\tau, \varphi)) = 0$, this means that x is a standard τ, and belongs to $A(\theta_\tau, \varphi)$ with no vagueness. Otherwise, we can get a number a, $0 < a \leq 1$, the vague degree of 'x belongs to $A(\theta_\tau, \varphi)$' is a.

Negative vague degree: to what extent an object x is not τ.

To what extent an object x is not τ, it is proportional to how many samples x is distinguishable from. Intuitively, if x is distinguishable from all samples, then 'x is not τ' is not vague. (If one is different from all samples of bald, he is not bald for sure, then the vague degree of 'he is not bald' is 0.)

Definition 6 $\quad \text{vd}^-(x, A(\theta_\tau, \varphi)) = 1 - \dfrac{\|\{\phi_{(\tau,\varphi)}y : x \notin \phi_{(\tau,\varphi)}y, y \in \theta_\tau\}\|}{\|\theta_\tau\|}$

When we make a judgement on whether an individual is τ, the greater the sample set is (that is, $\|\theta_\tau\|$ is greater), the more precise the vague degree is (whether positive or negative, it will acquire more decimal place). This indicates that, to get more precise vague degree, usually we need to find more samples.

3 Vague classes and a resolution of the Bald Paradox

3.1 The analysis on bald paradox

How do we judge whether a man is bald or not?

Whether an individual is bald or not is decided by an agent's visual impression. At the same time, being bald or not has a relation with the number of hairs. To judge whether an individual is bald, the number of hairs has some kind of power to decision power: up to a certain number, the individual must be bald, over another certain number, the individual cannot be bald.

Impressions are holistic and vague, while the number of hairs is discrete and can be accurately measured. So, there are two different perspectives: bald is a global 'macroscopic' impression, number (of hairs) is a result of close or 'microscopic' observation and count. In other words, impression is a cognitive outcome obtained in macroscopic perspective; the number of hairs is a cognitive outcome acquired by a cognitive measurement perspective. We will call these two perspectives whole or impression perspective versus measure or quantitative perspective, respectively.

Based on the above analysis, our views are these: the Bald Paradox is an epistemic problem, it should be discussed in an epistemic frame. It originates from the difference of two perspectives: impression and measure perspective. Repeated conversions between the two perspectives lead to puzzles.

Switching back and forth: *If a man with n hairs on his head is bald then a man with (n+1) hairs on his head is bald.* 'n hairs', '(n+1) hairs' come from a measure perspective, which focuses on hairs and sum of hairs. '. . . is bald' is an impression perspective, which focuses on the phenomenon of baldness.

If a man with n (number) hairs on his head is bald (impression) then a man with $(n+1)$ (measure) hairs on his head is bald (impression).

We judge whether one is bald or not by a macroscopic perspective, not by counting the number of hairs. If we only use a macroscopic perspective, vagueness is possible, but there is no paradox. When measure perspective is added to impression, and the two perspectives switch to each other repeatedly, the paradox arises.

To resolve the paradox, we need to get rid of the connections or break the switching. It is not reasonable to get rid of the connections, because the number of hairs has a basic relation with being bald. So we consider breaking the switch. Suppose the upper bound is m, then, when counting the hairs to m, switching should be stop. Now the problem is how to set the upper bound, and to us, it is based on the concept of a vague class.

3.2   Vague classes and a resolution of the Bald Paradox

Definition 7   Let $f(x)$ be the sum of x' s hairs. $\tau = B$ (bald), For any $x \in \kappa(A(\theta_B, \varphi))$, we set

$G(A(\theta_B, \varphi)) = Max(\{f(x) : x \in \kappa(A(\theta_B, \varphi))\})$

$H(A(\theta_B, \varphi)) = Max(\{f(x) : x \in A(\theta_B, \varphi))\})$

$G(A(\theta_B, \varphi))$ and $H(A(\theta_B, \varphi))$ are two boundaries around $A(\theta_B, \varphi)$ from a hair sum viewpoint. $G(A(\theta_B, \varphi))$ is the upper bound of $A(\theta_B, \varphi)$'s core. $H(A(\theta_B, \varphi))$ is the upper bound of $A(\theta_B, \varphi)$.

We make two quick remarks:

If $f(a) \geq G(A(\theta_B, \varphi))$, that is, $x \notin \kappa(A(\theta_B, \varphi))$, a cannot be a standard bald object.

If $f(a) \geq H(A(\theta_B, \varphi))$, that is, $x \notin A(\theta_B, \varphi)$, a cannot be bald with sample set θ_B under condition φ.

Something should be underlined here. First we get a bald class by sense and impression, after that, we measure this class and get two features of this class: $G(A(\theta_B, \varphi))$ and $H(A(\theta_B, \varphi))$. It is not true, though, that we get or define bald classes by both $G(A(\theta_B, \varphi))$ and $H(A(\theta_B, \varphi))$.

For any $A(\theta_B, \varphi)$, due to the upper bound on the hair sum, the induction on hair sums in the Bald Paradox does not go through.

Instead, the inductive argument should be rewritten to a conditional induction:

Condition 1: for any $x, y \in A(\theta_B, \varphi)$, $f(y) = x + 1$, $f(y) < G(A(\theta_B, \varphi))$.

Condition 2: for any $x, y \in A(\theta_B, \varphi)$, $f(y) = x + 1$, $f(y) < H(A(\theta_B, \varphi))$.

When condition 1 is satisfied, if x is bald, then $y(f(y) = x + 1)$ is bald, and it is standard bald.

When condition 2 is satisfied, if x is bald, then $y(f(y) = x + 1)$ is bald, and y also could be non-standard bald.

The restrictive conditions (resulting in the upper bound) ensure that we will not introduce new individuals by this induction. Thus, the paradox is dispelled.

3.3 Answers to some possible questions

In summary, the basic method to dispel the Bald Paradox is this. According to intuition, the induction should have an upper bound, if it goes beyond the upper bound, the induction is not valid any more.

This method seems to be right, but there are several questions we need to answer.

Suppose there is an upper bound n,

(1) Why is n so exact? How to determine this n (the problem of selecting the upper bound)?

(2) Since n is an upper bound, and for bald, $n + 1$ and n look no different, $n + 1$ (or $n - 1$) could also be upper bounds. Then, any n could be an upper bound. (The lifting of upper bound).

Here is our answer to these problems. n is determined by $A(\theta_B, \varphi)$, $A(\theta_B, \varphi)$ is determined by samples and distinguish power, n is not determined by information on the number of hairs. n is obtained by impression, not by measure. Under some conditions and impressions, we get n. We can also get $n + 1$ under other conditions and impressions, but that is because we are under different conditions. Under certain conditions, by impression we get only a certain n.

4 A formulation for vague classes

4.1 Language

Let $\mathcal{L}$ be a first order language, $\mathcal{L}^*$ is an extension of $\mathcal{L}$.

In $\mathcal{L}^*$, we add countably many new monadic predicate symbols Π (vague predicate) to $\mathcal{L}$, for any $P \in \Pi$, there are equality symbols $\approx_P$ and monadic predicates Θ_P. We add two types of formulas to the first-order formulas of $\mathcal{L}$: $x \approx_P y$ and $\Theta_P x$.

Definition 8: For any $P \in \Pi$,

$Px := \exists y(\Theta_P y \wedge x \approx_P y)$

$\odot Px := \forall y(\Theta_P y \to x \approx_P y)$

$\bigcirc Px := \exists y(\Theta_P y \wedge x \approx_P y) \wedge \exists y(\Theta_P y \wedge \neg(x \approx_P y))$

Here are the intuitive meanings for these formulas:

$x \approx_P y$: x is indistinguishable from y on P[3]

$\Theta_P x$: x is a sample of P.

$\odot Px$: x is a standard P.

$\bigcirc Px$: x is a vague P.

Px: x is P.

Now we will show how to express various baldness assertions in $\mathcal{L}^*$.

---

[3] Whether 2 individuals are distinguishable is a problem that depends on an agent's cognitive ability. To make our basic analysis simpler, we did not introduce explicit parameters for agents. Then the language and semantics can be seen as characterizing one single agent's vague classes.

$\Theta_B x$: x is a sample of bald.

$Bx := \exists y(\Theta_B y \wedge x \approx_B y)$

Bx: x is bald.

$\neg Bx$: x is not bald.

$\neg Bx \leftrightarrow \neg \exists y(\Theta_B y \wedge x \approx_B y) \leftrightarrow \forall y(\neg \Theta_B y \vee \neg(x \approx_B y)) \leftrightarrow \forall y(\Theta_B y \to \neg(x \approx_B y))$

$\odot Bx := Bx \wedge \forall y(\Theta_B y \to x \approx_B y)$

$\odot Bx$: x is standard bald.

$\neg \odot Bx$: x is not standard bald.

$\neg \odot Bx \leftrightarrow \neg(Bx \wedge \forall y(\Theta_B y \to x \approx_B y))$
$\qquad\qquad \leftrightarrow \neg Bx \vee \exists y \neg(\Theta_B y \to x \approx_B y) \leftrightarrow \neg Bx \vee \exists y(\Theta_B y \wedge \neg(x \approx_B y))$

$Bx \wedge \neg \odot Bx$:x is non-standard bald, x is bald, but not standard bald.

$Bx \wedge \neg \odot Bx \leftrightarrow Bx \wedge \exists y(\Theta_B y \wedge \neg(x \approx_B y))$

$\bigcirc Bx$: x is vague bald. (or x is bald vaguely)

$\bigcirc Bx := Bx \wedge \exists y(\Theta_B y \wedge \neg(x \approx_B y))$

$\bigcirc Bx \leftrightarrow Bx \wedge \neg \odot Bx$, A vague bald man is a non-standard bald man. Vice versa, a non-standard bald man is a vague bald man.

$\neg \bigcirc Bx$: x is not vague bald. (x is standard bald, or, x is not bald)

$\neg \bigcirc Bx \leftrightarrow \forall y(\Theta_B y \to x \approx_B y) \vee \forall y(\Theta_B y \to \neg(x \approx_B y))$

Bx, $\odot Bx$ and $\bigcirc Bx$ are abbreviations of complex formulas. Thus, all kinds of bald ($\odot Bx$, $\bigcirc Bx$ and even Bx, etc.) are not primitive notions, they are defined by samples and distinguishing power.

By definition 8, we can get:

$\bigcirc Bx \to Bx$, A vague bald man is a bald man.

$\odot Bx \to Bx$, A standard bald man is a bald man.

$\odot Bx \to \neg \bigcirc Bx$, A standard bald man is not a vague bald man (standard is not vague).

$\bigcirc Bx \to \neg \odot Bx$, A vague bald man is not a standard bald man (vague is not standard).

$\bigcirc Bx \leftrightarrow Bx \wedge \neg \odot Bx$, vague bald is equivalent with non-standard bald.

4.2 Model

The semantics for $\mathcal{L}^*$ extends the standard semantics for $\mathcal{L}$, composed of structure $\mathbb{A} = \langle D, \eta \rangle$, a valuation ρ and an interpretation $\sigma = \langle \mathbb{A}, \rho \rangle$. An interpretation is our notion of 'model'.

In $\mathcal{L}^*$, we add vague predicates set Π, and for any $P \in \Pi$, an equality symbol $\approx_P$ and monadic predicate Θ_P. In $\mathcal{L}^*$-semantic, we need some new elements to interpret these new symbols and the resulting new formulas.

Definition 9   A $\mathcal{L}^*$-structure is a tuple $\mathbb{B} = \langle D, \eta, T, \{\approx_\tau : \tau \in T\}, \theta, \xi \rangle$, a valuation ρ and an interpretation $\sigma = \langle \mathbb{B}, \rho \rangle$.

$\mathbb{B}$ is an expansion of $\mathbb{A} = \langle D, \eta \rangle$, adding the following components:

$T \neq \varnothing$ is a subject set (subject to distinguishing).

For any $\tau \in T$, $\approx_\tau$ is a reflexive and symmetry relation on D.

For any $\tau \in T$, θ is a sample choice function ($\theta : T \to \wp(D)$). For every subject τ, there is a sample set $\theta(\tau) \subseteq D$ with $\theta(\tau) \times \theta(\tau) \subseteq \approx_\tau$.

$\xi \colon \Pi \to T$. For any $P \in \Pi$, we interpret predicate P as a subject τ, or, assign a subject to every predicate, $\xi(P) = \tau$, ξ is similar to η.

Definition 10

(1) If α is a first-order $\mathcal{L}$-formula, then $\sigma(\alpha)$ is the same as in the definition for $\mathcal{L}$-semantic.

(2) $(\Theta_P(x))^\sigma = \begin{cases} 1, & if\, x^\sigma \in \theta_{\xi(P)} \\ 0, & otherwise \end{cases}$

(3) $x \approx_P y = \begin{cases} 1, & if\, x^\sigma \approx_{\xi(P)} y^\sigma \\ 0, & otherwise \end{cases}$

Definition 11 (Satisfiability and validity) Given a $\mathcal{L}^*$-formula α. If there is an interpretation $\sigma = \langle V, \rho \rangle$, such that $\sigma(\alpha) = 1$, then α is satisfiable. If for any σ, there is $\sigma(\alpha) = 1$, then α is valid, denoted as $\models \alpha$.

Propositions 2

(1) $\models \bigcirc Bx \to Bx$

(2) $\models \odot Bx \to Bx$

(3) $\models \odot Bx \to \neg \bigcirc Bx$

(4) $\models \bigcirc Bx \to \neg \odot Bx$

(5) $\models \bigcirc Bx \leftrightarrow Bx \wedge \neg \odot Bx$

(6) $\models \exists y \Theta_B y \to (\odot Bx \to Bx)$

The intuitions for clauses (1) - (5) were given in Section 4.1. The meaning of (6) is this: under the condition that a standard bald man exists, a standard vague bald man is a bald man. The proofs of (1) through (6) are routine.

4.3 Some explanations on truth value

When we make a judgement on whether a man is bald, samples and the indistinguishability relation are very important factors. When we choose samples, it is easy and not vague, but when we judge whether a man is bald or not from samples, vagueness arises. Our definition of baldness does not deal with vagueness in judgement, it only says that, when the judgement is done, we get a bald class. The dividing line between bald and non-bald is strict, and bald and not bald are contradictory to each other. Thus, from the view point of truth values, our treatment of these predicates is 2-valued. In the same way, standard bald and non-standard bald are 2-valued too.

Intuitively, bald and non-bald should be contradictory relations. We cannot say that an individual is both bald and non-bald. On the other hand, sometimes, it is difficult to distinguish bald and non-bald strictly, there is a vague area, which is also according to intuition. What we can say is that the former intuition comes from logic, the latter intuition comes from cognitive reasons. In this paper, what we discuss is the concept bald, in other words, we discuss how to define bald. Under the 2-valued definition of bald, the dividing line between bald and non-bald is clear, not vague. This is a theoretical definition. Our theory does not focus on how to decide whether an individual is bald or not based on samples, which is a cognitive problem. In a practical judgement, the boundary between bald and non-bald is still vague, and relies on our impressions. That is our cognitive reality. Vague class theory does not solve the cognitive problems for assigning bald.

5   Comparison with other approaches

5.1   Impressive versus measure perspective

To capture the nature of vague expressions, this paper introduced the concept of a vague class. As we have mentioned in Section 2.1, in fuzzy theories, there are fuzzy sets corresponding to vague expressions. The core concept for a fuzzy set is degree of membership, which depends on what we called a scientific measure perspective. In our opinion, to understand a vague expression, what we use is an impressive perspective, not a measure perspective. Based on this view, we introduced vague classes, an impressive perspective-based concept.

5.2   Indistinguishability versus indifference relations

In this paper, a vague class has two basic elements: the *sample* and the *indistinguishable relation*. In our extended first-order language $\mathcal{L}^*$ with equality symbols $\approx_P$ for indistinguishability, and monadic predicates Θ_P for samples, we defined:

Bald: $Px := \exists y(\Theta_P y \wedge x \approx_P y)$

Standard bald: $\odot Px := \forall y(\Theta_P y \to x \approx_P y)$

Vague bald: $\bigcirc Px := \exists y(\Theta_P y \wedge x \approx_P y) \wedge \exists y(\Theta_P y \wedge \neg(x \approx_P y))$

Here is another approach. In [1], a representative paper on the *tolerant approach* to vagueness, the concept of an *indifference relation* is introduced. The principle of tolerance can then be stated as:

$\forall x \forall y(P(x) \wedge x \smile_P y \to P(y))$

Here $\smile_P$ is an indifference relation which is reflexive and symmetric, but possibly non-transitive. In the opinion of the authors, if the semantics of vague predicate is made sensitive to such indifference relations, the tolerate principle can be validate. Intuitive said, x is tall *tolerantly* if there exists an individual y such that x is similar to y by way of how tall x looks, and y is tall *classically*. In the specific paper mentioned, the authors defined three distinct notions of truth to capture vague predicates: *tolerant truth* (t-truth), *classical true* (c-truth), *strict true* (s-truth). The relations between t-truth and c-truth, s-truth and c-truth can be represented as follows[4]:

---

[4]For more details, see the cited paper.

$M \vDash^t P(\underline{a})$ iff $\exists d \smallfrown_P a : M \vDash^c P(\underline{d})$

$M \vDash^s P(\underline{a})$ iff $\exists d \smallfrown_P a : M \vDash^c P(\underline{d})$.

Now, we compare the two theories. At first glance, they are very similar in their intuitive ideas, and we even could find a rough correspondence, using the authors' notions of truth and our Definition 8 above:

Indistinguishable relation   VS   Indifference relation,

$\bigcirc Px$(vague bald) is true   VS Px is t-true[5],

$\odot Px$ (standard bald) is true   VS Px is s-true,

$\ominus_P y$(a sample of bald) is true VS Px is c-true.

But when we go into details, the differences appear.

Firstly, in this paper, we define Px, $\bigcirc Px$ and $\odot Px$ in an object language $\mathcal{L}^*$, here as the tolerance approach uses three kinds of truth, and tries to capture their distinctions through model theory. Second, to define Px, $\odot Px$, and $\bigcirc Px$, crucially for us, samples $\ominus_P y$ are introduced in this paper, which are based on an individual's cognition. In the cited tolerance approach, the part corresponding to $\ominus_P y$ is Py which is classically true, there is no cognitive perspective. Finally, our approach has only one truth, making the system 2-valued, whereas the tolerance approach uses at least three truth values.

5.3   Related work

What are the truth values of borderline cases? Different approaches have different ideas. In fact, many approaches try to capture vagueness by changing classical valuations. Three-valued approaches introduce a new truth value I (indefinite/indeterminate) to capture the intermediate state in between T (true) and F (false). In fuzzy theories, sentences are true to some degree in $[0, 1]$[6]: for example, 'Tom is tall' may have a degree of 0.87. Supervaluationism introduces the technique of truth value gaps to vagueness. Fine [2] in particular proposed to account for the logical behavior of vague predicates by making a distinction between truth and 'supertruth'. A proposition is supertrue at a partial specification if it is classically true at all complete extensions. Epistemicists believe that there are strict boundaries between true and false. We cannot know where it is only because of agents' cognitive limitations: therefore, in epistemicism, classical 2-valuedness is preserved.

Let us briefly mention a few issues that face theories in the area. In technical aspects, many-valued approaches (including three-valued approaches like that of [1], fuzzy theories, and others, have many problems, both in validities [7] and in

---

[5]The 'true' in left part is given in definition 8 of this paper, the 't-true', 'c-true', 's-true' are concepts of [1].

[6]For fuzzy theories, tautologies are true to degree 1, contradictions are true to degree 0, and contingent propositions may be true to any degree corresponding to a real number between 0 and 1.

[7]$P(a) \vee \neg P(a)$ is not a tautology, and $P(a) \wedge \neg P(a)$ is not a contradiction.

the famous problem of 'penumbral connections'. [8] Moreover, higher-order vagueness is another prominent problem which many-valued approaches have to face. [9] The supervaluation approach can give a good analysis of penumbral connections, but it has problems with interpreting higher-order vagueness. Since 2-valued classical logic is preserved, epistemicism has no pressure when facing the technical problems mentioned above, but in a logical perspective, keeping 2-valued classical logic means that there is nothing new. If what we want to capture is how ordinary people (with imperfect logic) use vague expressions in daily life, classical 2-valued logic is not enough.

In this paper, there is a 2-valued definition of bald. From the perspective of technique, it is simpler than many-valued approaches and the supervaluation approach. On the other hand, different from the classical 2-valuedness of epistemicism, in this paper, vague predicates are defined based on samples and distinguishing power, and ordinary people's cognitive factors are considered and partly captured in our theory.

In addition to technical considerations, there is a deeper motive. We think that vague classes are based on human cognition, they do not primarily depend on truth. The concepts of truth and scientific measurement are acquired later on. To emphasize these ideas, we introduced the notion of sample Θ_{Py} which reflects our cognition, and we do not continue the traditional ways of emphasizing truth-value changes.

5.4   Sorites paradox

We add a few comments on solutions to the sorites paradox. The three valued approach ([13]) thinks that, because there is at least one not true (F or I) premise, we cannot get the conclusion by the standard inferences in the paradox. Fuzzy theories have two ways of dealing with the paradox. If the consequence relation requires keeping perfect truth, because the inductive premises have value almost 1 (but not perfectly true), we cannot get the conclusion. If the consequence relation itself requires maintaining true to a degree, then '→' is non-transitive in the analysis. In the supervaluation approach, the inductive premise is (super-)false under the interpretation, and the Sorites paradox is blocked. Epistemicists think that the inductive premises are not always true, and then the conclusion need not be true.

In short, there are mainly two ways to resolve the Sorites paradox: showing that

---

[8] As an illustration, consider the three-valued approach. (1) If a is indeterminate pink (P), red (R) and small (S), then $P(a) \wedge S(a)$ is indeed still indeterminate, but $P(a) \wedge R(a)$ should be false. Similarly with $P(a) \vee S(a)$ versus $P(a) \vee R(a)$: the former is indeterminate as desired, but we want the latter to be true. (2) Suppose that a and b are both borderline cases of tall individuals, though b is, in fact, a bit taller. Then $T(a) \wedge \neg T(b)$ should be false, but is predicated to be indeterminate. Fuzzy theories make similar wrong predictions for many other complex sentences as well.

[9] If we do not believe that there is a strict border between tall and not tall, then it is hard to believe that there are two strict borders between (i) the tall and the borderline cases, and (ii) the borderline cases and the not tall ones.

the inductive premise is not true, or cutting off the inference chains by denying transitivity. Every theory mentioned can resolve the paradox, but the key point is what they are based on (their interpretation for vagueness), and this should be able to make clear the mechanism underlying the paradox: why it is wrong, even though it looks reasonable.

In this paper, we give an analysis of the Bald Paradox, we judge whether one is bald by a macroscopic perspective, not by focusing on measuring numbers of hairs. If we only use a macroscopic perspective, vagueness is possible, but there is no paradox. When the measure perspective is added to that of impression, and the two perspectives switch to each other repeatedly, the paradox arises. The basic method to dispel the Bald Paradox our view is this: according to intuition, induction should have an upper bound, and if it goes beyond the upper bound, the induction is not valid any more.

6   Conclusion

In this paper, taking the simplest vague predicate "bald" as an example, we introduced the concept of a vague class. Based on vague classes, we gave a resolution to the Bald Paradox, and explained why it arises.

Vague classes are expressed by vague predicates. Vague predicates can be defined in an extended first-order language. Based on samples and distinguishing power, we defined bald, standard bald and vague bald, as an attempt to interpret vagueness in formal language. Constructing a new kind of predicate in a formal language to capture the nature of vague predication is a distinctive feature of this method. It is different from the usual truth-value changing approaches. Another special feature is our distinction between the impressive perspective and the measure perspective. In daily life, we usually use the impressive perspective when employing vague predicates. The Bald Paradox is a kind of sophistry making use of repeated switching between impression and measurement perspectives. The resolution method in this paper stops this conversion at a certain stage.

Acknowledgements   This research is supported by the Major Project of National Social Science Foundation of China (No. 12&ZD119) and The Project-sponsored by SRF for ROCS, SEM. We especially want to thank Johan van Benthem, Fenrong Liu for their very helpful comments and suggestions. We also thank Yanjing Wang for helpful exchange and comments at workshops. Further thanks go to various colleagues and audiences for their valuable feedback at conference and workshops held in Beijing and Shanghai.

BIBLIOGRAPHY

[1]   Cobreros, P., Ripley, D., Egre, P., van Rooij, R.A.M..: 'Tolerant, classical, strict'. *Journal of Philosophical Logic*, 41, 347-385, 2012.
[2]   Fine, K.: 'Vagueness, Truth and Logic', *Synthese* 30, pp. 265-300, 1975.

[3] Graff, D.: 'Shifting Sands: An Interest-Relative Theory of Vagueness', *Philosophical Topics* 28, pp. 45-81, 2000.
[4] Kamp, H.: 'Two Theories About Adjectives', in Keenan, E L (ed.) *Formal Semantics of Natural Language*, Cambridge: Cambridge University Press, 1975.
[5] Raffman, D.: 'Vagueness and Context-Relativity', *Philosophical Studies* 81, pp.175-92, 1996.
[6] Soames, S. *Understanding Truth*, New York: Oxford University Press,1999.
[7] Soames, S.: 'Precis of Understanding Truth and replies', *Philosophy and Phenomenological Research*, LXV: 429-452, 2002.
[8] Sorensen, R. A.: 'Ambiguity, Discretion and the Sorites', *Monist* 81, pp. 217-35, 1998.
[9] Sorensen, R. A.: *Vagueness and Contradiction*, Oxford: Oxford University Press, 2001.
[10] Stanley, J.: 'Context, interest-relativity, and the sorites', *Analysis*, 63: 269-80, 2003.
[11] Smith, N. J. J., *Vagueness and Degrees of Truth*, Oxford: Oxford University Press, 2008.
[12] Tye, M: 'Supervaluationism and the Law of Excluded Middle', *Analysis*, 49/3: 141-143, 1989.
[13] Tye. M.: 'Vague objects', *Mind*(99):535-57, 1990.
[14] van Rooij, R.A.M.: 'Vagueness, tolerance and non-transitive entailment'. In *Reasoning under Vagueness*. Berlin: Springer. 2011.
[15] Williamson, T.: *Vagueness*, London: Routledge, 1994.
[16] Williamson, T.: 'Soames on Vagueness', *Philosophy and Phenomenological Research* 65, pp. 422-8, 2002.
[17] Crispin, W.: 'The Epistemic Conception of Vagueness', *Southern Journal of Philosophy*, 33 (Supplement): 133-159, 1995.
[18] Zadeh, L.: 'Fuzzy Sets', *Information and Control* 8, pp. 338-53, 1965.

Learning to Live with Paradox

FRANK VELTMAN

1 Diagnosis[1]

Examples of the Sorites paradox are easy to come by. All you need is a vague predicate P, preferably one with a comparative form, the use of which is guided by the following principle:

> If two objects are observationally indistinguishable in the respects relevant to P, then either both satisfy P or else neither of them does.

Following [2], we will refer to this principle, which expresses the Equivalence of Observationally Indistinguishable objects, as EOI.[2] With such a vague predicate P, one can nearly always associate a domain $\mathcal{D}$ and a relation $>_P$ with the following properties:

(i) $\mathcal{D}$ is nonempty; the elements of $\mathcal{D}$ are similar in kind and (therefore) comparable in the respects relevant to P.

(ii) $>_P$ is the irreflexive and transitive relation on $\mathcal{D}$ consisting of the pairs $\langle d, e \rangle$ of which the first element d is observationally *more* P than the second element e.

(iii) There are objects d and e in $\mathcal{D}$ such that

  (a) P clearly applies to d;

  (b) P clearly does not apply to e;

  (c) There is a finite sequence of objects $d = d_0, d_1, \ldots, d_{n-1}, d_n = e$ in $\mathcal{D}$ such that for any two successive objects d_i and d_{i+1} in this sequence, neither $d_i >_P d_{i+1}$, nor $d_{i+1} >_P d_i$.

Let us write $x \approx_P y$ iff neither $x >_P y$ nor $y >_P x$. Whenever $x \approx_P y$, x and y are observationally indistinguishable in the respects relevant to the predicate P, and therefore it is tempting to read '$x \approx_P y$' as 'x is observationally *just as P as* y'. We will not resist this temptation, even though $\approx_P$ need not be an equivalence relation. That is, $\approx_P$ is reflexive and symmetric, but not in general transitive. It may very well occur that there is no discriminable difference between the objects x and y — x is observationally just as P as y — and no discriminable difference between the objects y and z — y is observationally just as P as z — whereas x and z *can* be discriminated — x is observationally more P than z. It is easy to see now how the paradox can

---

[1] Part of these comments are copied from notes for a course on logical analysis which were inspired by the work of Hans Kamp and Michael Dummett. See in particular [1] and [2].

[2] In [6], the principle EOI is referred to as the Principle of Tolerance.

arise. The principle EOI forces us to use the predicate P in a way that would only b
coherent if the relation $\approx_P$ were transitive. For, what else does EOI express but:

(EOI 1) For any $x, y \in \mathcal{D}$, if $x \approx_P y$ and P applies to x, P applies to y.

So, you cannot assign the predicate P to d_0 without having to assign it to $d_1, d_2, d_3, \ldots$
and, finally, to d_n as well. EOI forces you to do so, even though $\approx_P$ is not transitive
d_0 and d_n could be so far apart — d_0 has got no hairs, d_n has 150,000 hairs; the tem
perature of d_0 is 2° C, the temperature of d_n is 80° C — that it would seem perfectl
all-right to say that d_0 is P — bald, or cold — but that d_n is not. It looks like we ar
in a predicament: either we stick to the principle EOI, in which case we are forced t
conclude that the use of at least some observational predicates is intrinsically incor
sistent, or we give up this principle, in which case it seems to follow that there are n
truly observational predicates.

Modifying EOI

Most logicians choose the second horn of this dilemma and try to loosen the tie
between vague predicates and observation, without having to cut them. One way t
do this can already be found in in Russell [3]. Consider the observational relatic
$>_P$. The problem is that its resolution is too poor: an object x can *in fact* be mor
P than an object y, without *observationally* being so. Our powers of discriminatic
are limited; that is why observational equivalence is not transitive. We can, howeve
define a relation $>_P^*$ from $>_P$ which, as it were, increases the resolution:

Definition Let $>_P$ be an irreflexive and transitive relation on a domain $\mathcal{D}$. Define th
   sharpening $>_P^*$ of $>_P$ *relative to* $\mathcal{D}$ by: $x >_P^* y$ if and only if
   either (i) $x, y \in \mathcal{D}$ and there is some $z \in \mathcal{D}$ such that $z >_P y$ and not $z >_P x$
   or (ii) $x, y \in \mathcal{D}$ and there is some $z \in \mathcal{D}$ such that $x >_P z$ and not $y >_P z$.

Suppose you are faced with three objects d_1 and d_2, and d_3; you cannot see any di
ference in length between d_1 and d_2; you cannot see any difference in length betwee
d_2 and d_3 either; but you can see a difference in length between d_1 and d_3: d_1
observationally shorter than d_3. Wouldn't you then conclude that there *must* be a di
ference in length between d_1 and d_2 even though you cannot see it by the naked eye
Wouldn't you be inclined to say then that d_1 is *in fact* shorter than d_2, even thoug
this is not observationally so? It is easy to see that $>_P^*$ is irreflexive and transitiv
and that $>_P \subseteq >_P^*$. Hence, we can really think of $>_P^*$ as a sharpening of $>_P$. No
define a relation $\approx_P^*$ from $>_P^*$ in a way analogous to the way $\approx_P$ is defined from $>_P$
$x \approx_P^* y$ iff neither $x >_P^* y$ nor $y >_P^* x$. Then we find that even though $\approx_P$ is not ne
essarily transitive, $\approx_P^*$ is. Indeed, $\approx_P^*$ is an *equivalence* relation. So, if we replace th
formalization of the principle EOI given above by the following new formalization:

(EOI 2) For any $x, y \in \mathcal{D}$, if $x \approx_P^* y$ and P applies to x, then P applies
to y,

contradictions need no longer arise. Suppose you are presented with a series of o
jects as described in the beginning of this section. Given the first formulation of
principle of EIO, you cannot assign the predicate P to d_0 without having to assig
it to $d_1, d_2, d_3, \ldots$, and finally to d_n as well. But given the second formulation (E
2), you need not be led into contradicting yourself. (EOI 2) allows you to draw a li

·mewhere, while (EOI 1) does not. Clearly, there must be three objects d_i, d_{i+1}, and
₂ in the series such that $d_i >_P d_k$, but not $d_{i+1} >_P d_k$. Choose such d_i, d_{i+1}, and
₃, and you can deny the truth of the premise that if d_i is P, d_{i+1} is. Indeed, d_i and
₊₁ are *observationally* distinguishable in a respect relevant to P, since the purely
)servational predicate $x >_P d_k$ applies to d_i, but not to d_{i+1}. Are these new relations
$*_P$ and $\approx^*_P$ really observational? Well, they are clearly not *directly* observational like
₂ relation $>_P$. In many cases it is impossible to decide on the basis of *direct* ob-
rvation whether $>^*_P$ (or $\approx^*_P$) applies to a pair of objects or not. This is because in
₂ definition of $>^*_P$ (and hence in the definition of $\approx^*_P$) a quantification occurs. But,
hile $>^*_P$ and $\approx^*_P$ are not directly observational, they are certainly observational in
₂ sense that they are defined from a directly observational relation $>_P$ by means
˙logic alone. Every statement containing references to $>^*_P$ or $\approx^*_P$ can be translated
to an equivalent expression in which no reference is made to relations other than
$_P$. Therefore, if (EOI 1) is replaced by (EOI 2), the links between the predicate
and direct observation are not severed; the principle that the use of P should be
ided by *mere* observation is just replaced by the principle that it should be guided
˙ observation *aided by reason*. There are various reasons why this modification of
)I is not very attractive[3]. The most important one is that it presupposes that ones
·wers of discrimination are always the same. But it may very well happen that where
₂ observe some difference at one moment we don't see any difference the next. The
lations $>^*_P$ and $\approx^*_P$ are not stable. It's here that the theory of Zhou & Zhang comes
₎. I am pretty sure that they will reject the above solution to the paradox, because
ey are well aware that human perception constantly changes.[4] Still, their solution of
₂ paradox also amounts to a modification of EOI. On their account there is an upper
·und on the number of steps we can make using the principle of EOI starting from
mething that obviously has the property P – a *standard P* as they would call it –
₎d going step by step to things that get less and less P. Where exactly this upper
und is, depends on the indistinguishability relation at hand. Given that this is not a
nstant relation, the upper bound is not constant either. There is a line separating the
's and non-P's but we cannot exactly draw it. That's why these predicates are vague
the first place.

₄more pragmatic solution

₂e way Zhou & Zhang define this upper bound is quite ingenious. But this ingenuity
₎ its price. If we were to follow their advice we should be much more careful using
₃gue predicates than the unrestricted EOI allows. Yes, if people were prepared to
₂ vague adjectives the way Zhou & Zhang suggest, paradox could be avoided. But
ιy should they? Even if they're aware that the sloppy way they use vague adjectives
₃metimes leads to paradox, they might still refuse to change this way of using them
cause in the vast majority of cases this sloppy way serves its purpose just fine. This
₎ I think, the reaction a philosopher like Wittgenstein would have, not only to Zhou
Zhang's solution to the paradox but to any solution that proposes to modify EOI. In
₂ *Philosophical Investigations* Wittgenstein compares meaning rules with sign-posts
₎d he writes:

---

[3]See chapter 8 in [4] for an extensive discussion
[4]The *indistinguishable relation* $\approx_\tau$ plays much the same role in Zhou & Zhang's paper as the relation
₎ here, but they relativize this relation to a condition φ. The resulting relations $\approx_{\tau,\varphi}$ model the possible
₃nsions that the indistinguishability relation can have under various circumstances described by φ.

> The sign-post is in order – if, under normal circumstances, it fulfills its purpose.
> Wittgenstein (1957), PI § 87.

In other words, their is nothing wrong with the principle of EOI. After all, normal we are dealing with just a few objects, most of them very well distinguishable from each other. In those circumstances EOI does *not* give rise to inconsistency; normal it serves its purpose quite well. Only in exceptional situations — i.e. when we a confronted with sequences of objects as described above — things go wrong. B then, vague predicates like 'bald and also 'cold' and 'tall' are not meant to be used those situations; what we should use there is other, finer tools; we should no long talk in terms of 'cold', and 'short' etc., but in terms of numbers of hairs, degre centigrade or millimeters. Actually, Zhou & Zhang agree here. As they see it, tl Sorites Paradox typically arises when the coarse tools of vague predicates are us *side by side* with these finer tools, i.e. when one starts saying things like 'if someo with *23567* hairs on his head is not *bald*, then neither is someone with *234566* hairs his head'. They write that the Sorites is the result of a clash between on the one ha (i) a"macroscopic" perspective where the notion of observational indistinguishabil plays a central role, and on the other hand (ii) a "microscopic" perspective where w count hairs and measure length by millimeters.[5] Still, they go on proposing a seman solution rather than a pragmatic one. They propose to modify the coarse tools, whe Wittgenstein would claim that the Sorites merely reflects that one should not use the coarse tools in a situation where only other, more delicate, ones are applicable.

BIBLIOGRAPHY

[1] M. Dummett, M.: Wang's paradox. *Synthese*, 30:301-324(1975)
[2] Kamp, J.A.W.: The paradox of the heap. In U. Mönnich, editor, *Aspects of Philosophical Lo* (123-155). D. Reidel Publishing Company, Dord- recht, 1981.
[3] Russell, B.: An Inquiry into Meaning and Truth. George Allen and Unwin, London, 1940.
[4] Van Deemter, K.: *Not Exactly: In Praise of Vagueness*. Oxford University Press, 2010.
[5] Weiss, S.E.: The sorites fallacy: What difference does a peanut make. *Synthese*, 33:253-272(1976).
[6] Zhoe, B. and Zhang, L.: Vague Class and a Resolution to Bald Paradox. This volume, 2013.

---

[5] See [5] for a insightful discussion of this point.

Comments on Zhou and Zhang

WEN-FANG WANG

In "Vague Class and a Resolution to Bald Paradox", Zhou and Zhang propose an account of the genesis of the vague concept *bald* (and the extension of the predicate "bald") and a solution to the sorites paradox involving that concept. According to their account, the concept bald is formed (or learned) first by giving some "examples" or paradigm cases of bald heads and then by extending its applications to heads indistinguishable from these paradigm cases. It is emphasized in the account that paradigm cases and the relation of indistinguishability jointly determine a clear-cut boundary for the predicate "is bald". It is also emphasized that the property of being bald is attributed to a head, not on the ground of counting the number of its hairs, but on the ground that one's overall visual impression of the distribution of hairs on the head is indistinguishable from his/her impression of that of some paradigm cases. The problem of the bald paradox is then diagnosed by Zhou and Zhang as consisting of two parts. First, there is a repeated "switch" from the sensational concept (what they call the "impressive perspective") *bald* to the measurement concept (what they call the "measuring perspective") *the number of hairs*, which unduly connects them together. Second and more importantly, because the boundary of the concept *bald* is sharply determined by its paradigm cases and the relation of inditinguishability, the inductive premise in the bald paradox (i.e., "If a head with n hairs is bald, so is a head with n+1 hairs") "should have an upper bound" of its application, i.e., must not hold for some number m. After giving their account of the concept *bald* and the bald paradox, Zhou and Zhang go on to show how the predicate "is bald", "is standardly bald", and "is vaguely bald" can all be defined syntactically and determined semantically in a classical first-order language with a primitive one-place predicate "Θ_{bald}" ("is a paradigm case of being bald") and a primitive two-place predicate "x $\approx_{bald}$ y" ("x is indistinguishable from y w.r.t. baldness"). Even though they recognize that their account of the predicate "is bald" may not apply generally to all vague predicates, they still give their account in a very general way obviously with the hope that its application can at least be extended to a significant portion of vague predicates in our natural languages.

Though feeling interested, I am inclined to think that Zhou and Zhang's account is not on the right track and has less value than they think. To begin with, one should be reminded that offering an account of the *genesis* of a vague concept is not necessary the same thing as offering a logical analysis of that concept. The latter is, while the former may not be, about the logical forms and the truth-status of sentences involving predicates that express that concept and about the logical validity of arguments involving these sentences. To realize the difference between the two, one needs only to notice that an advocate of a non-classical solution to the bald paradox may (even though I doubt that s/he will) also accept Zhou and Zhang's account of the *genesis* of the vague concept *bald* while resolutely resists the semantics and the solution to the bald paradox that they offer. To be sure, Zhou and Zhang will retort that one needs not give up the good old classical logic in order to solve the bald paradox in particular and the sorites paradox in general, and that their account is just one of many attempts that try to save classical logic from the sorties paradox. But the problem is that, if one decides to stick with classical logic when dealing with the sorties paradox, s/he is bound to assert things offending many of our firm intuitions about vague concepts. These include the intuition that a vague concept is a concept without a clear-cut boundary, the intuition that there may well be higher-order vagueness of a vague concept as well as the first-order one, the intuition that a vague concept is a concept obeying some version of the so-called "tolerance principle" ([5], [2]), and the intuition that borderline sentences are indeterminate in the sense that there is no fact of the matter that can determine their truth-values. Many famous advocates of the classical approach, such as [4] and [3], have tried to answer to or explain away these intuitions, and I think that this is exactly what an advocate of a classical solution to the sorites paradox should have been doing. Sadly enough, Zhou and Zhang did little in this respect.

I am sure that Zhou and Zhang will protest that I have neglected the most important contribution of their paper, i.e., it has directly built the account of the genesis of the vague concept *bald* into the logical analysis of it and hence has offered a novel way to deal with vague concepts like the concept bald and associated paradoxes. I don't think, however, that the analysis of the concept *bald* presented in their paper is the right one, nor do I think that that analysis can in general be extended to other vague concepts. Two elements are especially emphasized by their account of the concept *bald*: paradigm cases and indistinguishability from a paradigm case. While I think that paradigm cases may be important for forming or learning the concept *bald*, the requirement of indistinguishability of our overall visual impression is certainly too much. I am inclined to think, though I don't have psychological evidences for this, that, as a matter of cognitive fact, people are inclined to apply the concept *bald* to people very similar to, though distinguishable from, a given paradigm case, and I assume that this is the most important cognitive fact that underlies the tolerance principle. But even if I were wrong about this, I would still argue that their analysis of the concept *bald* cannot be extended, in general, to other vague concepts for at least three obvious reasons. First, some vague concepts may never have any paradigm case. Consider the concept *being such a snail that walks much faster than most slow turtles* or the concept *being a much wiser man than any existent human being*. It is, I think, obvious that the former concept will never have an example while the latter cannot have one, though both concepts are still vague (after all, how fast must the snail be and how wise must the imagined man be?). These examples show that paradigm cases are not always necessary for forming a vague concept. They also show that there is an intensional aspect of a vague concept other than its extension. (No doubt the extensions of both concepts are the same, i.e., the empty set. They are nevertheless different vague concepts.) Second, some vague concepts may be such that they always require one to extend their applications to entities distinguishable from though very similar to paradigm cases. What I am thinking here are vague concepts about *kinds*, such as *board games, monsters in cartoon, talents, skills* and so on. Kinds are such entities that they cannot be indistinguishable from one another without being identical; consequently, applying a vague concept to a kind that is indistinguishable from a paradigm kind is simply to apply the same concept to the same paradigm kind again. So, if there is a genuine extended application of a vague concept about kinds, it must be an extended application to a kind that is distinguishable from any paradigm kind. This second consideration shows that indistinguishability from a paradigm case is also not necessary for forming or learning a vague concept. Finally, paradigm cases and indistinguishability from a paradigm case may not be sufficient for determining a reasonable extension of a vague concept. Consider again the case of the concept *bald*. Suppose I have only two paradigm cases of it: a_0 with no hair and a_{10} with 10 hairs. Suppose further that people with less than 5 hairs are indistinguishable from a_10 to me and people with 6 to 9 hairs are indistinguishable from a_{10} to me. Consequently, people with 0-10 hairs except the one with 5 hairs will be included, to me, in the extension of the concept *bald*, but this is not a reasonable concept, and probably not the concept that anyone of us actually has either. This example shows not only that paradigm cases and indistinguishability from a paradigm case are not sufficient, but also that something similar to Fine's notion of admissibility is needed ([1]). Similar considerations also show that something like Fine's admissibility is needed in order to promise the penumbral connection between two vague concepts, such as *red* and *pink*, in Zhou and Zhang's account.

There are also some technical and/or notional inadequacies in Zhou and Zhang's paper that I list below with only brief explanations:

1. Zhou and Zhang assert that all paradigm cases of P are standardly P in section 2.2, where P is a vague predicate. But this is not guaranteed by Definition 2 of their paper, nor is it guaranteed by the formal semantic that they offer unless they add (and they do add) the requirement to a model that every paradigm case of a concept P must be indistinguishable from every paradigm case of the

same concept. But this addition seems to me to be very unreasonable unless by "indistinguishable" they mean "similar enough" or something else. Two possible remedies are suggested here, where the second one probably will be preferred by the authors: (a) add a disjunct "either x is a paradigm P or" to the definiens of "standardly P" and drop the added requirement mentioned above; or (b) define the informal notion "x is indistinguishable* from y" as "either both x and y are paradigm cases or x is indistinguishable from y", then assert that what the formal notation "$\approx_p$" tries to capture is rather the relation of indistinguishbility*.

2. When defining "positive vague degrees" for objects, Zhou and Zhang appeal to the cardinal number of the set $\{\phi_{(\tau,\varphi)}(y) : x \in \phi_{(\tau,\varphi)}(y), y \in \theta_\tau\}$. This, however, does not capture their intention, for two paradigm cases y_1 and y_2 may determine the same indistinguishable set so that $\phi_{(\tau,\varphi)}(y_1) = \phi_{(\tau,\varphi)}(y_2)$. The issue will arise when we try to give the intuitively right vague degrees to things that are neither standardly P nor standardly not-P. The more appropriate way to define the same notion should appeal to the cardinal number of the set $\{y : x \in \phi_{(\tau,\varphi)}(y), y \in \theta_\tau\}$. Similar comments can be said about their definition of "negative vague degrees" for objects.

3. Several occurrences of the phrases "$f(y) = x + 1$" on pages 7-8 should be changed to "$f(y) = f(x) + 1$" to make the sentence meaningful.

4. The formal definition in Definition 8 of "$\copyright Px$" ("x is standardly P") does not guarantee that a standard P will be a P. Consequently, (2) in proposition 8 does not follow. To remedy this problem, the authors should add a conjunct "Px" to Definition 8 of "$\copyright Px$".

BIBLIOGRAPHY

[1] Fine, K.: 'Vagueness, truth and logic'. *Synthese*, 30: 265-300 (1975)
[2] Smith, N. J. J.: *Vagueness and Degrees of Truth*. Oxford: Oxford University Press, 2008.
[3] Sorensen, R.: *Vagueness and Contradiction*. Oxford: Clarendon Press, 2001.
[4] Williamson, T.: *Vagueness*. London: Routledge, 1994.
[5] Wright, C.: 'Further reflections on the sorites paradox'. *Philosophical Topics*, 15: 227-90 (1987).

Response to Veltman and Wang

BEIHAI ZHOU AND LIYING ZHANG

Response to Frank Veltman

We thank Frank Veltman for his careful reading, helpful summary, and considered comments in 'Learning to live with Paradox'. Veltman provided some systematic background on the Sorites paradox in 'A Diagnosis', summarized our work in 'Modifying EOI' from a historical perspective, and proposed his own view on how to face the Sorites paradox in 'A more pragmatic solution'. We would love to say many things in each of his points, but for now, we just say a little more on the last part of his comments, since it concerns the nature of the approach and claims in our theory. There are different kinds of reasoning, the reasoning of 'God' and the reasoning of ordinary people in actual world. The reasoning of God is perfect and ideal, while the reasoning of ordinary people is imperfect. On our understanding, the commentator thinks that perfect reasoning should be insisted on, and finer tools should replace coarse tools step by step. On the other hand, our work in this paper tries to represent and capture imperfect reasoning of ordinary people with logic tools. It is a kind of description rather than reform. So, we try to capture coarse tools with formal methods. The two viewpoints look quite different at first glance, but they are not contradictory at all. In our opinion, on the way to making people more logical, both description and reform are necessary, we need to understand human's reasoning first, and then try to improve them with perfect rules. Logicians can play a role in both parts.

Response to Wen-fang Wang

Wenfang Wang's comments on our paper consist of a summary and commentary, pointing out some potential problems, and adding advice for which we are grateful. We are happy to take this occasion for providing some clarifications of what we claim and achieve in the paper, triggered by what we take to be some major points in the commentary.

(1) *What makes our proposal different from existing systems such as fuzzy set theory?* Our main contribution is to put forward the new concept of a vague class. Fuzzy sets are based on a classical framework, which may not suit for cognition. Instead, vague classes depend on agents and cognitive factors. The Bald Man paradox is our trigger for introducing vague classes, and solving it is our test for the problem-solving ability of vague class theory. We hope that 'vague class' can become a useful notion and that research on vague class theory can start.

(2) *What is the status of boundaries in our theory, could they be uncertain?* Vague classes are based on agents and cognitive factors. Once a vague class is ascertained, a boundary will be established, but it may be only a theoretical one. For objects under the same subject, different agents may have different boundaries, and a variety of cognitive factors also can affect the formation of the border, all leading to uncer-

tainty. From this perspective, vague class theory is not against the intuition of uncer tain boundaries, it allows for refinement uncertainty. Second, people's intuitions ar not unchangeable rules we need to obey all the time, we also need to apply critic thinking, and consider where these intuitions come from. Finally, on boundary issue cognitivists such as Williamson think of vague predicates as having sharp boundarie but people do not know where they are. Now we have given a definition of boundar but we do not (and even cannot) use it directly to determine boundaries in daily li Like in geometry, whether or not there is a real circle in the world, we still need concept of theoretical circle.

(3) *Why do we use first-order logic?* Our theory offers a formalization of vagu classes in a first-order language with appropriate extensions that can express sor of the usual problems. We will continue our research to see what kind of logic really needed for the study of vague classes. This is not to save classical logic fro vagueness issues (though that might not be a bad idea), but we think we have alread introduced enough new ideas to not also change the logic immediately.

(4) *What is the range of application of our theory?* We are not saying that vag class theory can solve all problems of vague predicates. The commentator puts fc ward three kinds of vague concept as challenges to our account. This is not the pla to discuss these in detail, but here are two relevant thoughts. If one accepts that ba man is not identifiable with number of hairs, and is assigned first on the basis of in pressions, then we do not think the challenge examples are conclusive. We cou imagine a b2 having two hairs only, but the two hairs are very long, coiled on head, making him look indistinguishable with a10, or there is a c20 having 20 ve short hairs who also looks indistinguishable with a10, and then $b2 \in \theta_{(\tau,\varphi)}(a10)($ $b2 \in \phi_{(\tau,\varphi)}(a0)!)$, $c20 \in \phi_{(\tau,\varphi)}(a10)$. Such a $\phi_{(\tau,\varphi)}(x)$ is the 'equivalence class' our sense formed by the sample x and certain indistinguishability. As for a new obje a5, if someone feels a5 is bald, but a5 is not in the class generated by the existi samples, then a5 should be a new sample. There is no reason to limit to a fixed set samples. Behind this formal example, what we really want to emphasize is analysi issues from the perspective of cognition. In daily life, bald men are objects formed cognition based on visual impressions: related to objective physical features like number of hairs, but not entirely determined by them. To us, the latter emphasis see a deep-seated, but unwarranted presupposition of the literature on vagueness, whi hampers fruitful theorizing. But this also generates a question for us. If someone signs the predicate bald not based on counting hairs, then why can she say the pers described is bald? Obviously, there must be some other judgement method. Wha that method, what influences are important? We propose to analyse these questions being the fundamental ones, and thereby shift the focus in the area.

(5) *Does our theory contain 'technical and/or notional inadequacies'?* We ha taken note of the points raised by the commentator, we think that all of them ha convincing answers in our theory, but it would be tedious to go into formal deta here. We look forward to further discussion on these, and on all of the above matte

BIBLIOGRAPHY

[1] Fine, K. 1975. Vagueness, truth and logic, *Synthese* 30: 265–300.
[2] Sorensen, R. 2001. *Vagueness and Contradiction*. Oxford: Clarendon Press.
[3] Williamson, T. 1994. *Vagueness*. London: Routledge.

Probabilistic Semantics and Pragmatics: Uncertainty in Language and Thought. A Precis

Noah D. Goodman

In the following pages I provide a short summary of the soon-to-appear chapter "Probabilistic Semantics and Pragmatics: Uncertainty in Language and Thought" by Goodman and Lassiter [5], which itself is a fairly short sketch of an emerging theory of natural language. While there is a limit to what can be accomplished in a short exposition, I hope that both of these pieces will be useful in sparking further dialogue and new directions of work.

Our goal in [5] is to sketch a formal architecture for natural language understanding that grounds interpreted meaning in commonsense knowledge structures. This is a daunting task, which requires formalizing commonsense knowledge, understanding the relation of word meaning to knowledge, describing the composition of sentence meanings from word meaning, specifying the role of sentence meaning in belief update, and formalizing the inferential—pragmatic—aspects of interpreted meaning. Throughout this task we aim to understand language understanding as a whole, with complex interactions between interpretation, meaning, and knowledge. Of particular interest to us are the ways in which the context of an utterance affects its interpreted meaning, which have often been shunted from formal linguistics into the "wastebaskets" of pragmatics and general cognition.

Probabilistic methods have been very useful in capturing inductive learning and commonsense reasoning [8], and hence it is natural to use probability as one of the foundations of our system. This "Bayesian" approach describes knowledge of the world as generative models—structured probability distributions over possible situations. Probabilistic models provide a powerful formalization of knowledge of an uncertain world; probabilistic inference provides an idealized description of reasoning, which is inherently graded and leads to complex interactions. These properties are important in allowing the system to model aspects of human reasoning that can be vexing to standard logical systems, such as non-monotonicity. However the usual probability calculus is a relatively unstructured system, lacking the fine-grained compositionality needed to capture word meanings and interpretation. Instead, we build our system on stochastic lambda calculus—a simple extension of the untyped lambda calculus which inherits fine-grained compositionality but adds the ability to represent probabilistic uncertainty.

Stochastic lambda calculus (SLC) extends the ordinary, untyped lambda calculus with a uniform choice $\oplus$ operator, that forms an expression, $L \oplus R$, out of two options. A new reduction rule uniformly randomly reduces a choice expression, $L \oplus R$, to one of the two options, L or R. Complete reduction of an SLC expression thus specifies a (sub-)distribution over normal form expressions, rather than a single expression. The representational power of this system comes from its use of logical representations, while the inferential flexibility needed to deal with uncertainty comes from probabilistic inference. SLC is a fine-grained, universal [2] language for describing probability distributions, and hence well-suited to describing uncertain commonsense knowledge.

In the examples of [5] we use an extension of SLC into a full-featured probabilistic programming language, called Church [3]. We may use Church to define a set of new (stochastic) functions and symbols, and may then form an infinitude of expressions using these functions and symbols. That is, a set of Church definitions, say T, forms a domains specific language that extends SLC, call it SLC_T. The random-reduction (or *sampling*) semantics of Church is such that the definitions T induce a distribution over values of all the expressions in SLC_T (in fact each expression gets a countable set of values, in case it is encountered multiple times). In this sense, a set of definitions forms an intuitive theory of the domain: it describes which values are likely to be seen for all the "questions" that can be considered. We claim that such intuitive theories provide an appropriate target for lexical semantics.

With an intuitive theory as (common) background knowledge, we view the basic role of language as specifying a belief update. In probability theory belief update corresponds to conditioning, and hence we view the basic effect of language as conditioning the prior distribution over SLC_T. In order for this to be sensible, we need a meaning function which translates from natural language into the intuitive theory SLC_T. We assume that this meaning function first looks up meanings for words in a lexicon (mapping words to expressions in SLC_T) and then uses the hierarchical structure of the sentence to compose these meanings into the meaning of the sentence; it need not be a deterministic map, however. If a statement is translated into an expression that takes on Boolean values, then we may condition on this expression evaluating to *true*, resulting in a posterior distribution over SLC_T. However, rather than directly forming the infinite posterior distribution over SLC_T, we treat literal meaning as a map from the question under discussion (QUD) in SLC_T, to the posterior distribution over values of this expression.

This relatively simple model of literal meaning already gives rise to a number of important effects. In [5] we illustrate several of these, paying especially close attention to the effect of background knowledge on interpretation. Because we view interpretation as a joint probabilistic inference, there will be complex interactions between disparate aspects of world knowledge during interpretation of a sentence. For instance, the *explaining away* effect well-known in probabilistic modeling results in *a priori* independent propositions becoming dependent after hearing a sentence; this in turn can lead to non-monotonic effects when interpreting a sequence of sentences.

Another effect of treating interpretation as a joint inference is that world knowledge can influence the ambiguous mapping form surface from to meaning. For instance, a quantifier statement like "Most boys played in some game" has two different interpretations—in our systems the stochastic meaning function can produce two different SLC_T expressions. In some worlds, such as those in which games have only one player, one interpretation is more plausible than the other. This follows from a joint probabilistic inference of both the sentence meaning and the world (or QUD) state.

Compositionality is a key principle in both logic and natural language semantics, which also has key role in our model of language understanding. Indeed, in our architecture there are two interlocking "directions" of compositionality at work: compositionality of world knowledge, in which we use SLC to build both the domain theory T and lexical entries in SLC_T, and compositionality of linguistic meaning, in which the meaning function builds sentence meaning from the lexical meanings. However, it is important to note that these two directions of composition do not always coincide: this results in interactions during interpretation which may look non-compositional if only one direction is considered.

Compositionality in interpretation is further obfuscated when we consider the process of sentence understanding by a listener who does not take utterances literally, but instead infers meaning assuming a helpful speaker—that is, pragmatic interpretation. Building on recent work on formal models of pragmatics [1, 4] we describe a Bayesian listener who infers the value for the QUD under the condition that an informative speaker, who is considering the literal listener, would have made the given utterance. This model formalizes Gricean communicative pressures: utterances will tend to be relevantly informative, short, and so on. As a result, it captures quantity implicature (for instance, "some" implicates "not all") and a variety of other pragmatic phenomena. It also lends itself naturally to including incomplete knowledge—for instance [4] show that implicatures are weakened when the speaker does not have complete knowledge.

In standard formal approaches to semantics [6, 7], some context-sensitivity in meaning results from semantic *indices*: the meaning of a sentence is a truth-function on possible worlds only after fixing the values of a set of free, index, variables. However, the mechanism for setting these index variables given an utterance context is generally not specified. Within our framework it is natural to treat the index variables as additional random choices that must be inferred when interpreting a sentence. If this inference is performed as part of literal interpretation, there will be a pressure to set the indices such that the utterance is true. There is however no countervailing pressure to set the indices such that the sentence is informative, resulting in many utterances becoming vacuous (for instance, scalar adjectives as we explain in [5]). Instead, we lift inference about the index variables to the pragmatic listener. This allows Gricean principles—informativity, salience, and so on—to bear on the indices, resulting in a subtle interaction that sets indices to be useful given the context.

The treatment of indices as free variables filled in by pragmatic inference helps to capture several important aspects of natural language, including salience, reference, and vagueness. In [5] we describe a simple semantics for gradeable adjectives (such as "tall"), as comparison to a threshold which is such a free index. One consequence of this semantics is an intriguing result about the Sorites paradox:

1. Bob is tall.

2. Anyone who is 0.1cm shorter than someone who is tall is also tall.

3. Therefore, a midget is tall.

Under appropriate assumptions (about background statistics of heights) the second, inductive, premise has high probability, while the conclusion has low probability. If people endorse arguments according to their probabilistic strength, then this explains why people should endorse the first and second but reject the third statement—though mathematical induction seems to require the third given the first two.

To conclude, we sketch in [5] a probabilistic architecture for natural language understanding. This architecture builds lexical meaning from intuitive theories, builds sentence meaning from lexical meaning, and constructs interpreted meaning by probabilistic inference—literal and pragmatic. Because we reify background knowledge and inference as part of the interpretation process, we arrive at a very general treatment of the influence of context on conveyed meaning, and this is perhaps the most important contribution of our work.

BIBLIOGRAPHY

[1] Frank, M., Goodman, N.: Predicting pragmatic reasoning in language games. Science 336(6084), 998–998 (2012)
[2] Freer, C.E., Roy, D.M.: Computable de Finetti measures. Annals of Pure and Applied Logic 163(5), 530–546 (2012)
[3] Goodman, N.D., Mansinghka, V.K., Roy, D.M., Bonawitz, K., Tenenbaum, J.B.: Church: a language for generative models. Uncertainty in Artificial Intelligence (2008)
[4] Goodman, N., Stuhlmüller, A.: Knowledge and implicature: Modeling language understanding as social cognition. Topics in Cognitive Science (2013)
[5] Goodman, N.D., Lassiter, D.: Probabilistic semantics and pragmatics: Probabilistic semantics and pragmatics: Uncertainty in language and thought. In: Lappin, S., Fox, C. (eds.) Handbook of Contemporary Semantics. Wiley-Blackwell (to appear)
[6] Lewis, D.: General semantics. Synthese 22(1), 18–67 (1970)

[7] Montague, R.: The proper treatment of quantification in ordinary English. In: Hintikka, J., Moravcisk, J.M.E., Suppes, P. (eds.) Approaches to Natural Language, pp. 221–242. D. Reidel, Dordrecht (1973)

[8] Tenenbaum, J., Kemp, C., Griffiths, T., Goodman, N.: How to grow a mind: Statistics, structure, and abstraction. science 331(6022), 1279 (2011)

The Value of Semantics and Pragmatics: In the View of Linguistics, Psychology, and Cognitive Science

SHUSHAN CAI

1 Background

For the linguist De Saussure, semiology was 'a science which studies the role of signs as part of social life' ([12, 13]), while to the philosopher Peirce, this field of study, called semiotic, was the 'formal doctrine of signs', which was closely related to logic ([11]). Charles W. Morris defined semiotics (a reductive variant of Saussure's) as 'the science of signs' ([10]). Then, semiotics/semiology was divided into a three subfields, syntax, semantics and pragmatics, by Morris, Rudolf Carnap ([5]), and others. Since the 1950s, Chomsky, Montague, Austin, Searle and others then worked in the fields of syntax, semantics, and pragmatics separately, obtaining a very distinguished harvest ([6, 7]; [9]; [1]; [14, 15, 16]). As the results of these developments, a new synthetic interdisciplinary frame, that of cognitive science, was set up in the middle of 1970s, A new age of synthesis was begun, and interdisciplinary research was started ([4]). Syntax, semantics, and pragmatics are a basic frame for research in semiotics/semiology, , linguistics, as well as cognitive science ([3]). This reflects De Saussure's view that linguistics is only part of the general science of semiology. He said that the laws that semiology will discover will be laws applicable in linguistics ([13]).

2 Goodman and Lassiter's good points

Goodman and Lassiter's paper ([8]) has many good points. Some of them are as follows, with my comments and suggestions appended. In this paper, Goodman and Lassiter do significant work. In section 1, they provide background on probabilistic modeling and stochastic lambda calculus, and introduce a concrete scenario: the game of tug-of-war. In section 2, they provide a model of literal interpretation of natural language utterances and describe a formal fragment of English suitable for their running scenario. Using this fragment, they illustrate the emergence of nonmonotonic effects in interpretation and the interaction of ambiguity with background knowledge. In section 3, they describe pragmatic interpretation of meaning as probabilistic reasoning about an informative speaker, who reasons about a literal listener. This extended notion of interpretation predicts a variety of quantity implicatures and connects to recent quantitative experimental results. In section 4, the authors discuss the role of semantic indices in this framework and show that binding these indices at the pragmatic level allows us to deal with several issues in context sensitivity of meaning, such as the interpretation of scalar adjectives. Finally, they conclude with general comments about the role of uncertainty in pragmatics and semantics ([8]). I find these points impressive. In particular, the authors view the basic function of language understanding as belief update: moving from a prior belief distribution over worlds or situations to a posterior belief distribution given the literal meaning of a sentence. Then, they use some key concepts to build such a model. On the one hand they argue that uncertainty, formalized via probability, is a key organizing principle throughout language and cognition. On the other hand they use some vivid materials in natural language to test their model through experiments.

3   Comments and suggestions

This is a nice way of explaining the meaning of sentences and words in pragmatics. First, I find it significant that the authors try to get the full linguistic meaning needed in cognitive science meaning by the ideas and methods of pragmatics. Second, the authors use formal probability to describe uncertainty in language and thinking, and they build a model that can be tested. Third, where possible, the authors use materials in natural language to test their proposals. This is the value of pragmatics and semantics in language research; therefore, this is also the value of this paper. Pragmatics is the most complex and difficult field in linguistics, as it has many factors, speaker, hearer, time, place and context, which can all influence the meaning of language ([2]). Each of the factors is complex and difficult. In the comprehension of an utterance, many differences or biases in language, culture, logic, thinking, reasoning, psychology, intuition, heuristic, and experience will have effects on speakers and hearers, as they process information in their communicative speech acts. Therefore, interdisciplinary research in pragmatics and semantics is essential. I hope for, and look forward to, more discussions with the authors later.

BIBLIOGRAPHY

[1] Austin, J. L. (1962) *How to Do Things with Words*. Cambridge, Mass.: Harvard University Press.
[2] Cai, Shushan (1998) *Speech Acts and Illocutionary Logic*, Beijing: China Social Sciences Press.
[3] Cai, Shushan (2006) On the Trichotomy in Simiotics and its Influence on Philosophy of Language and Logic of Language. *Journal of Peking University*, Vol. 43, 2006(3): 50-58.
[4] Cai, S. The age of synthesis: From cognitive science to converging technologies and hereafter, Beijing: *Chinese Science Bulletin*, 2011, 56: 465-475, doi: 10.1007/ s11434- 010- 4005-7.
[5] Carnap, R. (1933)*Logische Syntax der Sprache*, English version, *The Logical Syntax of Language* (1937), translated by Amethe Smeaton.

[6] Chomsky, Noam (1957) *Syntactic Structure*. The Hague, Mouton.

[7] Chomsky, Noam (1968) *Language and Mind*. New York, Harcourt, Brace & World.

[8] Goodman, Noah D. and Daniel Lassiter (2013) Probabilistic Semantics and Pragmatics: Uncertainty in Language and Thought. To be published.

[9] Montague, Richard (1974) *Formal philosophy: Selected Papers*. Edited by Richmond H. Thomason. New Haven, Conn.: Yale University Press.

[10] Morris, Charles W. (1938) *Foundations of the Theory of Signs*. Chicago: Chicago University Press.

[11] Peirce, Charles Sanders (1931-58) *Collected Writings* (8 vols) (ed. Chrles Hartshorne, Paul Weiss and Arthur W. Burks). Cambridge, MA: Harvard University Press.

[12] Saussure, Ferdinand de (1916/1974) *Course in General Linguistics* (trans. Wade Baskin). London: Fontana/Collins.

[13] Saussure, Ferdinand de (1916/1983) *Course in General Linguistics* (trans. Roy Harris). London: Duckworth.

[14] Searle, John R. (1969) *Speech Acts: An Essay in the Philosophy of Language*. London: Cambridge University Press.

[15] Searle, John R. (1979) *Expression and Meaning: Studies in the Theory of Speech Acts*. Cambridge University Press.

[16] Searle, John R. (1983) *Intentionality: An Essay in the Philosophy of Mind*. Cambridge and New Youk: Cambridge University Press.

The Place of Logic in a Probabilistic semantics: Comments on Goodman and Lassiter

Thomas F. Icard III

The paper by Goodman and Lassiter presents a line of research that represents some of the most exciting recent developments in the study of natural language meaning. To my knowledge, this work offers the first credible attempt to combine serious compositional semantics with a serious story about pragmatics and the dynamics of conversation, all integrated into a single, principled formal framework. This particular paper focuses on the process of interpreting an utterance, which involves a subtle negotiation on the part of a listener between general world knowledge, specific knowledge about the language, and speaker interpretation, the latter amounting to a special case of intentional action understanding. The coverage of topics is ambitious: gradable adjectives, three different kinds of ambiguity (syntactic, lexical, and compositional), quantity implicatures, semantic indices, vagueness, and more. But in each case, and especially when taken as a whole, I believe genuinely new insight is gained. In particular, I think the observation about the "two interlocking directions" of compositionality, as described in §2.4, is deep and well worth noting for anyone interested in developing compositional probabilistic semantics. The work described in the second half of the paper fulfills the Gricean dream of reducing pragmatic reasoning to well motivated principles of general rationality.

Since its modern inception, model-theoretic semantics has had something of a split-personality, existing somewhere in between empirical psychology and cognitive science on the one hand, and anti-psychologistic logic and philosophy of language on the other.[1] Very roughly speaking, the latter tradition sees meaning as involving relations between abstract symbolic systems and the world, while the former takes meaning to be a matter of what happens in the minds of speakers and interpreters. Goodman and Lassiter take an unabashedly psychological perspective, interpreting models as "intuitive theories" in people's heads, and subjecting their work to empirical data. This has the advantage of making clear what the goals

---

[1] See [16] for an early discussion. According to Thomason's preface to the collected works [14], Montague's conception of syntax and semantics, presumably following Frege, "renders statistical evidence about, say, the reactions of a sample of native speakers to 'Mary is been my mother' just as irrelevant to syntax as evidence about their reactions to '7 + 7 = 22' would be to number theory" (2).

of the project are, and how we are to interpret and evaluate a given proposal. It is an interesting question how this kind of approach to semantics should or could relate to a more externalist, "objective" approach to semantics, but I will not speculate on that here (though see the section below on "Good Arguments").

Given the occasion, I would like to focus on how logic fits into the program as described in the paper. In some ways logic is central, with the use of stochastic lambda calculus and especially the Church programming language. While using untyped lambda calculus as the basis of a consistent logical system is notoriously problematic [10], Scheme is (latently) typed, including a primitive Boolean type, and therefore supports a standard notion of entailment, inherited by Church. That said, logical inference *per se* plays a relatively small role in the project as described. Logic is used more for its representational power in building complex generative models than for reasoning, which is thoroughly probabilistic. To a large extent I think this is justified. At the same time, I also want to suggest that, now with the framework up and running, many central themes from logic return, and the project only stands to gain from assimilating some of these ideas. After all, notions from logic such as "valid inference", "sound argument", "deduction", "knowledge", "strategic reasoning", and so on, are important parts of ordinary people's intuitive theories, and have been for at least two and a half centuries.[2] It would be surprising if they did not show up in a theory of how people think, speak, and reason.

My comments will be split up into those pertaining to the "Semantics" part of the paper—that is, the part of the project that can be described already at the level of Goodman and Lassiter's `literal-listener`—and those pertaining to the "Pragmatics" part, where multi-agent reasoning takes center stage.

Semantics and Inference

A subject's intuitive theory is formalized as a Church program π, which, simplifying matters slightly, we can think of as defining a space of expressions, $\mathcal{E}$, with each $e \in \mathcal{E}$ taking on values in $\mathcal{V}(e)$. If we let W stand for the set of full valuations, i.e., functions assigning to each $e \in \mathcal{E}$ an element of $\mathcal{V}(e)$, then W can be thought of as a set of (maximally specific) *possible worlds*, and π can be seen as defining a distribution P_π over W.[3] Running π forward will generate a given evaluation history with probability determined by the product of the probabilities for random choices made along the history. Since in general a given execution will not assign values to some elements of $\mathcal{E}$, we might take $\mathcal{V}(e)$ always to include the "undefined" value $\bot$. The probability of a particular world $w \in W$ is then the sum

---

[2]Indeed, surprisingly modern discussions of these topics can be found in China as far back as the fourth century, B.C. See [13] for an interesting discussion.

[3]I will assume here we are only dealing with programs that halt with probability 1. I am also ignoring the generalization to densities.

of probabilities of possible executions of π resulting in the assignments as given by w.

The language we can use to query and condition this underlying probability space is extremely rich. It includes all of Church (and thus of pure Scheme), in addition to any new expressions (variables, procedures, etc.) defined by π. Let $\mathcal{L}_\pi$ refer to a well-defined class of Boolean statements expressible in this language.[4] Then P_π also gives us probabilities for elements $\varphi \in \mathcal{L}_\pi$, which roughly play the role of "events" or "propositions":[5]

$$P_\pi(\varphi) = \sum_{w \in W:\ w \models \varphi} P_\pi(w) .$$

Conditional distributions $P_\pi(\varphi | \psi)$ are also well-defined for all $\varphi, \psi \in \mathcal{L}_\pi$, provided $P_\pi(\psi) > 0$. We are to think of P_π as encoding the *beliefs* of an agent, at least in some idealized sense.

It is therefore natural to model Bayesian inference in this setting. To make a prediction about the world, an agent might sample (some part of) π, possibly conditional on some assumption or observation. The literal-listener does just that: conditions its model (program) on an utterance, and makes a probabilistic inference about the world.

More specifically, there is assumed to be a question-under-discussion (QUD), which in principle could be any expression partitioning W. An utterance is simply a string of English, and it is interpreted by a meaning procedure (not strictly speaking a function, since it is noisy), which assigns a Church expression to each well-formed string of English.[6] Upon hearing an utterance, the subject will make a joint inference about the meaning of the utterance and the answer to the QUD, which may interact in subtle ways. See the paper for illustrative examples and discussion (§2.2-§2.4).

It is important to point out that the only inferential relation considered between (the meaning of) the utterance and possible answers to the QUD is conditional probability. That is, holding fixed a meaning $\psi \in \mathcal{V}(\text{meaning}(u))$ for utterance string u, an answer φ to the QUD is good to the extent that $P_\pi(\varphi | \psi)$ is higher than alternative answers. Indeed, a subject is predicted to infer φ with some probability proportional to $P_\pi(\varphi | \psi)$, since the basic inference model has the subject sampling from π, conditioned on ψ being true. The more samples she takes before guessing, the more likely she is to guess the answer with highest posterior probability, according to her own model of the world.

This is an elegant set-up insofar as it makes the inference process totally uniform. Everything is conditional sampling. In some ways, the picture is a Bayesian

---

[4] Strictly speaking, I mean expressions that are always typed Boolean.

[5] In the following, $w \models \varphi$ means the Boolean statement φ holds in w.

[6] Thus, the meaning function and the lexicon are formally part of the subject's model.

version of model-checking: guiding inference by constructing a few models of the premises and checking whether the conclusion holds, albeit in a principled, rational way. But this uniformity, as it stands, seems to leave no room for distinctively *logical inference*, e.g., perceived entailments that hold on the basis of form alone. To be sure, if as a matter of logic ψ entails φ, then it easy to see $P_\pi(\varphi|\psi) = 1$, and so the subject is predicted always to infer φ from ψ anyway. Why, then, should we bother worrying about logic at all?

I would like to suggest two possible reasons. The first is simply that constructing models online is not free, and there may be cases where detecting logical patterns is cheaper and more reliable. It is of course an empirical question whether it actually is cheaper, and whether it is a strategy people do use. But consider an example like the following.

EXAMPLE 1. Our subject, Rita, is picking up her son, Carl, from daycare. She is concerned about whether he had fun, and wonders whether he played any games. She also knows of 10 other children who were there today, and she knows their individual ages. She overhears one of the employees say, "All the children played at least two games with three other children of different ages", and concludes that indeed Carl played some games.

It is highly doubtful that Rita simulates even one scenario in which that sentence holds, even though in principle she has the knowledge required to do so. I doubt anyone would propose that. If the inference is to proceed by sampling, the problem must be simplified. But how will the algorithm know how to simplify the problem ahead of time—that is, before even the first sample—without detecting some general pattern? Logics based on "surface reasoning" systematize patterns like those used in Example 1, which can be used to simplify an inference problem [17, 9]. In some cases, such as in this example, the surface patterns are actually *sufficient* to reach the appropriate conclusion. One is tempted to suggest that in such cases we may be able to bypass sampling or model-checking altogether.

Relatedly, in some situations a subject may be unable to interpret part of some utterance, not because of the complexity of building the representation, but because her meaning function is undefined on some substring, or some part of the utterance was not perceived. One can imagine a number of coping strategies in such situations. But depending on the QUD, we may be able simply to ignore that part of the utterance. Following Example 1, if Rita instead heard, "All the children and quadragenarians played games together", she would still be able to resolve her question, even though she has no idea what a quadragenarian is.[7]

One of the distinctive features of logical entailment is that it is subject-neutral. That is, if φ is entailed by ψ, this is so no matter what the world is like. In terms of the formalism sketched above, we have $P_\pi(\varphi|\psi) = 1$ for any π whatsoever,

---

[7] A quadragenarian is someone between the ages of 40 and 49, inclusive.

provided φ and ψ are well-defined by π. This makes logic a potentially useful tool, portable across belief states even as an agent changes its view of the world. Moreover, for the specific class of entailment patterns pertaining to monotonicity and related notions—concepts that are particularly well suited to higher-order functional languages like Church—there are independent (psycho)linguistic reasons for thinking people are implicitly paying attention to them, viz. the grammar of negative polarity items.[8]

There are obviously a number of questions about how this is to work in concert with the framework in the paper:

1. Are these patterns supposed to be detected at the level of logical form, or directly between strings of English and QUD expressions? Or ought even QUDs be thought of as sometimes quasi-linguistic?

2. On any given occasion of communication, how does the interpretation algorithm decide whether to sample states of the world or search for logical patterns? Are there natural interpolations between these two?

3. How are these patterns to be learned? Could they be acquired as a natural result of experience with reasoning?

On the last question, one can imagine a system caching certain patterns that are observed to be useful in inference, and applying those without performing unnecessary computation.[9] It would be intriguing to work out the details of this. I do not believe logical patterns can completely drive inference by themselves, but I do believe they play an important role, even for relatively automatic, unconscious language comprehension.

Good Arguments

The second broad reason for taking logical entailment seriously is the fact that people do sometimes care explicitly about reasoning well and making good arguments. Logical arguments are not the only kind people care about, but they make up an important class. Furthermore, arguments are often couched in natural language, so one might take it to be a goal of semantics to be able to say something about which arguments are good and why. Characterizing good logical argument has been more closely associated with "objective" approaches to semantics, but it seems no less important for the psychological approach. Indeed, Goodman and Lassiter take on these issues by addressing the Sorites paradox.

---

[8]See [8, 9] for a logical take on NPIs, and for references to the literature. The exact nature of the generalization is controversial.

[9]Something akin to Fragment Grammar [15] suggests itself.

They present their analysis as a version of the "deny one of the premises" strategy. However, I think their framework and analysis instead suggest an instance of the "deny validity of the argument" strategy. Explaining why, I hope, will illustrate how the framework might relate to more traditional logical semantic approaches to entailment.

The meaning of a gradable adjective like 'tall' is assumed to be a predicate `tall` with a free degree variable, say,

$$(\lambda(x) \ (>= \ (\text{height } x) \ d),$$

where `height` is a memoized procedure assigning numbers to individuals, say,

$$(\text{mem } (\lambda(x) (\text{gaussian } m \ s))),$$

for some mean m and standard deviation s. As developed further in other work[11], Goodman and Lassiter suggest that under this setup the second Sorites premise,

($*$) Anyone who is 0.000000001mm shorter than someone who is tall is tall.

is actually false, although an agent will assent to it with high probability, since most samples of two individuals together with a threshold d will verify it. Thus, while people might reasonably accept it after only thinking about it for a moment, it is strictly speaking not true. Thus the argument does not go through.

But recall what 'true' means in this framework. It is not a statement directly about the world. To say that ($*$) is not true is to say that in the agent's intuitive model of the world it does not have probability 1. But is it actually false? That is, does it have probability 0? This depends on the model. Suppose our subject believes there are only finitely many people, with the number drawn from some distribution. Then on a given execution of the agent's program, that is, in a given possible world in the subject's model, ($*$) may well turn out to be true. In fact, no matter what (the distribution over) d is, ($*$) may be true.

To dramatize the point even more, if our subject thinks it highly unlikely that two people differ by less than 0.000000001mm in height, which might not be unreasonable, then she will accord extremely high probability to ($*$). If we even allow that the subject assigns probability 0 to the possibility that they differ by that little,[10] then ($*$) will receive probability 1. It is possible to enforce that ($*$) comes out having probability 0 by supposing that there are infinitely many individuals in the subject's model. But this seems to me implausible. One might also take "anyone" to quantify over merely possible individuals, and allow that we can compare heights across worlds, so to speak. I will also ignore this possibility, since it is not the analysis given.

---

[10]For this possibility we may want to make the value even smaller.

So what of the Sorites argument if we do not want to say the second premise is false? It is standardly assumed that a person makes a *mistake* in accepting the argument. But assenting to something that has very high probability is surely not a mistake. Once a person accepts (∗), are they forced by logic to accept the Sorites conclusion? Consider what would be necessary in order for the Sorites to be a good *logical* argument. It would have to be the case that any model of the premises is also a model of the conclusion. Suppose the first premise and the conclusion are:

(#) Patrick, whose height is 2m, is tall.

($) Anyone whose height is 1m is tall.

Suppose moreover we are assuming the framework in the paper, and the semantics of the relevant expressions as sketched. Then it is easy to see that the inference to ($) from (∗) and (#) is not logically valid. Happily, this comes right out of the semantic framework, for there are plenty of worlds where (∗) and (#) are true but ($) is false. In fact, any model that includes a 'Patrick' of 2m together with a 1m tall person, and a threshold value of, say, 1.5m, is very likely to make (∗) true as well. Thus, the probability of ($) conditional on (#) and (∗) will typically not be any higher than the prior probability of ($), and depending on the model and whether (#) was learned by testimony, it could well be lower. Thus, the argument is neither logically valid, nor probabilistically compelling in any reasonable sense.

Whether this captures what is wrong with the Sorites I will not venture to say. Rather, I would like to use it as an illustration of how one could in principle evaluate arguments couched in natural language in this framework, using assumptions about the semantics of certain expressions. Traditionally, testing argument patterns in this way has been one of the main sources of "data" for logical semantics. If an intuitive entailment is not predicted, or if an intuitive non-entailment is predicted to be an entailment, this would be taken to show a given semantic account is off the mark. As witnessed by the Montague quotation above (Footnote 1), many have thought we should be concerned with *genuinely* good arguments, rather than argument patterns that people merely typically assent to or produce. Where these come apart, a choice must be made.[11] At any rate, the example of the Sorites illustrates that the framework can profitably be used to understand genuinely good arguments as couched in natural language, whether or not that is one of the authors' aims. This is particularly attractive since we can evaluate complex arguments simultaneously from a logical and from a probabilistic point of view. For instance, there will certainly be arguments that give a conclusion probability 1 while not making it a logical entailment. I think this aspect of the project should be especially attractive to philosophers interested in these topics.

---

[11] See [7] for a recent discussion of this issue in the case of English epistemic modals.

Common Knowledge

Many have observed that intelligent communication often seems to require reasoning about others' minds. An effective speaker should be sensitive to how a listener is likely to respond to what she says, and a smart listener will reason about what the speaker is intending to convey. The second half of the paper presents an explanation of pragmatic reasoning from the perspective of a "level-1" listener, listener, who models an idealized speaker, who in turn has a model of the literal-listener sketched above. The account is very powerful, explaining a number of well recognized implicatures and other pragmatic effects, and predicting new ones in addition. I will not go into the details here, though I do want to raise a question about the role of common knowledge, and belief therein.

In the current version common knowledge does not play an explicit role, however its effects can be seen in various places. First, in the typical case considered here, there will be a number of assumptions about the world and about the language that the listener must assume are believed by the speaker to be shared by the listener. Of course, this cannot include the answer to the QUD, since the listener is assumed not to know that, but to believe that the speaker does know it, and implicitly the speaker is assumed to know this about the listener. In terms of the standard common knowledge hierarchy [12], this takes us only about three steps up. Informally, for certain φ, e.g., meaning statements and background assumptions, we have $B_L(\varphi)$ and $B_L(B_S(B_L((\varphi))))$, whereas for each possible answer ψ to the QUD, we have $\neg B_L(\psi)$ and $B_L(B_S(\neg B_L(\psi)))$. These claims do not have a precise formal meaning in terms of the listener's model—I think one of the interesting challenges is to make this connection more rigorous—but this roughly captures the extent of the assumptions required to explain, e.g., basic quantity implicatures.

A second effect of common knowledge, mentioned in this paper but explored more thoroughly elsewhere [5], stems from the idea that in some situations certain objects, people, or topics are more salient than others, and would therefore more likely be the object of conversation. This again is reflected only three or so levels up, with the speaker assuming a prior on what is most likely to be spoken about and assuming that the speaker knows she makes that assumption. Nonetheless, the relevant knowledge seems to be available at even higher levels. To take an example from [4], if Ann and Bob see each other at the movies, then in fact Ann knows that Bob knows that Ann knows ... that they saw each other at the movies. If asked about any level in the hierarchy, she would surely have no hesitation in assenting.

What if we wanted to make full common knowledge of some fact an explicit assumption of the listener? This is of course perfectly expressible in Church, using the familiar fixed point definition:[12]

---

[12] See, e.g., [2]. I am forgoing the opportunity to define this in Church. I am also switching from B_L and B_S to K_L and K_S—from L's perspective the relevant facts might look like knowledge.

$$CK_{\{L,S\}}(\varphi) := \varphi \wedge K_L(CK_{\{L,S\}}(\varphi)) \wedge K_S(CK_{\{L,S\}}(\varphi)).$$

Adding a version of this assumption by itself to the listener's intuitively theory is of course completely inefficacious. Why would we care to include it? Once again, I would like to offer two reasons.

The work described in the paper so far does not tell us how the listener's prior view of the world and of her interlocutor generate the appropriate cascade of decreasingly sophisticated conversational agents on which to base her inference about the QUD. Presumably the information that is assumed to be shared all the way down this cascade was already taken by the listener to be commonly known, or at least commonly believed up to the requisite level. Ideally one would like some way of going directly from an observed speech act and the listener's view of the speaker, to the appropriate inference on the part of the listener as laid out in the paper. It is conceivable that a general, explicit representation for what might be assumed common knowledge would be helpful in this regard.

Second, it is a common theme of other work on multiagent interaction, notably dynamic epistemic logic [1, 18], that common knowledge is also a typical *effect* of successful communication. The outcome of a successful speech act will not merely include the listener resolving the QUD, it will also result in the listener (and the speaker, for that matter) taking it to be common knowledge that it has been resolved. This is interesting in itself, but it may also be crucial for what is to follow, if, for example, correctly interpreting the next utterance depends on their taking the previous utterances as common knowledge. Again, if the interpretation process is to be totally automated given the subject's mental state, then these higher-order beliefs should perhaps be made explicit.

Working this out in the context of the framework of the paper may be challenging. But I also think it opens up some exciting possibilities. A few examples of relevant issues, beyond those already mentioned, might be:

1. It is well known that communication does not always result in common knowledge. For instance, if the listener takes the speaker to be insincere, but the listener believes the speaker does not know this, then she will not take her resolution of the QUD to be common knowledge, even if she believes they both know the answer. Is there some way of characterizing in this setup when the answer to a QUD will result in common knowledge?

2. More generally, there is a huge space of possible strategic interactions that could interface in interesting ways with a more complex and realistic model of what can be expressed in language. We should be able to study the inconspicuous details of word choice that uphold the art of persuasion and deception.

3. The standard way of dealing with higher-order probabilistic beliefs is with "type spaces" [6], though these are notoriously complex. Analogous to the taming of large, uncountable probability spaces with probabilistic programs (see above), is there some way of using Church to reason explicitly, while tractably, about higher-order probabilistic beliefs? This is related to the question of how these beliefs should be encoded.

Answering these questions may help create new links between important themes from dynamic epistemic logic and epistemic game theory, and the semantics and pragmatics of natural language.

Conclusion

I would like to conclude with a general remark about the range of "meanings" we might like to be able to express on a semantic theory, and relatedly what should be the space of possible "objects of belief". I would argue that all three of the examples discussed above—surface reasoning, the Sorites, and common knowledge—require going beyond mere representation of the world. Surface reasoning requires some recognition *that* one form follows from another, not just the ability to infer one thing from another. For the Sorites, to decide whether the argument is valid, one must be able to ponder what the relevant words mean, and determine whether the conclusion in fact follows. This is something people do, and it seems to require representing meanings and entailment relations as such. The question about common knowledge is whether in addition to modeling the effects of common knowledge, it may also be useful to model interlocutors' assumptions *about* common knowledge explicitly.

Moreover, these are not isolated instances. Language is full of references to more abstract notions and entities, closer to the realm of logic. For example, as is familiar from the literature on situation semantics [3], sentences like

($\star$) Nathaniel saw Delilah kiss the chef.

seem to entail reference to something like a scene, event, or situation. This is reinforced by the observation that an utterance of ($\star$) can be followed by "It made him angry", and it seems clear that "it" is referring to the situation. In principle, I think the framework in the paper (perhaps uniquely) has the resources to account for examples like this. There is already a powerful pragmatic story about anaphora resolution, as a special case of resolving semantic indices (§4). Moreover, "partial" worlds, in the spirit of situation semantics, are already identified as important in limiting the belief change triggered by an utterance (§2). The natural next step is to make these partial worlds "first-class citizens" in subjects' intuitive theories, so that they may serve as the object of reference or the value of a variable. One of the advantages of a rich representation language like Church is that this should be possible.

Indeed, one of the most attractive features of the framework presented in this paper is that we can explore these questions formally and rigorously. Goodman and Lassiter's paper already tells a very compelling story about why psychologists and linguists should find these developments exciting. I hope to have given some indication for why philosophers and logicians should as well. In the other other direction, I hope some more of what we have learned in logic and philosophy might also find its way back into the framework, with fruitful interactions in both directions, as its coverage of phenomena expands.

Acknowledgements

I would like to thank Noah Goodman, Wesley Holliday, Daniel Lassiter, and Shane Steinert-Threlkeld for helpful discussions, and Johan van Benthem and Fenrong Liu for the opportunity to take part in this discussion.

BIBLIOGRAPHY

[1] Alexandru Baltag, Lawrence S. Moss, and Slawomir Solecki. The logic of public announcements, common knowledge, and private suspicions. Technical Report SEN-R9922, CWI, Amsterdam, 1999.

[2] Jon Barwise. Three views of common knowledge. In Moshe Y. Vardi, editor, *Proceedings of the Second Conference on Theoretical Aspects of Reasoning About Knowledge (TARK)*, pages 365–379. Morgan Kaufmann, 1988.

[3] Jon Barwise and John Perry. *Situations and Attitudes*. MIT Press, 1983.

[4] Herbert H. Clark and Catherine R. Marshall. Reference diaries. In D. L. Waltz, editor, *Theoretical issues in natural language processing*, volume 2, pages 57–63. Association for Computing Machinery, 1978.

[5] Michael C. Frank and Noah D. Goodman. Predicting pragmatic reasoning in language games. *Science*, 336, 2012.

[6] John C. Harsanyi. Games with incomplete information played by Bayesian players, I. *Management Science*, 14(3):159–183, 1967.

[7] W. H. Holliday and T. F. Icard. Measure semantics and qualitative semantics for epistemic modals. In *Proceedings of 23rd Semantics and Linguistics Theory (SALT) Conference*, 2013.

[8] Thomas F. Icard. Inclusion and exclusion in natural language. *Studia Logica*, 100(4):705–725, 2012.

[9] Thomas F. Icard and Lawrence S. Moss. Recent progress on monotonicity. *Linguistic Issues in Language Technology*, 2013. forthcoming.

[10] Stephen C. Kleene and J. B. Rosser. The inconsistency of certain formal logics. *Annals of Mathematics*, 36(3):630–636, 1935.

[11] Daniel Lassiter and Noah D. Goodman. Context, scale structure, and statistics in the interpretation of positive-form adjectives. In *Proceedings of 23rd Semantics and Linguistics Theory (SALT) Conference*, 2013.

[12] David K. Lewis. *Convention*. Blackwell, 1969.

[13] Fenrong Liu, Jeremy Seligman, and Johan van Benthem. Models of reasoning in Ancient China. *Studies in Logic*, 43(3):57–81, 2011.

[14] Richard Montague. *Formal Philosophy*. Yale University Press, 1974.

[15] Timothy O'Donnell, Jesse Snedeker, Noah D. Goodman, and Joshua B. Tenenbaum. Productivity and reuse in language. In *Proceedings of the Thirty-Third Annual Conference of the Cognitive Science Society*, 2011.

[16] Barbara Partee. Semantics–mathematics or psychology? In Rainer Bäuerle, Urs Egli, and Arnim von Stechow, editors, *Semantics from Different Points of View*, pages 1–14. Springer, 1980.

[17] Johan van Benthem. Meaning: Interpretation and inference. *Synthese*, 73:451–470, 1987.

[18] Johan van Benthem, Jan van Eijk, and Barteld Kooi. Logics of communication and change. *Information and Computation*, 207(11):1620–1662, 2007.

Response to Icard and Cai

NOAH D. GOODMAN

1 Response to Icard

I would first like to thank Thomas Icard for his extremely thoughtful comments on my contribution, they are both insightful about the architecture itself and tremendously useful in building a bridge to the logic community. I would next like to thank the organizers of this occasion, Fenrong Liu and Johan van Benthem, for their effort, vision, and patience in curating many diverse (and somewhat intransigent) academics. And now let me turn to a consideration of Icard's comments.

Let me reiterate and extend several points about our methodology that Icard notes in his commentary. We do, indeed, take an "unabashedly psychological perspective." In my view this psychological perspective is both liberating and clarifying. For instance, it frees us from the need to establish semantic reference independent of the individuals communicating, and it brings psychological principles of knowledge representation and inference into our tool chest for theory building. Our psychological perspective also means that the methodological approach of cognitive science pervades our architecture. In particular we have implicitly appealed to the *levels of analysis* methodology [4], establishing our computational-level model of language understanding (essentially a competence theory) without making strong claims about the algorithmic process of inference. This is an important point to clarify, as it affects the conclusions we should draw about the relationship between sampling and logical reasoning.

The semantics of stochastic lambda calculus (SLC) can be viewed either in terms of sampling (the random reduction or evaluation process that we focus on) or directly in terms of induced distributions (on normal forms or return values). They are equivalent, so, for instance, the sampling semantics of an SLC expression induces a probability distribution. It is important to note that while sampling semantics can be used in this way to specify a probabilistic model, and even to specify conditional inference on such a model, there is no claim intended that inference must be carried out by any (or any particular) sampling algorithm. A computer or a brain could consistently approximate the distributions implied by a SLC model using exact enumeration, rejection sampling, Markov chain Monte Carlo, etc. That is, a model in SLC specifies which probabilistic inferences should be drawn, but not how they should be arrived at—the sampling *semantics* is used to build computational-level models, though sampling *algorithms* could also be used to implement them.

I believe that this methodological distinction clarifies the potential relationship between our model of language understanding and logical reasoning, which Icard flags as an important point for consideration. If we take logical inferences to be high-level regularities that follow from the form of expressions without needing to consider their content, then there is no inconsistency with more model-based inference strategies (like straightforward sampling). At the level of inference algorithms there is every reason to leverage any information which can help to draw good conclusions. As Icard notes "detecting logical patterns is cheaper and more reliable" than building complete situations, and hence is a good candidate to help with inference. Imagine that we had a "natural logic" system that could cheaply establish some entailment relationships between expressions in some domain language SLC_T. We could imagine using this capability to guide a sampling-based inference procedure, for instance constructing situations by sampling from entailments of premises rather than premises themselves. (Indeed, this is reminiscent of the class of *lifted inference* algorithms developed recently in the stochastic relational learning community.) We could even imagine using entailments to "bypass sampling or model-checking altogether." When we join Icard in making this move, considering logical inference as a collection of strategies or heuristics for guiding probabilistic inference, a number of interesting possibilities are opened. For instance, Icard indicates that the heuristic rules of logical inference within a particular domain SLC_T could be learned from inference experience—and they could be highly likely, rather than certain relations. This is an intriguing engineering suggestion for making more efficient probabilistic inference systems, as well as an important research direction in the psychology of reasoning and language comprehension. If we allow that both logical, form-based and probabilistic, model-based reasoning processes are available, we must still answer the empirical questions of when they are used and how they relate.

I will note before going on that I have joined Icard in a somewhat controversial view of logic within psychology and language semantics: logical representation as a computational-level foundation, logical inference as an algorithmic-level heuristic. I believe that the flexibility of human thought, and especially human language, demands the fine-grained compositionality that logical representations provide. I also believe that much of human thinking happens through imagination of particular situations, rather than by form-level re-write rules; though it may be that such rules can be learned as useful strategies.

This view of logical inference as a set of sometimes-useful heuristics is especially intriguing in the Sorites paradox. We show in [3] how the inductive premise of the Sorites comes out as probabilistically strong, while the conclusion comes out as probabilistically weak. Icard points out that in our set-up the inductive premise is also not logically valid in a mathematical sense (true in all models). However, there may be a third "logical heuristic" sense: we agree to the inductive premise because it is probabilistically strong, then use a logical heuristic (mathematical induction) to draw the conclusion from form alone—leading us to endorse the conclusion—while also sensing that the conclusion is wrong.

Later in his comments Icard brings up the issue of common knowledge within our approach and the relation to explicit approaches such as dynamic epistemic logic. I generally agree very much with his two points: that the role of common ground is an important avenue for future work, and that logical approaches may have much to offer in motivating these extensions. In our framework common knowledge is generally strong and implicit—it is a result, for instance, of using the same intuitive theory as background knowledge for the pragmatic listener, the informative speaker, and the literal listener. One can extend the model to remove particular facts from common knowledge. For instance, [2] removes the knowledgeability of the speaker from common knowledge, and show that implicatures are weakened when the speaker does not have complete knowledge. In many ways this is reminiscent of the treatment of epistemic states in dynamic epistemic logic. Clark [1] argues that it is not cognitively plausible for humans to represent the recursive cascade of common knowledge assumptions ("I believe that you believe that I believe..."). In our approach, rather than explicitly adding assumptions to put facts into common knowledge, common ground is complete unless otherwise noted—we must add modifications to *remove* facts from common knowledge. This may provide a useful formal mechanism for common knowledge that also satisfies Clark's challenge of physiologically plausibility.

Icard has insightful comments about several other aspects of our approach. I will rest by simply reiterating his call to understand the relationship between logics, such as natural logic and dynamic epistemic logic, and probabilistic approaches to language and cognition.

2   Response to Cai

I would like to thanks Shushan Cai for his kind words and useful summary of our approach to language understanding. Cai uses his comments to renew a call for interdisciplinary work and communication in the cognitive sciences. This is a call that I agree with completely. Our small advances are made possible largely because they draw on and unify established programs in formal semantics, statistics, logic, and psychology. It is not always easy to dialog across such disciplinary boundaries, but the result is usually worth the effort.

In opening his comments Cai reminds us of the origins of linguistic research in the quest to understand the role of "signs" in our mental and social lives. It is important to remember this history, and to remember the intuitions that existed before any discussion of truth-functions or informativity. We encounter signs every day in our lives, and some signs are the result of natural processes—a broken twig may be the sign of game passing, or a skid mark may be the sign of a car accident. These *natural signs* get their meaning through an inferential process that depends on a belief of causal connection between the signs and the event that led to it. Thus natural signs have as their "semantics" knowledge of causal processes in the world. They are interpreted through a (Bayesian) inference: from the observation of the sign to the (posterior distribution over) the events that generated it. A natural sign can be used for simple communication, as when we make the sounds of a lion to warn another of a nearby predator.

The genius of human language is generalizing from natural signs to *arbitrary signs*. Arbitrary signs are used similarly in interpretation: they are complex observations that constrain events in the world by Bayesian conditioning (as described in Goodman and Lassiter, to appear). However the semantics of arbitrary signs does not follow immediately from the natural world; we construct the semantics as a collective social act. Arbitrary signs are fundamentally *useful* for communication and coordination.

Let us take as granted that the origins of communication are simple Bayesian inference from natural signs and the end point is the Bayesian model we have described, in which arbitrary signs are interpreted through a complex social inference process. There are then several immediate and fascinating questions: How did this change happen? That is, through what process of cultural (or biological) innovation did we shift from natural to arbitrary signs. Was the shift to arbitrary signs immediate and unified or were their several steps? If there were several steps, did pragmatic inference or complex semantics come first? We may find traces of answers in many places including historical studies of language change and onto-genetic studies of language development in children.

As this meditation makes clear, many of the most important questions in cognitive science, as well as their answers, come from a dialog between disciplines and generations. I am grateful to Cai for framing our work in this light.

BIBLIOGRAPHY

[1] Clark, H.H.: *Using language*. Cambridge University Press Cambridge (1996)
[2] Goodman, N., Stuhlmüller, A.: Knowledge and implicature: Modeling language understanding as social cognition.*Topics in Cognitive Science* (2013)
[3] Goodman, N.D., Lassiter, D.: Probabilistic semantics and pragmatics: Probabilistic semantics and pragmatics: Uncertainty in language and thought. In: Lappin, S., Fox, C. (eds.) *Handbook of Contemporary Semantics*. Wiley-Blackwell (to appear)
[4] Marr, D.: *Vision*. Freeman Publishers (1982)

The Medium or the Message? Communication Relevance and Richness in Trust Games

CRISTINA BICCHIERI, AZI LEV-ON, AND ALEX CHAVEZ

Subjects communicated prior to playing trust games; the richness of the communication media and the topics of conversation were manipulated. Communication richness failed to produce significant differences in first-mover investments. However, the topics of conversation made a significant difference: the amounts sent were considerably higher in the unrestricted communication conditions than in the restricted communication and no-communication conditions. Most importantly, we find that first-movers' expectations of second-movers' reciprocation are influenced by communication and strongly predict their levels of investment.[1]

[1]The original paper appeared in *Synthese* 176:125–147, 2010. Owing to unforeseen circumstances, the authors have not been able to write a response to the commentators at this stage. However, the discussion will be continued at the Beijing conference, and beyond.

Comments on Bicchieri, Lev-On, and Chavez

MIEYUN GUO AND JIANYING CUI

Many experimental findings reveal that communication is conducive to cooperation and trust among agents in social settings. [1] report experimental results about the effects of communication relevance (Relevant versus Irrelevant) and the communication medium (Face to Face, FtF, vs. Computer-Mediated, CMC). Generally speaking, they show that the topics of conversation made a significant difference, while media richness had no effects on trust. The authors use the act of promise-making to explain why some communications are conducive to positive social behavior such as trust and cooperation.

In what follows, we put forward four questions from a logical point of view. Our concern is to stimulate further interdisciplinary research in this interesting field.

Question 1: What is the precise sense of the minimal rationality stipulated in the paper? How can we explain the effects on trust or reciprocity of the different acts of promise-making?

In their introduction, the authors say the following:

The culprit here is not rationality per se, but instead the common auxiliary hypothesis that players only care about their own material payoffs. ...We assume players are rational, in the minimal sense of maximizing their expected utilities, but drop the hypothesis that they only care about their material payoffs. In particular, we show how communication among players, when accompanied by promising, leads them to trust and reciprocate even in difficult circumstances.

This is a bit ambiguous. What does the minimal sense mean? Is it that maximizing expected utility is not the only goal for players in trust games? Or, should we say that players could neglect, or even abandon, the goal of maximizing utility in *some* situations? We can even imagine that some first movers may not care much about an endowment of \$6 (after all, a small sum of money), so they are inclined to be altruistic, or to take risks and invest some large portion of the endowment. But that means the players may be irrational in the standard sense.

If the agents also take reputation as one of their goals, it will make things much more complicated. An alternative interpretation would be that we still assume players are rational in the standard sense of maximizing expected payoffs, but they have incomplete information on the types (cooperative or non-cooperative) of the other players.

In fact, in their paper, the authors explore the act of promise-making to clarify why and how it is conducive to trust (p.130). Both in Experiment 2b (CMC-Relevant communication) and in Experiment 3b (FtF-Relevant communication), the first mover is allowed to make promises about her investment. However, unlike with FtF-Relevant communication, the first mover is not allowed to reveal her identity to her counter player in the case of CMC-Relevant communication. Hence the social norms of promise-keeping may not be activated for either the first or the second mover. Here the condition of promise-making seems to become entangled with the condition of media richness. The authors did not provide much explanation for their results from the perspective of different acts of promise-making, so we see this as a natural question.

Question 2: Could the authors give an explanation for the different results with CMC-Irrelevant and No-communication?

As stated in the paper, relevant communication or FtF communication may include indicators such as visual, verbal, and social cues to reveal the types of the pair of agents. That is why communication can have a significant effect on fostering trust and cooperation.

However, intuitively, the No-Communication case is very similar to CMC-Irrelevant, since the designed questions do not reflect the characteristics or the personality of the other player at all. Likewise, the CMC-Irrelevant communication lacks these indicators. Thus, it seems that the distribution of trust for No-Communication should be very close to that of CMC-Irrelevant. But in reality, significant differences are found in the experiment data of the paper (Fig.2 in p.136). Interestingly, the distribution of reciprocity did not make a big difference in those two cases (Fig.3 in p.136) .

Question 3: How to connect your results on trust games with experimental studies on Centipede games?

In order to compare with the Centipede Game, let us simplify the experiment in [1] a little bit. Assume the first mover (agent A) only has 1 dollar, and can make a decision to invest or not. If not, she will keep the 1 dollar. If she invests, there are 4 possibilities as the 1 dollar gets tripled, that is, she may be reciprocated in the form of 3, 2, 1, or 0 dollars from the second mover (agent E). This can be depicted in an extensive form game as in Figure 1.

The original version of the centipede game involved a hundred moves and linearly increasing payoffs (Rosenthal 1982), but the term has been extended to related versions. To simplify matters, we give the following two-round and three-move Centipede game:

In this way, the trust game in [1] can be taken as a centipede game. By using Backward Induction, the game will stop in the first round and the first mover will choose not to invest at all. However, in reality, many findings show that people

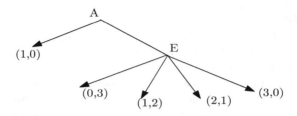

Figure 1. Trust Game

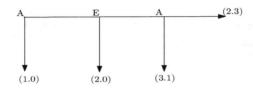

Figure 2. Centipede Game

deviate from the Nash outcome predicted by the Backward Induction argument. In fact, one of the authors proposed in Bicchieri (1988) to modify a theory of game so as to include a theory of belief revision to solve this paradox from a theoretical point of view. [3] was the first to make an experimental study of this phenomenon. In their paper, models based on errors, small amounts of altruistic behavior, and different learning rules were proposed to explain their data. Hence, a natural question arises here is to compare your results with experimental studies for centipede games.

Question 4: Does a player also engage in probabilistic reasoning based on empirical studies?

The experimental results by the authors show that two-thirds of second movers sent back exactly half (in our simplified case, 1.5 dollar), while the remaining one-third sent nothing ([1], p.134 and Fig.1).

Now suppose that the agent is a probabilistic reasoner who knows these experimental results (based on personal experience, or by consulting Google). The first mover may then calculate her expected utility as: $(2/3)*1.5+0*(1/3)) = 1$. Surprisingly, it is the same as embarking on no adventure at all. So the first mover can choose randomly. This seems connected with our earlier point about incomplete information on agent types.

Firstly, to invest or not depends on the types of investor (here, the first mover). Investors can be adventurous or cautious, optimistic or pessimistic. An adventurer may think there is a possibility of getting more. A conservative investor may think that it is safer to keep the 1 dollar, instead of having an expected utility.

Secondly, things also depend on beliefs on the types of the other agent (cooperative or non-cooperative). The investors can get that kind of information through observation in FtF-Irrelevant communication, which may help explain why FtF-Irrelevant communication has a higher frequency of trust when compared with No-communication.

Discussion and conclusion

Trust also plays an important role in research on effective interaction and cooperation in multi-agent systems (MAS), with key topics like witness information and the history of interaction in temporal settings.

[5] introduced an influential cognitive theory of social trust that is based on the concepts of goal, belief, opportunity, ability and willingness. Modal logics have been recently developed to study different types of trust. [6] have studied reasoning about trust from a logical point of view by combining the languages of temporal, epistemic and dynamic logic. [7] uses a dynamic testimonial logic extending a static conditional doxastic logic to capture agents' epistemic trust in other agents' testimony, for instance, in bandwagon effects.

Moreover, as we argued above, knowledge or belief about agent types, as well as probabilistic reasoning, are important to understanding social behavior. The probabilistic epistemic logic in [8] suggests an interesting theoretical approach to these phenomena.

In fact, the experimental data in [1] offer scientific evidence for all these matters. Hence, empirical studies and a logic approach should be complementary in helping us understand and foster trust and cooperative relationships in our society. More generally, it seems that a Theory of Play, as suggested in [4], should also be based on experimental studies, rather than remaining restricted to game theory and logic. It is time to have more communication to build trust and cooperative relationships between researchers in all these fields.

BIBLIOGRAPHY

[1] C.Bicchieri, A. Lev-on and A. Chavez, The medium or the message? Communication relevance and richness in trust games. *Synthese*, 176:125-147, 2010.
[2] C.Bicchieri, Common Knowledge and Backward Induction: A Solution to the Paradox. *Proceedings of TARK*,381-393,1988.
[3] R.D. McKelvey and T.R. Palfrey, An experimental study of the centipede game. *Econometrica*, Vol.60, 4:803-836, 1992.
[4] J. van Benthem, *Logic in Games*. The MIT Press, Cambridge (Mass.), 2013.
[5] C. Castelfranchi and R. Falcone, *Trust Theory: A Socio-Cognitive and Computational Model*. John Wiley and Sons, Chichester, UK, 2010.
[6] A. Herzig, E. Lorini, J. F. Hubner, and L. Vercouter, A logic of trust and reputation. Logic. *Journal of the IGPL*, 18(1):214-244, 2010.
[7] W.Holliday, Dynamic Testimonial Logic. *Proceeding of the Second International Workshop on Logic, Rationality and Interaction*, Chongqing, China, October,161-179, 2009.
[8] R.Fagin and J.Halpern, Reasoning about knowledge and probability. *Journal of the Association for Computing Machinery*, 41:340367, 1994.

Communication and Cooperation

JAKUB SZYMANIK

Communication has a positive effect on collaboration. Bicchieri, Lev-On, and Chavez [1] provide an experimental evidence in favor of this claim. In this short note I will summarize their findings, discuss possible extensions, and argue for the need of a more general theory.

1 The experiment

The authors focus on trust games in the context of two aspects of communication: *content* (relevant vs. irrelevant communication) and *media-richness* (face to face, FtF, vs. computer-mediated communication, CMC). In the trust game they studied, there are two players: first-mover and second-mover. The first-mover received 6$ and she could decide to send any discrete amount of dollars to the second-mover. The amount second-mover received was tripled by the experimenter. Then, from this new amount, the second-mover could send any discrete dollar amount back to the first-mover. Participants were paired randomly and played 3 games in the following order: no-communication game (base condition), relevant or irrelevant CMC communication game, and lastly, relevant or irrelevant FtF communication game. In the two latter cases the communication preceded the execution of the moves. The researchers were interested in 3 dependent variables: *trust* – defined as the amount of dollars sent by the first-mover, *reciprocity* – the amount returned by the second-mover, and *expectation* – the amount the first-mover expected to get back. In general, across all conditions relevance and FtF had positive effects on all dependent variables. Looking into the interactions within conditions, the statistical analysis returned the following main findings (see Fig. 1 for the graph of dependencies among the variables):

(1) Trust was positively influenced by relevance.

(2) Reciprocity increased with trust.

(3) Reciprocity increased in FtF communication.

(4) Trust increased with expectations.

(5) Expectations predicted reciprocity better than trust itself did.

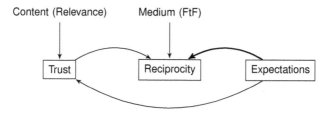

Figure 1. Dependencies between experimental variables as discovered by statistical analysis.

2 Discussion

Let us explore the possibility of a simplified picture of the observed interdependencies. The experimental findings suggest that reciprocity is better predicted by expectations than by trust. On the other hand, trust is influenced by the act of promising when relevant communication is allowed. Intuitively, the act of promising should trigger higher expectations of the first-mover and in turn the expectations should influence trust. In other words, one could hypothesize that in relevant communication, the act of promising influences only expectations that in turn imply various levels of trust. Therefore, one could expect that Figure 2 correctly depicts the dependencies in a simpler way. The authors do not discuss the relationship between content and expectations. Is such an interpretation consistent with the experimental findings?

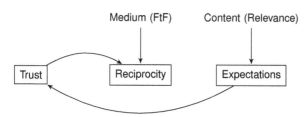

Figure 2. An alternative (simpler?) model.

Moreover, reciprocity differed significantly by medium. Namely, in the FtF conditions subjects engaged in a positive reciprocation to the highest extent. This supports the authors hypothesis that promises focus participants on social norms and motivate them to reciprocate. I find this hypothesis intuitively very convincing. However, looking at Figure 1 we still see that reciprocity is also dependent on trust and expectations. However, the dependency on expectation may be only statistical in nature. The second-movers took their decision without knowing the first-movers' expectations. Taking into account agents' information processing capabilities, the reciprocity should rather depend only on trust (that signals expectations). The amount of money send by the first-mover signals the strength of the social norm of 'honoring the deal' and motivates the second-mover's level of reciprocation (see Fig. 2).

FtF communication seems to encourage a more 'social' interpretation of the game on at least two levels. Firstly, in such games the players see each other, they can draw conclusions about the social standing of their opponent. For instance, the age or gender can play a role of the 'peer-defining' category. We know that gender turned out to be insignificant across the experimental conditions. But was it also an insignificant factor within the FtF condition? Secondly, note that in the experiment the ordering of conditions was fixed. All participants first played no-communication games, then CMC, and finally FtF. Therefore, they started with a scenario promoting strategic (selfish) behavior and then gradually moved towards a more 'social' setting where other factors except pay-off maximization could play a bigger role. The authors mention the problem but they claim that there should be no ordering effect as pairings for each game were unique. Still the players had to first develop a strategy for no communication condition. Arguably, they could then use the same strategy throughout the whole experiment. What would happen if the players started with FtF condition? The latter resembles to a lager extend our everyday strategic interactions that are rarely played once and never again. It seems that we are evolutionary programmed to think about social situations in terms of long-term exchanges. This can explain the positive levels of reciprocity observed in trust games [2]. When starting with FtF communication subjects should reciprocate even more.

3   The need of cognitive models

To better understand the above issues it would be helpful to have some theory of how subjects arrive at their decisions. What is the reasoning behind the formation of certain expectations and trust? What is the second-movers' reasoning leading to a particular reciprocation decision? Having a theory like that could help in designing experiments looking more into the dynamics of cognitive processes involved in trust games.

The discipline of behavioral game theory could profit from a general framework that would allow to *a priori* predict the dependencies between various variables and to compare different models. In such a framework we could answer, for instance, the questions like: Is the model from Fig. 2 simpler than model from Fig. 1? Do they predict different cognitive processes? Will changing the order of games influence subjects' reasoning?

At first glance such a framework seems improbable, as behavioral game-theory deals with an extremely complicated facet of our cognition. However, cognitive processes related to language and communication do not seem significantly simpler, and yet psycholinguistics offers many formal and computational cognitive models. These models usually draw from logic, (evolutionary) game-theory, probability theory, and computer science – yet another facet of 'communication effect' between disciplines. Recently, the network of collaborations between game-theory, logic, and computer science has been constantly growing – this volume is a very witness to it. Also, the interest in cognitive science among logicians and game-theorists is slowly increasing (see, e.g., the recent papers where techniques from logic, game-theory, and computer science are combined to predict subjects' behavior in various games [3, 4]). Hopefully within coming years we will observe an increase in communication between researchers doing formal work and cognitive scientists. As the paper by Bicchieri and colleagues exemplifies, this communication can result in increased trust and collaboration between the communities, maybe even leading to a cognitive turn in logic and game-theory.

BIBLIOGRAPHY

[1] Bicchieri, C., Lev-On, A., Chavez, A.: The medium or the message? Communication relevance and richness in trust games. *Synthese* 176(1) (2010) 125–147
[2] Camerer, C.: *Behavioral Game Theory: Experiments in Strategic Interaction.* Princeton University Press, New Jersey (2003)
[3] Szymanik, J., Meijering, B., Verbrugge, R.: Using intrinsic complexity of turn-taking games to predict participants' reaction times. In: *Proceedings of the 35th Annual Conference of the Cognitive Science Society.* (2013)
[4] Gierasimczuk, N., van der Maas, H., Raijmakers, M.: An analytic tableaux model for deductive mastermind empirically tested with a massively used online learning system. *Journal of Logic, Language and Information* (2013)

Part IV

LOGIC AND SOCIAL INTERACTION

The last group of papers in this book reflects a new development, the emerging contacts between logic and the social sciences. Many activities of interest to modern logicians, such as argumentation, communication, social deliberation and decision making, irreducibly involve interaction between many agents. A typical case in point are current logics for multi-agent information update and belief revision for individuals and groups, or the logical analysis of games and game theory. While existing tools of logic may still apply at this interface, the agenda of research themes changes considerably.

In 'Where do preferences come from?', *Franz Dietrich* (CNRS and University of East Anglia) and *Christian List* (London School of Economics) develop the idea that a rational choice is based on good reasons, a notion that is mostly absent from standard decision theory. The central idea is that an agent's preferences are based on salient properties of the alternatives over which the preferences are held, encoded in an agent's motivational state. This is made precise in a formal model with a weighting relation on the salient properties, which then forms the basis for a new rich account of preference formation and preference change. *Alexandru Baltag* (University of Amsterdam) scrutinizes the proposed model of motivational attitudes and weights, asking what its elaboration in terms of formal postulates adds to the semantic model itself, and wonders whether the account really captures all forms of preference change (down to falling out of love), or even just those that have a rational basis. *Patrick Girard* and *Shaun White* (University of Auckland) discuss how modal logics might be used formalize Dietrich and List's interaction between an agent's weighing relation and state-based preferences, and draw comparisons with various lines of related work on reasons in the current philosophical and logical literature.

Alexandru Baltag, Zoë Christof (University of Amsterdam), *Jens Ulrik Hansen* (Lund University), and *Sonja Smets* (University of Amsterdam) tackle a major social phenomenon today, fanned by information technology, in 'Logical models of informational cascades'. Using techniques from dynamic-epistemic logic, they investigate scenarios where agents form beliefs on the basis of their own information and observed beliefs of others. This can lead to epistemic failures at a group level. Amongst other things, the analysis shows that cascades are sometimes unavoidable by rational means: the group's failure to reach the truth may stem from each agent's rational attempt at individual truth-tracking, even when agents have unbounded higher-order reasoning powers about other agents' minds and the current protocol. *Hu Liu* (Sun Yat-sen University, Guangzhou) asks a number of technical questions about the probabilistic-logical machinery used, and also wonders how helpful the analysis is as an internal account of agents, since the complexity of the logic may go well beyond what agents are capable of doing in informational cascades. *Eric Pacuit* (University of Maryland) questions the specific rules in the paper for counting evidence by the agents, and also asks what more general underlying principle would lead agents to 'follow the crowd'. Furthermore, he stresses that informational cascades are a diverse phenomenon, making it imperative to separate peculiarities of the urn example in the paper from general insights of the proposed logical approach.

In 'Knowledge, friendship and social announcements', *Jeremy Seligman* (University of Auckland), *Fenrong Liu* (Tsinghua University), and *Patrick Girard* (University of Auckland) present a formal logic for reasoning about the changing patterns of knowledge and friendship in social networks. The mechanism is social announcements from a sender transmitting a message with information to one or more receivers, treated in a dynamic-epistemic logic. The approach is applied to concrete scenarios, and some general issues are highlighted, including the special syntactic form of bona fide epistemic friendship-based model transformations. *Carlo Proietti* (Lund University) wonders about connections with already available hybrid logics in the literature, and asks whether there should not also be a dynamic of changing friendship links. He points out that the special network patterns studied by social scientists, such as circles, wheels, or stems, might have a systematic reflection in the logic. *Katsuhiko Sano* (Japan Advanced Institute of Science and Technology, Kanazawa) points at the indexical character of many relevant announcements in the authors' area of study, and suggests that existing work on indexical modal logics from the computer science literature might be relevant. He also asks whether the proposed epistemic friendship logic is decidable, and whether such a fact might be shown by embedding it into existing logics for communication in the presence of channels.

Yun Xie and *Minghui Xiong* (Sun Yat-sen University, Guangzhou) investigate a concrete area of reasoning-driven social life in 'Logics for litigation argumentation'. Contacts between logic and law go back to the Sophists in antiquity, but in the 20th century, there has been a split between formal logic and informal logic in the area of legal studies. The authors integrate insights from both these traditions in a precise game-based model of litigation involving three agents, informed by insights from informal logic, but allowing for exact treatment by modern dynamic logical systems. *Davide Grossi* (University of Liverpool) relates the authors' three-person games to formal studies of argumentation networks, games played over these, and recent systems of dynamic logic that systematize them. *Henry Prakken* (Utrecht University and University of Groningen) points out the significance for legal studies of the extensive work under the heading of AI and Law, and the existing rich logical research program of defeasible reasoning.

Where do Preferences Come From? A Summary[1]

FRANZ DIETRICH AND CHRISTIAN LIST

1 Introduction

The paper to be presented at this conference ([3]) sketches some basic ideas underlying a broader, ongoing decision-theoretic project. In that project, we aim to develop a new, general approach to decision theory which

(i) improves upon standard decision theory in both its idealized, 'rational' and its more psychologically informed, 'behavioural' variants (the former are associated with classical rational choice theory, the latter with behaviourial economics and economic psychology);

(ii) is widely applicable in the social sciences and in philosophy; and

(iii) provides a framework for expressing some key philosophical debates about the relationship between reasons and rational decisions, which are not adequately captured by standard formal decision theory.

One important task is the development of a theory of preference formation and preference change, since standard decision theory says very little about where an agent's preferences come from and when they might change. Here, we introduce a simple formal framework for modelling preference formation and preference change.[1]

2 Informal summary

The idea that a rational choice is (among other things) a choice based on reasons – perhaps subjective reasons – is a very natural one, but the notion of a reason is more

---

[1]This summary is based on [3] and presents, in abridged and adjusted form, material from that paper, supplemented by some more general remarks about our ongoing research. We would kindly request that, in any references, the present summary be cited together with the original paper. Christian List wishes to acknowledge support from the Leverhulme Trust.

[1]Some important related contributions include work by logicians on the logic of preference and preference change. See, e.g., [5], [1], [4], [6], and the historical contribution by [8]. More detailed references to related literature, including some related work in economics, are included in [3].

or less absent from standard decision theory. In standard models, an agent has beliefs and preferences, formally modelled as subjective probabilities and utilities, and acts so as to satisfy his (or her) preferences according to his (or her) beliefs. While beliefs may be updated in light of new evidence, the agent's preferences – at least with respect to fully described outcomes – are typically assumed to be *fixed* and *exogenously given* (in [3], we cite a few exceptional works, many of them from outside mainstream economics). An agent's preferences are simply taken to be an essential but inexplicable feature of his personal identity. On this picture, preferences cannot be rationally assessed or criticized (provided they satisfy some minimal internal consistency constraints), and we cannot capture the idea that an agent's preferences may be the product of something more fundamental, such as the agent's reasons and his weighing of these reasons.

To overcome these limitations, we propose a 'property-based' account of preference formation. The central idea is that an agent's preferences are based on certain 'motivationally salient' properties of the alternatives over which the preferences are held. An agent's preferences may change as new properties of the alternatives become salient or previously salient properties cease to be salient. The motivationally salient properties serve as the reasons for the agent's preferences.

More precisely, the agent ranks different alternatives according to the way he 'weighs' the motivationally salient properties of these alternatives. This works as follows. For each alternative, the agent considers the set of motivationally salient properties of that alternative. An underlying 'weighing relation', defined as a binary relation over sets of properties, is then used to rank these property combinations relative to each other.

For example, when a consumer forms his preferences over different goods in a supermarket, such as different yoghurts, he could in principle consider a very large number of properties (characteristics) of these goods. In practice, however, he will only consider a small subset of these properties: the motivationally salient properties. This may include whether a yoghurt is cherry-flavoured, low-fat, and free from chemical additives, but exclude whether the yoghurt has an odd number of letters on its label (a totally irrelevant property) or whether it has been produced in an environmentally sustainable manner (something only an ethically oriented consumer will pay attention to). The consumer then determines his preferences over different yoghurts on the basis of his weighing relation over property combinations. He will most prefer the yoghurt with the 'highest-ranked' combination of motivationally salient properties. The consumer's preferences can change when new properties of the alternatives become motivationally salient, for example when he starts caring about environmental sustainability, or when previously salient properties cease to be salient, for example when he becomes less diet-conscious.

The theory also permits a normative reinterpretation. Under this reinterpretation, the focus is no longer on the properties the agent is *actually* motivated by, i.e., the *motivationally salient* properties, but instead on the properties the agent *ought ideally* to be motivated by, i.e., the *normatively relevant* properties. The appropriate weighing relation then captures, not the agent's *actual disposition* to weigh different property combinations relative to each other, but the way they *ought* to be weighed, according to some normative background theory.

3 The formal framework

In what follows, we give a brief exposition of the central formal framework and a simple axiomatic characterization result.

3.1 Preferences and properties

Let X be a non-empty set of fundamental objects of preference (e.g., fully described outcomes or consequences of actions, possible worlds, social states, bundles of goods, or policy platforms). The elements of X are mutually exclusive and jointly exhaustive. We call them *alternatives*.

We represent the agent's *preferences* by some order $\succsim$ on X (a complete and transitive binary relation), where $x \succsim y$ means 'the agent weakly prefers x to y'. As usual, $\succ$ and $\sim$ denote the strict and indifference parts of $\succsim$.

To address the question of how $\succsim$ is formed and when it may change, we introduce the idea that the agent's preferences depend on certain properties of the alternatives. Informally, a *property* is a characteristic that an alternative may or may not have. For example, being vegetarian is a property that a meal may or may not have. (For simplicity, we set aside non-binary properties; see, e.g., [2].) Formally, a *property* is an abstract object, P, which *picks out* a subset of X. (It need not be *identified* with that subset; two more distinct properties could in principle have the same extension of alternatives in X satisfying them.) Let $\mathcal{P}$ denote the set of all properties.

In forming his preferences, the agent focuses on some, but not necessarily all, properties of the alternatives. We call the properties that the agent focuses on the *motivationally salient* ones, and call the set of such properties, M, the agent's *motivational state*. Formally, $M \subseteq \mathcal{P}$. When a property is in M, this simply means that the agent pays attention to it; it does not imply that the property is satisfied by any particular alternative under consideration. Also, inclusion of a property in M does not mean that the agent is always positively, or always negatively, disposed towards alternatives with that property. It only means that whether or not an alternative has the property may sometimes make a difference to what the agent's preference in relation to that alternative is.

Motivational salience is a primitive notion of our framework. Which properties are motivationally salient for an agent in any context is a psychological question that our formalism alone cannot answer (here, empirical work is required). We write $\mathcal{M}$ to denote the set of all motivational states that are deemed psychologically possible for the agent. Formally, $\mathcal{M}$ is a non-empty set of sets of properties.[2]

To indicate notationally that the agent's preference order $\succsim$ depends on his motivational state M, we append the subscript M to the symbol $\succsim$. So, $\succsim_M$ denotes the agent's preference order in motivational state M. A full model of the agent requires the ascription of a *family* $(\succsim_M)_{M \in \mathcal{M}}$ of preference orders to the agent, consisting of one preference order $\succsim_M$ for each motivational state $M \in \mathcal{M}$.

How exactly does $\succsim_M$ depend on M? The family of preference orders $(\succsim_M)_{M \in \mathcal{M}}$ is *property-based* if there exists a binary relation $\geq$ over property combinations (consistent sets of properties[3]) such that, for any motivational state $M \in \mathcal{M}$ and any alternatives $x, y \in X$,

$$x \succsim_M y \Leftrightarrow \{P \in M : x \text{ satisfies } P\} \geq \{P \in M : y \text{ satisfies } P\}.$$

When this definition applies, we say that x's having the properties in $\{P \in M : x \text{ satisfies } P\}$ and y's having the properties in $\{P \in M : y \text{ satisfies } P\}$ are the agent's *motivating reasons* for preferring x to y in state M. We call $\geq$ the agent's *weighing relation* over property combinations. The weighing relation ranks different property combinations relative to each other, indicating which property combinations – if salient – are 'preferable to' or 'better than' which others for the agent.

3.2 An example

A simple example illustrates our framework. Suppose an agent faces a choice between the following four alternatives:

S&H: a sweet and healthy cake,   nS&H: a non-sweet and healthy cake,
S&nH: a sweet and unhealthy cake,   nS&nH: a non-sweet and unhealthy cake.

For simplicity, suppose the only properties that may become motivationally salient are:

S: sweetness;   H: healthiness.

Suppose further that any set of properties can in principle be motivationally salient, so that the set of all possible motivational states is

$$\mathcal{M} = \{\{S,H\}, \{S\}, \{H\}, \varnothing\}.$$

---

[2] By stating which specifications of M are included in $\mathcal{M}$, we can capture different assumptions about which properties can simultaneously become motivationally salient for the agent. This could include assumptions about 'crowding out' or 'crowding in' effects, whereby the motivational salience of some properties either rules out, or brings about, the motivational salience of others.

[3] A set of properties is *consistent* if there exists an alternative $x \in X$ which satisfies all of them.

Now the agent's preferences across different $M \in \mathcal{M}$ might be as follows:

In state $M = \{S,H\}$:   S&H $\succ_M$ nS&H $\succ_M$ S&nH $\succ_M$ nS&nH.
In state $M = \{S\}$:   S&H $\sim_M$ S&nH $\succ_M$ nS&H $\sim_M$ nS&nH.
In state $M = \{H\}$:   S&H $\sim_M$ nS&H $\succ_M$ S&nH $\sim_M$ nS&nH.
In state $M = \varnothing$:   S&H $\sim_M$ nS&H $\sim_M$ S&nH $\sim_M$ nS&nH.

We must emphasize that this is just one example of what the agent's family of preference orders across different motivational states might be. (In general, the motivationally salient properties in the different states only *constrain* the agent's preferences in those states; they do not by themselves *determine* those preferences. The preferences are determined only together with the underlying weighing relation.)

The family of preference orders in the present example can be verified to be property-based, with respect to the following weighing relation:

$$\{S,H\} > \{H\} > \{S\} > \varnothing.$$

The example illustrates that, when the agent's preferences are property-based, a single weighing relation over property combinations suffices to induce the agent's entire family of preference orders across different motivational states.

3.3   An axiomatic characterization

We now offer an axiomatic characterization of property-based preferences. The following two axioms constrain the relationship between motivationally salient properties and preferences.

Axiom 1. The agent is indifferent between any two alternatives whose motivationally salient properties are the same. Formally, for any two alternatives $x, y \in X$ and any motivational state $M \in \mathcal{M}$,

if $\{P \in M : x$ satisfies $P\} = \{P \in M : y$ satisfies $P\}$, then $x \sim_M y$.

Axiom 2. If the agent's motivational state changes, in that additional properties become motivationally salient, the agent's preference between any alternatives satisfying none of the newly added properties remains unchanged. Formally, for any two alternatives $x, y \in X$ and any two motivational states $M, M' \in \mathcal{M}$ with $M' \supseteq M$,

if neither x nor y satisfies any $P \in M'\backslash M$, then $x \succsim_M y \Leftrightarrow x \succsim_{M'} y$.

In the main paper, we prove that, if the set of possible motivational states $\mathcal{M}$ satisfies a suitable closure condition, Axioms 1 and 2 characterize the class of property-based families of preference orders. Call $\mathcal{M}$ *intersection-closed* if, for all $M_1, M_2 \in \mathcal{M}$, we have $M_1 \cap M_2 \in \mathcal{M}$.

THEOREM 1. *Suppose $\mathcal{M}$ is intersection-closed. Then the agent's family of preference orders $(\succsim_M)_{M \in \mathcal{M}}$ satisfies Axioms 1 and 2 if and only if it is property-based.*

Thus the two axioms guarantee that the agent's family of preference orders across motivational states can be represented by a single underlying weighing relation over property combinations. The present result is just one of several theorems than can be obtained in our framework.

3.4 The basic implication

According to our theory, the stable feature characterizing an agent is not the agent's preference order over the alternatives in X, but the agent's weighing relation over property combinations. On this picture:

- An agent *forms* his or her preferences by adopting a particular motivational state, i.e., by focusing – consciously or otherwise – on certain properties of the alternatives as the motivationally salient ones (and by adopting a weighing relation in the first place).

- An agent may *change* his or her preferences when the motivational state changes, i.e., when new properties of the alternatives become motivationally salient or others cease to be salient.

In the main paper, we consider in detail whether our theory is empirically testable and present some game-theoretic applications.

4   Concluding remarks

We conclude this summary with a few general remarks on the kinds of issues our theory is intended to shed light on.

4.1   Non-informational preference change

Standard decision theory has no difficulty explaining how an agent's preferences over some uncertain prospects (as opposed to fully described outcomes) can change as the agent changes his beliefs about the likelihood of various possible outcomes of those prospects. However, standard decision theory is unable to explain preference changes of a different kind: those driven by a change in (i) the agent's normative priorities, (ii) the salience of certain considerations, or (iii) the agent's conceptual scheme or mental representation of the decision alternatives. The proposed theory of property-based preference formation points towards a unified treatment of such phenomena.

4.2   Framing and nudging effects

Since Tvserky and Kahneman's seminal work (e.g., [7]), it is well known that people's choices are often influenced by subtle changes in how the decision options are described or framed. Standard decision theory struggles to explain such phenomena, which are sometimes interpreted simply (but incorrectly) as violations of rationality. The proposed framework allows us to go beyond this standard picture, by capturing the way in which different frames or descriptions activate different motivating reasons in an agent's preference-formation process.

4.3   Deliberation, beyond the exchange of information

While economists often think of deliberation just in terms of the exchange of information, philosophers, political scientists, and others hold that deliberation is much richer: it has many other aspects, from the consideration of arguments and the weighing of reasons to the hypothetical assumption of other agents' perspectives. The proposed theory provides a better language for capturing the content of deliberative processes. It is able to do so by (i) explicitly modelling the mechanisms by which reasons constrain preferences, (ii) permitting a formalization of the notions of motivating and normative reasons, and (iii) capturing the possibility that an agent's set of motivating reasons may change in response to normative reflection.

4.4   Ethical considerations in decision theory

While standard decision theory rests on a purely formal account of rationality that provides no resources for ethically assessing or criticizing an agent's preferences, philosophers and others are often interested in a more substantive account of rationality, under which we can assess an agent's motivations and distinguish between attitudes that are formally but not substantively rational, and attitudes that meet stronger substantive or moral constraints. The proposed theory is intended to capture such broader concerns and thereby to build a bridge between formal and substantive approaches to thinking about rationality.

BIBLIOGRAPHY

[1]   De Jongh, D., Liu, F. (2009) Preference, priorities and belief. In T. Grüne-Yanoff and S. O. Hansson (eds.) *Preference Change: Approaches from Philosophy, Economics and Psychology*. Dordrecht (Springer): 85-108.

[2]   Dietrich, F., List, C. (2011) A model of non-informational preference change. *Journal of Theoretical Politics* 23(2): 145-164.

[3]   Dietrich, F., List, C. (2013) Where do preferences come from? *International Journal of Game Theory* 42(3): 613-637.

[4]   Grüne-Yanoff, T., Hansson, S. O. (2009) *Preference Change: Approaches from Philosophy, Economics and Psychology*. Dordrecht (Springer).

[5]   Hansson, S. O. (2001) Preference Logic. In D. Gabbay and F. Guenthner (eds.), *Handbook of Philosophical Logic*, 2nd ed., vol. 4. Dordrecht (Kluwer): 319-393.

[6]   Liu, F. (2010) Von Wright's '*The Logic of Preference*' revisited. *Synthese* 175(1): 69-88.

[7] Tversky, A., Kahneman, D. (1981) The framing of decisions and the psychology of choice. *Science* 211(4481): 453-458.
[8] Von Wright, G. H. (1963) *The Logic of Preference*. Edinburgh (Edinburgh University Press).

Comments on Dietrich and List

ALEXANDRU BALTAG

The paper proposes a setting for preference formation and preference change. Essentially, a more fundamental relation (than the preference order on outcomes) is postulated, namely a *weighting relation*: this is a kind of "preference order on sets of attributes" (or "relevant properties"). An agent's "motivational state" specifies which of these attributes are taken as *salient* in a given context. So the same underlying weighting relation is typically consistent with many motivational states. But for a given motivational state, a given weighting relation gives rise to a unique preference order on outcomes: for any two states, the order is given by the weighting order between the sets of all the salient properties of the two states. In fact, by varying the motivational state, we obtain a whole family of preference orders (having the same underlying weighting relation)

The authors make two remarks that are in fact in a mutual tension, conceptually speaking: first, that even if the preference order is complete and transitive (as is usually assumed), the weighting relation does not have to be so; and second, that the weighting relation underlying a given family of preferences is "essentially unique". The tension arises from the fact that to consider the weighting relation as "unique", we have to neglect (as irrelevant) the sets of salient properties that do not match the set of properties of any given state. But, if we do that, then we can also conclude that the weighting is "essentially complete and transitive": its restriction to the "relevant" sets (i.e. the sets of all salient properties of any state) IS complete and transitive! The first remark simply amounts then to noticing that the weighting relation is not uniquely determined on the "irrelevant" sets (and hence can be taken to be non-complete or non-transitive on some of those sets).

This is a beautiful, innovative, far-reaching, and indeed very ambitious, paper. The proposed setting looks simple and natural enough in itself, but then the authors also attempt to give it axiomatic justifications (Theorems 1 and 2), in terms of supposedly more intuitive postulates connecting the weighting of motivationally salient properties with the preference order. These results are mildly interesting, though the proofs seem very easy. But they do not seem to me to give much extra insight, or add any further justifying evidence in favor of the proposed setting: the connecting postulates do not seem to me to be significantly more natural or more intuitive than the basic setting (with the preference order defined as above in terms of the weighting relation).

The authors argue that this framework is empirically meaningful and (in partic-ular) falsifiable[1], by exhibiting some specific consequences that seem empirically falsifiable. They go on to relate this setting to a Kantian notion of "substantive ra-tionality"[2] (opposed to a Humean notion of "formal rationality", that is shared by most authors in classical Economics, Decision Theory and Game Theory). They apply this theory (in section 7) to a game-theoretic example (Prisoner's Dilemma). They extensively (and convincingly) argue (cf section 6) that alternative strategies going along the classical lines *cannot explain all* the types of preference changes captured by the account proposed here (although they may explain *some* of these preference changes).

What is left unclear is whether the same objection cannot be raised against the newly proposed theory: can it really explain all (natural) preference changes? To recall, the essence of the new theory is that the weighting relation is taken to be the underlying stable feature characterizing an agent at all moments, and that both preference formation and preference change can be *completely* explained in terms of adopting a particular motivational state, or changing it to a different motivational state. According to this view, all preference change is generated by a change in the *salience* of various attributes (since, in different contexts, different properties are salient), i.e. a change of *focus* or of *motivation* on the side of the agent.

However, I am not convinced that this view is correct. While such changes in focus and motivation affecting the attributes' salience do happen, and when they happen they do typically induce preference changes, I see no reason to assume that all preference changes are produced in this way. The authors give no good reasons in favor of this thesis. Well, in fact, it does seem likely that, at a purely formal level, one can always artificially choose the attributes and the weighting relation in such a way that one can simulate any preference change in this framework. But such a formal move doesn't address the underlying (empirical) problem: given a family of preferences and a family of "natural" attributes (together with a family of "natural" motivational states), can one simulate any "natural" preference change in this framework (as a change of motivational state)? I have no reason to believe that the answer is yes. Intuitively, preference change and salience change are different things, and one can easily imagine one without the other. I can easily think of an agent whose set of salient attributes remains invariant, despite her change of preferences.

In general, I think that any attempt to fully *explain* or *justify* preference change in purely "rational" terms (and in particular the present proposal) arise from an

---

[1] Although why is this really important remains unexplained: I guess that the authors betray here their ingrained Popperian bias.

[2] The name may be unfortunately chosen. There is no relation to Aumann's well-known concept of substantive rationality: on the contrary, Aumann's notion is an example of what the authors call "formal rationality"!

unjustified belief in a type of "hyper-rationality", that would explain away all the irrational (in particular, preferential) components that shape our formal rationality. But rationality (the formal kind, at least) is defined (and justified) in terms of preferences, and not vice-versa. There have to be more fundamental, "irrational" or rather *infra-rational* choices underlying (and circumscribing) any meaningful notion of rationality, or otherwise an infinite explanatory cycle (of the vicious type) would arise. In the standard theory, preferences play the role of such infra-rational factors. I am not at all against attempts at replacing preferences by even more fundamental factors. But these factors will have to remain (at least partially) infra-rational. Even if some preferences (and preference changes) are indeed "rationally" motivated (maybe in terms of what the authors call "substantive rationality"), it does not follow that all preferences and preference changes are motivated in this way. On the contrary, it seems very likely that an agent may sometimes simply change her preference "irrationally", i.e. for no reason whatsoever, even when all her "reasons" (her motivational state, her set of salient attributes etc.) stay the same. A person can fall in love, or fall out of love, just so, purely and simply, for no deeper reasons. There is no need for (and no good evidence in favor of) assuming that this person has suddenly switched her focus from some set of desirable attributes to another. She may still appreciate the same qualities (e.g, charm, beauty, intelligence etc.) over others, and she may still recognize those same qualities in the person she used to love. She just doesn't love him anymore.

Comments on Dietrich and List

PATRICK GIRARD AND SHAUN WHITE

Dietrich & List introduce a new model of preference formation and preference change. We would like to pose some questions about a central feature of the model, the agent's weighing relation. In Section 1 we discuss how modal logic might formalise the interaction between an agent's weighing relation and their state-based preferences. In Section 2 we compare Dietrich & List's weighing relation with the 'priority' order that Horty uses in his formalisation of the concept of reason. In Section 3 we ask when a weighing relation can be deduced from an order on single properties; we also observe that a weighing relation can sometimes 'coincide' with a preference relation. In Section 4 we ask if changes in motivational states need to be restricted, and, if so, how any restrictions can be made consistent with the weighing relation. In Section 5 we briefly pose a question on collective motivational states and weighing relations.

1 Motivations, the weighing relation, and modal logic

The logical challenge is to formalise the interaction between motivational states $M \in \mathcal{M}$, the motivated preferences $\succeq_M$ for each $M \in \mathcal{M}$, and the weighing relation $\geq_{\mathcal{M}}$. Given a set of motivational states $\mathcal{M}$, models for motivated preference logic have two kinds of relations, a weighing relation $\geq_{\mathcal{M}}$ defined over sets of sets of formulas (i.e., over motivational states), and a motivativated preference relation $\succeq_M$ for each $M \in \mathcal{M}$. For the analysis of motivated preference relations, one could use a modality of the form $[P_M]\varphi$. Several semantic definitions for this modality are conceivable, for instance:

$$N, w \models [P_M]\varphi \text{ iff } \forall v \in W : N, v \models \varphi \Rightarrow w \succeq_M v.$$

This definition states that the alternative w is better according to $\succeq_M$ than every φ-alternative v (Following Liang & Seligman ([3]), this semantic definition is one for a window operator, cf., Blackburn et al, [1], p.424). Correspondingly for the weighing relation, we would introduce an operator $[\mathcal{M}]\varphi$ defined over $\geq_{\mathcal{M}}$. The first step is to understand precisely how the semantics of this modality works. We need guidance from Dietrich & List, but we presume a semantic definition might look like this:

$$N, w \models [\mathcal{M}]\varphi \text{ iff } \forall v \in W : N, v \models \varphi \Rightarrow$$
$$\forall M \in \mathcal{M} : \{\psi \in M \mid N, w \models \psi\} \geq_{\mathcal{M}} \{\psi \in M \mid N, v \models \psi\}.$$

This says that the alternative w is better than any φ-alternative v according to each motivational state. Perhaps the following would provide a better semantics?

$$N, w \models [\mathcal{M}]\varphi \text{ iff } \forall v \in W : N, v \models \varphi \Rightarrow$$
$$\{\psi \in P \mid N, w \models \psi\} \geq_{\mathcal{M}} \{\psi \in P \mid N, v \models \psi\}.$$

This latter definition says that the set of properties that alternative w satisfies is better than the set of properties satisfied by any φ-alternative v.

Regardless of the semantic definition adopted for $[\mathcal{M}]\varphi$, an important technical challenge is to formalise the interaction between $[P_M]\varphi$ and $[\mathcal{M}]\varphi$. Is one definable in terms of the other? If not, what kind of interaction principles between the modalities will be required for a completeness proof? To answer this question we need a better understanding of $\geq_{\mathcal{M}}$.

To make this more precise, consider Dietrich & List's Theorem 1, which states that an intersection-closed collection of motivational states can be represented by a single weighing relation iff the collection satisfies their axioms 1 and 2. In such cases, there is a weighing relation $\geq_{\mathcal{M}}$ such that

$$w \succeq_M v \text{ iff } \{\psi \in M \mid N, w \models \psi\} \geq_{\mathcal{M}} \{\psi \in M \mid N, v \models \psi\}$$

Assuming that the collection of motivational states that satisfy the conditions of Theorem 1 are good, is there a set of axioms complete for the class of good models, i.e., those with an $\mathcal{M}$ that is closed under taking intersections or subsets, or some other constraints?

2 Motivations, the weighing relation, and Horty's reasons

Dietrich & List write that "to mark the interpretation of motivationally salient properties as *reasons* for the agent's preferences, we sometimes call property-based preferences also *reason-based*." John Horty recently proposed a formalisation of the concept of 'reason' using the non-monotonic framework of default logic (c.f., [2]). In this framework, "reasons are identified with the premises of triggered defaults" ([2], p27). For example, the default $\delta_1 : B \rightarrow F$ asserts that birds fly. Upon learning that Tweety is a bird, default δ_1 is triggered, and the premise B becomes a reason for concluding that Tweety flies. But several defaults may get triggered at the same time. For instance, that $\delta_2 : P \to \neg F$, "penguins do not fly". Upon learning that Tweety is a penguin, we have two conflicting triggered defaults. To resolve conflicts Horty uses a primitive (strict partial) order $>_{Horty}$ over the set of defaults. In the Tweety example, the ordering of defaults would be $\delta_2 >_{Horty} \delta_1$. So when both defaults are triggered, δ_2 takes precedence over δ_1, and we conclude that Tweety does not fly (our reason being that Tweety is a penguin) even though Tweety is a bird.

Is a similar non-monotonic formalism of motivations for preference conceivable? Perhaps motivations for preferences would emerge as defaults that apply unless superseded by other more important defaults. So, containing chocolate may be a motivation to prefer ice cream x to ice cream y, but containing nuts is a motivation to prefer y over x – if I'm allergic to nuts. How does Horty's notion of reason compare to Dietrich & List's? In general, what is the difference between motivation and reason?

More specific questions arise when we more closely compare Horty's default priority order $>_{Horty}$ to Dietrich & List's weighing relation $>_{\mathcal{M}}$. In Horty's set-up, if $\delta_1 >_{Horty} \delta_2$ then whenever δ_1 and δ_2 are amongst the salient defaults, the agent gives δ_1 priority over δ_2. In Dietrich & List's framework, if $\{p_1\} >_{\mathcal{M}} \{p_2\}$ then when X's only salient property is p_1 and Y's only salient property is p_2, the agent prefers X to Y; can this ever mean the agent 'prioritises' p_1 over p_2? In the next section we conjecture that some weighing relations can be interpreted as being, in some sense, 'priority' relations.

Horty's fundamental relation $>_{Horty}$ orders single formulas while Dietrich & List's fundamental relation $>_{\mathcal{M}}$ orders sets of formulas. Could we extend $>_{Horty}$ into a relation on sets of formulas (sets of defaults) in such a way that the extended relation is comparable to Dietrich & List's weighing relation?

3   Interpretations and properties of the weighing relation

Imagine an agent in a bookshop, browsing the shelves, contemplating making a purchase. Two properties of books are potentially salient: the format (paperback or hardback) and the content (fiction or non-fiction). All four possible varieties of book are on sale. Let p denote the property of being a paperback, and let f denote the property of being a work of fiction. The agent's weighing relation is

$$\{p, f\} >_{\mathcal{M}} \{p\} >_{\mathcal{M}} \{f\} >_{\mathcal{M}} \emptyset.$$

There are two interesting things about this weighing relation. First, the relation $\geq_{\mathcal{M}}$ can be condensed down to a relation R over properties, and R might be interpretable as a priority relation over properties. Ceteris paribus, this agent always prefers a paperback to a hardback, always prefers a work of fiction to a work of non-fiction, and, if he or she considers both form and content salient, he or she always considers form 'more important' than content. The agent's weighing relation can be deduced from the order

$$p \, R \, f \, R \, 0$$

where we use 0 to stand for absence of whatever it is compared with; $p \, R \, 0$, for example, indicates that a book with property p is always preferred to a book without property p. Of course, not all weighing relations can be derived from a relation on single properties.

Here's the other reason the bookworm's weighing relation is interesting: it coincides with the agent's preferences when the agent considers all properties to be salient. When the agent considers both format and content salient, they will, for instance, prefer a non-fiction paperback to a hardback work of fiction. More generally, if $M^* = \bigcup_{M \in \mathcal{M}} M$ is a plausible motivational state, and if every combination of properties is realised in some alternative, then $\succeq_{M^*}$ is derivable from $\succeq_{\mathcal{M}}$, and vice versa.

So we're led to think about two questions. In what types of practical choice situations will a weighing relation boil down to an order on properties? And in which situations will an agent's weighing relation necessarily coincide with one of their state-based preference relations?

4   Rational changes between motivational states

By giving agents a stable weighing relation, Dietrich & List allow agents to change their preferences while retaining their rationality. If we want to fully describe a rational agent, we might need to restrict the kinds of changes in motivational state that the agent can undergo. In some situations, an agent could have fully rational preferences in each motivational state but be capable of making irrational changes between motivational states.

Here is an extreme example. It's based on the idea of the money pump. Helen has a pen and a pencil, and John needs to borrow one of them. When John is in a confident state, he prefers to write with a pen, and when he lacks confidence, he prefers to write in pencil. Helen knows this. Initially John is feeling confident, and asks to borrow Helen's pen. Helen agrees. Later, she manipulates John's state, and causes him to lose confidence. John then agrees to pay $1 to exchange the pen for a pencil. Later still, Helen manipulates John's state again, and restores his confidence. He then agrees to pay $1 to swap the pencil for a pen. And so on. In any particular state, John's preferences appear rational. His weighing relation could be perfectly reasonable mathematically – suppose the pen has properties

a and b, the pencil has properties c and d, his motivational states are $\{a, c\}$ and $\{b, d\}$, and his weighing relation has $\{a\} > \{c\}$ and $\{d\} > \{b\}$. If we want to go into more detail about what it means for an agent to be rational, will we need to restrict the kind of changes in motivational state that are acceptable? If so, how could we ensure that the kind of changes we accept as rational will be consistent with the agent's weighing relation?

5 Motivations, the weighing relation, and communities

Do motivational states and weighing relations only ever belong to individual agents? Social choice theorists sometimes refer to the aggregated preferences of a community as 'the community's preferences'. After reading Dietrich & List's paper, a social choice theorist might ask: if a community can have preferences, can it also have a motivational state? And, if a community can have motivational states, can it have a weighing relation? Do the authors have any thoughts about the idea of a collective motivational state, or a collective weighing relation?

BIBLIOGRAPHY

[1] Blackburn, Patrick, Maarten de Rijke, and Yde Venema (2001) *Modal Logic*. Cambridge University Press, UK.
[2] Horty, John F. (2012) *Reasons as Defaults*. Oxford University Press, USA.
[3] Zhen, Liang and Jeremy Seligman (2011) "A Logical Model of the Dynamics of Peer Pressure", *Electronic Notes in Theoretical Computer Science*, Volume 278, 3 November 2011, Pages 275-288.

Response to Girard and White, and Baltag

FRANZ DIETRICH AND CHRISTIAN LIST

1 Introduction

The aim of this short note is to offer some brief responses to the comments on [1] that we have received from Alexandru Baltag and, in a coauthored piece, Patrick Girard and Shaun White. We would like to begin by thanking Baltag, Girard, and White for their thoughtful and generous comments. Given space constraints, unfortunately, we are not able to do full justice to all their interesting suggestions, thoughts, and questions here, but we hope to address at least some of their central points.

Although Baltag on the one hand and Girard and White on the other raise a number of distinct issues, there is some thematic overlap between them. Several of the comments concern the formal properties and substantive interpretation of the 'weighing relation', a central concept in our reason-based model of preference formation. In what follows, we first recapitulate the concepts of reason-based preferences and the underlying weighing relation and then address some of the questions Baltag, Girard, and White raise about those concepts and about our approach more generally.

2 Reason-based preferences and the weighing relation revisited

The aim of the formal framework developed in [1] is to model the relationship between an agent's preferences and his or her 'reasons' for holding those preferences. Preferences are represented by some (complete and transitive) order $\succsim$ on a set of alternatives X. Crucially, each alternative in X is conceptualized by the agent, not as a primitive object, but as a bundle of properties. A *property*, for our purposes, is a binary characteristic that an alternative may or may not have. At any given time, the agent is in a particular *motivational state*, defined by the set M of properties that the agent focuses on. Inclusion of a property in M only means that the agent cares about, or pays attention to, that property. It does not mean that he or she always likes, or always dislikes, the property; the property simply makes some difference to the agent's preferences in motivational state M. To indicate that the agent's preference order depends on M, we write $\succsim_M$ to denote the agent's preference order in motivational state M.

Assuming that $\mathcal{M}$ is the (non-empty) set of all motivational states deemed possible (e.g., compatible with the agent's psychology), we call the family $(\succsim_M)_{M \in \mathcal{M}}$ of preference orders across $M \in \mathcal{M}$ *property-based* if there exists an underlying binary relation $\geq$ over consistent sets (combinations) of properties such that, for any motivational state $M \in \mathcal{M}$ and any alternatives $x, y \in X$,

$$x \succsim_M y \Leftrightarrow \{P \in M : x \text{ satisfies } P\} \geq \{P \in M : y \text{ satisfies } P\}.$$

Property-basedness means that the agent's entire family of preference orders across different possible motivational states can be represented in terms of a single underlying binary relation $\geq$ over combinations of properties. We call this a *weighing relation*. It captures how good, or preferable, different combinations of properties are relative to each other, from the perspective of the agent. Any family of preference orders that satisfies two basic axioms (discussed in [1] is representable in this way.

3   Relation-theoretic properties of the weighing relation

Baltag notes that, in [1], we make two claims about the relation-theoretic properties of the weighing relation. Our first claim is that, for any property-based family of preference orders, the underlying weighing relation is *essentially unique*, meaning that it is unique on all pairs of property combinations that 'matter' for the agent's preferences, i.e., all pairs that co-occur in some set $X_M = \{\{P \in M : x \text{ satisfies } P\} : x \in X\}$ for $M \in \mathcal{M}$. Our second claim is that, even though the agent's preference order $\succsim_M$ in every motivational state $M \in \mathcal{M}$ is transitive, the underlying weighing relation can still be intransitive.

Baltag wonders whether these two claims are in tension. Specifically, he observes that the weighing relation restricted to each X_M is transitive, and suggests that our claim about the relation's possible intransitivity 'simply amounts ... to noticing that the weighing relation is not uniquely determined on the "irrelevant" sets (and hence can be taken to be non-complete or non-transitive on some of those sets).'

We agree with Baltag's first observation (that the relation $\geq$ restricted to each X_M separately is transitive), but disagree that the possible intransitivity is just due to the residual non-uniqueness. To explain this point, we recall an example from a different paper [2].

Consider a consumer choice over three alternatives (different cars):

a Monster Hummer, which is fast, big, but not environmentally friendly $(FB\neg E)$;

a Sports Beetle, which is fast, not big, but environmentally friendly $(F\neg BE)$;

a Family Hybrid, which is not fast, but big and environmentally friendly $(\neg FBE)$.

Suppose that any of the three properties of a car, 'fast', 'big', and 'environmentally friendly', could serve as reasons for or against preferring it. Formally,

- property F (fastness) is satisfied by the Hummer $(FB\neg E)$ and the Beetle $(F\neg BE)$,

- property B (big) is satisfied by the Hummer $(FB\neg E)$ and the Hybrid $(\neg FBE)$,

- property E (environmentally friendly) is satisfied by the Beetle $(F\neg BE)$ and the Hybrid $(\neg FBE)$.

Suppose, further, that any subset of $\{F, B, E\}$ can potentially constitute a motivational state M. So $\mathcal{M}$ is the powerset of $\{F, B, E\}$. It can be checked that the following family of preference orders across $M \in \mathcal{M}$ is property-based (the left column shows the motivational state; the right column shows the corresponding preference order $\succsim_M$ in that state, with $\succ_M$ and $\sim_M$ denoting strict preference and indifference, respectively).

$$
\begin{aligned}
M = \{F, B, E\} \quad &: \quad \text{Hummer} \sim_M \text{Beetle} \sim_M \text{Hybrid,} \\
M = \{F, B\} \quad &: \quad \text{Hummer} \succ_M \text{Beetle} \succ_M \text{Hybrid,} \\
M = \{B, E\} \quad &: \quad \text{Hybrid} \succ_M \text{Hummer} \succ_M \text{Beetle,} \\
M = \{F, E\} \quad &: \quad \text{Beetle} \succ_M \text{Hybrid} \succ_M \text{Hummer,} \\
M = \{F\} \quad &: \quad \text{Hummer} \sim_M \text{Beetle} \succ_M \text{Hybrid,} \\
M = \{B\} \quad &: \quad \text{Hummer} \sim_M \text{Hybrid} \succ_M \text{Beetle,} \\
M = \{E\} \quad &: \quad \text{Beetle} \sim_M \text{Hybrid} \succ_M \text{Hummer,} \\
M = \varnothing \quad &: \quad \text{Hummer} \sim_M \text{Beetle} \sim_M \text{Hybrid.}
\end{aligned}
$$

Importantly, the underlying weighing relation $\geq$ must have the following features (otherwise it would not generate the family of preference orders just shown):

$${F, B} \equiv {B, E} \equiv {F, E},$$

$${F, B} > {F} > {B},$$
$${B, E} > {B} > {E},$$
$${F, E} > {E} > {F},$$

$${F} > \varnothing,$$
$${B} > \varnothing,$$
$${E} > \varnothing.$$

Here $>$ and $\equiv$ denote the strict and indifference components of $\geq$. The intransitivity of $\geq$ is a direct consequence of the second, third, and fourth rows of the displayed list.

It is important to note that this intransitivity is not due to non-uniqueness. In the present example, the agent's family of preference orders is representable *only* by an intransitive weighing relation. This remains true even if the weighing relation is chosen in the 'sparsest' possible way, i.e., with the smallest number of relata with which the given family of preference orders can be generated. Formally, there may have to be cycles within the set $\bigcup_{M \in \mathcal{M}} (X_M \times X_M)$ of pairs that matter for the agent's preferences.

4 A further analysis of the weighing relation

We now move on to some of Girard and White's main questions about the weighing relation. First of all, they suggest developing an explicit logic to formalise the relationship between an agent's motivational states, his or her preferences in those states, and the underlying weighing relation. They also briefly describe some ingredients of such a logic. We agree that the development of a logic of this kind would be useful for a number of purposes, and, unfortunately, our paper does not offer one (instead, it offers a more 'ordinary', decision-theoretic framework in the tradition of the representation theorems of von Neumann and Morgenstern, Savage, and others). We leave this issue as a challenge for future work, though we would like to draw attention to Osherson and Weinstein's related contributions (2012a,b).

We next turn to another important question raised by Girard and White: in which cases can the weighing relation, which is a binary relation over property *combinations*, be reduced to a binary relation over *individual properties*? Girard and White note one such case: the case in which the weighing relation has a *lexicographic structure* (though Girard and White do not use this term). In the lexicographic case, the weighing relation $\geq$ is generated by a *priority order*, R, over properties (a linear order). For notational simplicity, let $R(1), R(2), R(3)$ etc. denote the highest-ranked, second-highest-ranked, third-highest-ranked etc.

properties with respect to R. (For simplicity, we assume that the total number of properties is finite.) For any two property combinations S_1 and S_2, we then define $S_1 \geq S_2$ if and only if either $S_1 = S_2$ or there is some n such that

- $R(n) \in S_1$ and $R(n) \notin S_2$, and

- for all $m < n$, $[R(m) \in S_1$ if and only if $R(m) \in S_2]$.

This yields a weighing relation $\geq$ induced by the priority order R. Interpretationally, properties here play the role of 'good-making features', and the priority order represents their order of importance. An approach along these lines is suitable for representing choices by checklists (e.g., [5]) or take-the-best heuristics (e.g., [4]).

In Dietrich and List (2013a), we consider another case in which the weighing relation can be represented in terms of further primitives: the *additive* case. Here we introduce a *weighing function* w over individual properties, which assigns to each individual property P a real number, $w(P)$, interpretable as the 'weight' of P. The weighing relation $\geq$ is now induced by the weighing function as follows. For any two property combinations S_1 and S_2, we define $S_1 \geq S_2$ if and only if

$$\sum_{P \in S_1} w(P) \geq \sum_{P \in S_2} w(P).$$

What do the lexicographic and additive cases have in common? The answer is relevant to Girard and White's question about when a weighing relation is reducible to either an ordering or a function *over individual properties*. Both the lexicographic and additive cases involve a *separable* weighing relation. Intuitively, in those cases, the 'valence' of a property for the agent – whether it counts in favour of or against an alternative when motivating – does not depend on which other properties are present. In the additive case, an even stronger condition of *additive separability* is met. Given space constraints, we set the formal details aside.

Separable weighing relations are certainly interesting and important, but it would be a loss of generality to assume that an agent's preference formation is always based on them. In this sense, Girard and White's comment draws attention to an important, but nonetheless special case.

5   The bigger picture

Finally, we would like to respond to some of Baltag's and Girard and White's comments about the 'bigger picture' underlying our approach. According to our framework, the stable feature of an agent is no longer the agent's preference order over the alternatives in X, as in standard rational choice theory, but the agent's weighing relation over property combinations. The variable feature is the agent's motivational state. Both Baltag and Girard and White raise some questions about this picture. These include the following:

(1) Is this picture sufficiently general? Does it need to be generalized further?

(2) Can the criticisms that are normally directed at the fixed-preference assumption of rational choice theory also be directed at our assumption of a fixed weighing relation?

(3) Are all changes between motivational states genuinely 'rational'?

Regarding question (1), there is indeed some scope for further generalization. The paper under discussion here [1] allows the motivationally salient properties (those in M) to vary across different motivational states M, and thereby across different decision-making contexts that the agent might be in, but takes the properties themselves to be completely context-independent. This means that whether an alternative satisfies a given property does not depend at all on the context or situation in which the agent is confronted with that alternative. In ongoing work [3], we discuss the possibility that an agent's motivationally salient properties (those in M) may not just vary across different decision-making contexts, but that they may also include properties that refer to the context itself. To illustrate, consider Amartya Sen's famous example of a polite dinner party guest. This guest never chooses the largest piece of fruit offered to him or her, in order to avoid being greedy. So, at a superficial level, the agent seems to display different preferences over pieces of fruit in different situations. (Whether a particular apple is chosen – revealed-preferred in a given context – depends on which other pieces of fruit are also on offer.) However, the best explanation of what is going on here involves, not varying the motivational state M, but rather including the property of 'politeness' in M. Crucially, 'politeness' is a *relational* property: whether an alternative is 'politely choosable' depends not only on the alternative itself, but also on which other alternatives are available (a feature of the decision-making context). In [3], we argue that these observations point towards two very different ways in which the decision-making context may make a difference: *context-variance* (here, the agent has different Ms in different contexts) and *context-regardingness* (here, some Ms may include properties that refer to the context, such as relational properties). It should be evident that introducing both kinds of context-dependence opens up a more general picture, of which the framework in [1] is a special case.

Regarding question (2), a critic is of course right to note that assuming a fixed weighing relation imposes a certain restriction. Methodologically, we believe, however, that a model of individual choice should not involve too many free variables; otherwise it would run the risk of becoming unfalsifiable. In addition, even when there are many free variables, each agent needs to be specified in terms of *some* fixed characteristics; otherwise it is unclear what explanatory role the ascription of agency plays (there still needs to be a sense in which we are dealing with a single agent). How restrictive or permissive the assumption of a fixed weighing relation is depends significantly on how rich the set of properties is that we invoke for explanatory purposes. With sufficiently many properties – possibly allowing non-separable interaction effects between them – we may indeed be able to explain even complex shifts in the agent's evaluative dispositions. In Baltag's comment, he gives the example of someone who 'can fall in love, or fall out of love, just so, purely and simply, for no deeper reasons'. If, by 'reasons', we mean 'substantively rational and fully conscious reasons', we would of course agree; such is human psychology. However, if an agent's preferences change, we may still be able to attribute that change to some *cause*, not necessarily a substantively rational or conscious *reason*. We think that especially the more general version of our framework, in which contexts can influence an agent in two very different ways, can accommodate apparently non-rational preference changes, without having to suggest that they are 'uncaused'.

Finally, regarding question (3), we wish to note that the technical framework presented in [1] only allows us to assess the formal relationship between an agent's motivational state M and the corresponding preference order $\succsim_M$. (For example, does that relationship satisfy the two axioms characterizing property-basedness?) We do not wish to suggest that changes from one state M to another state M' are always substantively rational. Indeed, Girard and White provide a nice 'money pump' example of what can go wrong when an agent keeps vacillating between different motivational states. The more general framework in [3] that we have briefly described can be used to draw a distinction between (i) those forms of context-dependence that can count as rational in some sophisticated sense (certain kinds of norm-following might fall into this category, as in Sen's politeness example) and (ii) those forms of context-dependence that are boundedly or 'sub-'rational. Girard and White's example would fall into the latter category.

We conclude by thanking Baltag, Girard, and White once again for their comments.

BIBLIOGRAPHY

[1] Dietrich, F., List, C. (2013) Where do preferences come from? *International Journal of Game Theory* 42(3): 613-637.
[2] Dietrich, F., List, C. (2013a) A reason-based theory of rational choice. *Nous* 47(1): 104-134.
[3] Dietrich, F., List, C. (2013b) Reason-based rationalization. Working paper, London School of Economics.
[4] Gigerenzer, G., Todd, P. M., ABC Research Group (2000) *Simple Heuristics that Make Us Smart.* New York (Oxford University Press).
[5] Mandler, M., Manzini, P., Mariotti, M. (2012) A million answers to twenty questions: Choosing by checklist. *Journal of Economic Theory* 147: 71-92.

Logical Models of Informational Cascades

ALEXANDRU BALTAG, ZOÉ CHRISTOFF, JENS ULRIK HANSEN,
AND SONJA SMETS

ABSTRACT. In this paper, we investigate the social herding phenomenon known as *informational cascades*, in which sequential inter-agent communication might lead to epistemic failures at group level, despite availability of information that should be sufficient to track the truth. We model an example of a cascade, and check the correctness of the individual reasoning of each agent involved, using two alternative logical settings: an existing probabilistic dynamic epistemic logic, and our own novel logic for counting evidence. Based on this analysis, we conclude that cascades are not only likely to occur but are sometimes unavoidable by "rational" means: in some situations, the group's inability to track the truth is the direct consequence of each agent's rational attempt at individual truth-tracking. Moreover, our analysis shows that this is even so when rationality includes unbounded higher-order reasoning powers (about other agents' minds and about the belief-formation-and-aggregation protocol, including an awareness of the very possibility of cascades), as well as when it includes simpler, non-Bayesian forms of heuristic reasoning (such as comparing the amount of evidence pieces).

Social knowledge is what holds the complex interactions that form society together. But how reliable is social knowledge: how good is it at tracking the truth, in comparison with individual knowledge? At first sight, it may seem that groups should be better truth-trackers than the individuals composing them: the group *can "in principle" access* the information possessed by each agent, and in addition it *can access whatever follows* from combining these individual pieces of information using logic.

And indeed, in many situations, this "virtual knowledge" of a large group is much higher than the knowledge of the most expert member of the group: this is the phenomenon known as *wisdom of the crowds* [32]. Some examples of the wisdom of the crowds are explainable by the logical notion of *distributed knowledge*: the kind of group knowledge that can be realized by inter-agent communication. But most examples, typically involving no communication, are of a different, more "statistical" type, and they have been explained in Bayesian terms by Condorcet's Jury Theorem [26, 19, 25], itself based on the Law of Large Numbers. In particular, some political scientists used (variants and generalizations of) the Jury Theorem, to provide epistemic arguments in favour of deliberative democracy: this is the core of the so-called "epistemic democracy" program [26, 19]. Roughly speaking, the conclusion of this line of research is that groups are more reliable at tracking the truth than individuals and that the larger the group, the more probable it is that the majority's opinion is the right one.

However, the key word in the above paragraph is *virtual* (as in "virtual knowledge" of a group). Most explanations based on (variants of) the Jury Theorem seem to rely on a crucial condition: agents do not communicate with each other, they only *secretly* vote for their favorite answer. Their opinions are therefore taken to be completely *independent* of each other.[1] In contrast, the logical notion of distributed knowledge is tightly connected to communication: by sharing all they know, the agents can convert their virtual knowledge into actual knowledge. But without communication, how do the agents get to actualize the full epistemic potential of the group? Or do they, ever?

There seems to be a tension between the two ingredients needed for maximizing actual group knowledge: independence (of individual opinions) versus sharing (one's opinions with the group). Independence decreases when inter-agent communication is allowed, and in particular when agents are making *public and sequential* guesses or decisions. In such cases, some agents' later epistemic choices might very well be influenced by other agents' previous choices. Being influenced in this way may be perfectly justifiable on rational grounds at an individual level. After all, this is what discussion and deliberation are all about: exchanging information, so that everybody's opinions and decisions are better informed, and thus more likely to be correct. So, at first sight, it may seem that communication and rational deliberation can only be epistemically beneficial to each of the agents, and hence can only enhance the truth-tracking potential of the group. But in fact the primary consequence of communication, at the group level, is that the agents' epistemic choices become *correlated* (rather than staying independent). This correlation undermines the assumptions behind positive theoretical results (such as the Jury Theorem), and so the conclusion will also often fail. The group's (or the majority's) actual knowledge may fall way behind its "virtual knowledge": indeed, the group may end up voting unanimously for the wrong option!

Informational cascades are examples of such "social-epistemic catastrophes" that may occur as a by-product of sequential communication. By observing the epistemic decisions of the previous people in a sequence, an individual may rationally form an opinion about the information that the others might have, and this opinion may even come to outweigh her other (private) information, and thereby affect her epistemic decision. In this way, individuals in a sequence might be "ra-

---

[1] But see e.g. [19] for majority truth-tracking in conditions that allow for some very mild forms of communication within small subgroups.

tionally" led to ignore their own private evidence and to simply start *following the crowd*, whether the crowd is right or wrong. This is *not mindless imitation*, and it is *not due to any irrational social-psychological influence* (e.g. group pressure to conform, brainwashing, manipulation, mass hysteria etc). Rather, this is the result of rational inference based on partial information: it is not just a cascade, but an "informational" cascade. A classical example is the choice of a restaurant. Suppose an agent has some private information that restaurant A is better than restaurant B. Nevertheless, when arriving at the adjacent restaurants she sees a crowded restaurant B and an empty restaurant A, which makes her decide to opt for restaurant B. In this case our agent interprets the others' choice for B as conveying some information about which restaurant is better and this overrides her independent private information. However, it could very well be that all the people in restaurant B chose that restaurant for the exact same reason. Other examples of informational cascades include bestseller lists for books, judges voting, peer-reviewing, fashion and fads, crime etc [13].

While models of such phenomena were independently developed in [12] and [6], the term *informational cascades* is due to [12]. A probabilistic treatment of cascades, using Bayesian reasoning, can be found in [15]. Traditionally investigated by the Social Sciences, these social-informational phenomena have recently become subject of philosophical reflection, as part of the field of *Social Epistemology* [17, 18]. In particular, [21] gives an excellent philosophical discussion of informational cascades (and the more general class of "info-storms"), their triggers and their defeaters ("info-bombs"), as well as the epistemological issues raised by the existence of these social-epistemic phenomena.

In a parallel evolution, logicians have perfected new formal tools for exploring informational dynamics and agency [8], and for modeling public announcements and other forms of distributed information flow. An example is the fast-growing field of Dynamic-Epistemic Logic (DEL for short), cf. [8], [4], [14] etc. More recently, variants of DEL that focus on *multi-agent belief revision* [3, 7, 5] and on the *social dynamics of preferences* [10, 27] have been developed and used to investigate social-epistemic phenomena that are closely related to cascades: epistemic bandwagonning [22], mutual doxastic influence over social networks [31, 30], and pluralistic ignorance [29, 20].

The time seems therefore ripe for an epistemic-logical study of informational cascades. In this paper, we take the first step in this direction, by modeling "rational" cascades in a logical-computational setting based on (both probabilistic and more qualitative) versions of Dynamic-Epistemic Logic.

When the total sum of private information possessed by the members of a group is in principle enough to track the truth, but nevertheless the group's beliefs fail to do so, one might think that this is due to some kind of "irrationality" in the formation and/or aggregation of beliefs (including unsound reasoning and misinterpretation of the others' behavior, but possibly also lack of cooperation, lack of relevant communication, lack of trust etc). However, this is not always the case, as was already argued in the original paper [12]. One of the standard examples of an informational cascade (the "Urn example" which will be discussed in section 1), has been used to show that cascades can be "rational". Indeed, in such examples, the cascade does seem to be the result of correct Bayesian reasoning [15]: each agent's opinion/decision is perfectly justified, given the information that is available to her. A Bayesian model of this example is given in [15] and reproduced by us in section 1. The inescapable conclusion seems to be that, in such cases, *individual rationality may lead to group "irrationality"*.

However, what is typically absent from this standard Bayesian analysis of informational cascades is the agents' higher-order reasoning (about other agents' minds and about the whole sequential protocol in which they are participating). So one may still argue that by such higher-order reflection (and in particular, by becoming aware of the dangers inherent in the sequential deliberation protocol), "truly rational" agents might be able to avoid the formation of cascades. And indeed, in *some* cases the cascade can be prevented simply by making agents aware of the very possibility of a cascade.

In this paper, we prove that this is *not* always the case: there are situations in which no amount of higher-order reflection and meta-rationality can stop a cascade. To show this, we present in section 2 a formalization of the above-mentioned Urn example using Probabilistic Dynamic Epistemic Logic [9, 24]. This setting assumes perfectly rational agents able to reflect upon and reason about higher levels of group knowledge: indeed, epistemic logic takes into account all the levels of mutual belief/knowledge (beliefs about others' beliefs etc) about the current *state* of the world; while dynamic epistemic logic adds also all the levels of mutual belief/knowledge about the on-going *informational events* ("the protocol"). The fact that the cascade can still form proves our point: cascades cannot in general be prevented even by the use of the most perfect, idealized kind of individual rationality, one endowed with unlimited higher-level reflective powers. Informational cascades of this "super-rational" kind can be regarded as "epistemic Tragedies of the Commons": *paradoxes of (individual-versus-social) rationality*. In such contexts, a cascade can only be stopped by an external or "irrational" force, acting as *deus ex machina*: an "info-bomb", in the sense of [21]. This can be either an intervention from an outside agent (with different interests or different information that the agents engaged in the cascade), or a sudden burst of "irrationality" from one of the participating agents.

In section 3 we address another objection raised by some authors against the Bayesian analysis of cascades. They argue that real agents, although engaging in cascades, do it for non-Bayesian reasons: instead of probabilistic conditioning, they seem to use "rough-and-ready" qualitative heuristic methods, e.g. by simply *counting* the pieces of evidence in favor of one hypothesis against its alternatives. To model cascades produced by this kind of qualitative reasoning (by agents who still maintain their higher-level awareness of the other agents' minds), we introduce a new framework – a *multi-agent logic for counting evidence*. We use this setting to show that, even if we endow our formal agents only with a heuristic way of reasoning which is much less sophisticated, more intuitive and maybe more realistic than full-fledged probabilistic logic, they may still "rationally" engage in informational cascades. Hence, the above conclusion can now be now extended to a wider range of agents: *as long as the agents can count the evidence, then no matter how high or how low are their reasoning abilities* (even if they are capable of full higher-level reflection about others' minds, or dually even if they can't go beyond simple evidence counting), *their individual rationality may still lead to group "irrationality"*.

1   An Informational Cascade and its Bayesian Analysis

We will focus on a simple example that was created for studies of informational cascades in a laboratory [1, 2]. Consider two urns, respectively named U_W and U_B, where urn U_W contains two white balls and one black ball, and urn U_B contains one white ball and two black balls. One urn is randomly picked (say, using a fair coin) and placed in a room. This setup is common knowledge to a group of agents, which we will denote $a_1, a_2, ..., a_n$ but they do not know which of the two urns is in the room. The agents enter the room one at a time; first a_1, then a_2, and so on. Each agent draws one ball from the urn, looks at it, puts it back, and leaves the room. Hence, only the person in the room knows which ball she drew. After leaving the room she makes a guess as to whether it is urn U_W or U_B that is placed in the room and writes her guess on a blackboard for all the other agents to see. Therefore, each individual a_i knows the guesses of the previous people in the sequence $a_1, a_2, ..., a_n$ before entering the room herself. It is common knowledge that they will be individually rewarded if and only if their own guess is correct.

In this section we give the standard Bayesian analysis of this example, following the presentation in [15]. Let us assume that in fact urn U_B has been placed in the room. When a_1 enters and draws a ball, there is a unique simple decision rule she should apply: if she draws a white ball it is rational to make a guess for U_W, whereas if she draws a black one she should guess U_B. We validate this by calculating the probabilities. Let w_1 denote the event that a_1 draws a white ball and b_1 denote the event that she draws a black one. The proposition that it is urn U_W which is in the room will be denoted similarly by U_W and likewise for U_B. Given that it is initially equally likely that each urn is placed in the room the probability of U_W is $\frac{1}{2}$ ($P(U_W) = \frac{1}{2}$), and similarly for U_B. Observe that $P(w_1) = P(b_1) = \frac{1}{2}$. Assume now that a_1 draws a *white* ball. Then, via Bayes' rule, the posterior probability of U_W is

$$P(U_W|w_1) = \frac{P(U_W) \cdot P(w_1|U_W)}{P(w_1)} = \frac{\frac{1}{2} \cdot \frac{2}{3}}{\frac{1}{2}} = \frac{2}{3}.$$

Hence, it is indeed rational for a_1 to guess U_W if she draws a white ball (and to guess U_B if she draws a black ball). Moreover, when leaving the room and making a guess for U_W (resp. U_B), all the other individuals can infer that she drew a white (resp. black) ball.

When a_2 enters the room after a_1, she knows the color which a_1 drew and it is obvious how she should guess if she draws a ball of the same color. If a_1 drew a white ball and a_2 draws a white ball, then a_2 should guess U_W. Formally, the probability of U_W given that both a_1 and a_2 draw white balls is

$$P(U_W|w_1, w_2) = \frac{P(U_W) \cdot P(w_1, w_2|U_W)}{P(w_1, w_2)} = \frac{\frac{1}{2} \cdot \frac{2}{3} \cdot \frac{2}{3}}{\frac{5}{18}} = \frac{4}{5}.$$

A similar reasoning applies if both drew black balls. If a_2 draws an opposite color ball of a_1, then the probabilities for U_W and U_B become equal. For simplicity we will assume that any individual faced with equal probability for U_W and U_B will guess for the urn that contains more balls of the color she saw herself: if a_1 drew a white ball and a_2 draws a black ball, a_2 will guess U_B.[2] Hence, independent of which ball a_1 draws, a_2 will always guess for the urn matching the color of her privately drawn ball. We assume that this tie-breaking rule is common knowledge among the agents too. In this way, every individual following a_2 can also infer the color of a_2's ball.

When a_3 enters, a cascade can arise. If a_1 and a_2 drew opposite color balls, a_3 is rational to guess for the urn that matches the color of the ball she draws. Nevertheless, if a_1 and a_2 drew the same color of balls (given the reasoning previously described, a_3 will know this), say both white, then no matter what color of ball a_3 draws the posterior probability of U_W will be higher than the probability of U_B (and if a_1 and a_2 both drew black balls the other way around). To check this let us calculate the probability of U_W given that a_1 and a_2 drew white balls and a_3 draws a black one:

$$P(U_W|w_1, w_2, b_3) = \frac{P(U_W) \cdot P(w_1, w_2, b_3|U_W)}{P(w_1, w_2, b_3)} = \frac{\frac{1}{2} \cdot \frac{2}{3} \cdot \frac{2}{3} \cdot \frac{1}{3}}{\frac{1}{9}} = \frac{2}{3}.$$

It is obvious that $P(U_W|w_3, w_2, w_1)$ will be even larger, thus whatever ball a_3 draws it will be rational for her to guess for U_W. Hence, if a_1 and a_2 draw the

---

[2]This tie-breaking rule is a simplifying assumption but it does not affect the likelihood of cascades arising. Moreover, there seems to be some empirical evidence that this is what most people do and it is also a natural tie-breaking rule if the individuals assign a small chance to the fact that other people might make errors [2].

same color of balls a cascade will start from a_3 on![3] The individuals following a_3 should therefore take a_3's guess as conveying no new information. Furthermore, everyone after a_3 will have the same information as a_3 (the information about what a_1 and a_2 drew) and their reasoning will therefore be identical to the one of a_3 and the cascade will continue.

If U_B is, as we assumed, the urn actually placed in the room and both a_1 and a_2 draw white balls (which happens with probability $\frac{1}{9}$) then a cascade leading to everyone making the *wrong* guess starts. Note, however, that if both a_1 and a_2 draw black balls (which happens with probability $\frac{4}{9}$), then a cascade still starts, but this time it will lead to everyone making the *right* guess. Thus, when a cascade happens it is four times more likely in this example that it leads to right guesses than to the wrong guesses. This already supports the claim that rational agents can be well aware of the fact that they are in a cascade without it forcing them to change their decisions. The general conclusion of this example is that even though informational cascades can look irrational from a social perspective, they are not irrational from the perspective of any individual participating in them.

The above semi-formal analysis summarizes the standard Bayesian treatment of this example, as given e.g. in [15]. However, as we mentioned in the introduction, several objections can be raised against the way this conclusion has been reached above. First of all, the example has only been partially formalized, in the sense that the public announcements of the individuals' guesses are not explicitly present in it, neither is the reasoning that lets the individuals ignore the guesses of the previous people caught in a cascade. Moreover, the Bayesian analysis given above does not formally capture the agents' full higher-order reasoning (i.e. their reasoning about the others' beliefs and about the others' higher-order reasoning about their beliefs etc). So one cannot use the above argument to completely rule out the possibility that some kind of higher-order reflection may help prevent (or break) informational cascade: it might be the case that, after realizing that they are participating in a cascade, agents may use this information to try to stop the cascade.

For all these reasons, we think it is useful to give a more complete analysis, using a model that captures both the public announcements and the full higher-order reasoning of the agents. This is precisely what we will do in the next section, in the framework of Probabilistic Dynamic Epistemic Logic [9, 24].

---

[3]Note that the cascade will start even if we change the tie-breaking rule of a_2 such that she randomizes her guess whenever she draws a ball contradicting the guess of a_1. In this case, if a_1 and a_2 guess for the same urn, a_3 will not know the color of a_2's ball, but she will still consider it more likely that a_2's ball matches the ball of a_1 and hence consider it more likely that the urn which they have picked is in the one in the room.

2   A Probabilistic Logical Model

In this section we will work within the framework of Probabilistic Dynamic Epistemic Logic [9, 24]. Our presentation will be based on a simplified version of the setting from [9], in which we assume that agents are *introspective as far as their own subjective probabilities are concerned* (so that an agent's subjective probability assignment does not depend on the actual state of the world but only on that world's partition cell in the agent's information partition). We also use slightly different graphic representations, which make explicit the *odds* between any two possible states (considered pairwise) according to each agent. This allows us to present directly a *comparative* treatment of the rational guess of each agent and will make obvious the similarity with the framework for "counting evidences" that we will introduce in the next section. We start with some definitions.

DEFINITION 1 (Probabilistic Epistemic State Models). A probabilistic multi-agent epistemic state model $\mathcal{M}$ is a structure $(S, \mathcal{A}, (\sim_a)_{a \in \mathcal{A}}, (P_a)_{a \in \mathcal{A}}, \Psi, \| \bullet \|)$ such that:

- S is a set of states (or "worlds");

- $\mathcal{A}$ is a set of agents;

- for each agent a, $\sim_a \subseteq S \times S$ is an equivalence relation interpreted as agent a's epistemic indistinguishability. This captures the agent's hard information about the actual state of the world;

- for each agent a, $P_a : S \to [0, 1]$ is a map that induces a probability measure on each $\sim_a$-equivalence class (i.e., we have $\sum \{P_a(s') : s' \sim_a s\} = 1$ for each $a \in \mathcal{A}$ and each $s \in S$). This captures the agent's subjective probabilistic information about the state of the world;

- Ψ is a given set of "atomic propositions", denoted by $p, q, \ldots$. Such atoms p are meant to represent *ontic "facts'* that might hold in a world.

- $\| \bullet \| : \Psi \to \mathcal{P}(S)$ is a "valuation" map, assigning to each atomic proposition $p \in \Psi$ some set of states $\|p\| \subseteq S$. Intuitively, the valuation tells us which facts hold in which worlds.

DEFINITION 2 (Relative Likelihood). The *relative likelihood* (or "odds") of a state s against a state t according to agent a, $[s : t]_a$, is defined as

$$[s : t]_a := \frac{P_a(s)}{P_a(t)}.$$

We graphically represent probabilistic epistemic state models in the following way: each state is drawn as an oval, having inside it the name of the state and the facts p that are "true" at the state (i.e. the atomic sentences p having this state in their valuation $\|p\|$); and for each agent $a \in \mathcal{A}$, we draw a-labeled arrows going from each state s towards all the states in the same a-information cell to which a attributes equal or higher odds (than to state s). Therefore, the qualitative arrows represent both the hard information (indistinguishability relation) and the probability ordering relative to an agent, pointing towards the indistinguishable states that she considers to be at least as probable. To make explicit the odds assigned by agents to states, we label these arrows with the quantitative information (followed by the agents' names in the brackets). For instance, the fact that $[s : t]_a = \frac{\alpha}{\beta}$ is encoded by an a-arrow from state s to state t labeled with the quotient $\alpha : \beta(a)$. For simplicity, we don't represent the loops relating each state to itself, since they don't convey any information that is specific to a particular model: in every model, every state is a-indistinguishable from itself and has equal odds $1 : 1$ to itself.

To illustrate probabilistic epistemic state model with odds, consider the initial situation of our urn example presented in Section 1 as pictured in Figure 1. In this initial model $\mathcal{M}_0$, it is equally probable that U_W or U_B is true (and therefore the prior odds are equal) and all agents know this. The actual state (denoted by the thicker oval) s_B satisfies the proposition U_B, while the state s_W satisfies the proposition U_W. The bidirectional arrow labeled with "1:1 (all a)" represents the fact that all agents consider both states equally probable.

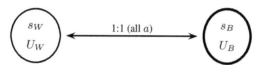

Figure 1. The initial probabilistic state model $\mathcal{M}_0$ of the urn example

DEFINITION 3 (Epistemic-probabilistic language). As in [9], the "static" language we adopt to describe these models is the epistemic-probabilistic language due to Halpern and Fagin [16]. The syntax is given by the following Backus-Naur form:

$$\varphi := p \mid \varphi \mid \varphi \wedge \varphi \mid K_a\varphi \mid \alpha_1 \cdot P_a(\varphi) + \ldots + \alpha_n \cdot P_a(\varphi) \geq \beta$$

where $p \in \Psi$ are atomic propositions, $a \in \mathcal{A}$ are agents and $\alpha_1, \ldots, \alpha_n, \beta$ stand for arbitrary rational numbers. Let us denote this language by $\mathcal{L}$.

The *semantics* is given by associating to each formula φ and each model $\mathcal{M} = (S, \mathcal{A}, (\sim_a)_{a \in \mathcal{A}}, (P_a)_{a \in \mathcal{A}})$, some *interpretation* $\|\varphi\|_\mathcal{M} \subseteq S$, given recursively by the obvious inductive clauses[4]. If $s \in \|\varphi\|_\mathcal{M} \subseteq S$, then we say that φ is *true* at

---

[4]It is worth noting that, when checking whether a given state s belongs to $\|\varphi\|$, every expression of the form $P_a(\psi)$ is interpreted conditionally on agent a's knowledge at s, i.e. as $P_a(\|\psi\| \cap \{s' \in S :$

state s (in model $\mathcal{M}$).

In this language, one can introduce strict inequalities, as well as equalities, as abbreviations, e.g.:

$$P_a(\varphi) > P_a(\psi) \quad := \quad \neg(P_a(\psi) - P_a(\varphi) \geq 0),$$

$$P_a(\varphi) = P_a(\psi) \quad := \quad (P_a(\varphi) - P_a(\psi) \geq 0) \wedge (P_a(\psi) - P_a(\varphi) \geq 0)$$

One can also define an expression saying that *an agent a assigns higher odds to φ than to ψ* (given her current information cell):

$$[\varphi : \psi]_a > 1 \quad := \quad P_a(\varphi) > P_a(\psi)$$

To model the incoming of new information, we use *probabilistic event models*, as introduced by van Benthem *et alia* [9]: these are a probabilistic refinement of the notion of *event models*, which is the defining feature of Dynamic Epistemic Logic in its most widespread incarnation [4]. Here we use a simplified setting, which assumes introspection of subjective probabilities.

DEFINITION 4 (Probabilistic Event Models). A probabilistic event model $\mathcal{E}$ is a sextuple $(E, \mathcal{A}, (\sim_a)_{a \in \mathcal{A}}, (P_a)_{a \in \mathcal{A}}, \Phi, pre)$ such that:

- E is a set of possible events,

- $\mathcal{A}$ is a set of agents;

- $\sim_a \subseteq E \times E$ is an equivalence relation interpreted as agent a's epistemic indistinguishability between possible events, capturing a's hard information about the event that is currently happening;

- P_a gives a probability assignment for each agent a and each $\sim_a$-information cell. This captures some new, independent subjective probabilistic information gained by the agent during the event: when observing the current event (without using any prior information), agent a assigns probability $P_a(e)$ to the possibility that in fact e is the actual event that is currently occurring.

- Φ is a finite set of mutually inconsistent propositions (in the above probabilistic-epistemic language $\mathcal{L}$), called *preconditions*;

- pre assigns a probability distribution $pre(\bullet|\phi)$ over E for every proposition $\phi \in \Phi$. This is an "occurrence probability": $pre(e|\phi)$ expresses the prior probability that event $e \in E$ might occur in a(ny) state satisfying precondition ϕ;

$s' \sim_a s\}$. See [16], [9] for other details.

As before, the probability P_a can alternatively be expressed as probabilistic odds $[e : e']_a$ for any two events e, e' and any agent a. Our event models are drawn in the same fashion as our state models above: for each agent a, a-arrows go from a possible event e towards all the events (of a's information cell) to which a attributes equal or higher odds. As an example of an event model, consider the first observation of a ball in our urn case, as represented in the model $\mathcal{E}_1$ from Figure 2. Here a_1 draws a white ball from the urn and looks at it. According to all the other agents, two events can happen: either a_1 observes a white ball (the actual event w_1) or she observes a black one (event b_1). Moreover, only agent a_1 knows which event is the actual one. The expressions $pre(U_W) = \frac{2}{3}$ and $pre(U_B) = \frac{1}{3}$ depicted at event w_1 represents that the prior probabilities $pre(w_1 \mid U_W)$ that event w_1 occurs when U_W is satisfied is $\frac{2}{3}$ while the probability $pre(w_1 \mid U_B)$ that event w_1 happens when U_B is satisfied is $\frac{1}{3}$ (and vice versa for event b_1). The bidirectional arrow for all agents except a_1 represents the fact that agent a_1 can distinguish between the two possible events (since she *knows* that she sees a white ball), while the others cannot distinguish them and have (for now) no reason to consider one event more likely than the other, i.e., their odds are $1 : 1$.

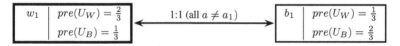

Figure 2. The probabilistic event model $\mathcal{E}_1$ of agent a_1 drawing a white ball

To model the evolution of the odds after new information is received, we now combine probabilistic epistemic state models with probabilistic event models using a notion of product update.

DEFINITION 5 (Probabilistic Product Update). Given a probabilistic epistemic state model $\mathcal{M} = (S, \mathcal{A}, (\sim_a)_{a \in \mathcal{A}}, (P_a)_{a \in \mathcal{A}}, \Psi, \| \bullet \|)$ and a probabilistic event model $\mathcal{E} = (E, \mathcal{A}, (\sim_a)_{a \in \mathcal{A}}, (P_a)_{a \in \mathcal{A}}, \Phi, pre)$, the updated state model $\mathcal{M} \otimes \mathcal{E} = (S', \mathcal{A}, (\sim'_a)_{a \in \mathcal{A}}, (P'_a)_{a \in \mathcal{A}}, \Psi', \| \bullet \|')$, is given by:

$$S' = \{(s, e) \in S \times E \mid pre(e \mid s) \neq 0\},$$

$$\Psi' = \Psi,$$

$$\|p\|' = \{(s, e) \in S' : s \in \|p\|\},$$

$$(s, e) \sim'_a (t, f) \text{ iff } s \sim_a t \text{ and } e \sim_a f,$$

$$P'_a(s, e) = \frac{P_a(s) \cdot P_a(e) \cdot pre(e \mid s)}{\sum \{P_a(t) \cdot P_a(f) \cdot pre(f \mid t) : s \sim_a t, e \sim_a f\}},$$

where we used the notation

$$pre(e \mid s) := \sum \{pre(e \mid \phi) : \phi \in \Phi \text{ such that } s \in \|\phi\|_{\mathcal{M}}\}$$

(so that $pre(e \mid s)$ is either $= pre(e|\phi_s)$ where ϕ_s is the unique precondition in Φ such that ϕ_s is true at s, or otherwise $pre(e \mid s) = 0$ if no such precondition ϕ_s exists).

This definition can be justified on Bayesian grounds: the definition of the new indistinguishability relation simply says that the agent puts together her old and new hard information[5]; while the definition of the new subjective probabilities is obtained by multiplying the old probability previously assigned to event e (obtained by applying the conditioning rule $P_a(e) = P_a(s) \cdot P_a(e \mid s) = P_a(s) \cdot pre(e \mid \phi_s)$) with the new probability independently assigned (without using any prior information) to event e during the event's occurrence, and then renormalizing to incorporate the new hard information. The reason for using multiplication is that the two probabilities of e are supposed to represent two independent pieces of probabilistic information.[6]

Again, it is possible, and even easier, to express posterior probabilities in terms of posterior relative likelihoods:

$$[(s,e) : (t,f)]_a = [s : t]_a \cdot [e : f]_a \cdot \frac{pre(e \mid s)}{pre(f \mid t)}.$$

The result of the product update of the initial state model $\mathcal{M}_0$ from Fig. 1 with the event model $\mathcal{E}_1$ of Fig. 2 is given by the new model $\mathcal{M}_0 \otimes \mathcal{E}_1$ of Fig. 3. The upper right state is the actual situation, in which U_B is true, but in which the first ball which has been observed was a white one. Agent a_1 *knows* that she observed a white ball (w_1), but she does not know which urn is the actual one, so her actual information cell consists of the upper two states, in which she considers U_W to be twice as likely as U_B. The other agents still cannot exclude any possibility.

This is going to change once the first agent announces her guess. To model this announcement we will use the standard *public announcements* of [28], where a (truthful) public announcement $!\varphi$ of a proposition φ is an event which has the effect of deleting all worlds of the initial state model that do not satisfy φ. Note that, public announcements $!\varphi$ can be defined as a special kind of probabilistic event models: take $E = \{e_{!\varphi}\}$, $\sim_a = \{(e_{!\varphi}, e_{!\varphi})\}$, $\Phi = \{\varphi\}$, $pre(e_{!\varphi} \mid \varphi) = 1$, $P_a(e_{!\varphi}) = 1$.

---

[5]This is the essence of the "Product Update" introduced by Baltag et alia [4], which forms the basis of most widespread versions of Dynamic Epistemic Logic.

[6]In fact, this feature is irrelevant for our analysis of cascades: no new non-trivial probabilistic information is gained by the agents during the events forming our cascade example. This is reflected in the fact that, in our analysis of cascades, we will use only event models in which the odds are 1 : 1 between any two indistinguishable events.

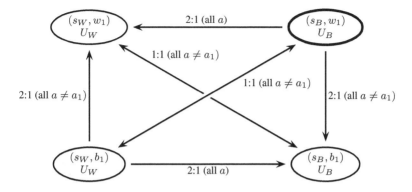

Figure 3. The updated probabilistic state model $\mathcal{M}_0 \otimes \mathcal{E}_1$ after a_1 draws a white ball

Now, after her private observation, agent a_1 publicly announces that she considers U_W to be more likely than U_B. This is a public announcement $!([U_W : U_B]_{a_1} > 1)$ of the sentence $[U_W : U_B]_{a_1} > 1$ (as defined above as an abbreviation in our language), expressing the fact that agent a_1 assigns higher odds to urn U_W than to urn U_B. Since all agents know that the only reason a_1 could consider U_W more likely than U_B is that she drew a white ball (her announcement can be truthful only in the situations in which she drew a white ball), the result is that all agents come to know this fact. This is captured by our modelling, where her announcement simply erases the states (s_W, b_1) and (s_B, b_1) and results in the new model $\mathcal{M}_1$ of Fig. 4.

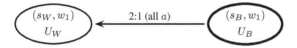

Figure 4. The updated probabilistic state model $\mathcal{M}_1$ after a_1's announcement

By repeating the above very reasoning, we know that, after another observation of a white ball by agent a_2 (the event model is as above in Fig. 2 but relative to agent a_2 instead of agent a_1) and a similar public announcement of $[U_W : U_B]_{a_2} > 1$, the resulting state model $\mathcal{M}_2$, depicted in Fig. 5, will be such that all agents now consider U_W *four* times more likely than U_B.

Let us now assume that agent a_3 enters the room and privately observes a black ball. The event model $\mathcal{E}_3$ of this action is in Figure 6, and is again similar to the earlier event model (Fig. 2) but relative to agent a_3 and this time, since a *black* ball

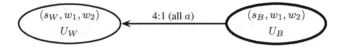

Figure 5. The updated probabilistic state model $\mathcal{M}_2$ after a_2's announcement

is observed, the actual event is b_3.

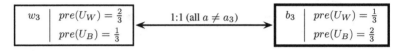

Figure 6. The probabilistic event model $\mathcal{E}_3$ of a_3 drawing a black ball

The result of a_3's observation is then given by the updated state model $\mathcal{M}_2 \otimes \mathcal{E}_3$ shown in Figure 7.

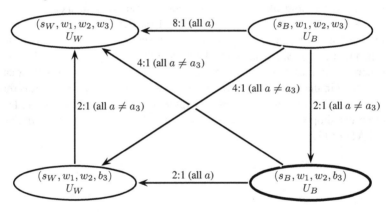

Figure 7. The probabilistic state model $\mathcal{M}_2 \otimes \mathcal{E}_3$ after a_3 draws a black ball

Since only agent a_3 knows what she has observed, her actual information cell only contains the states in which the event b_3 has happened, while all other agents cannot distinguish between the four possible situations. Moreover, agent a_3 still considers U_W more probable than U_B, *irrespective* of the result of her private observation (w_3 or b_3). So the fact that $[U_W : U_B]_{a_3} > 1$ is now *common knowledge* (since it is true at all states of the entire model). This means that announcing this fact, via a new public announcement of $[U_W : U_B]_{a_3} > 1$ will *not* delete any state: the model $\mathcal{M}_3$ after the announcement is simply *the same as before* (Fig. 7).

So the *third agent's public announcement bears no information whatsoever*: an informational cascade has been formed, even though all agents have reasoned correctly about probabilities. From now on, the situation will keep repeating itself: although the state model will keep growing, *all agents will always consider U_W more probable than U_B in all states* (irrespective of their own observations). This is shown formally by the following result.

PROPOSITION 6. *Starting in the model in Fig. 1 and following the above protocol, we have that: after $n-1$ private observations and public announcements $e_1, !([U_W : U_B]_{a_1} > 1) \ldots, e_{n-1}, !([U_W : U_B]_{a_{n-1}} > 1)$ by agents $a_1, \ldots, a_{n-1}$, with $n \geq 3$, $e_1 = w_1$ and $e_2 = w_2$, the new state model $\mathcal{M}_{n-1}$ will satisfy*

$$[U_W : U_B]_a > 1, \text{ for all } a \in \mathcal{A}.$$

Proof. To show this, we prove a stronger
   CLAIM: after $n-1$ private observations and announcements as above, the new state model $\mathcal{M}_{n-1}$ will satisfy

$$[U_W : U_B]_{a_i} \geq 2, \text{ for all } i < n, \text{ and}$$

$$[U_W : U_B]_{a_i} \geq 4, \text{ for all } i \geq n.$$

From this claim, the desired conclusion follows immediately.
   PROOF OF CLAIM:
   We give only a sketch of the proof, using an argument based on partial descriptions of our models. The base case $n = 3$ was already proved above. Assume the inductive hypothesis for $n - 1$. By lumping together all the U_W-states in $\mathcal{M}_{n-1}$, and similarly all the U_B-states, we can represent this hypothesis via the following partial representation of $\mathcal{M}_{n-1}$:

Note that this is just a "bird's view" representation: the actual model $\mathcal{M}_{n-1}$ has 2^{n-2} states. To see what happens after one more observation e_n by agent n, take the update produce of this representation with the event model $\mathcal{E}_n$, given by:

The resulting product is:

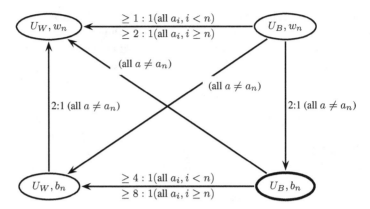

where for easier reading we skipped the numbers representing the probabilistic information associated to the diagonal arrows (numbers which are not relevant for the proof).

By lumping again together all indistinguishable U_W-states in $\mathcal{M}_{n-1}$, and similarly all the U_B-states, and reasoning by cases for agent a_n (depending on her actual observation), we obtain:

Again, this is just a bird's view: the actual model has 2^n states. But the above partial representation is enough to show that, in this model, we have $[U_W : U_B]_{a_i} \geq 2$ for all $i < n + 1$, and $[U_W : U_B]_{a_i} \geq 4$ for all $i \geq n + 1$. Since in particular $[U_W : U_B]_{a_n} > 1$ holds in all the states, this fact is common knowledge: so, after publicly announcing it, the model stays the same! Hence, we proved the induction step for n. ∎

So, in the end, all the guesses will be wrong: the whole group will assign a higher probability to the wrong urn (U_W). Thus, we have proved that *individual Bayesian rationality with perfect higher-level reflective powers can still lead to "group irrationality"*. This shows that in some situations there simply is no higher-order information available to any of the agents to prevent them from entering the cascade; not even the information that they *are* in a cascade can help in this case. (Indeed, in our model, after the two guesses for U_W of a_1 and a_2, it is already common knowledge that a cascade has been formed!)

3   A Logical Model Based on Counting Evidence

A possible objection to the model presented in the previous section could be that it relies on the key assumption that the involved agents are perfect *Bayesian* reasoners. But many authors argue that *rationality* cannot be identified with *Bayesian rationality*. There are other ways of reasoning that can still be deemed rational without involving doing cumbersome Bayesian calculations. In practice, many people seem to use much simpler *"counting" heuristics*, e.g. guessing U_W when one has more pieces of evidence in favor of U_W than in favor of U_B (i.e. one knows that more white balls were drawn than black balls).

Hence, there are good reasons to look for a model of informational cascades based on counting instead of Bayesian updates. In this section we present a formalized setting of the urn example using a notion of rationality based on such a simple counting heuristic. The logical framework for this purpose is inspired by the probabilistic framework of the previous section. However, it is substantially simpler. Instead of calculating the probability of a given possible state, we will simply count the evidence in favor of this state. More precisely, we label each state with a number representing the strength of all evidence in favor of that state being the actual one. This intuition is represented in the following formal definition:

DEFINITION 7 (Counting Epistemic Models). A counting multi-agent epistemic model $\mathcal{M}$ is a structure $(S, \mathcal{A}, (\sim_a)_{a \in \mathcal{A}}, f, \Psi, \| \bullet \|)$ such that:

- S is a set of states,

- $\mathcal{A}$ is a set of agents,

- $\sim_a \subseteq S \times S$ is an equivalence relation interpreted as agent a's epistemic indistinguishability,

- $f : S \to \mathbb{N}$ is an "evidence-counting" function, assigning a natural number to each state in S,

- Ψ is a given set of atomic sentences,

- $\| \bullet \| : \Psi \to \mathcal{P}(S)$ is a valuation map.

We can now represent the initial situation of the urn example by the model of Figure 8. The two possible states s_W and s_B correspond to U_W (resp. U_B) being placed in the room. The notation $U_W ; 0$ at the state s_W represents that $f(s_W) = 0$ and that the atomic proposition U_W is true at s_W (and all other atomic propositions are false). The line between s_W and s_B labeled by "all a" means that the two states are indistinguishable for all agents a. Finally, the thicker line around s_B represents that s_B is the actual state.

We now turn to the issue of how to update counting epistemic models. However, first note that, at this stage there is not much that distinguish counting epistemic models from probabilistic ones. In the case the models are finite, one can simple sum the values of $f(w)$ for all states w in a given information cell and

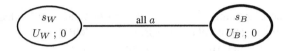

Figure 8. The initial counting model of the urn example

rescale $f(w)$ by this factor thereby obtaining a probabilistic model from a counting model. Additionally, assuming that all probabilities are rational numbers one can easily move the other way as well. In spite of this, when we move to dynamic issues, the counting framework becomes much simpler as we do not need to use multiplication together with Bayes' rule and renormalization, we can simply use addition. Here are the formal details:

DEFINITION 8 (Counting Event Models). A counting event model $\mathcal{E}$ is a quintuple $(E, \mathcal{A}, (\sim_a)_{a \in \mathcal{A}}, \Phi, pre)$ such that:

- E is a set of possible events,

- $\mathcal{A}$ is a set of agents,

- $\sim_a \subseteq E \times E$ is an equivalence relation interpreted as agent a's epistemic indistinguishability,

- Φ is a finite set of pairwise inconsistent propositions,

- $pre : E \to (\Phi \to (\mathbb{N} \cup \{\bot\}))$ is a function from E to functions from Φ to the natural numbers (extended with $\bot$)[7]. It assigns to each event $e \in E$ a function $pre(e)$, which to each proposition $\phi \in \Phi$ assigns the strength of evidence that the event e provides for ϕ.

As an example of a counting event model, the event model of the first agent drawing a white ball is shown in Figure 9. In this event model there are two events w_1 and b_1, where the actual event is w_1 (marked by the thick box). A notation like $pre(U_W) = 1$ at w_1 simply means that $pre(w_1)(U_W) = 1$.[8] Finally, the line between w_1 and b_1 labeled "all $a \neq a_1$" represents that the events w_1 and b_1 are indistinguishable for all agents a except a_1.

A counting epistemic model is updated with a counting event model in the following way:

---

[7]Here, $\bot$ essentially means "undefined": so it is just an auxiliary symbol used to describe the case when pre is a *partial* function.

[8]To fit the definition of counting event models properly, U_W and U_B must be pairwise inconsistent, however, this claim fits perfectly with the example where only one of the urns is placed in the room and we could simple replace U_W by $\neg U_B$.

w_1;	$\mathsf{pre}(U_W) = 1$		all $a \neq a_1$		b_1;	$\mathsf{pre}(U_W) = 0$
	$\mathsf{pre}(U_B) = 0$					$\mathsf{pre}(U_B) = 1$

Figure 9. The counting event model of a_1 drawing a white ball

DEFINITION 9 (Counting Product Update). Given a counting epistemic model $\mathcal{M} = (S, \mathcal{A}, (\sim_a)_{a \in \mathcal{A}}, f, \Psi, \| \bullet \|)$ and a counting event model $\mathcal{E} = (E, \mathcal{A}, (\sim_a)_{a \in \mathcal{A}}, pre)$, we define the product update $\mathcal{M} \otimes \mathcal{E} = (S', \mathcal{A}, (\sim'_a)_{a \in \mathcal{A}}, f', \Psi', \| \bullet \|)$ by

$$S' = \{(s, e) \in S \times E \mid pre(s, e) \neq \bot\},$$
$$\Psi' = \Psi,$$
$$\|p\|' = \{(s, e) \in S' : s \in \|p\|\},$$
$$(s, e) \sim_a (t, f) \text{ iff } s \sim_a t \text{ and } e \sim_a f,$$
$$f'((s, e)) = f(s) + pre(s, e), \quad \text{for } (s, e) \in S',$$

where we used the notation $pre(s, e)$ to denote $pre(e)(\phi_s)$ for the unique $\phi_s \in \Phi$ such that $s \in \|\phi_s\|_{\mathcal{M}}$, if such a precondition $\phi_s \in \Phi$ exist, and otherwise we put $pre(s, e) = \bot$.

With this definition we can now calculate the product update of the models of the initial situation (Fig. 8) and the first agent drawing a white ball (Fig. 9). The resulting model is shown in Figure 10.

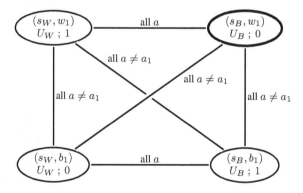

Figure 10. The updated counting model after a_1 draws a white ball

We need to say how we will represent the action that agent a_1 guesses for urn U_W. As in the probabilistic modeling we will interpret this as a public announcement. A public announcement of ϕ in the classical sense of eliminating all non-ϕ

states, is a special case of a counting event model with just one event e, $\Phi = \{\phi\}$, $\sim_a = \{(e, e)\}$ for all $a \in \mathcal{A}$, and $pre(e)(\phi) = 0$. Setting $pre(e)(\phi) = 0$ reflects the choice that we take public announcements not to provide any increase in the strength of evidence for any possible state, but only revealing hard information about which states are possible. In the urn example it is the drawing of a ball from the urn that increases the strength of evidence, whereas the guess simply convey information about the announcer's hard information about the available evidence for either U_W or U_B. Similar to the previous section, we will interpret the announcements as revealing whether their strength of evidence for U_W is smaller or larger than their strength of evidence for U_B.

We therefore require a formal language that contains formulas of the form $\phi <_a \psi$, for all formulas ϕ and ψ. The semantics of the new formula is given by:

$$\|\phi <_a \psi\|_{\mathcal{M}} = \{s \in S \mid f(a, s, \|\phi\|_{\mathcal{M}}) < f(a, s, \|\psi\|_{\mathcal{M}})\},$$

where for any given counting model $\mathcal{M} = (S, (\sim_a)_{a \in \mathcal{A}}, f, \Psi, \| \bullet \|)$ and any set of states $T \subseteq S$ we used the notation

$$f(a, s, T) := \sum \{f(t) : t \in T \text{ such that } t \sim_a s\}.$$

Now, the event that agent a_1 announces that she guesses in favor of U_W will be interpreted as a public announcement of $U_B <_{a_1} U_W$. This proposition is only true at the states (s_W, w_1) and (s_B, w_1) of the above model and thus the states (s_W, b_1) and (s_B, b_1) are removed in the resulting model shown in Figure 11.

Figure 11. The counting model after a_1 publicly announces that $U_B <_{a_1} U_W$

Moreover, the event that a_2 draws a white ball can be represented by an event model identical to the one for agent a_1 drawing a white ball (Fig. 9) except that the label on the line should be changed to "all $a \neq a_2$". The updated model after the event that a_2 draws a white ball will look as shown in Figure 12. Note that in this updated model, $U_B <_{a_2} U_W$ is only true at (s_W, w_1, w_2) and (s_B, w_1, w_2), thus when a_2 announces her guess for U_W (interpreted as a public announcement of $U_B <_{a_2} U_W$) the resulting model will be the one of Figure 13.

Assuming that agent a_3 draws a black ball this can be represented by an event model almost identical to the one for agent a_1 drawing a white ball (Fig. 9). The only differences are that the label on the line should be changed to "all $a \neq a_3$)" and the actual event should be b_3. Updating the model of Figure 13 with this event will result in the model of Figure 14.

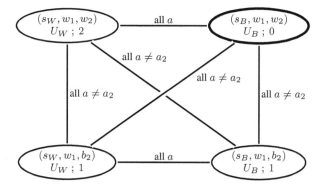

Figure 12. The updated counting model after a_2 draws a white ball

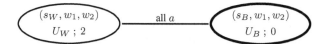

Figure 13. The counting model after a_2 publicly announces that $U_B <_{a_2} U_W$

Note that in Figure 14 the proposition $U_B <_{a_3} U_W$ is true in the entire model. Hence, agent a_3 has more evidence for U_W than U_B and thus, no states will be removed from the model when she announces her guess for U_W (a public announcement of $U_B <_{a_3} U_W$). If a_3 had drawn a white ball instead, the only thing that would have be different in the model of Figure 14 is that the actual state would be (s_B, w_1, w_2, w_3). Therefore, this would not change the fact that a_3 guesses for U_W and this announcement will remove no states from the model either. In this way, none of the following agents gain any information from learning that a_3 guessed for U_W. Subsequently whenever an agent draws a ball, she will have more evidence for U_W than for U_B. Thus, the agents will keep guessing for U_W. However, these guesses will not delete any more states. Hence, the models will keep growing exponentially reflecting the fact that no new information is revealed. In other words, *an informational cascade has started*. Formally, one can show the following result:

PROPOSITION 10. *Let $\mathcal{M}_n$ be the updated model after agent a_n draws either a white or a black ball. Then, if both a_1 and a_2 draw white balls (i.e. we are in the model of Fig. 12), then for all $n \geq 3$, $U_B <_{a_n} U_W$ will be true in all states of $\mathcal{M}_n$.*

In words: after the first two agents have drawn white balls all the following agents will all have more evidence for U_W than U_B (no matter which color ball

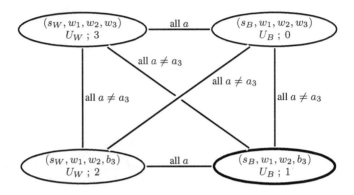

Figure 14. The updated counting model after a_3 draws a black ball

they draw) and will therefore guess for U_W, however, these guesses will be unin-formative to the subsequent agents as the public announcement of $U_B <_a U_W$ will delete no worlds.

Before we can prove this proposition we need some definitions. For all $n \geq 3$, let $\mathcal{E}_n$ be the event model that agent a_n draws either a white ball (w_n) or a black ball (b_n)[9], for instance $\mathcal{E}_1$ is shown in Figure 9. Furthermore, let $\mathcal{M}_n$ denote the model obtained after updating with the event $\mathcal{E}_n$, hence $\mathcal{M}_n = \mathcal{M}_{n-1} \otimes \mathcal{E}_n$. The model $\mathcal{M}_3$ is shown in Figure 14. We will denote the domain of $\mathcal{M}_n$ by $dom(\mathcal{M}_n)$. For a proposition ϕ, we will by $f^n(\phi)$ denote $\sum \{f(s) \mid s \in \|\phi\|_{\mathcal{M}_n}\}$. Note that $f^2(U_W) = 2$, $f^3(U_W) = 5$, $f^2(U_B) = 0$, and $f^3(U_B) = 1$.

Now Proposition 2 follows from the following lemma:

LEMMA 11. *For all $n \geq 3$ the following hold:*

(i) *Let $[w]_n := dom(\mathcal{M}_{n-1}) \times \{w_n\}$ and $[b]_n := dom(\mathcal{M}_{n-1}) \times \{b_n\}$. Then $[w]_n$ and $[b]_n$ are the only two information cells of agent a_n in $\mathcal{M}_n$, $dom(\mathcal{M}_n) = ([w]_n \cup [b]_n)$, and $|[w]_n| = |[b]_n| = 2^{n-2}$. Additionally, for all $k > n$, $\mathcal{M}_n$ contains only one information cell for agent a_k, namely the entire $dom(\mathcal{M}_n)$. Furthermore, U_W is true in 2^{n-3} states of $[w]_n$ and 2^{n-3} states of $[b]_n$, and similar, U_B is true in 2^{n-3} states of $[w]_n$ and 2^{n-3} states of $[b]_n$.*

(ii) *For $s \in [w]_n$, $f(a_n, s, U_W) = f^{n-1}(U_W) + 2^{n-3}$ and $f(a_n, s, U_B) = f^{n-1}(U_B)$. For $s \in [b]_n$, $f(a_n, s, U_W) = f^{n-1}(U_W)$ and $f(a_n, s, U_B) = f^{n-1}(U_B) + 2^{n-3}$.*

---

[9]Which color ball a_n draws does not matter as it only affect which state will be the actual state.

(iii) $f^n(U_B) + 2^{n-2} < f^n(U_W)$.

(iv) $U_B <_{a_n} U_W$ is true at all states of $\mathcal{M}_n$.

Proof. The proof goes by induction on n. For $n = 3$ the statements $(i) - (iv)$ are easily seen to be true by inspecting the model $\mathcal{M}_3$ as shown in Figure 14 of section 3. We prove the induction step separately for each of the statements $(i) - (iv)$.

(i): Assume that (i) is true for n. Then, for agent a_{n+1} the model $\mathcal{M}_n$ consists of a single information cell with 2^{n-1} states where U_W is true in half of them and U_B in half of them. Considering the event model $\mathcal{E}_{n+1}$ it is easy to see that updating with this will result in the model $\mathcal{M}_{n+1}$, where there will be two information cells for agent a_{n+1} corresponding to the events w_{n+1} and b_{n+1}, i.e. $[w]_{n+1}$ and $[b]_{n+1}$, and each of these will have 2^{n-1} states. It is also easy to see that for all $k > n + 1$ there will only be one information cell for a_k. Finally, it is also easy to see that U_W will be true in 2^{n-2} states of $[w]_n$ and 2^{n-2} states of $[b]_n$ since U_w where true in 2^{n-2} states of $\mathcal{M}_n$ and likewise for U_B.

(ii): Assume that (ii) is true for n. Assume that $s \in [w]_{n+1}$. Then using (i), the definition of the product update, and (i) again, we get:

$$
\begin{aligned}
f(a_{n+1}, s, U_W) &= \sum \{f(t) \mid t \in [w]_{n+1} \cap U_W\} \\
&= \sum \{f((t, w_{n+1})) \mid t \in \mathcal{M}_n, t \in U_W \text{ (in } \mathcal{M}_n)\} \\
&= \sum \{f(t) + pre(w_{n+1})(U_W) \mid t \in \mathcal{M}_n, t \in U_W \text{ (in } \mathcal{M}_n)\} \\
&= \sum \{f(t) + 1 \mid t \in \mathcal{M}_n, t \in U_W \text{ (in } \mathcal{M}_n)\} \\
&= f^n(U_W) + 2^{n-3}.
\end{aligned}
$$

Similarly for U_B we get

$$
\begin{aligned}
f(a_{n+1}, s, U_B) &= \sum \{f(t) \mid t \in [w]_{n+1} \cap U_B\} \\
&= \sum \{f(t) + pre(w_{n+1})(U_B) \mid t \in \mathcal{M}_n, t \in U_B \text{ (in } \mathcal{M}_n)\} \\
&= \sum \{f(t) + 0 \mid t \in \mathcal{M}_n, t \in U_B \text{ (in } \mathcal{M}_n)\} \\
&= f^n(U_B).
\end{aligned}
$$

If $s \in [b]_{n+1}$ then,

- $f(a_{n+1}, s, U_W) = f^n(U_W)$ and

- $f(a_{n+1}, s, U_B) = f^n(U_B) + 2^{n-3}$,

follow by reasoning in similar manner.

(iii): Assume that $f^n(U_B) + 2^{n-2} < f^n(U_W)$. Note that from (i) and (ii) we have that

- $f^{n+1}(U_W) = 2f^n(U_W) + 2^{n-2}$,

- $f^{n+1}(U_B) = 2f^n(U_B) + 2^{n-2}$.

But, then

$$
\begin{aligned}
f^{n+1}(U_B) + 2^{n-1} &= 2f^n(U_B) + 2^{n-2} + 2^{n-1} \\
&= 2(f^n(U_B) + 2^{n-2}) + 2^{n-2} \\
&< 2f^n(U_W) + 2^{n-2} \\
&= f^{n+1}(U_W).
\end{aligned}
$$

(v): Now assume that $U_B <_{a_n} U_W$ is true at all states of $\mathcal{M}_n$. Consider agent a_{n+1}, we then want to prove that $U_B <_{a_{n+1}} U_W$ is true at all states of $\mathcal{M}_{n+1}$. That is, we need to prove that

$$
f(a_{n+1}, s, U_B) < f(a_{n+1}, s, U_W),
$$

for all $s \in dom(\mathcal{M}_{n+1})$. By (i) and the definition of f, we only have to consider two cases, namely when $s \in [w]_{n+1}$ and when $s \in [b]_{n+1}$. Moreover, by (ii) we just need to prove that

a) $f^n(U_B) < f^n(U_W) + 2^{n-2}$;

b) $f^n(U_B) + 2^{n-2} < f^n(U_W)$.

It is clear that $a)$ follows from $b)$ and $b)$ follows directly from (iii).

This completes the proof.   ∎

4   Conclusion and Further Research

We provided two logical models of the same example to argue for the somewhat counterintuitive claim that informational cascades are the direct consequence of individual rationality, even when rationality includes full higher-order reasoning or/and is non-Bayesian.

On the question whether real humans are Bayesian reasoners, a variation of the urn example was conducted as an experiment [2] to test whether people were using a simple counting heuristic instead of Bayesian update. In this variation, urn U_W contained 6 white and 1 black ball, whereas U_B contained 5 white and 2 black balls (in this way, a black ball provides more information than a white ball). Even though more decisions were consistent with Bayesian update than with a simple counting heuristic, the counting heuristic could not be neglected. For critique of the conclusion that most individuals use Bayesian updating, see [23]. In fact, these experimental results seem to rule out only the simplest counting heuristic where one compares the number of white balls against the number of black balls. However, one could argue that in this variation of the experiment another counting heuristic would be more appropriate (for instance counting one black ball several times, to reflect the higher weight of this piece of evidence) and that the experimental results are consistent with such an alternative. In general, our logic for counting evidence (presented in Section 3) can be used to capture various forms of "weighting" heuristics.

Another aim for future research is to generalize both the probabilistic logic of Section 2 and the logic for counting evidence of Section 3 to obtain a general logic of evidence that can capture both quantitative and qualitative approaches to reasoning about evidence, where an example of the latter is [11].

A deeper aim is to look at variations of our example, in order to investigate ways to *stop or prevent the cascade*. It is easy to see (using Condorcet's Jury Theorem) that, if we change the protocol to forbid all communication (thus making individual guesses private rather public), then by taking a poll at the end of the protocol, the majority vote will match the correct urn with very high probability (converging to 1 as the number of agents increases to infinity).

This proves that examples such as the one analyzed in this paper are indeed cases of "epistemic Tragedies of the Commons": situations in which *communication is actually an obstacle to group truth-tracking*. In these cases, a cascade can be stopped only in two ways: either by "irrational" actions by some of the in-group agents themselves, or else by outside intervention by an external agent with different information or different interests than the group. An example of the first solution is if some of the agents *simply disregard* the information obtained by public communication and make their guess solely on the basis of their own observations: in this way, they lower the probability that their guess is correct (which is "irrational" from their own individual perspective), but they highly increase the probability that the majority guess will be correct. An example of the second solution is if the protocol is modified (or only disrupted) by some external agent with regulative powers (the "referee" or the "government"). Such a referee can simply *forbid communication* (thus returning to the protocol in the Condorcet's Jury Theorem, which assumes independence of opinions). Or she might require *more communication*; e.g. require that the agents should announce, not only their beliefs about the urns, but also their *reasons* for these beliefs: *the evidence* supporting their beliefs. This evidence might be the number of pieces of evidence in favor of each alternative (in the case that they used the counting heuristics); or it might be the subjective probability that they assign to each alternative; or finally, it might be all their available evidence: i.e. the actual color of the ball that they observed (since all the rest of their evidence is already public). Requiring agents to share any of these forms of evidence is enough to stop the cascade in the above example.

One may thus argue that *partial communication* (sharing opinions and beliefs, but not sharing the underlying reasons is evidence) is the problem. More (and better) communication, more true deliberation based on sharing arguments and justifications (rather than simple beliefs), *may* sometimes stop the cascade. However, there are other examples, in which communicating some of the evidence is not enough: cascades can form even after each agent shares some of her private evidence with the group. It is true that a "total communication", in which everybody shares *all* their evidence, all their reasons, all the relevant facts, will be an effective way of stopping cascades (provided that the agents perfectly trust each other and they are justified to do so, i.e. nobody lies). In our toy example, this can be easily done: the relevant pieces of evidence are very few. But it is unrealistic to require such total communication in a real-life situation: the number of facts that might be of relevance is practically unlimited, and moreover it might not be clear to the agents themselves which facts are relevant and which not. So in practice this would amount to asking the agents to publicly share all their life experiences! With such a protocol, deliberation would never end, and the moment of decision would always be indefinitely postponed.

So in practice the danger remains: no matter how rational the agents are, how well-justified their beliefs are, how open they are to communication, how much time they spend sharing their arguments and presenting their evidence, there is always the possibility that all this rational deliberation will only lead the group into a cascading dead-end, far away from the truth. The only practical and sure way to prevent cascades seems to come from the existence of a significant number of "irrational" agents, who simply ignore or refuse to use the publicly available information and rely only on their own observations. Such extreme, irrational skeptics will very likely get it wrong more often than the others. But they will perform a service to society, at the cost of their own expected accuracy: due to them, in the long run society might correct its entrenched errors, evade its cascades and get better at collectively tracking the truth.

The conclusion is that communication, individual rationality and social deliberation are not absolute goods. Sometimes (and especially in Science, where the aim is the truth), it is better that some agents effectively isolate themselves and screen off some communication for some time. Allowing (and in fact encouraging) some researchers to "shut themselves up in their ivory tower" for a while, pursuing their independent thinking and tests without paying attention to the received knowledge in the field (and preferably without access to any Internet connections), may actually be beneficial for the progress of Science.

Acknowledgments.

The research of Sonja Smets and Zoé Christoff leading to these results has received funding from the European Research Council under the European Communitys Seventh Framework Programme (FP7/2007-2013)/ERC Grant agreement no. 283963.

Jens Ulrik Hansen is sponsored by the Swedish Research Council (VR) through the project "Collective Competence in Deliberative Groups: On the Epistemological Foundation of Democracy".

BIBLIOGRAPHY

[1] L.R. Anderson and C.A. Holt. Classroom games: Information cascades. *The Journal of Economic Perspectives*, 10(4):187–193, 1996.

[2] L.R. Anderson and C.A. Holt. Information cascades in the laboratory. *The American Economic Review*, 87(5):847–862, 1997.

[3] G. Aucher. A combined system for update logic and belief revision. Master's thesis, ILLC, University of Amsterdam, Amsterdam, the Netherlands, 2003.

[4] A. Baltag, L. Moss, and S. Solecki. The logic of common knowledge, public announcements and private suspicions. In *TARK Proceedings*, pages 43–56, 1998.

[5] A. Baltag and S. Smets. A qualitative theory of dynamic interactive belief revision. *Logic and the Foundations of Game and Decision Theory (LOFT7)*, (1–3):11–58, 2008.

[6] A.V. Banerjee. A simple model og herd behavior. *The Quarterly Journal of Economics*, 107(3):797–817, 1992.

[7] J. van Benthem. Dynamic logic for belief revision. *Journal of applied and non-classical logics*, 17(2):129–155, 2007.

[8] J. van Benthem. *Logical Dynamics of Information and Interaction*. Cambridge Univ. Press, 2011.

[9] J. van Benthem, J. Gerbrandy, and B. Kooi. Dynamic update with probabilities. *Studia Logica*, 93(1), 2009.

[10] J. van Benthem and F. Liu. Dynamic logic of preference upgrade. *Journal of Applied Non-classical Logics*, 17(2):157–182, 2007.

[11] J. van Benthem and E. Pacuit. Dynamic logics of evidence-based beliefs. *Studia Logica*, 99(1–3):61–92, 2011.

[12] S. Bikhchandani, D. Hirshleifer, and I. Welch. A theory of fads, fashion, custom, and cultural change as informational cascades. *Journal of Political Economy*, 100(5):992–1026, 1992.

[13] S. Bikhchandani, D. Hirshleifer, and I. Welch. Learning from the behavior of others: Conformity, fads, and informational cascades. *Journal of Economic Perspectives*, 12(3):151–170, 1998.

[14] H. van Ditmarsch, W. van der Hoek, and B. Kooi. *Dynamic Epistemic Logic*. Synthese Library, Springer, 2007.

[15] D. Easley and J. Kleinberg. *Networks, Crowds, and Markets. Reasoning about a Highly Connected World*. Cambridge University Press, 2010.

[16] R. Fagin and Halpern J.Y. Reasoning about knowledge and probability. *Journal of the ACM*, 41(2):340–367, 1993.

[17] A. Goldman. *Knowledge in a Social World*. Oxford Univ Press, 1999.

[18] A. Goldman and D. Whitcomb, editors. *Social Epistemology, Essential Readings*. Oxford University Press, 2011.

[19] R.E. Goodin. *Reflective Democracy*. Oxford Univ. Press, 2003.

[20] J.U. Hansen. A logic-based approach to pluralistic ignorance. In Jonas De Vuyst and Lorenz Demey, editors, *Future Directions for Logic – Proceedings of PhDs in Logic III*, volume 2 of *IfColog Proceedings*, pages 67–80. College Publications, 2012.

[21] V.F Hendricks, P.G Hansen, and R. Rendsvig. Infostorms. *Metaphilosophy*, 44:301–326, 2013.

[22] W. Holliday. Trust and the dynamics of testimony. In *Logic and Interactive Rationality, Dynamics Yearbook 2009*, pages 147–178. Institute for Logic, Language and Computation, Universiteit van Amsterdam, 2010.

[23] S. Huck and J. Oechssler. Informational cascades in the laboratory: Do they occur for the right reasons? *Journal of Economic Psychology*, 21(6):661–671, 2000.

[24] B.P. Kooi. Probabilistic dynamic epistemic logic. *J. of Logic, Lang. and Inf.*, 12(4):381–408, September 2003.

[25] C. List. Group knowledge and group rationality: A judgment aggregation perspective. In *Social Epistemology, Essential Readings*, pages 221–241. Oxford University Press, 2011.

[26] C. List and R.E. Goodin. Epistemic democracy: Generalizing the condorcet jury theorem. *Journal of Political Philosophy*, 9(3):277–306, 2001.

[27] F. Liu. *Changing for the Better: Preference Dynamics and Agent Diversity*. PhD thesis, University of Amsterdam, 2007.

[28] J. Plaza. Logics of public communications. In M.L. Emrich, M.S. Pfeifer, M. Hadzikadic, and Z.W. Ras, editors, *Proceedings of the 4th International Symposium on Methodologies for Intelligent Systems*, pages 201–216. Oak Ridge National Laboratory, 1989.

[29] C. Proietti and E. Olsson. A ddl approach to pluralistic ignorance and collective belief. *Journal of Philosophical Logic*, 2013.

[30] J. Seligman, P. Girard, and F. Liu. Logical dynamics of belief change in the community, under submission. 2013.

[31] J. Seligman, F. Liu, and P. Girard. Logic in the community. In Mohua Banerjee and Anil Seth, editors, *Logic and Its Applications*, volume 6521 of *Lecture Notes in Computer Science*, pages 178–188. Springer Berlin Heidelberg, 2011.

[32] J. Surowiecki. *Wisdom of Crowds*. Anchor books USA, 2004.

Comments on Baltag, Christoff, Hansen and Smets

Hu Liu

The paper by Baltag, Christoff, Hansen, and Smets (Henthforth, BCHS paper) analyzed the phenomena of informational cascades with the help of two types of logical models: One is the model of probabilistic dynamic epistemic logic [9][24], and the other is a novel model, which resembles the first one, but is based on counting evidences instead of calculating probabilities. (References in this comment refer to BCHS paper's bibliography.)

The two models have the same logical structure. Analysis of informational cascades within both models show the same result. Informational cascades cannot be eliminated even if agents are endowed with unlimited reasoning powers: They can reason about each other's epistemic status and about the underlying sequential protocol with unlimited higher level.

The problem of informational cascade was originally analyzed by probabilistic reasoning [12]. There are also critiques of the Bayesian approach (See the footnote of page 14, BCHS paper). BCHS paper provided two models to reflect the two schools. This comment applies to both models. I will use the counting evidence model for representation because this one is new in literature.

The paper is neat in technique and convictive in intuition. It is especially interesting to see how techniques of logic, like product models of epistemic modles and event models, can be used for a special problem outside the logic society. I shall discuss their framework from two perspectives. First, is the model they give "good" from the point of view of logic? Second, how well does their model helps us to understand the problem of informational cascades?

Dynamics of the sequential actions is realized in BCHS paper by the known technique of a product of an epistemic model and an event model. The novel part is how they deal with the number of evidences. This is done by this clause: $f'((s, e)) = f(s) + pre(s, e)$. The reading of the clause is natural: After an action e has been applied, the number of evidences attached to state s ($f'((s, e))$) is equal to the number of evidences attached to s before the action ($f(s)$) plus the number of evidences that the action provides for s.

The model is not a standard logical model. The valuation is restrictive in that at each state only one proposition (U_W or U_B in BCHS paper) can be true. It is different from hybrid logic because a proposition is allowed to be true at more than one state. Readers may wonder if it is possible to generalize their model to full logical strength and still keep the natural interpretation. If so, the model may be helpful in analyzing agents' behavior in a general background.

This is not direct as can already be seen in BCHS paper. They allowed states at which no proposition is true. This is a "noisy" generalization in the sense that it is not useful for the problem of informational cascades. They introduce a special symbol $\bot$ which will be assigned to states with no true proposition. These states are eliminated in the updated model.

For a general logical model, the problem is, the function *pre* gives the strength of evidence that an action provides for a *proposition*, whereas in a model, numbers of evidences are attached to *states*. It is not a problem with the restriction in BCHS model, because each state is labeled by a single proposition. Intuitively, the strength of evidence should be on propositions. How do we calculate the number of evidences for a state at which more than one proposition is true? A simple sum of the numbers of evidences for each labeled propositions seems insufficient, because in this way, states with less propositions being true are inclined to have less evidences, and states with no proposition being true have no support.

Another thing that was missed in the paper is a full definition of the logic. It is definitely helpful, since the main claim of the paper refers to "all the level of mutual belief/knowledge". Even though epistemic logic defined on epistemic models is intrinsically with iterative knowledge operators, without an explicit syntactic representation, it is not easy to see the role of high-level knowledge reasoning in the formation of a cascade.

The authors did introduce a notation $U_W <_a U_B$ (or $U_B <_a U_W$) into the language, which is read as that there are more evidences for U_B than for U_W from the point of view of agent a. From the context it may stand for the formula $K_a(N(U_W) < N(U_B))$, which is read as that a knows/believes that U_B has more evidences than U_W. I wonder why they did not use the direct epistemic language.

Now for the second perspective, it makes sense to have a formal investigation of the problem of informational cascades, because it is agents' reasoning that makes a cascade happening. The question is, what can we gain from such a formal framework. As in BCHS paper, "what is typically absent from the standard Bayesian models of informational cascades is the agents' high-order meta-reasoning about other agents' minds and about the whole sequential protocol in which they are participating. The paper fills in the gap, and makes a correct claim that no amount of high-order reasoning can eliminate the problem.

In the standard treatment of informational cascades, the power of agents' reasoning is not an explicit part of models. They focused on how agents calculate probabilities. Knowledge reasoning is informally used in models as granted. Also, there is an implicit hypothesis that agents can reason about the protocol. In the running example of BCHS paper, this means actions of the third and further agents provide no information about the true situation, because agents know that they are in a cascade. BCHS paper investigated the problem in a formal background and highlighted the role of reasoning in a cascade. This itself is a nice contribution to literature.

The new result in BCHS paper concerns higher-order reasoning. Because no syntactic representation is given, it is not clear from their proof why higher-order reasoning cannot eliminate the problem.

As far as I understand the problem, agents' limited level of epistemic reasoning is sufficient for a cascade happening. Higher-order reasoning does not change anything. An agent a_n knows what previous agents did $(K_n(did(a_1)) \wedge \ldots \wedge K_n(did(a_{n-1})))$. From this the agent may infer what previous agents knows about the true situation $(K_n K_1 P_1 \wedge \ldots \wedge K_n K_{n-1} P_{n-1})$. Also the agent should know whether or not a cascade has occurred. $(K_n(Cas))$. It seems that two-level reasoning is enough for this purpose. This observation can only be clear with the help of a well-defined formal language.

Whether or not Cas is taken as a primary in the language will change the needed level of reasoning. I am not sure if we can define a cascade in an epistemic language (with the mechanism of either calculating probability or counting evidence). In the running example, the following statement indicates a formation of a cascade.

$$K_3 K_1(N(U_W) < N(U_B)) \wedge K_3 K_2(N(U_W) < N(U_B)) \wedge \ldots \wedge K_n K_1(N(U_W) < N(U_B)) \wedge K_n K_2(N(U_W) < N(U_B)).$$

It is not direct how do we define a general notion of cascades.

Besides the "limited level" restriction, reasoning in information cascades is restrictive in another way. Since only sequential protocols are in question, agents do not really reason about each other. Agents only reason about those agents acting before them. These restrictions on epistemic reasoning suggest that we may consider a simpler logic than the standard epistemic logic for the problem of informational cascades.

Comments on Baltag, Christoff, Hansen and Smets

Eric Pacuit

1. Overview. We often rely on other people's opinions when making decisions or forming our beliefs. For example, suppose that you are deciding whether to eat dinner at restaurant A or restaurant B. You have some information about the restaurants suggesting that A is better (e.g., you have looked at the menus of both restaurants). However, when you arrive at restaurant A, you notice that it is nearly empty while restaurant B (which happens to be next door) is almost completely full. Naturally, you take this as evidence that restaurant B is in fact better and decide to eat there. There is nothing particularly surprising or troubling about the diner's behavior: upon receiving new evidence that more people are eating at restaurant B, she updates her beliefs and makes her decision where to eat accordingly. However, situations such as this, in which people form their beliefs using information obtained by observing the behavior or opinions of others, can bring about a so-called *information cascade*. This is a situation in which "it is optimal for an individual, having observed the actions of others ahead of him, to follow the behavior of the preceding individual without regard to his own information." [2]. In this fascinating and rich paper, Alexandru Baltag, Zoé Christoff, Jens Ulrik Hansen, and Sonja Smets develop a logical model that can be used to analyze information cascades.

There are two main contributions of this paper. The first contribution is to use a probabilistic dynamic epistemic logic to explicitly represent the reasoning of the agents involved in an information cascade. To that end, the authors focus on a simple example of a information cascade that has been discussed in the literature. In this example, there are two urns: the first, denoted S_W, contains two white balls and one black ball and the second, denoted S_B, contains two black balls and one white ball. One of the two urns is randomly chosen and placed in a room. The agents enter the room one at a time, select a ball from the urn, observe the color of the ball, and then place the ball back into the urn. The agents then guess whether the urn is S_W or S_B and write their guess on a board. So, the agents form their beliefs based on two pieces of evidence: the color of the ball that the agent (privately) observes and a public signal listing all the guesses of the agents that have entered the room earlier. I will not rehearse the analysis again here, but the key idea is as follows: Suppose that the first two agents observe a black ball. A straightforward Bayesian analysis shows that, in this case, it is optimal for all subsequent agents to simply guess S_B regardless of the color of the ball that is observed. The logical framework used in this paper goes beyond this simple Bayesian analysis by explicitly representing the agents' "higher-order" reasoning. That is, it represents what the agents know and believe about what the other agents know and believe about the situation, including whether the agents know that they are involved in an information cascade. The second contribution of this paper is a novel dynamic epistemic logic that incorporates the agents' ability to "count" the evidence in favor of each state. This logical framework is much simpler than probabilistic dynamic epistemic logic, but, arguably, still faithfully represents the agents' reasoning in an information cascade.

2. Framing the issues: The *Wisdom of the Crowds*. I want to start with a general comment about the discussion of the *Wisdom of the Crowds*. The authors say that the *Wisdom of the Crowds* is "explainable in logical terms by the notion of distributed knowledge." The classic example of the "wisdom of the crowds" runs as follows: Each agent is asked to guess how many jelly beans are in a jar. The surprising fact is that the group's estimate, calculated by averaging the individual guesses, is typically much closer to the actual value than any individual. Strictly speaking, this is not an instance of "distributed knowledge". I take distributed knowledge to be what the group "knows" if they "combine" their knowledge. For example, if agent i knows that $p \rightarrow q$ and agent j knows that p, then there is distributed knowledge in the group consisting of i and j that q is true. Even if all the agents combine their knowledge in this way, the group will not "know" the number of jelly beans in the jar. The agents need to *aggregate* their estimates (in this case, by averaging) in addition to simply combining their knowledge. But, I do not want to make too much of this criticism. The general point is simply that "distributed knowledge" is only one aspect of the Wisdom of the Crowds, for a more in-depth analysis, see [4].

3. Logical analysis of information cascades. The logical analysis of information cascades presented in this paper is best understood as an extension of the standard Bayesian analysis (or an alternative in the case of the counting evidence models). This is a very interesting and useful contribution to the literature. However, the focus on the specific urn example left me wondering about the scope of this logical analysis. Commenting on the different examples of the information cascades and their models, Easley and Kleinberg note that:

> ... the general conclusions from these models tend to be qualitatively similar: when people can see what others do but not what they know,

there is an initial period when people rely on their own private information, but as time goes on, the population can tip into a situation
where people—still behaving fully rationally—begin ignoring their
own information and following the crowd. [3]

Can we develop a logical framework that makes explicit what exactly is "qualitatively similar" between the different instances of information cascades? More
generally, what is the underlying logic of an information cascade? This requires
enhancing the logical systems presented in this paper along the following lines:

1. What exactly is the evidence that the agents use to form their beliefs? Making this explicit will allow us to describe when the agents are "ignoring"
   their own information. In the models in this paper, the agents use all available evidence to form their beliefs (in the case of the urn, this includes the
   guesses of the other agents and the private observation of the ball). The
   sense in which agents ignore their own information is that their guess will
   be the same regardless of the color of the ball that they observe.

2. What is the underlying principle that drives the agents' decision to follow
   the crowd? Again, the agents do not explicitly decided to imitate the crowd.
   Rather, they form their beliefs based on all the available evidence, but the
   rational guess for certain agents turns out to be the one that imitates the
   crowd.

   I will illustrate this idea with a principle that has played an important role in
   qualitative models of Aumann's classic agreeing to disagree theorem [5, 1].
   Let D be a set of decisions (e.g., "write S_W on the board" or "write S_B on
   the board"), and for each each i, let $d_i : S \to D$ be a function specifying
   the action that agent i chooses at each state. Finally, let $[d_i = d]$ denote the
   set of states where agent i chooses d. The *sure-thing principle* is (here, I am
   treating the knowledge modalities as set functions[1] mapping a set of states
   E to the set of states where the agent knows E):

$$\left(\bigcap_{E \subseteq S} K_i(K_i E \to K_j E) \ \cap \ K_i([d_j = d]) \right) \subset [d_i = d]$$

   This says that if agent i knows that j is at least as knowledgable as she is
   and i knows that j chooses d, then i also chooses d. This is not the principle
   driving the reasoning in information cascades. Since each agent i has private
   information about the color of the ball, she does not *know* that the agents

---

[1]This means, in particular, that $E \to F$ is defined to be the set $\overline{E} \cup F$, where $\overline{E}$ is the complement
of E.

that entered the room earlier are at least as knowledgeable as she is. Perhaps some variant of this principle is driving the agents' decision to follow the crowd in an information cascade.

4. Counting evidence. The counting evidence models extends a standard epistemic model with a function $f : S \to \mathbb{N}$ specifying the "weight" of evidence for each state. This is an interesting model, but I was left with a few questions about the functions counting evidence functions f.

- Does f represent cardinal or ordinal information about the weight of evidence for each state? It seems that you are only using ordinal information about the evidence weights. So for example, if $f(s) = 4$ and $f(t) = 2$, then there is more evidence for state s than t, but does it make sense to say that there is "twice as much evidence for s as t".

- The weight of a proposition P is defined by summing over the weights of the states in the proposition that the agent has not ruled out. This means that a proposition that has a lot of states, each with relatively small weights, may be considered more likely than a proposition consisting of a few states each having relatively large weights. For example, suppose that $P = \{s_1, s_2, \ldots, s_{10}\}$ and $Q = \{t_1, t_2\}$ with $f(s_i) = 1$ for $i = 1, \ldots, 10$ and $f(t_1) = f(t_2) = 4$. Then (assuming the agent has not ruled-out any of the states), $Q <_a P$. This is not necessarily a problem, but there are other options for defining the $<_a$ relation: for example, comparing the maximum weights in the propositions or the product of the weights in the propositions.

- It also makes sense to include operations that *decrease* the weight of evidence for a state. This would allow agents to react to agents that they know may be lying or misrepresenting their evidence.

BIBLIOGRAPHY

[1] Bacharach, M. (1985). Some extensions of a claim of aumann in an axiomatic model of knowledge. *Journal of Economic Theory 37*(1), 167 – 190.

[2] Bikhchandani, S., D. Hirshleifer, and I. Welch (1998). Learning from the behavior of others: Conformity, fads and informational cascades. *Journal of Economic Perspectives 12*(3), 151 – 170.

[3] Easley, D. and J. Kleinberg (2010). *Networks, Crowds and Markets. Reasoning about a Highly Connected World.* Cambridge University Press.

[4] Lyon, A. and E. Pacuit (2013). The wisdom of crowds: Methods of human judgement aggregation. In *Handbook of Human Computation.*

[5] Samet, D. (2010). Agreeing to disagree: The non-probabilistic case. *Games and Economic Behavior 69*, 169 – 174.

Response to Pacuit and Liu

ALEXANDRU BALTAG, ZOÉ CHRISTOFF, JENS ULRIK HANSEN AND SONJA SMETS

We thank our commentators for their careful and insightful analysis of our paper. We only reply here to a selected set of points raised in their comments.

1 Reply to the comments by Eric Pacuit

We fully agree with Eric Pacuit that "distributed knowledge" is only one aspect of the "Wisdom of the Crowds". We actually think that it is important to clearly distinguish between different types of group knowledge, applicable to different situations. The explanation in terms of distributed knowledge applies to situations in which unlimited inter-agent communication is allowed, and thus the agents can pool together all their "hard" information. In contrast, the explanation in terms of (variants and generalizations of) Condorcet's Jury Theorem applies to forms of "statistical group knowledge", in which no communication (or only communication in very small disjoint subgroups) is allowed, and in which a given yes/no issue is to be decided, based only on private, independent observations giving "soft" (fallible) information about the issue under consideration. In such situations, the judgment aggregation method does *not* involve inter-agent communication, but rather counting of the agents' votes followed by some type of preference aggregation (e.g. majority rule etc).

Unlike statistical group knowledge, distributed knowledge *can, in principle,* be converted into actual individual knowledge by means of simple inter-agent communication. Of course, this possibility might in fact be realizable only in very idealized scenarios, while in practice the group's dynamics, the structure of the social network, the individuals' different epistemic interests and agendas, and their lack of mutual trust would prevent the full actualization of distributed knowledge. But even from a purely logical perspective, taking these factors into account and thus developing a more accurate formalization of a group's potential knowledge is a fascinating open problem. Ongoing work by A. Baltag and S. Smets, presented at two recent conferences, shows that in these more realistic conditions, even if we restrict to the scenarios allowing inter-agent communication as their aggregation method, true group knowledge turns out to be different from both distributed knowledge and from common knowledge (lying instead somewhere in between these extremes).

Pacuit's question about the underlying logic of informational cascades is of primary importance. Indeed, one would like to have a more sophisticated logical analysis of the cascade-forming mechanisms. Such an analysis would reach beyond the model-theoretic formalization presented in our paper. A good qualitative syntax should make the distinction between different types of evidence, coming from different sources (such as the second-hand evidence coming from the guesses of others versus the direct evidence based on private observations). The model should allow for different treatments of such different types of evidence, for instance by attaching more weight to private observations and less to the second-hand evidence. In particular, it is clear that agents who trust *only* their own private information can avoid the cascade (though, at an individual level, they might be more likely to make the wrong guess than the ones who take into account the others' evidence). But this is only one extreme of the spectrum; the "ideal" (but cascade-generating) Bayesian reasoner formalized in our paper is at the other extreme. Of course, the most promising evidence-assessment policies lie in between these extremes. It would be extremely interesting to give a comparative analysis of the cascade-forming (or cascade-preventing) potential of these various policies.

Another good question posed by Pacuit is about the underlying principle driving the agent's decision to follow the crowd. He speculates on a possible analogy with the sure-thing principle. The best way we can understand the analogy is to claim that the third agent (and any subsequent one) reasons that the *group* of the first two agents has collectively *more* relevant evidence (about the state of the urn) than the evidence that is privately available to her. However, the analogy is imperfect, since in the sure-thing principle the agent i does not have access to the specific evidence/information used by the (more knowledgeable) agent j; she only has information about j's decision (and about the fact that j has more information than herself). So, agent i has to follow j's decision without being able to justify her choice except by appealing to j's higher knowledge. In contrast, the agent in the Urn example has indirect access to the actual evidence possessed by the first two agents: assuming that those agents are rational, she can infer their private evidence from their decisions (guesses). She can then use the evidence extracted in this way (balancing it against her own private evidence) to form her decision. This means that she can justify (and that we can explain) her decision of following the crowd by appealing only to her individual reasoning, her private evidence and the evidence extracted from the others' behavior)combined with her belief in the others' rationality). It seems that the sure-thing principle is not needed here. This is similar to the difference between adopting your college teacher's expressed belief (say, in the truth of Darwin's theory of evolution) because you understand the reasons behind her belief (say, the arguments that she presented to you, the available evidence that she explicitly or implicitly gave you access to) versus adopting it simply because you believe/know that she is smarter and more knowledgeable than you. Arguably, the first kind of adoption can be a source of real "knowledge", while the second kind is at best a source of true beliefs. So it seems to us that the sure-thing principle can be at best a source of true beliefs; while the rational Bayesian reasoning formalized in our paper can (and typically is) a source of knowledge (though occasionally it may also lead to errors, e.g. non-truth-tracking cascades such as in the Urn example).

Moving on to the comments on the section on counting evidence, Pacuit asks if the function f in our counting models represents cardinal or ordinal information, and suggests that in fact only the ordinal information is actually used. This does not seem quite right, as it can be seen from Pacuit's next comment: addition of (strengths of) evidence plays an important role in our analysis; in particular, the evidence for U_W is obtained by adding the (new and old) evidence in support of each of the states satisfying U_W. Even if each of these states is weakly supported, the total proposition (set of states) can be strongly supported. Pacuit rightly points out that there are other possible choices on how to define the weight of a proposition. Such alternatives are indeed worth exploring, but given the choice made here, our counting models use both the *order* structure and the *additive* structure of natural numbers. But it is true that we do *not* use the *multiplicative* structure: as Pacuit correctly notes in his commentary, "having twice as much evidence for s as for t " is not a relevant issue in this framework. This is in fact one of the main differences from the Bayesian reasoning used in the probabilistic DEL framework. Nevertheless, the counting-evidence heuristic, as formalized in our paper, cannot be reduced to purely ordinal information.

Eric Pacuit's last suggestion is to include operations that decrease the weight of evidence for a state. This is certainly a natural move, an evidential analogue of the *contraction* operation (in the AGM theory of belief revision). In this paper, we tried to limit our operations to the minimum required for the analysis of the Urn cascade. But, once we turn our ideal agents into suspicious ones, equipped with the abilities to lie and misrepresent their information, operations that decrease evidence weight are definitely worth exploring.

2   Reply to Hu Liu

In this commentary two points are raised:1) the question whether our model is "good" from the point of view of logic and 2) how well our model helps us to understand the problem of informational cascades.

With respect to the first point above, Hu Liu claims that our model is not a "standard" logical model because supposedly the valuation is restrictive in each state to only one atomic proposition so that only U_W or U_B can be true. First let us mention that the notion of "model" that we adopt in our paper is standard. In the counting case our models are simply labeled transition systems equipped with a function that assigns a natural number to each state. This means that if one starts from a given epistemic language consisting of a set of atomic sentence and building up a set of well-formed formulas in the usual way, these formulas can be interpreted in our models. We have added a note to our paper to make this point more explicit. In general, different atomic propositions can be true at the same state. It is true that in our Urn example, we used atomic sentences U_W and U_B which happened to be mutually exclusive. But this is just an example!

Hu Liu remarks that "for a general logical model, the problem is, the function pre gives the strength of evidence that an action provides for a *proposition*, whereas in a model, numbers of evidence are attached to *states*. ... Intuitively, the strength of evidence should be on propositions." Hence he asks how we can calculate the number of evidence for a state at which more than one proposition is true? According to Hu Liu, "the sum of the numbers of evidences for each labeled proposition seems insufficient, because in this way, states with less proposition being true are inclined to have less evidences, and states with no proposition begin true have no support". As far as we understand this remark, we think it is based on a confusion of how our event models work. Our event models (see definition 7) have a build-in precondition function pre which assigns to every event and every proposition coming from the set Φ a weight. The set of propositions Φ is given as part of the event model and consists of pairwise inconsistent propositions (hence there are no states in which both such pairs of formulae can be true). This set of propositions Φ is not necessarily a set of atomic sentences, it can consist of complex epistemic formulae. Indeed a formula in Φ is meant to express the precondition (or proposition) that has to be true for an evidence-changing event to occur. It now happens to be the case in our example that the set Φ coincides with the set of atomic sentences Ψ, however in general that need not be the case. Besides this mechanism, our model does have an operation for evidence to be assigned to propositions, that is exactly what $f(a, s, T)$ was introduced for. However, we do not see any problem with the computation of $f(a, s, T)$ in case a state has more than one true atomic sentence.

We agree with Hu Liu that a full definition of the logic, providing not only the semantics but also a fully developed axiom system for an epistemic language will be helpful to express the mechanism of a cascade. The axiom system will have to be worked out in future work.

With respect to the second question of Hu Liu mentioned above, we refer to the issue of "higher-order reasoning". Indeed, the agent's limited level of epistemic reasoning is sufficient for a cascade to happen. Two levels of higher-order reasoning can already get agents stuck into a cascade, the only thing we need is for these agents to have knowledge of their own private observations and knowledge about the other's (knowledge of) their private observations. Exploring a setting with a "limited level" restriction can be interesting in its own respect, certainly if one aims to have the simplest possible logical setting that can still prove a cascade. This is however not what we aimed for this in this paper. Our attention went to the surprising result that, contrary to what one would expect, when giving agents an unrestricted capacity to reason about all levels of knowledge and allowing them to conclude that they are part of a cascade will still not be enough for them to break a rational cascade. In this paper we paid no attention to the "limited level" restriction, our aim was to equip agents with all the possible reasoning powers available in epistemic logic so that these agents are fully rational, fully logically omniscient and never make any mistake. Now the surprising result is that even in this ideal case, fully rational agents can get stuck into a cascade with no rational way out.

Knowledge, Friendship and Social Announcements

JEREMY SELIGMAN, FENRONG LIU AND PATRICK GIRARD

ABSTRACT. We present a formal language for reasoning about the changing patterns of knowledge and friendship in social networks. A social announcement consists of an agent (the sender) transmitting some information (the message) to one or more other agents (the receivers) within a network and each of these three components can be described in different ways, from different perspectives. We discuss a number of conceptual issues that arise in such social communication and illustrate our ideas with a number of examples about cold-war spy networks and office gossip.

> Love comes from blindness; friendship from knowledge.
> —Comte De Bussy-Rabutin

1 Introduction

We will be interested in analysing scenarios in which there is a significant interaction between knowledge and social relationships. For example:

> *Berlin 1978. A spy network has recently been uncovered by the Stasi, who are rounding up the spies and their associates. Bella is friends with Charlie and Erik, neither of whom are friends with each other. Unknown to the others is that Erik is a spy. The others are not spies, and Erik knows that because all spies know who else is a spy (we suppose). Bella knows that Charlie is not a spy, but Charlie does not know about her. After the network is exposed all the spies and their friends will be interrogated by the police. But just before this happens a message is relayed to all agents revealing whether or not they are in danger, that is, whether they are a spy (which they would know in any case) or a friend of a spy. Who now knows that Erik is a spy?*

To answer questions of this kind in such a scenario we need to reason about knowledge on the basis of social relationships. The purpose of this paper is to discuss a number of conceptual issues that arise when considering communication between agents in such networks, both from one agent to another, and broadcasts to socially-defined groups of agents, such as the group of my friends. We aim to provide a precise language for exploring 'logic in the community' [13] and reasoning in social networks in general. The framework we are going to propose is a combination of epistemic logic, dynamic logic and hybrid-like logic. In what follows, we will first review what has been developed in those fields and then position our contribution.

Epistemic logic (EL), with Hinttika's pioneering work [5] in philosophy and further developments in computer science and AI such as [7], is used to reason about knowledge. When more agents are involved, to talk about other agents' knowledge, EL was extended to multi-agent epistemic logic (multi-EL), with little change in the logic. A more significant extension was to include operators for common knowledge, in [3]. This produced a far richer agenda of topics and techniques for epistemic logic, in which such propositions as the following can be formalised:

- If I don't know that I'm in danger then I know that I don't know it.   (EL)

- Bella knows that Erik knows that Charlie doesn't know that Erik is in danger. (multi-EL)

- It is not common knowledge that Erik is in danger. (EL^C)

Since the 1980s, when the information-driven dynamics of knowledge and belief came to the fore, interest shifted to how knowledge and beliefs change in response to new information. The AGM-paradigm [1] proposed rational postulates governing belief revisions. A slightly different framework, PAL, the logic of public announcement [9], analysed how an agent's information is updated in response to concrete action-like announcements. Typically public announcement results in common knowledge of the announced message, because each agent knows not only the content of the message, but that every other agent has received it. In dynamic epistemic logic, DEL, following [2], private announcements, in which a message is received by a limited set of agents are also considered. This extended the analysis of knowledge and communication to include such propositions as:

- Were it to be publicly announced that Erik is a spy, Charlie and Bella would know that.   (PAL)

- If Bella does not know that she is in danger, and Charlie were told this in private, Bella would still not know it.   (DEL)

For uniformity of presentation, we will summarise the work in this area using the following partially specified formal language, based on a set Prop of propositional variables, a set A of agents, and a set $\mathcal{D}$ of dynamic operators:

$$\varphi ::= \rho \mid \neg\varphi \mid (\varphi \wedge \varphi) \mid K_a\varphi \mid C_G\varphi \mid \Delta\varphi \quad (\rho \in \text{Prop},\ \alpha \in A,\ G \subseteq A,\ \Delta \in \mathcal{D})$$

Here, $K_a\varphi$ means that agent a knows that φ, $C_G\varphi$ means that it is common knowledge among the agents in the group G that φ, and $\Delta\varphi$ (for $\Delta \in \mathcal{D}$) means that were action/event Δ to occur, φ would be true. Details of the class $\mathcal{D}$ vary and some approaches are based on belief rather than knowledge, but the above will suffice for illustrative purposes.

With such a language, we can represent epistemic propositions such as the following:

$K_e(p \to q)$ Erik (e) knows that if (p) the network has been exposed then (q) there is a mole.

$K_b\neg K_c p$ Bella (b) knows that Charlie (c) does not know the network has been exposed.

$C_{bc}p$ It is common knowledge between Bella and Charlie that the network has been exposed.

$[p!]K_e q$ If it were publicly announced that the network has been exposed, Erik would know there is a mole.

$(\neg K_e q \wedge$ $\neg K_e q)$ Erik doesn't know that there is a mole, and were it to be announced privately to the others that the network has been exposed, he would still not know.

The last example uses the dynamic operator which represents the action of announcing p to all agents other than e. The announcement of p is represented by the node marked $[p!]$, which is highlighted as the action actually performed. But there is another possible action I, the 'identity' action in which nothing changes. Agent e's ignorance about whether or not the announcement has taken place is represented by the line marked e. This is what makes the announcement, in some sense, private. Operators like these are called 'event models' or 'action models' in the literature, because of their Kripke-model-like appearance, but they are really part of the syntax of the language, not its semantics.

For semantics, a model for the language is a standard Kripke model with a binary accessibility relation for each modal operator: $M = \langle W, k, V \rangle$, where W is a set of epistemic states and for each propositional variable $\rho \in \text{Prop}$, the function V assigns a set $V(\rho) \subseteq W$ of states in which ρ holds. Then, for each agent $a \in A$, there is an equivalence relation k_a on W, representing that agent's ignorance about

Jeremy Seligman, Fenrong Liu and Patrick Girard

which state she is in. If $k_a(u, v)$ then agent a cannot determine of u or v is the actual state: for her they are 'epistemically indistinguishable'. For a group G of agents, we define k_G to be the transitive closure of $\bigcup_{a \in G} k_a$, so that two states u and v are related by k_G if there is some path $u_0, \ldots, u_n$ connecting $u = u_0$ to $v = u_n$, such that for each link in the path, say from u_i to u_{j+1}, there is at least one agent in the group who cannot distinguish between u_i and u_{j+1}. Each dynamic operator $\Delta \in \mathcal{D}$ is associated with a transformation, mapping each model M and state w to a new model ΔM and state Δw, representing the result of performing some action. The details depend, of course, on the operator itself. Formulas are evaluated as follows:

$$
\begin{array}{lll}
M, w \models \rho & \text{iff} & w \in V(\rho) \\
M, w \models \neg\varphi & \text{iff} & M, w \not\models \varphi \\
M, w \models (\varphi \wedge \psi) & \text{iff} & M, w \models \varphi \text{ and } M, w \models \psi \\
M, w \models K_a\varphi & \text{iff} & M, v \models \varphi \text{ for all } v \in W \text{ such that } k_a(w, v) \\
M, w \models C_G\varphi & \text{iff} & M, v \models \varphi \text{ for all } v \in W \text{ such that } k_G(w, v) \\
M, w \models \Delta\varphi & \text{iff} & \Delta M, \Delta w \models \varphi
\end{array}
$$

Consider the model M depicted below, with six epistemic states, among them the actual state u is shown as larger than the others. Two propositions p and q are being considered and their truth value varies from state to state; states in which p holds are marked with a 'p'. The relations k_a are represented as paths in the model, labelled with the name of the agent a. (There is a path from any point to itself, and if two or more lines marked 'a' are joined, they form a path.) The relation k_G for a group G is therefore given by paths made up of lines that are labelled by at least one agent in G.

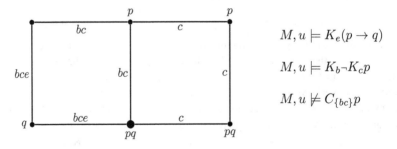

$$M, u \models K_e(p \to q)$$

$$M, u \models K_b \neg K_c p$$

$$M, u \not\models C_{\{bc\}}p$$

The classic example of a dynamic operator is that of publicly announcing some proposition. The public announcement $[p!]$ of p transforms the model M into the model $[p!]M$ in which all not-p worlds are eliminated, as shown below. The actual state $[p!]u$ is just u itself. We can then evaluate formulas $K_e q$ and $C_{\{bce\}}p$ in the new model and they are both true.

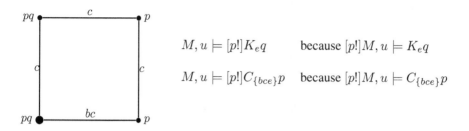

$$M, u \models [p!]K_e q \qquad \text{because } [p!]M, u \models K_e q$$

$$M, u \models [p!]C_{\{bce\}}p \quad \text{because } [p!]M, u \models C_{\{bce\}}p$$

For private announcement, things are trickier. We review the main idea by considering the evaluation of our example formula:

$$(\neg K_e q \wedge \boxed{\substack{\boxed{p!} \quad e \quad \boxed{I}}} \neg K_e q)$$

The dynamic operator acts on our model M by combining the two models $[p!]M$ and IM, which is just M itself, by adding links representing e's ignorance about which model he is in. The actual state of the new model is the pair $\langle u, d \rangle$ where u is the actual state of M and d is the designated node of the operator, representing the action actually performed.

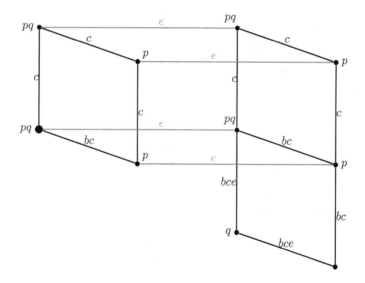

In this new model, we can see that q is false at the right bottom state, so $\neg K_e q$ is true at the actual state. Namely, we have that $M, u \models (\neg K_e q \wedge \boxed{\substack{\boxed{p!} \quad e \quad \boxed{I}}} \neg K_e q)$ because $M, u \models \neg K_e q$ and $\boxed{\substack{\boxed{p!} \quad e \quad \boxed{I}}} M, \langle u, d \rangle \models$

$\neg K_e q$. After an announcement to the other agents that the network is compromised, Erik would still not know it. Similarly, readers are invited to check the truth value of the following two formulas concerning common knowledge in the above model, too. It is not hard to see that in M at u, the formula $\boxed{\begin{array}{ccc} \fbox{$p!$} & \overset{e}{\rule{1.5em}{0.4pt}} & \fbox{I} \end{array}} C_{bc}p$

is true, whereas $\boxed{\begin{array}{ccc} \fbox{$p!$} & \overset{e}{\rule{1.5em}{0.4pt}} & \fbox{I} \end{array}} C_{be}p$ is false, which is to say that were it to be announced to all agents other than Erik that the network is compromised, then this would be common knowledge among Bella and Charlie, but not among Bella and Erik.

The main concern of this paper is to extend this analysis to reasoning about knowledge and communication defined implicitly in terms of social relationships, most simply the binary relation of 'friendship', which we interpret in a minimal way as any symmetric irreflexive relation between agents on the basis of which exchange of information can occur. Some examples of propositions of this kind are the following:

- I know that all my friends are in danger but not all my friends know they are.

- Were Bella to tell all her friends whether or not they are in danger, she would know that Charlie knows he is not in danger.

- Were Erik to tell all his friends that he is a spy, Charlie would not know whether he is in danger.

- Were I to announce to all my friends that they are my friends, they would know this.

We will base our analysis on agent-indexical propositions, such as the first and last of the above, using names to refer to specific agents, and treating expression such as 'all my friends' as modal operators. The analysis of communication in such a setting involves careful attention to the perspective from which the communication is described. All three components of the communication (the sender, the message, and the receivers) can be specified in a variety of ways that need to be distinguished. In particular, the receivers of a message may be listed explicitly, or described as 'Bella's friends', or even as 'my friends', so that a sender may not know exactly who receives his message. Likewise, the content of the message may be about the sender or the receiver. For example, Charlie may tell Bella 'you are in danger' (about the receiver, Bella) or 'I am not a spy' (about the sender, Charlie). He may broadcast to all 'my friends are in danger', which if Bella is a friend, and does this, will enable her to infer that she is in danger, or send a message only to her friends that they are in danger. All such possibilities, together with their epistemic consequences, will be studied in subsequent sections.

The technical framework for this work is general dynamic dynamic logic (GDDL) [4], which provides a method for extending modal logics with dynamic operators for reasoning about a wide range of model-transformations, starting with those definable in propositional dynamic logic (PDL) and extended to allow for the more subtle operators involved in, for example, private communication, as represented in dynamic epistemic logic (DEL) and related systems. We provide a hands-on introduction to GDDL, introducing elements of the formalism as we go, showing how GDDL can be employed in a two-dimensional setting, but leave the reader to consult [4] for further details.

2 A language of social knowing

To represent the logical structure of propositions and reasoning about knowledge in a social context, specifically those involving friendship, we extend epistemic logic to EFL, an *epistemic logic of friendship*. The language is based on atoms of two types: propositional variables $\rho \in$ Prop representing indexical propositions such as 'I am in danger', and (a finite set of) agent nominals $n \in$ ANom which stand for indexical propositions asserting identification: 'I am n'. The language is then inductively defined as:

$$\varphi ::= \rho \mid n \mid \neg\varphi \mid (\varphi \land \varphi) \mid K\varphi \mid F\varphi \mid A\varphi \mid \downarrow n\, \varphi$$

We read K as 'I know that', F as 'all my friends', A as 'every agent', $\downarrow n\, \varphi[n]$ as 'φ (me) holds from my perspective'.

As usual in modal logic we can define the duals of the operators, which we write inside angle brackets: $\langle K \rangle = \neg K\neg$ 'it is epistemically possible for me that', $\langle F \rangle = \neg F\neg$ 'I have a friend who', and $\langle A \rangle = \neg A\neg$ 'there is someone who'. We also use abbreviations for the hybrid-logic-like operators $@_n\varphi = A(n \to \varphi)$ (equivalently, $\langle A \rangle(n \land \varphi)$).[1] So, for example, if $\underline{n}$ is Charlie then the operator $@_n$ simply shifts the indexical subject to Charlie, so that $@_n d$ means 'Charlie is in danger'.

The English glosses are not so exact and require some manipulation to get proper translations because of the way pronouns work in English. For example, if d represents 'I am in danger' then $\langle F \rangle K d$ means 'I have a friend who knows that he is in danger' rather than 'I have a friend who I know that I am in danger' which

---

[1] Although reminiscent of hybrid logic, the 'agent nominals' n, binder $\downarrow n$ and now the operator $@_n$ are not exactly the same as their hybrid-logic namesakes, but are rather some sort of two-dimensional cousins. A true nominal, for example, is a proposition that is logically compelled to be satisfied by exactly one evaluation index, which in the case of our models, would have to be the pair $\langle w, a \rangle$.

$\neg K @_e s$	I don't know that Erik (e) is a spy (s)
$F d$	All my friends are in danger
$K F d$	I know that all my friends are in danger
$@_e K F d$	Erik knows that all his friends are in danger
$F K d$	All my friends know they are in danger
$\langle F \rangle d$	Some of my friends are in danger
$\langle F \rangle c$	Charlie (c) is my friend
$\langle F \rangle K @_e d$	Some of my friends know that Erik is in danger
$\downarrow n \, \langle F \rangle K @_n d$	I have a friend who knows that I am in danger

Figure 1. Some statements of EFL

is not even grammatically correct! The hybrid feature of the language enables us to express indexical propositions. Finally, $\downarrow n \, FK \langle F \rangle n$ says 'all my friends know they are friends with me'. This provides a way of referring to 'me' inside the scope of other operators, by shifting the referent of n to the current agent. To illustrate, we give more examples in Figure 1.

Models for this language are two-dimensional Kripke models of the form $M = \langle W, A, k, f, V \rangle$, where W is a set (of epistemic states), A is a set (of agents), and

1. k is a family of equivalence relations k_a for each agent $a \in A$, representing the ignorance of a in distinguishing epistemic possibilities (as for standard S5 epistemic logic)

2. f is a family of symmetric and irreflexive relations f_w for each $w \in W$, representing the friendship relation in state w.

3. g is a function mapping each agent nominal $n \in$ ANom to the agent $g(n) \in A$ named by n. We abbreviate $g(n)$ to $\underline{n}$ when the model is clear from the context.

4. V is a valuation function mapping propositional variables Prop to subsets of $W \times A$, with $(w, a) \in V(p)$ representing that the indexical proposition p holds of agent a in state w.

In order to interpret the $\downarrow$ operator, we introduce a slightly different mapping: For $m \in$ ANom,

$$g_{[a]}^{[n]}(m) = \begin{cases} a & \text{if } m = n \\ g(m) & \text{otherwise} \end{cases}$$

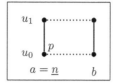

Figure 2. A simple EFL model

Models $\langle W, A, k, f, g[^n_a], V \rangle$ based on this mapping is denoted as $M[^n_a]$, which is the result of changing M so that n now names a.

For example, Figure 2 illustrates a simple model for a language in which there is only one propositional variable p and one agent name n. The set of states is $W = \{u_0, u_1\}$ and the set of agents is $A = \{a, b\}$, with $g(n) = a$, n naming agent a. Both agents are ignorant about which state they are in, so $k_a = k_b$ is the universal relation. These are indicated by the two columns of the diagram. The left column displays the k_a relation with a thick line; the right column displays the k_b relation, similarly. The lines are non-directional because the relations are assumed to be symmetric. In more complex diagrams, we will assume that the relations depicted are the reflexive, transitive closures of what is shown explicitly. The rows of the diagram show the relations f_{u_0} (first row) and f_{u_1} (second row) with dotted lines. This represents the two agents being friends in both states of W. Again these are non-directional because we assume symmetry. But for these lines we do *not* take the reflexive, transitive closure, since we assume that f_w is irreflexive and may or may not be transitive. Finally, that p holds only of agent a in state u_0, i.e., that $V(p) = \{(u_0, a)\}$ is shown by labelling the lower left node of the diagram with p.

Models are used to interpret $\mathcal{L}$ in a double-indexical way, as follows:

$M, w, a \models \rho$	iff	$(w, a) \in V(\rho)$, for $\rho \in$ Prop
$M, w, a \models n$	iff	$g(n) = a$, for $n \in$ ANom
$M, w, a \models \neg\varphi$	iff	$M, w, a \not\models \varphi$
$M, w, a \models (\varphi \wedge \psi)$	iff	$M, w, a \models \varphi$ and $M, w, a \models \psi$
$M, w, a \models K\varphi$	iff	$M, v, a \models \varphi$ for every $v \in W$ such that $k_a(w, v)$
$M, w, a \models F\varphi$	iff	$M, w, b \models \varphi$ for every $b \in A$ such that $f_w(a, b)$
$M, w, a \models A\varphi$	iff	$M, w, b \models \varphi$ for every $b \in A$
$M, w, a \models \downarrow n \, \varphi$	iff	$M[^n_a], w, a \models \varphi$.

We say that M is a *named agent* model, if every agent in M has a name, i.e., for each $a \in A$, there is an $n \in$ ANom such that $g(n) = a$. The model depicted in Figure 2 is *not* a named agent model because agent b has no name. Our discussion in this context will not be restricted to named agent models.

2.1 Transforming models with PDL

We will define a class of operators $\mathcal{D}$ and corresponding actions on models such that for each $\Delta \in \mathcal{D}$ and each M model for $\mathcal{L}$, there is an $\mathcal{L}$ model ΔM, and for each state w of M, a state Δw of ΔM. We then extend $\mathcal{L}$ to a language $\mathcal{L}(\mathcal{D})$ of *dynamic epistemic friendship logic* (DEFL) by adding the elements of $\mathcal{D}$ as propositional operators and defining

$$M, w, a \models \Delta \varphi \qquad \text{iff} \quad \Delta M, \Delta w, a \models \varphi$$

To define $\mathcal{D}$, we use the language of propositional dynamic logic (PDL) with basic programs K, F and A, given by

$$\begin{aligned}
\mathcal{T} \quad & \pi ::= K \mid F \mid A \mid \varphi? \mid (\pi; \pi) \mid (\pi \cup \pi) \mid \pi^* \\
\mathcal{F} \quad & \varphi ::= \rho \mid n \mid \neg\varphi \mid (\varphi \vee \varphi) \mid \langle \pi \rangle \varphi
\end{aligned}$$

for $\rho \in \mathsf{Prop}$ and $n \in \mathsf{ANom}$. The denotation of program terms $\pi \in \mathcal{T}$ and formulas $\varphi \in \mathcal{F}$ in a model M are defined in the manner shown in Table 1. Note

$$\begin{aligned}
\llbracket \rho \rrbracket^M &= V(\rho), \text{ for } \rho \in \mathsf{Prop} \\
\llbracket n \rrbracket^M &= W \times \{g(n)\}, \text{ for } n \in \mathsf{ANom} \\
\llbracket (\varphi \wedge \psi) \rrbracket^M &= \llbracket \varphi \rrbracket^M \cap \llbracket \psi \rrbracket^M \\
\llbracket \neg\varphi \rrbracket^M &= W \setminus \llbracket \varphi \rrbracket^M \\
\llbracket \langle \pi \rangle \varphi \rrbracket^M &= \{w \in W \mid w\llbracket \pi \rrbracket^M v \text{ and } v \in \llbracket \varphi \rrbracket^M \text{ for some} \\
& \quad v \in W \} \\
\llbracket K \rrbracket^M &= \{\langle (w, a), (v, a) \rangle \mid k_a(w, v)\} \\
\llbracket F \rrbracket^M &= \{\langle (w, a), (w, b) \rangle \mid f_w(a, b)\} \\
\llbracket A \rrbracket^M &= \{\langle (w, a), (w, b) \rangle \mid a, b \in A, w \in W\} \\
\llbracket \varphi? \rrbracket^M &= \{\langle w, w \rangle \mid w \in \llbracket \varphi \rrbracket^M\} \\
\llbracket \pi_1; \pi_2 \rrbracket^M &= \{\langle w, v \rangle \mid w\llbracket \pi_1 \rrbracket^M s \text{ and } s\llbracket \pi_2 \rrbracket^M v \text{ for some } s \in \\
& \quad W \} \\
\llbracket \pi_1 \cup \pi_2 \rrbracket^M &= \llbracket \pi_1 \rrbracket^M \cup \llbracket \pi_2 \rrbracket^M \\
\llbracket \pi^* \rrbracket^M &= \{\langle w, v \rangle \mid w = v \text{ or } w_i \llbracket \pi \rrbracket^M w_{i+1} \text{ for some } n \geq 0, \\
& \quad w_0, \ldots, w_n \in W, w_0 = w \text{ and } w_n = v\}
\end{aligned}$$

Table 1. Semantics of PDL terms and formulas

in particular, the clauses for K, F and A, in which these program terms refer to the accessibility relations of the corresponding operators of EFL, when interpreted two-dimensionally. Complex program terms are built up in the usual way: $(\pi_1; \pi_2)$ for the relational composition of π_1 and π_2, $(\pi_1 \cup \pi_2)$ for their union (or choice), $\varphi?$ for the 'test' consisting of a link from (w, a) to itself iff $M, w, a \models \varphi$, and π^* for the reflexive, transitive closure of π, which is understood as a form of iteration.

Note also that we have abused notation so that formulas φ of EFL, written with existential operators $\langle K \rangle$, $\langle F \rangle$ and $\langle A \rangle$, are also programs formulas (in F). This is justified by the obvious semantic equivalence:

$$M, w, a \models \varphi \text{ iff } (w, a) \in [\![\varphi]\!]^M$$

Now the class of dynamic operators will be defined using the theory of General Dynamic Dynamic Logic (GDDL) given in [4], which provides an extension to any language of PDL. The simplest of these operators are called PDL-*transformations*. These consist of assignment statements which transform models by redefining the basic programs. For example, the operator $[K := \pi]$ acts on model M to produce a new model $[K := \pi]M$ such that

$$[\![K]\!]^{[K:=\pi]M} = [\![\pi]\!]^M$$

On states, there is no change: $[K := \pi]w = w$, so the resulting DEFL operator has the following semantics:

$$M, w, a \models [K{:=}\pi]\varphi \quad \text{iff} \quad [K{:=}\pi]M, w, a \models \varphi$$

We must be a little careful in the choice of π so as to ensure that the resulting model $[K := \pi]M$ is still a model for EFL. For example, consider the program term $n?; K$. In M, this relates (u_0, a) to (u_1, b) in case $(u_0, a) \in [\![n]\!]^M$ and $(u_0, a)[\![K]\!]^M(u_1, b)$, which only holds when $g(n) = a$, $a = b$, and $k_a(u_0, u_1)$. Then $[K := n?; K]M$ is the structure $\langle W, A, k', f, V \rangle$ in which $k'_a = k_a$ and $k'_b = \emptyset$, for $b \neq a$. This is *not* a model for EFL. To make it into a model for EFL, we need to make each k_a reflexive. This can be done with the program term $\top?$, since $[\![\top?]\!]^M$ is the identity relation. Thus taking π to be $(n?; K) \cup \top?$ we get the model $[K := (n?; K) \cup \top?]M$ which is the structure $\langle W, A, k'', f, V \rangle$ in which $k''_a = k_a$ and k''_b is the identity relation for all $b \neq a$. The application of $[K := (a?; K) \cup \top?]$ to a particular model is illustrated in Figure 3. Here, M is a named agent model, so we allow ourselves the abuse of notation involved in writing a for the name of a. In this model there are two friends, a and b, who are both ignorant about whether they are in state u_0 or u_1. p holds only of agent a in state u_0, so in particular, $M, u_0, b \models (K\neg p \wedge \neg K \langle F \rangle p)$, which means that agent b knows that she is not p but does not know whether she has a friend who is p. After the action $[K := (n?; K) \cup \top?]$ we get the model shown on the right, in which k_a is as before but now k_b is the identity relation. In the transformed model, agent b now knows that she has a friend who is p. Thus we get the dynamic fact:

$$M, u_0, b \models [K := (n?; K) \cup \top?]K \langle F \rangle p$$

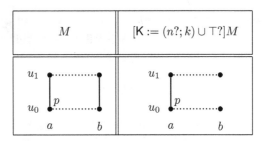

Figure 3. A simple PDL-transformation.

In effect, the PDL-transformation, $[K := (n?; K) \cup \top?]$ is the action of revealing everything to every agent other than $\underline{n}$. We will consider more subtle forms of epistemic change in subsequent sections. Now it is time for an analysis of the spy network example in the Introduction. We repeat the story here.

> *Berlin 1978. A spy network has recently been uncovered by the Stasi, who are rounding up the spies and their associates. Bella (b) is friends with Charlie (c) and Erik (e), neither of whom are friends with each other. Unknown to the others is that Erik is a spy (s). The others are not spies, and Erik knows that because all spies know who else is a spy (we suppose). Bella knows that Charlie is not a spy, but Charlie does not know about her. After the network is exposed, all the spies and their friends will be interrogated by the police. But just before this happens a message is relayed to all agents revealing whether or not they are in danger, that is, whether they are a spy (which they would know in any case) or a friend of a spy. Who now knows that Erik is a spy?*

A model M of the initial situation is depicted in Figure 4, with u_0 representing the actual state. In EFL we can state pertinent facts such as $@_b(K\neg s \land \neg K \langle F \rangle s)$ 'Bella knows that she is not a spy but doesn't know if a friend of hers is a spy'. We will write d 'I am in danger' as an abbreviation for $(s \vee \langle F \rangle s)$ 'either I'm a spy or I have a spy as a friend', and, for convenience, we have labelled those state-agent pairs at which d holds. Thus we can read that $@_b(d \land \neg Kd)$ 'Bella is in danger but doesn't know it', whereas $@_b K @_c \neg d$ 'Bella knows that Charlie is not in danger'.

Now consider the PDL-term $\text{cut}_K(\varphi)$ defined by

$$(\varphi?; K; \varphi?) \cup (\neg\varphi?; K; \neg\varphi?)$$

Before M	After $[K := \mathsf{cut}_K(d)]M$

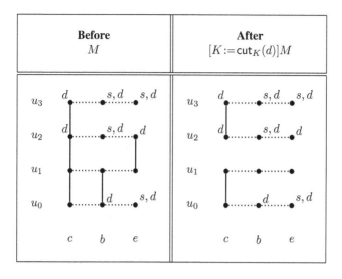

Figure 4. Spy Network

This relates $\langle w, a \rangle$ to $\langle v, b \rangle$ iff $a = b$, $k_a(w, v)$, and either φ is true of a in both states w and v or false of a in both states. Thus the operator $[K := \mathsf{cut}_K(\varphi)]$ produces a new model $[K := \mathsf{cut}_K(\varphi)]M$ from M by removing the k_a links between states with conflicting values for φ (about a). Effectively, this 'reveals' to each agent whether or not φ holds (for them).[2]

In our example, the situation after the revelation of d 'you are in danger' is given by the model $[K := \mathsf{cut}_K(d)]M$, shown in the right part of Figure 4. Notice that the k_c link between u_1 and u_2 are cut because $M, u_1, c \not\models d$ but $M, u_2, c \models d$; Charlie finds out that he is not in danger. Similarly, the k_b link between u_0 and u_1 is cut because Bella finds out that she *is* in danger ($@_b Kd$). Finally, the k_e link between u_1 and u_2 is cut because everyone now knows that Erik knows whether he is in danger (although only Bella knows which). Reasoning about such changes can be represented in the language of DEFL such as the valid schema

$$[K := \mathsf{cut}_K(\varphi)]A(K\varphi \vee K\neg\varphi)$$

which states (for non-epistemic facts φ such as $d = \langle F \rangle s$) that after φ is revealed, everyone knows whether φ or not.

After the same update we can also ask who then knows that Eric is a spy (and did not know before)? Formally put in DEFL,

---

[2]This operator was first introduced in [15].

For which n, $@_n(\neg K @_e s \wedge [K := \mathsf{cut}_K(d)] K @_e s)$?

Again reading from the right part of the Figure 4, we see that it is Bella who knows that Eric is a spy, as Charlie is still uncertain.

2.2 GDDL **operators**

More complicated operators can be constructed from finite relational structures whose elements are each associated with a PDL transformation, and whose combined effect on the model is calculated by 'integrating' them according to a further transformation. A GDDL operator Δ is something that looks like this:

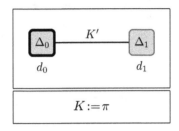

This represents an action d_0 (highlighted as the action that is actually performed) whose effect on the model is given by the PDL-transformation Δ_0. There is also an action d_1 with associated PDL-transformation Δ_1, and the relationship between d_0 and d_1 is marked as K'.[3] The effect of the operator on an EFL model M with domain W is computed by forming a product model M' (in the manner of [2]) whose domain is $W \times \{d_0, d_1\}$, in which the elements (w, d_i) represent the state resulting from action d_i when the initial state is w. The model M' consists of copies of two models $[\Delta_0]M$ with domain $W \times \{d_0\}$ and $[\Delta_1]M$ with domain $W \times \{d_1\}$, and a duplication of the model occurring in Δ itself, with, in this case, $(w, d_0)[\![K']\!]^{M'}(w, d_1)$ for each $w, v \in W$. Finally, the model $[\Delta]M$ is computed by applying the 'integrating' transformation $[K := \pi]$ to M'. This uses a PDL program term π to compute the new value for K from a combination of relations in the copied models $[\Delta_0]M$ and $[\Delta_1]M$ and the new relation K' from Δ itself.[4]

This somewhat complex operation is best explained by looking at a simple example. Consider the case in which Δ_0 is the PDLtransformation $[K := (a?; K) \cup \top?]$ considered earlier, and Δ_1 is the identity transformation, I. We will also take π to be $(K \cup a?; K')^*$.

---

[3] In the general case, as explained in [4], there may be many actions and many new relation symbols; also, propositional variables.

[4] Again, the general case is more flexible, allowing any of the basic expressions K, F, agent nominal and propositional variable to be reinterpreted at the integrating stage.

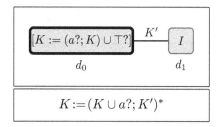

The action of this GDDL operator on the model M considered earlier, is shown in Figure 5. It represents a situation in which an action d_0 gives complete information to all agents other than a. The occurrence of d_0 is known to all agents other than a, who stays completely in the dark. Not only is k_a unchanged in both $[\Delta_0]M$ (the top half of the diagram) and $[\Delta_1]M$ (the bottom half), but a is also ignorant about which of these two submodels she is in, as represented by the vertical lines in connecting the two halves of the a column: $(w, d_0)k_a(w, d_1)$ for all $w \in W$. Once

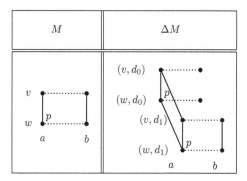

Figure 5. A simple GDDL operator in action.

again, we must check that the resulting model is an EFL model. In this case, it is. The k_a and k_b relations are transitive thanks to the application of the $*$ operator in the integrating transformation $[K := (K \cup a?; K')^*]$.

We'll say that a GDDL-transformation Δ is a *general* EFL *dynamic operator* if it is in the language of PDL terms defined above, possibly augmented with internal relations such as K' and also preserves the property of being a EFL-model: whenever M is an EFL-model, so is ΔM. Further work is needed to characterise this class syntactically.

3 Social announcements

It is time to consider direct communications, or 'announcements', within a social network. As we stated before, an announcement in a social network consists of three components (i) the sender (ii) the receiver(s) and (iii) the dynamic transmitting of some message.[5] In this section, we will study each component and try to capture the subtleties of social communication, in particular, when the indexical propositions are involved.

As a starting point, we ignore the sender and define a basic act of communication in which a message ψ is sent (anonymously, we suppose) to a group of agents described by formula θ by

$$\mathsf{send}_\theta(\psi) = [K := (\theta?; \mathsf{cut}_K(\psi)) \cup (\neg\theta?; K)]$$

The action $\mathsf{send}_\theta(\psi)$ reveals the truth or falsity of ψ (which may be different for different agents) to all agents satisfying θ, and leaves the k_a relation unchanged for agents a not satisfying θ.

To see how this works, consider $\mathsf{send}_{\langle F\rangle b}(d)$ in the case of our spy network. This is an anonymous announcement to the friends of Bella (but not to Bella herself) whether or not they are in danger. The effect of this action is shown in Figure 6. The formula θ describing the receivers of the message is $\langle F\rangle b$, which is satisfied by Charlie and Erik in the actual state u_0. Thus only the relations k_c and k_e are changed; k_b remains the same. This is by no means our final analysis of communication. For one thing, actions of this sort are only 'semi-private', i.e., directed at particular individuals, but with others not involved in the communication still aware that it has occurred. Later, we will need to make the analysis more complex to cope with a great degree of privacy, in which only the sender and receivers are aware that the communication has occurred. For example, after the communication to Bella's friends, Bella knows something that she didn't know before: before she knew that Charlie was not in danger, now she knows that Charlie knows this:

$$M, u_0, b \models [\mathsf{send}_{\langle F\rangle b}(d)] K @_c K \neg d$$

Yet before we get to the issue of privacy, we will bring the sender into our model, and explore some subtle distinctions about the nature of the message itself.

---

[5]We are aware of the attempts by others in this respect. [10] analysed specific types of communication network (i.e., communications that take place between one agent and another, or between an agent and a group of agents) when considering the issue of how distributed knowledge can be established by a group of agents through communication. Communication graphs were adopted by [8] to study communication between agents. Agent i directly receiving information from agent j is represented by an edge from agent i to agent j in such graph. Neither approach considers groups of agents described in terms of social relations.

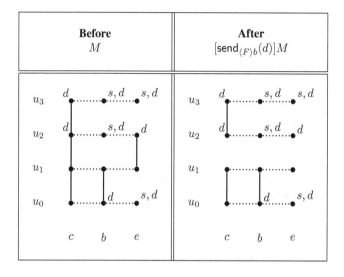

Figure 6. Restricting to Bella's friends

3.1 Announcements about the sender

The first case is that of a message sent by agent n to *agents described by* θ with a message ψ, which is understood to be about the sender, e.g. 'I am in danger'. We define $[n \lhd \psi! : \theta]\varphi$, the statement that φ would hold after such a communication, as

$$(@_n K\psi \rightarrow [\mathsf{send}_\theta (@_n \psi)]\varphi)$$

To make sense of this, we will look at a progression of simpler cases. First, with $\theta = \top$, the formula $[n \lhd \psi! : \top]\varphi$ means that φ would hold were agent n publicly to announce that ψ, noting that it simplifies to $(@_n K\psi \rightarrow [K := \mathsf{cut}_K(@_n \psi)]\varphi)$.

We make the rather strong assumption that the message is known by the sender.[6] Suppose, for example, that Erik, unable to keep his secret any longer, told everyone that he is a spy. After this, everyone would know that he is a spy (and Bella, his friend, would know that she is in danger). This follows from the validity of $[e \lhd s! : \top]AK@_e s$.[7] Note that $[b \lhd s! : \top]AK@_b s$ is also true (since it is valid!). This says that everyone would know that Bella is a spy were she to announce it. But the reason is quite different: Bella could not announce that she is a spy,

---

[6]The standard assumption of PAL that announcements are true is thus equivalent to supposing that they are made by God, or some other omniscient entity. [6] studied different types of agent (truth-teller, liar and bluffer), how they make announcements, and are subsequently interpreted in communication.

[7]In fact, the information that Erik is a spy becomes common knowledge.

because she knows that she isn't.[8]

The second case is an announcement to *a particular agent*. In this case, θ is an agent nominal m and the formula $[n \lhd \psi! : m]\varphi$ means that φ would hold were agent n to announce to m that ψ. For example, Erik may be more cautious in his admission, telling only Bella, after which she, but not Charlie would know: $[e \lhd s! : b]@_b K@_e s$ and $(\neg(b \lor K@_e s) \rightarrow [e \lhd s! : b]\neg K@_e s)$ are both valid, and the latter says that an agent who is neither Bella nor (already) knows that Erik is a spy, still doesn't know this after he announces it to Bella.

In the most general case, θ is a description of a group of agents. For example, $[b \lhd \neg s! : \langle F \rangle b]\varphi$ states that φ would hold after Bella tells her friends that she is not a spy. Again we have a useful validity: $[b \lhd \neg s! : \langle F \rangle b]@_b FK@_{b}\neg s$, which says that if Bella were to tell her friends that she is not a spy then they would all know that she isn't a spy.

3.2 Announcements about the receivers

Announcements that are indexical about the receiver such as 'you are in danger' (announced to Bella by Erik) or 'you are my friends' (announced by Bella to her friends) can be expressed with a slight change that captures the different preconditions for announcements of this kind. We define $[n : \psi! \rhd \theta]\varphi$, the statement that φ holds after agent n announces message ψ (about θ) to agents satisfying θ as

$$(@_n KA(\theta \rightarrow \psi) \rightarrow [\mathsf{send}_\theta(\psi)]\varphi)$$

Again, we first consider the simple case of public announcement, represented by $[n : \psi! \rhd \top]\varphi$, which can be seen to be equivalent to $(@_n KA\psi \rightarrow [K := \mathsf{cut}_K(\psi)]\varphi)$. Consider, for example, my announcing to everyone 'you are in danger'. The precondition that I know everyone is in danger is captured by the antecedent KAd, and after the announcement everyone knows that she is in danger, as is represented by the validity of $\downarrow n\, [n : d! \rhd \top]AKd$.

The case of agent-to-agent announcement displays a nice symmetry between the two kinds of indexical message. Agent n announcing 'you are in danger' to agent m is equivalent to announcing (again to m) that m is in danger. More generally, the following equivalences are valid

$$[n : \psi! \rhd m]\varphi \leftrightarrow [n \lhd @_m \psi! : m]\varphi$$
$$[n \lhd \psi! : m]\varphi \leftrightarrow [n : @_n \psi! \rhd m]\varphi$$

This symmetry between announcements is more delicate when announcing to groups. Announcing 'you are in danger' to each of my friends is only the same as

---

announcing to them 'all my friends are in danger' on the assumption that each friend knows only that she is my friend, and knows nothing about the others. Without this assumption,

$$[n : \psi! \rhd \langle F \rangle n] \varphi \ \leftrightarrow \ [n \lhd @_n F \psi! : \langle F \rangle n] \varphi$$

is not always valid.[9]

For announcement to friends, an interesting new phenomenon arises. Consider the case of my announcing 'you are my friend' to my friends. That φ holds after such an announcement is represented by $[n : \langle F \rangle n! \rhd \langle F \rangle n]$. The message is the same as the description of the set of receivers, so when this is expanded, we find that the precondition for the announcement is $\downarrow n \, KA(\langle F \rangle n \to \langle F \rangle n)$, which is valid, so the announcement can always be made, by anyone. But nonetheless, it can be informative, as can be seen from the validity of $\downarrow n \, [n : \langle F \rangle n! \rhd \langle F \rangle n] FK \langle F \rangle n$, which says that after my making this announcement, my friends all know that they are my friends, something they may not have known before.

Finally, we note that any sender-indexical announcement to a group θ is equivalent to a receiver-indexical announcement to the same group θ in the case that there is at least one receiver ($A \neg \theta$ is false). The trick is that the statement ψ about n (the sender) is then equivalent to the statement $@_n \psi$ about any (every) receiver. More formally, the following is valid:[10]

$$(\neg A \neg \theta \to [n \lhd \psi! : \theta] \varphi \ \leftrightarrow \ [n : @_n \psi! \rhd \theta] \varphi)$$

3.3 Private announcements

Communications of the form $[n \lhd \psi! : \theta]$ and $[n : \psi! \rhd \theta]$ are only semi-private. Their effect on the model ensures that every agent will know that the announcement has occurred, if the sender satisfies the precondition, so, for example,

$$\downarrow n \, [n \lhd d! : m] AK(@_n Kd \to @_m K @_n d)$$

---

[9]For a simple counterexample, consider ψ to be d and the model M (shown left).

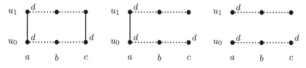

The precondition of $[b : d! \rhd \langle F \rangle b]$ is $@_b KA(\langle F \rangle b \to d)$, which is equivalent to the precondition $@_b KFd$ of $[b \lhd @_b Fd! : \langle F \rangle b]$ which is satisfied in M, and the resulting two models are shown middle and right. Yet these are easily distinguished, by taking φ to be $@_a K @_c d$.

[10]The key observation here is that the precondition for the sender-indexical announcement is $@_n K \psi$, which is equivalent to $@_n K U_A (\theta \to @_n \psi)$ when $U_A \neg \theta$ is false.

is valid: after I announce to m that I am in danger, everyone will know that if I know I am in danger then m also knows it. This is (typically) an unjustified violation of the privacy of the communication between me and m.

To make the action $\mathsf{send}_\theta(\psi)$ private, we embed it in a GDDL operator similar to the one given in our earlier example. Thus, for the sender-indexical[11] version, that φ would hold after the private announcement of ψ by n to agents θ is be represented as

$$(@_n K\psi \quad \rightarrow$$ 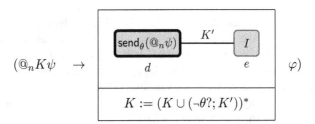 $$\varphi)$$

Call this formula $[n \triangleleft \psi! : \theta]\varphi$. Inside the GDDL operator, the internal relation K' represents ignorance about whether the communication $\mathsf{send}_\theta^n(\psi)$ has occurred or not, the latter possibility represented by the identity transformation, I. The integrating transformation $[K := (K \cup (\neg\theta?; K'))^*]$ restricts ignorance of the K' kind to agents other than θ and factors this in to the new epistemic relation. The $*$ is needed to ensure that the result is an equivalence relation. We will see an example of this operator in action at the end of the next section.

4   Knowing your friends

So far, the friendship relation in our models has been relatively tame, remaining fixed across epistemic states. We have used it to determine which group of agents receive a message, and even to specify the content of a message, but we have not yet considered ignorance about who is friends with whom. This is where it gets really interesting. We will explore some of the possibilities with an everyday example of infidelity and gossip.

> *Peggy (p) knows that Roger (r) is cheating (c) on his wife, Mona (m). What's more, Roger knows that Peggy knows, because they met accidentally while he was with his mistress. Mona does not know about the affair, and both Peggy and Roger know this. The situation (for Roger) deteriorates when he discovers that Peggy is a terrible*

---

[11] The receiver-indexical version is obtained by changing the message and the precondition as in the simple semi-private case.

gossip. She is bound to have told all her friends about his affair. What Roger does not know is whether Mona is a friend of Peggy (she is).

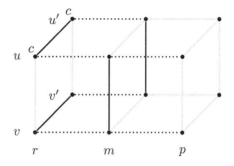

Figure 7. Roger's Quandry

We can represent the epistemic state of this network before Peggy's announcement with the model depicted in Figure 7, assuming that married couples are also friends. (The grey construction lines are only included to make the diagram easier to read; they have no epistemic or social significance.) Note that the friendship relations are now different in different states. At u (the actual state) for Roger r, the statements listed in Table 2 are all true. As a result, we can compute that at w in the original model for Roger r, the formula

$$\downarrow n\, [p \triangleleft @_n c! : \langle F \rangle p] @_m K @_n c$$

is true, i.e., "I don't know that Mona will know about my cheating after Peggy tells her friends about it." That some proposition φ holds after the announcement 'Roger is cheating!' that Peggy makes to her friends is given by $[p \triangleleft @_r c! : \langle F \rangle p]\varphi$, which expands and simplifies to

$$(@_p K @_r c \rightarrow [K := (\langle F \rangle p?; \mathsf{cut}_K (@_r c)) \cup (\neg \langle F \rangle p?; K)]\varphi)$$

When evaluated at u, the presupposition that Peggy knows that Roger is cheating is satisfied, and so the formula φ is evaluated in the transformed model shown in Figure 8. (Note the missing vertical line in the middle.)

This is all very well, but Roger needs a little more privacy.

Before returning home to face Mona, Roger is uneasy. He would really like to know whether or not she knows about his affair. He

c	I'm cheating
$\downarrow n\ K(@_pK@_nc\ \wedge$ $@_m\neg K@_nc)$	I know that Peggy (but not Mona) knows I am cheating.
$\downarrow n\ @_pK@_nK@_pK@_nc$	Peggy knows I know she knows I am cheating
$\neg K@_m\langle F\rangle p\ \wedge$ $\neg K@_m\neg\langle F\rangle p$	I don't know whether Peggy and Mona are friends.
$\downarrow n\ @_pK@_n\neg K@_m\langle F\rangle p$	Peggy knows I don't know whether she and Mona are friends.

Table 2. Facts about Roger

already knows that she knows if and only if she is friends with Peggy. So if Peggy told him that they are friends, he would be prepared for Mona's fury. But for his planned excuses to be convincing, Mona must not know that he knows she knows (about the affair). It is therefore very important that Peggy tells him in private.

Now let us suppose that the ever-loquacious Peggy announces to Robert privately that Mona is her friend, represented as $[p \triangleleft \langle F\rangle m! : r]$. Now, whether the crucial proposition φ

$$(@_rK@_mK@_rc \wedge \neg@_mK@_rK@_mK@_rc)$$

(that Roger knows Mona knows he has been cheating but Mona doesn't know that he knows) holds must be determined by evaluating it in the model obtained by transforming the one in Figure 8 using the following GDDL operator, call it Δ:

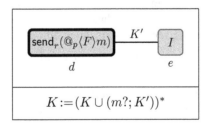

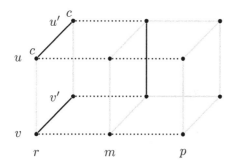

Figure 8. After Peggy's gossip

The result is shown in Figure 9.

The upper half of the diagram represents the result of action d, Peggy telling Roger that she is friends with Mona ($\text{send}_r(@_p\langle F\rangle m)$), whereas the lower half represent the result of action e, nothing (I); it is just a copy of the model in Figure 8. Mona is the only one of the three who doesn't know which action has taken place, and her ignorance is represented by the lines connected corresponding states in the upper and lower halves (in the m column). We see that $K@_m K@_r c$ holds of r in state (u, d), so Roger can meet Mona prepared.[12]

We may wonder about the accuracy of the model in representing Roger and Mona as friends after Peggy's announcement. This will probably lead to changes to the social network, Roger and Mona may no longer be friends. We leave such issues for other occasions.

5   Conclusions

What has emerged from this study is an appreciation of the diversity of subtle logic distinctions when combining epistemic and social relations, especially when allowing indexical propositions, as are very common in the social setting. The patterns of inference we commonly use when reasoning about everyday social situations have been shown to be more intricate that one might first have thought. In this paper we have not touched on the topic of common knowledge in the social setting. Some initial observations on this are given in our [14]. Work by Ruan and Thielscher [11] and by Sano and Tojo [12] adopt a similar approach to the formalisation of 'logic in the community' [13]. Yet, although [11]

---

[12]Even the additional level of privacy offered here is still not perfect, as it involves some change in Mona's knowledge. She goes from knowing that Roger doesn't know that she is friends with Peggy to not knowing this. However, one may just think that privacy is a matter of degree.

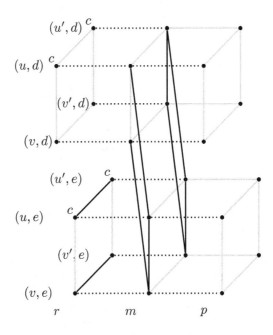

Figure 9. Peggy to Roger, privately.

includes a common knowledge operator, it captures only common knowledge of an enumerated set of agents (as in the traditional setting of multi-agent, non-social epistemic logic) and so does not capture the perspectival distinctions we have been emphasising here. The more recent [12] focusses on belief rather than knowledge, with a more limited range of dynamic operators but a more direct axiomatisation of the logic. All these approaches are compatible and point to future developments studying the interaction between propositional attitudes, indexicality and communication within social networks.

BIBLIOGRAPHY

[1] Carlos E. Alchourrón, Peter Gärdenfors, and David Makinson. On the logic of theory change: Partial meet contraction and revision functions. *Journal of Symbolic Logic*, 50:510–530, 1985.

[2] Alexandru Baltag, Lawrence S. Moss, and Slawomir Solecki. The logic of public announcements, common knowledge and private suspicions. Technical Report SEN-R9922, CWI, Amsterdam, 1999.

[3] Ronald Fagin, Joseph Y. Halpern, Yoram Moses, and Moshe Y. Vardi. *Reasoning about Knowledge*. Cambridge, MA: The MIT Press, 1995.

[4] Patrick Girard, Jeremy Seligman, and Fenrong Liu. General dynamic dynamic logic. In Thomas Bolander, Torben Braüner, Silvio Ghilardi, and Lawrence S Moss, editors, *Advances in Modal Logics Volume 9*, pages 239–260, 2012.

[5] Jaakko Hinttika. *Knowledge and Belief*. Ithaca: Cornell University Press, 1962.

[6] Fenrong Liu and Yanjing Wang. Reasoning about agent types and the hardest logic puzzle ever. *Minds and Machines*, 23(1):123–161, 2013.
[7] John-Jules Ch. Meyer and Wiebe van der Hoek. *Epistemic Logic for Computer Science and Artificial Intelligence*. Cambridge: Cambridge University Press, 1995.
[8] Eric Pacuit and Rohit Parikh. Reasoning about communication graphs. In Dov Gabbay Johan van Benthem, Benedikt Loewe, editor, *Interactive Logic*, pages 13–60. Amsterdam University Press, 2007.
[9] Jan A. Plaza. Logics of public announcements. In *Proceedings of the 4th International Symposium on Methodologies for Intelligent Systems*, 1989.
[10] Floris Roelofsen. Exploring logical perspectives on distributed information and its dynamics. Master's thesis, ILLC,The University of Amsterdam, 2005.
[11] Ji Ruan and Michael Thielscher. A Logic for Knowledge Flow in Social Networks. In Dianhui Wang and Mark Reynolds, editors, *AI 2011: Advances in Artificial Intelligence*, pages 511–520. Springer Berlin Heidelberg, 2011.
[12] Katsuhiko Sano and Satoshi Tojo. Dynamic epistemic logic for channel-based agent communication. In Kamal Lodaya, editor, *Logic and Its Applications*, volume 7750 of *Lecture Notes in Computer Science*, pages 109–120. Springer Berlin Heidelberg, 2013.
[13] Jeremy Seligman, Fenrong Liu, and Patrick Girard. Logic in the community. In Mohua Banerjee and Anil Seth, editors, *ICLA*, volume 6521 of *Lecture Notes in Computer Science*, pages 178–188, 2011.
[14] Jeremy Seligman, Fenrong Liu, and Patrick Girard. Facebook and the epistemic logic of friendship. In Burkhard C Schipper, editor, *Proceedings of the 14th Conference on Theoretical Aspects of Rationality and Knowledge*, pages 229–238, Chennai, India, 2013.
[15] Johan van Benthem and Fenrong Liu. Dynamic logic of preference upgrade. *Journal of Applied Non-Classical Logic*, 17:157–182, 2007.

Comments on Seligman, Liu and Girard

CARLO PROIETTI

Knowledge, friendship and social announcements continues a research line in-augurated by the same authors with *Logic in the community* and carried on with a series of papers on *Epistemic Friendship Logic* (EFL) whose general aim is to foster the analysis of "reasoning about knowledge and communication defined im-plicitly in terms of social relationships, most simply the binary relation of friend-ship". *Friendship* is intended here in its most general and abstract sense as any symmetric and irreflexive relation between agents building up a network – i.e. a graph where the latter are the points and the former constitute the edges. The formal framework on which the present paper builds upon is General Dynamic Dynamic Logic (GDDL), which consists of "a combination of epistemic logic, dy-namic logic and hybrid-like logic" (p. 2) and is meant to enable to reason about a wide range of model tranformations induced by information exchange and com-munication in multi-agent systems.

Reasoning about knowledge and social relationships is relevant for the analysis of security protocols of information exchange, which is one of the most signifi-cant applications for Dynamic Epistemic Logics (DEL) since [3]. The work pro-vides interesting countributions to this agenda by disclosing some specific model-theoretic dynamics of information exchange, in particular by specifying the model tranformations induced by

- private messages that carry information about the sender vs. messages that carry information about the receiver (sect. 3)

- two levels of information privacy: one where (a) only receivers acquire in-formational content but everybody is aware of the information exchange and the other where (b) only receivers acquire informational content and are aware of the exchange (sect. 3)

- providing information about the structure of the social network to some of the agents belonging to it (sect. 4).

Informational protocols are by no means the unique area of possible applica-tion for dynamic EFL to the extent that this logic has a special cross-disciplinary interest. Indeed, its potential lies in enabling to reason about *Social Networks*, i.e.

a concept of joint interest for Behavioural Economics and Social Psychology at least since [5]. The emergence and the dynamics of many collective phenomena are indeed explained by considerations on the features of abstract networks – one may think about Schelling's explanation of *segregation* and *herding* but also about specifically epistemic phenomena such as *informational cascades*. The study of social and economic networks constitutes an interesting and expanding discipline - as witnessed by the number of publications and the diffusion of extensive textbooks such as [4] and [2]- merging countributions and insights from different formal areas such as game theory, graph theory, experimental social psychology and probability theory. Research on EFL may constitute a relevant toolkit in this wide toolbox.

In view of these considerations I shall divide my comments into more "technical" ones (points 1 and 2) and more general questions and motivational issues for future research (points 3 and 4).

1. The language of EFL is constituted by indexical propositions $(p, q, \ldots)$ and by nominals $(i, j \ldots)$ and operators from hybrid logic $(\downarrow n, @_n)$. The model-theoretic interpretation of the latters is here a non-standard one, e.g. nominals name different points in the model (see fn. 3 at p. 7). Semantics thus formulated are intuitive, yet it would be interesting to explore – if the authors haven't done it elsewhere – the *first-order correspondence theory* of this language (see [1]. chap. 3.5). In the case of standard hybrid logic nominals are translated as constants, but that cannot be the case here. How should then one think about first-order correspondence for EFL?

2. An interesting technical question is hinted at and left for future exploration: "We'll say that a GDDL-transformation Δ is a *general EFL dynamic operator* if it is in the language of PDL terms defined above[...] and also preserves the property of being an EFL-model: whenever M is an EFL-model, so is ΔM. Further work is needed to characterize this class syntactically" (p. 14). A syntactic characterization would indeed provide an interesting limitative result discerning which program constructions of PDL are allowable and which ones are not. A further point is to ask what is "wrong" with the latters (as in the examples from the paper): are non-EFL transformations still interpretable as informational updates in the general sense of the term? If so, what type of updates do they represent? If not, why so?

3. As shown in section 4 with the "infidelity and gossip example" the network structure of friendship relations may be very relevant: indeed Peter does not know if he is part of a three-node path $m - p - b$ or of a triangle $m - p - b - m$ and this has a number of consequences. Social network theorists are typically interested in some prototypical structures such as the 'circle', the 'wheel', the 'star' or the 'stem'. It would be useful for a logic

of the community to be able to discriminate among different topologies via defining formulas, i.e. formulas that are valid in a model if and only if the model has a specific network structure – or else via formulas that are satisfied at a state if and only if this state is part of such a structure. It seems to me that at the actual state of the art general properties as 'being in a circle' are hardly definable with the EFL language and this may be a direction for future extensions.

4. Another issue of interest concerns the dynamic of friendship links among agents. In the examples of the paper friendship links are assumed to be stable across the different model transformations. This can be viewed as a limit of the actual EFL formalism. Indeed, a relevant topic in social network analysis is constituted by the study of the impact on global welfare of small structural changes in a network. *Ripple effects* (see [2] chap. 12) are a typical example of that and the"infidelity and gossip" example of section 4 also witnesses the importance of such structural modifications. Would it be possible to implement similar network transforming operations on the extant EFL framework? How so?

BIBLIOGRAPHY

[1] Blackburn P., de Rijke M. and Venema Y. (2001). *Modal Logic*. Cambridge (Uk): Cambridge University Press

[2] Easley, D. and Kleinberg, J. (2010). *Networks, Crowds and Markets*. Cambridge: Cambridge University Press Press.

[3] Fagin, R., Halpern, J. Y., Moses Y. and Vardi, M. Y. (1995). *Reasoning about Knowledge*. Cambridge: MIT Press.

[4] Jackson, M. O. (2008). *Social and Economic Networks*. Princeton: Princeton University Press.

[5] Schelling, T. (1978). *Micromotives and Macrobehaviour*. New York: Norton.

Comments on Seligman, Liu and Girard

Katsuhiko Sano

The authors' paper opens up a new perspective on logical dynamics of social knowledge over agents' network (e.g. Facebook). The paper contains intuitive and nice examples on spy network and office gossip, and also provides a general technical framework for dynamic operators of several types of social announcements. The paper is also clearly written. Therefore, I would like to supply references related to the authors' indexical reading of propositions (section 1). Moreover, I would like to give three technical questions (sections 2 and 3), all of which are not explained explicitly in the paper.

1   Relevant Works on Indexical Reading of Propositions

The authors read a propositional letter, say 'd', as an indexical sentence such as 'I am in danger'. As far as I know, such indexical reading traces back to Prior [5], where he used the term 'egocentric' for indexical sentences and regarded egocentric sentences as gerunds (see also descriptions by Blackburn [1]). Thus, Prior read d as '"Being in danger" is true of me'. Later in [3, ch.5.3], Lewis rewrote the information of 'me' into a free variable x and simplified Prior's reading of d into "x is danger" is true of me'. Of course, neither Prior and Lewis considered indexical sentences to deal with social knowledge, but we can situate the authors' indexical reading in such historical literatures on modal logic.

Lewis' notation may remind some readers of description logic, where we can describe roles and properties of individuals. In particular, Wolter and Zakharyaschev [8] proposed a *modalized* version of description logic (modal description logic) so that we can also handle modalized roles and properties of individuals. If we borrow (and generalize a bit) the notations of modal description logic from [8], we may symbolize some of the authors' examples as follows:

- Fd (all of my friends are in danger) $\rightsquigarrow$ $\forall$friends.Danger.

- FKd (all of my friends knows that they are in danger) $\rightsquigarrow$ $\forall$friends.KDanger.

- $@_eKFd$ (Eric knows that all of his friends are in danger) $\rightsquigarrow$
  $K\forall$friends.Danger(ERIC).

However, the authors' framework is still different from [8] in the following three respects. First of all, Wolter and Zakharyaschev did not consider any dynamic operators. Second, their modal description logic cannot express the dependence of an individual (say 'Eric') to the notion of modal operator (say 'Eric knows that ...'). Because of this, the syntax of Wolter and Zakharyaschev needs to have the individual symbol ERIC for Eric and the knowledge operator K_e for Eric independently. However, the authors' syntax and its two dimensional semantics can handle this dependence with the help of hybrid logic. In particular,

$$\mathfrak{M}, (w, a) \models @_{\text{ERIC}} K \varphi \quad \text{iff} \quad \text{for all } w' \text{ such that } k_{\underline{\text{ERIC}}}(w, w'), \mathfrak{M}, (w', a) \models \varphi,$$

where $@_{\text{ERIC}} K \varphi$ is read as 'Eric knows that he is φ' or 'Eric knows that φ is true of him' and $\underline{\text{ERIC}}$ is the corresponding individual (or value) of the name ERIC and $k_{\underline{\text{ERIC}}}$ is an agent-indexed binary relation on worlds. This idea seems relevant also to Priest's work [4] on intentional operators, where we have an intentional operator $t \Psi \varphi$ (an individual t Ψs that φ, where Ψ is an intentional verb) for each term t of FOL. Third, the notion of individual in [8] is world-dependent. That is, the names of individuals may change from worlds to worlds. On the other hand, the individuals in the authors' framework is rigid, i.e., the names of individuals are constant through worlds.

2 Recursive Axiomatizability of EFL and its PDL-expansion

One of the authors' main contributions is to develop a general framework to capture several kinds of social actions by agents. This is done by the following three steps:

EFL $\longrightarrow$ PDL expansion of EFL $\longrightarrow$ GDDL expansion of 'PDL expansion of EFL'.

First, the authors set up the two-dimensional syntax (EFL) capturing both knowledge and friendship relation between agents. This syntax can be regarded as a combination of epistemic logic (for world-axis) and hybrid logic with the downarrow binder (for agent-axis) and, as a result, its semantics employs a two dimensional frame

$$(W, A, (k_a)_{a \in A}, (f_w)_{w \in W}),$$

where W is a set of worlds, A is a set of agents, $k_a \subseteq W \times W$ is an equivalence relation, and $f_w \subseteq A \times A$ is irreflexive and symmetiric. Let us define the set of all the valid formulas on the class of all the two-dimensional frames above as the *logic* of EFL.

Second, with the help of the notion of PDL transformation, the authors provided a general method for extending two-dimensional syntax with dynamic operators such as announcements about the sender to his/her friends or announcements to the receivers. In particular, section 2.1 amounts to propose propositional dynamic logic (with Kleene star and test) over the two-dimensional models (in this sense, it would be nice to give the two-dimensional denotations to *all* the terms and formulas of Table 1 in the paper).

Third, by employing the authors' previous technique of GDDL [2], they expand PDL-expansion of EFL to cover dynamic operators such as private announcements. This enables us to consider action models in two-dimensional setting. In this third step, GDDL technique seems to allow us to obtain the *reduction* axioms for the obtained dynamic operators also on two-dimensional models. This implies that, if we can axiomatize the logic of the second step, then we can automatically obtain the complete axiomatization of the logic of the third step with the help of reduction axioms.

Now, I can state the following two questions to the authors.

(1) Can we axiomatize the logic of EFL?

(2) Can we axiomatize the authors' PDL-expansion of the logic of EFL?

Affirmative answers to the questions (1) and (2) will provide more solid logical basis with the authors' work and also give us a possibility of studying inference patterns of logical dynamics of social knowledge from proof-theoretic viewpoints. As for the question (1), if we hybridize also the world-axis with nominals and satisfaction operators, then the result of [6, Remark 4.16] (or a generalized version [7, Theorem 1]) allows us to obtain the axiomatization the (hybrid) logic of EFL *for free*, though I am not sure if hybridization of the world-axis have any effect on the third step above.

3 Decidability

Compared to FOL, one of the merits of modal logic consists in decidability, i.e., we can recursively check if a given formula is satisfiable or not. Since EFL contains the downarrow binder of hybrid logic, it may lead us to the undecidability result, provided we impose no assumptions on models. Then, one natural question is the following.

(3) Is the satisfiability problem of the logic of EFL without the downarrow binder decidable? If it is decidable, what is the computational complexity?

Wolter and Zakharyaschev [8] showed that the satisfiablity problem of their modal description logic is decidable when we interpret a given modal operator as S5-modality. They also have an undecidability result when they expand their logic with the common knowledge operator and the number of agents is more than 1. As a possible direction, we may try to apply the method of [8] for the question (3).

BIBLIOGRAPHY

[1] P. Blackburn. Arthur Prior and hybrid logic. *Synthese*, 150(3):329–372, 2006.
[2] P. Girard, J. Seligman, and F. Liu. General dynamic dynamic logic. In S. Ghilardi T. Bolander, T Braüner and L. S. Moss, editors, *Advances in Modal Logics*, volume 9, pages 239–260. College Publications, 2012.
[3] D. Lewis. *Counterfactuals*. Blackwell Publishing, 1973. 2nd edition was pubulished in 1986.
[4] G. Priest. *Towards Non-Being: The Logic and Metaphysics of Intentionality*. Oxford University Press, 2007.
[5] A. N. Prior. Egocentric logic. *Nôus*, 2:191–207, 1968.
[6] K. Sano. Axiomatizing hybrid products: How can we reason many-dimensionally in hybrid logic? *Journal of Applied Logic*, 8(4):459–474, December 2010.
[7] K. Sano and S. Tojo. Dynamic epistemic logic for channel-based agent communication. In K. Lodaya, editor, *Logic and Its Applications*, volume 7750 of *Lecture Notes in Computer Science*, pages 109–120, Belrin, Heidelberg, 2013. Springer.
[8] F. Wolter and M. Zakharyaschev. Modal description logics: modalizing roles. *Fundamenta Informaticae*, 39:411–438, 1999.

Response to Proietti and Sano

JEREMY SELIGMAN, FENRONG LIU AND PATRICK GIRARD

We would like to begin our response with grateful acknowledgements to the two commentators, Katsuhiko Sano and Carlo Proietti, for their careful reading of our paper and insightful comments. To align with the aims of the Tsinghua Logic Conference, the paper was written as a general introduction to our recent work on 'logic in the community,' introduced in [11], without too many technical details. Here we will keep to that spirit, focusing primarily on conceptual issues raised by the commentators. For the equally interesting technical questions, which we take seriously and will use to direct future work (perhaps even in collaboration!), we will here provide only a few brief comments and conjectures rather than complete solutions.

The two sets of comments allow us to put our research in a broader context, both conceptually and technically. We had not realised, for example, the nice connection of our work with modal description logic, or with the older work on indexical propositions mentioned by Sano. Sano's technical question about the axiomatisation of EFL and its PDL extensions is not something we are able to answer in this brief response, except to say that he is correct to suggest that our logic can be understood more abstractly as the product of two modal logics for the operators K and F given by the Gabbay-Shehtman construction in [4]. Given two independent relations R_1 and R_2 defined over two different domains, this construction yields a two-dimensional model defined, following [10], by:

$$(x, y)R_h(z, u) \quad \text{iff} \quad xR_1z \text{ and } y = u,$$
$$(x, y)R_v(z, u) \quad \text{iff} \quad x = z \text{ and } yR_2u.$$

In this setting, the questions of axiomatisation, first-order correspondence and decidability of our logic have already been partially answered, for the products of modal logics and hybrid modal logics, some by Sano himself in [9], and we suspect that the axiomatisation of EFL can be studied with similar methods. With that in place, its extension with GDDL using the results of our [5] should be straightforward. The task of specifying syntactically exactly which GDDL operators preserve the structure of our models requires a bit more work.

As pointed out by Proietti, this paper 'continues a research line inaugurated by the same authors with *Logic in the community*' in which we aim to investigate logical aspects of social interaction. Our research started with the simple observation that social relationships between agents play an important role in communication, by providing informational pathways, or channels. We reason about other people's states of mind, their beliefs, preferences and so forth, on the basis of their social connections, and because of the high value of social connections, we also reason about them explicitly. So the main concern for us has been how to extend the now well-established DEL framework to account for social communication and information flow in social contexts. This project, as noted by Proietti, also fits nicely with research in Behavioural Economics and Social Psychology, to which it may be seen as a partial formalisation. This observation opens the possibility for more technical studies, for instance by comparing our approach to research on social networks in economics, such as that in Jackson's [6].

We will now elaborate a bit more on two themes that came up in the comments. A lot more would need to be said to address all comments we received, but we have a limited amount of space and time.

Proietti suggested that it would be useful to be able to describe social network structures such as the 'circle', the 'wheel' or the 'star'. Although modal logic is not so good at defining structures that involve counting or loops, the addition of nominals improves the situation, at least in regard to frame-definability, especially when combined with the resources of PDL. For example, the formula

$$D_2 \quad = \quad (\langle F \rangle i \wedge \langle F \rangle j \wedge \langle F \rangle k) \rightarrow (\langle F \rangle (i \wedge j) \vee \langle F \rangle (i \wedge k) \vee \langle F \rangle (j \wedge k))$$

is valid on a frame just in case each agent has no more than 2 friends, or, in network terminology, that the maximal degree of the social network is ≤ 2.[1] Likewise, that the diameter of the network is no more than 3, say, is defined by

$$\neg(k \wedge \langle F \rangle (j \wedge \langle F \rangle ((i \wedge \neg k) \wedge \langle F \rangle ((h \wedge \neg j \wedge \neg k) \wedge \langle F \rangle (\neg i \wedge \neg j \wedge \neg k)))))$$

Circles are networks consisting of agents that we can imagine sitting at a round table so that each agent has all her friends sitting next to her. On the corresponding frames, D_2 is clearly valid. But also, given any three agents i, j and k, we can divide the rest into three classes: those sitting between i and j, those between j and k and those between i and k. This property can be defined by first noting that the formula $(\neg i \wedge \langle (F; (\neg j \wedge \neg k))^* \rangle i)$ is satisfied by those agents who are connected to agent i by a sequence of friends that includes neither j nor k. (This is

---

[1] This can clearly be generalised to give a formula D_n that is valid on all and only frames of maximal degree $\leq n$. Specifying that the maximal degree is at least n, or even that its minimal degree is at least n, is not possible, as a simple hybrid bisimulation argument will show. Other extensions to modal logic such as the 'counting modalities' of Areces et al [2] might be put to service here.

of course important if one wants to relay a message to i without alerting j and k). Abbreviate this as p_i. We can define similar formulas p_j and p_k, and so an agent will be 'between' i and j, say, iff she satisfies $(p_i \wedge p_j)$. Then the required formula is

$$B \quad = \quad (p_i \wedge p_j) \vee (p_j \wedge p_k) \vee (p_i \wedge p_k)$$

A social network is a circle iff both D_2 and B are valid on the corresponding frame.

This is just a starting point. As suggested, it would be interesting to look at reasoning patterns, both static and dynamic, that result from additional assumptions about network structure. It would also be interesting to go in the opposite direction: look at classes of structures that are easily definable in our language and explore them from the perspective of network studies. In particular, the prospect of investigating games on networks using these languages, in the manner of Agotnes and van Ditmarsch [1] is an appealing one.

But instead of characterising shapes of networks, recent research on 'logic in the community' has been directed at characterising the dynamics of social interactions, mostly concerning the influence of agents on other agents with whom they are socially connected. [8, 7] and more recently [3] propose models of influence within a social network and apply them to the dynamics of preference and belief, focusing on such topics as peer pressure, social belief revision and pluralistic ignorance.[2]

The model works by defining some rules for the way in which agents are influenced by their 'friends'. In the above papers, there are two rules of influence: strong and weak. In each case, a triggering condition is identified, together with an operation that changes the agent's mental state. For example, one rule that is considered is that an agent is strongly influenced to believe some proposition p when all of her friends believe it (Bp), but is only weakly influenced when none of her friends disbelieve it ($\neg B \neg p$). The effect of being influenced is for the agent to change her belief. This allows us to study the dynamic properties of the distribution of beliefs across the network. Here's a simple example of doxastic influence (Ip) regarding p in action:[3]

---

[2]Pluralistic ignorance is 'a situation where a majority of group members privately reject a norm, but assume incorrectly that most others accept it' (Wikipedia).

[3]See [8] for the precise statement of the rules of influence and further discussion of examples like these.

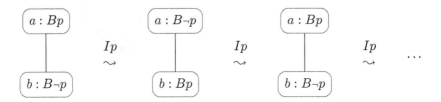

The community pictured above is in 'flux': beliefs will continue to change in an alternating pattern. By contrast, other communities stabilise fast:

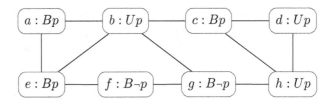

The difference between communities that remain in flux and those that eventually become stable can be described precisely in logical terms. [7] provides a formula to do so and a theorem to prove that it does. The method used is to show that the dynamics can be characterised by a cellular automaton, whose states and transitions are described using formulas. This is a fairly general technique and we suspect that it can be applied to study many different kinds of logical dynamic behaviour in social networks.

Another question we have started to think about (in [12]) is how to describe changes to the social network itself. On Facebook, for example, friends are frequently added and sometimes deleted. Those simple actions have an impact on information flow which we would hope to understand better. Consider the following continuation of the office gossip story (from our current paper):

> Roger, scared of the possibility that Mona will find out about his affair
> from Peggy, does all that he can to distance them. His smear campaign
> is designed to break their friendship and so protect his information.

To model deleting a friendship link, we define the result of cutting the friendship link between agents n and m in one direction:

$$\mathsf{cut}_F(n, m) = (\neg n?; F) \cup (F; \neg m?)$$

Since the friendship relation is symmetric, we need to cut the relation in both directions:

$$[-F_{n,m}] = [F := \mathsf{cut}_F(n, m)][F := \mathsf{cut}_F(m, n)]$$

It is easy to show that $[\![F]\!]^{[-F_{nm}]M} = [\![F]\!]^M \setminus \{\langle n,m\rangle, \langle m,n\rangle\}$, as required. This follows from the fact that $a[\![F]\!]^{[F:=\mathsf{cut}_F(n,m)]M}b$ iff $a[\![F]\!]^M b$ and $\langle a,b\rangle \neq \langle n,m\rangle$. So how is this going to help Roger? Well, after the application of $[-F_{mp}]$ to the model of Figure 7, Peggy's announcement to her friends that Roger is cheating has no effect. In fact, she has no friends to receive the message. So the model is unchanged. Back in the original model it is true for Roger that

$$[-F_{mp}] \downarrow n \; [p \triangleleft @_n c!:\langle F\rangle p]@_m \neg K @_n c$$

'after Peggy loses Mona as a friend, even after she tells her friends that I am cheating, Mona won't know.'

Next, think of adding a friend. In the basic case, we can define the operation $[+F_{n,m}]$ by analogy with deletion, but more simply, as

$$[F =: F \cup (n?; A; m?)]$$

But a more interesting model of adding friends follows the protocol of Facebook and other online social networks, whereby one must first request friendship. To capture this aspect of network change, we need to represent whether or not an agent *wants* to be friends with another agent. This could be done with a preference order, showing that the agent prefers states in which they are friends to those in which they are not. Again, we would like to explore this issue in the near future.

BIBLIOGRAPHY

[1] Thomas Ågotnes and Hans van Ditmarsch. What will they say?—Public announcement games. *Synthese*, 2011.

[2] Carlos Areces, Guillaume Hoffmann, and Alexandre Denis. Modal logics with counting. In A Dawar and R de Queiroz, editors, *WoLLIC 2010*, pages 98–109. Springer, 2010.

[3] Zoé Christoff and Jens Ulrik Hansen. A two-tiered formalization of social influence. In *Proceedings of LORI-4*, 2013.

[4] Dov M Gabbay and Valentin B Shehtman. Products of modal logics, part 1. *Logic journal of IGPL*, 6(1):73–146, 1998.

[5] Patrick Girard, Jeremy Seligman, and Fenrong Liu. General dynamic dynamic logic. In Thomas Bolander, Torben Braüner, Silvio Ghilardi, and Lawrence S Moss, editors, *Advances in Modal Logics Volume 9*, pages 239–260, 2012.

[6] Matthew O. Jackson. *Social and Economic Networks*. Princeton University Press, August 2008.

[7] Zhen Liang and Jeremy Seligman. A logical model of the dynamics of peer pressure. *Electronic Notes in Theoretical Computer Science*, 278:275–288, 2011.

[8] Fenrong Liu, Jeremy Seligman, and Patrick Girard. Logical dynamics of belief change in the community. To appear.

[9] Katsuhiko Sano. Axiomatizing hybrid products: How can we reason many-dimensionally in hybrid logic? *Journal of Applied Logic*, 8(4):459–474, 2010.

[10] Darko Sarenac. *Products of topological modal logics*. PhD thesis, Stanford University, 2006.

[11] Jeremy Seligman, Fenrong Liu, and Patrick Girard. Logic in the community. In Mohua Banerjee and Anil Seth, editors, *Logic and Its Applications - 4th Indian Conference, ICLA 2011, Delhi, India, January 5-11, 2011. Proceedings*, volume 6521 of *Lecture Notes in Computer Science*, pages 178–188, 2011.

[12] Jeremy Seligman, Fenrong Liu, and Patrick Girard. Facebook and the epistemic logic of friendship. In Burkhard C Schipper, editor, *Proceedings of the 14th Conference on Theoretical Aspects of Rationality and Knowledge*, pages 229–238, Chennai, India, 2013.

Logics for Litigation Argumentation

MINGHUI XIONG AND YUN XIE

ABSTRACT. The discussion of logics for legal argumentation has a very long history in the studies of law and philosophy. It stems from Protagoras, the first sophist in Ancient Greece who presented the paradox of the court. Since then the search for logic in law has attracted considerable attention from both philosophers and jurists. From the middle of last century, the study on logics for legal argumentation has been carried out in a systematic way, along two paths — formal and informal. The former is based on formal logic, whereas the latter is grounded in informal logic and argumentation theory. Scholars endorsing the formal approach try to conceive of legal logic as only application of traditional or modern logics, whereas researchers in the informal camp attempt to study legal argumentation from a perspective of non-formal logic. Although scholars of these two camps sometimes seem to be able to understand each other, for most of the time they don't. However, we think it is possible to integrate these two lines in a way to better promote mutual understanding and promising cooperation. In this paper, we aim to show this kind of possibility for litigation argumentation.

1 Introduction

The discussion of logics for legal/litigation argumentation has a long history in the studies of law and philosophy. It stems from Protagoras, the first sophist in Ancient Greece who presented the paradox of the court. Then Aristotle, in his treatise *On Sophistical Refutations*, discussed the issues of refuting sophist's argumentation and detecting their fallacies. In 1588, Abraham Fraunce, who continued the search for logic in law and published the *Lawier Logike-exemplifying the praecepts Logike* by *the practise of the common Lawe*, claimed in his poem that "I see no reason, why that Law and Logike should not bee. The nearest and the dearest friends, and therefore best agree...... I sought for Logike in our Law, and found it as I thought." In 1881, Jr. Oliver Wendell Holmes asserted in his works that "The life of the law has not been logic; it has been experience" ([8], p.1), a misleading claim which has induced many of his followers to decline the important role of logic in law. However, they have in fact all misunderstood what Holmes really wanted to say, since a few years later Holmes had further stressed explicitly that the training of lawyers was indeed a training in logic, and the language of judicial decision was mainly the language of logic ([9]).

More recently, the systematic study on logics for legal argumentation started from the middle of last century, i.e., after the publication of Ulrich Klug's *Juristische Logik* (1951) and Lee Loevinger's *An Introduction to Legal Logic* (1952). Since then, studies of logics for legal argumentation have been carried out along two different approaches — formal and informal. The former is based on formal logic, whereas the latter is grounded in informal logic and argumentation theory. Within the formal approach, legal logic is regarded as only applied logic. Many scholars urged to identify legal logic as just some applications of traditional formal logic in law, while Klug and Loevinger refined it as the applied modern logic. For scholars endorsing the informal approach, they have been greatly influenced by works of Chaim Perelman and Stephen Toulmin, namely their *Traité de l'argumentation* (1958) and *The Uses of Argument* (1958). They have developed their studies of logic for legal argumentation from a perspective of non-formal logic, emphasizing the substantial features of argument in legal context. For many years studies of legal logic in these two fields have just been developed in parallel, without any interaction between. And for most of the time, scholars endorsing these different approaches even don't understand each other. However, we think it is possible to integrate these two lines of studies in a promising way to better promote mutual understanding and cooperation.

2   Protagoras Paradox and Legal/Litigation Argumentation

The *Protagoras Paradox*, also known as the *Paradox of the Court*, is a very old issue in logic which could be dated back to ancient Greece. It is said that the famous sophist, Protagoras, once accepted a pupil, Euathlus, with the agreement that the student will pay Protagoras for his instructions after he wins his first case in court. However, Euathlus, with no intention to pay for the amount owed, chose not to get any case. So Protagoras decided to sue Euathlus, asking specifically for the payment for his instruction. Accordingly, Protagoras has presented his legal argument (the proponent' argument), believing that he could definitely get his payment: "If I win the case, I will be paid by the decision of the court. Or if you win the case, I will be paid according to the original contract, because you win your first case. Anyway, I will be paid no matter what the court's ruling is." While Euathlus, interestingly, has also put forward a legal argument (the opponent's argument) against Protagoras' claim: "If I win this case, then by the court's decision I will not pay you. Or if you win, then I haven't yet won my first case, so I won't pay you either. In a word, I won't pay you in any case."

From this interesting story, we could reveal at least four features of legal/litigation argumentation. First, legal argumentation consists of at least two sub-arguments. One is the proponent's legal argument and the other is the opponent's legal argument. Second, the conclusions of those two arguments are incompatible, or even inconsistent with each other. This is the so-called *legal dispute*, which is similar to the *difference of opinion* in the Pragma-dialectical framework developed by van Eemeren and Grootendorst ([20]). Third, there are at least two or more agents who take part in the legal argumentation. One of them is the suitor, who is the plaintiff in a civil trial or the prosecutor in a criminal trial. Another one is the respondent, who is the defendant in a civil trial or the defense in a criminal trial. Fourth, in legal argumentation there also exists a trier (including a judge or a jury) who takes the role of a referee or of an arbiter. As we understand it, the role of the trier is important and indispensable in legal argumentation, because the trier must be an agent of some legal argument when they make a judicial decision in a trial. But the role of a trier in litigation games could be different, for example, a judge only acts as a referee in the common law system, while he/she can act as an officer in the continental law system.

3 Logic for Legal Reasoning Based-on Zero-agent Logic

Since Aristotle, logic has been regarded as a discipline providing basic tools to analyze, evaluate and construct a real argument or argumentation in everyday life. In the context of law, legal logic is expected to explain and to guarantee the rational ground for legal reasoning, argument, and argumentation. In other words, legal reasoning or arguments are always seen to be based on some kind of logic, such as syllogistic logic, propositional logic, predicate logic, defeasible logic, or dynamic logic. However, the choice of the type of logic to be employed is dependent on the mainstream of logical studies at that time.

At the very beginning, the logic for legal reasoning is a zero-agent logic. It is the logic that assures the legal certainty. As Alder has indicated, the science of law in discourse is a purely formal science; its only instrument is formal logic, which deals with certainty and nothing else ([1]). Many legal scholars have identified formal logic with the Aristotelian syllogism, so studies in legal logic become simply the studies of how the Aristotelian logic or traditional formal logic could be applied to the field of law. Accordingly, legal syllogism has been treated for a long time as the basic model of legal argument. A legal syllogism always consists of three statements, i.e., the major premise, the minor premise and the conclusion. The major premise concerns with the question of law, the minor premise represents the question of fact, while the conclusion contains the statement which follows from the two premises [18]. It is due to this three-part-structure that a legal syllogism is regarded as a kind of syllogism. But Aristotelian logic is a zero-agent logic, which could only, according to van Benthem, deal with the mathematical relationships between static propositions ([19]). As a result, legal logic based on zero-agent logic cannot explain the logical foundation of legal reasoning in a very convincing way.

Alternatively, MacCormick conceives of the legal syllogism in a more subtle way. He identifies its logical structure as the following form

Minghui Xiong and Yun Xie

(1) In any case, if p then q

(2) In the instant case p

(3) ∴ in the instant case, q ([13], p. 25)

Moreover, he explains the justificatory power of legal reasoning by recognizing the above structure as an embodiment of the Modus Ponens in propositional logic,

(1) If p then q

(2) p

(3) ∴ q ([13], p. 87; [7], p.2)

However, from the perspective of modern logic, it is easy to see that these two forms are apparently different from each other. Therefore, it appears as though MacCormick's position is grounded on some unfortunate confusion.

Similarly, Alexy ([2], p.222) also suggests to understand legal syllogism in a different way. According to him, the simplest form of internal justification (of legal argument) has the following structure,

(1) $(x)(Tx \rightarrow ORx)$

(2) Ta

(3) ∴ ORa   (1), (2)

At the first sight, Alexy seems to have introduced into legal reasoning the modal or deontic operator O. But within the structure he proposed, this operator is not working in the process from the premises to the conclusion. In fact, we cannot logically infer the conclusion (3) simply from the premises (1) and (2). Its validity could only be established through predicate logic, in the following way,

(1) $(x)(Tx \rightarrow ORx)$ Premise

(2) Ta   Premise

(3) Ta $\rightarrow$ ORa   (1) Universal Instantiation (U. I.)

(4) ∴ ORa   (3), (2) Modus Ponens (M.P.)

Obviously, MacCormick constructs his theory of legal argumentation based on propositional logic, whereas Alexy's theory of legal argumentation is based on predicate logic. But no matter which logic is taken, their frameworks are both based on some zero-agent logic, since in their analysis the role of arguer has been completely ignored, let alone that of the audience. Therefore, it is impossible for their theories to be insensitive to the role of agent and the audience in legal reasoning. As a result, it is apt to identify legal logic with studies in syntax and semantics, thus to exclude from its sphere all the pragmatic investigations ([10], p.19). Moreover, in their zero-agent logical frameworks, only one conclusion is allowed to be correct in legal reasoning. Hence the conclusion of legal reasoning will become unique and universal. But this is just an undesirable consequence, because there will be no way for us to explain why different judges and juries can make different judicial decisions on the same case. To sum up, theories of legal reasoning based on zero-agent logic will turn out to be unadvisable for analyzing our legal reasoning in a trial [1].

4 Logic for Legal Argument Based-on One-agent Logic

A legal argument always exists in some context, such as legal argumentations or legal discussions, so the concept of legal argument could be broad in scope. According to Alexy ([2], p. 211), there are different types of legal discussions, such as discussions in legal science, judicial deliberations, debates in the courts of law, legislative treatments of legal question, and discussions of legal questions among students and among jurists or lawyers. Following Alexy's idea, Feteris claims that anyone who presents a legal thesis, such as a lawyer, a judge, a legislator, or a legal scholar, could make a legal argument ([3], p.1). Accordingly, the agents of legal argument would include not only lawyers, but also judge and juries in a trial. First of all, judges and juries must make some decision, even in the toughest cases. No judge can say that this case is too tough or too hard for him so that he is not going to rule on it. In other words, the judge must make a legal argument in any legal case, and usually in a relatively short period of time, because throwing the case out of court will just amount to ruling in favor of the defendant or the defense [4].

According to Perelman (1982, p.9), a demonstration is one form of proof that conforms to the rules which are made explicit in a formalized system, while an argumentation flows out of natural language. The aim of argumentation is not to deduce consequences from some given premises, but is to elicit or to increase the adherence of the members of an audience to the thesis that is presented for their consent. Thus, the speaker as the arguer, and the audience, are very important for the analysis and evaluation of arguments. On that basis, Perelman states that the rational force of an argument will depend on some non-formal conditions, and will leave room for "material" rules of inference. Consequently, he further claims that a legal argument is non-conclusive and non-formal (cf. [10], pp.105-107). In his book *The Uses of Argument*, Toulmin relies on the legal process to establish the acceptability of his model of argument. In *An Introduction to Reasoning* (1984), together with Rieke and Janik, he also specifies how the argument model could be applied into the legal contexts. Particularly, s/he indicates that if an arguer wants to give a sufficient justification of her/his position, s/he must respond to all those questions put forward by an antagonist who challenges the claim.

However, for both Perelman and Toulmin, they all neglect considering different roles of agents in legal argumentation, and the interactions between agents. Therefore, their logic for legal argument could be regarded as based on one-agent logic. According to van Benthem ([19]), the one-agent logic focuses only on such activ-

---

[1]This paper focuses only on studies in logic of legal reasoning and argument. Actually, in the literature on legal logic, there also exists another line of research in which specific legal notions, such as notions of norm and right, are characterized and formalized with zero-agent logics, especially with modal logic systems (cf. [26, 27])

ity of drawing the conclusion from the premises, hence the role of the arguer is recognized as very important for evaluating an argument. Nevertheless, although reasoning could be aimless, argument is a social and verbal means of trying to resolve, or at least to contend with, a conflict of difference that has arisen or exists between two or more parties ([23]). This means that no argument is aimless. The goal of an argument is the arguer's intention or goal. So any arguer who is making a legal argument will always have his/her own intention or goal in the legal case. Specifically, the judge has an intention to make a fair, rational judicial decision. The jury's goal is to make a reasonable fact-finding. And the aim of the lawyer on behalf of the plaintiff or prosecutor is to make judges and juries to support his/her own legal claim, while the goal of the defendant or defense is to persuade judges and juries of not accepting the opponent's claim.

Similar to the zero-agent logic frameworks, legal logics based on one-agent logic still cannot explain that in reality different judges can make conflicting judgments on the same case, because, in their frameworks, only the conclusion which is normally that of the judges' or of the juries' legal argument will be regarded as correct. Moreover, we also cannot explain why there sometimes exist two different judicial decisions between the first stage and the second stage in a trial.

5   Logic for Legal Argumentation Based-on Two-agent Logic

According to van Benthem, many-agent logic takes into consideration the interactive process of argumentation in our analysis and evaluation of arguments. This interactive process could happen between different agents and their moves in argumentation. Most of the logical models for legal argumentation developed in AI are based on two agents ([15]; [5]; [12]; [22]). Two agents model is the simplest form of multi-agent logical model. Normally, two agents will mainly refer to the suitor and the respondent in legal argumentation. And arguments and counterarguments are seen as moves in this interactive process, thus they are well taken into considerations in those AI frameworks ([22]; [5], [15]).

Gordon ([5]) develops a framework called the Pleadings Game, in which there are two players, the plaintiff and defendant. The allocations of rights and obligations of a player are dependent to a limited extent on the role taken. For example, the plaintiff has the burden of proving the main claim. However, both judge and juries are not included as possible players, so this framework can only describe a civil pleading, but not a criminal pleading.

In Argumentation System (AS) proposed by Prakken ([15]), arguments are regarded as one of the most important core concepts. This framework has been developed out from the work of Dung's abstract argumentation theory. Prakken tries to define the relationships between different moves, such as attack (p.157), defeat (p.162), and rebut (p.174). There are also two agents in his dialogue game, one is called the proponent, and the other is the opponent (p.165). In the Argumentation

System, the judge or jury seems to be treated as one player, because they have a duty to judge whether or not one argument has successfully attacked, defeated, or rebutted another argument.

Verheij ([22]) develops the ARGUMED Family of Argument Assistants in the field of law. His basic assumption is that argumentation is a process during which arguments are constructed and counterarguments are adduced. Actually, this framework is based on the idea of dialectical arguments, which consists of statements which have two different relationships: a statement can support another, or a statement can attack another (p.55). In his first prototype: ARGUE! system (p.19), the arguer who makes argument is not the same agent to the other arguer who makes counterarguments. And it seems that both the suitor and the respondent are required to play this litigation game in collaboration. As a program, it is designed for computer users. Therefore, anyone who could take part in a lawsuit can take an argument move according to this program. Apparently, the logical foundation of this kind of game is also some kind of two-agent logic.

Lodder ([12]) puts forward a legal logic game, *DiaLaw*, for modeling the legal justification in AI and Law. In his framework, there exist two players, the plaintiff or the prosecutor, and the defendant or the defense. Although in reality the judges or juries perform the role of the third party in the court of law, they are not modeled in DiaLaw, because Lodder doubts that there could be some substantial criterion with which the judge can use to settle conflicts or to make decisions. However, unlike the other frameworks, the moves modeled in Lodder's DiaLaw are not arguments, but are illocutionary acts, such as claims, questions, acceptance and withdrawals.

6   Logic for Legal Argumentation Based-on Three-agent Logic

According to the Pragma-dialectic theory, a legal trial is a civilized but arbitrary way of settling a disagreement, because it is to lay the matter before a third part who serves as judge and decides who is right ([21], p.24). Following this idea, [3] (p.171-172) conceives of legal argumentation as a part of critical discussion aiming at the resolution of a dispute. The dispute covers not only the disagreements between the protagonist (the plaintiff in a civil process/the prosecutor in a criminal process) and the antagonist (the defendant in a civil process/the accused in a criminal process), but also the disagreements between these parties and the judge. For this reason, Feteris introduces the judge into legal argumentation as the third agent. However, it is a pity that she overlooks the importance of the jury as a legal arguer in a lawsuit. Moreover, Feteris also appears to be incoherent in her articulation of legal argumentation. She suggests that anyone who presents a legal thesis can make a legal argument ([3], p.1). But in some of these cases, it is not clear who could be the protagonist or the antagonist, or who will be the judge or jury. For some cases, obviously there is no judge or jury to take part in a legal argumentation in the discussions between legal scholars, or in the legislative debates in parliament.

Besides, some might also doubt that Feteris' framework of legal argumentation could be qualified as logical studies of legal argumentation. However, we agree with Sartor that there have been two "logics" of legal reasoning, namely the formal symbolic logic and argumentation theory ([16]), so Feteris does provide us a possible type of legal logic, since the Pragma-Dialectical theory is undoubtedly a global theory of argumentation.

7   Litigation Argumentation Based-on Three-agent Game

A legal logic for litigation argumentation should be based on three agents. [25] suggests that the traditional concept of legal argumentation is both too broad and too narrow. On the one hand, it is too broad because legal argumentation exists not only in a court but also outside the court. In other words, legal argumentation covers the arguments occurred both in the process of legislation, law-enforcement and justice. On the other hand, the traditional concept of legal argumentation is too narrow, because it is customary for many researchers to limit the scope solely to adjudication, i.e., to the application of law to concrete cases for the purpose of reaching a verdict and passing a sentence. As Klug has claimed, legal logic is some theory about applying the rules of formal logic within the framework of adjudication (cf. [10], p.21).

Accordingly, we propose firstly a new concept of litigation argumentation, which is an argument-based game between and among three agents: the suitor (including the plaintiff and the prosecutor), the respondent (including the defendant and the defense), and the trier (including the judge and the jury) ([24], pp. 72-76). They all take the role of a legal arguer, and interact with each other.

Secondly, we develop a five-part structure of legal argument, expanding the traditional structure of legal syllogism ([24], p.144). By this new model of legal argument, we can reveal both the inference relationship from the norm to the interpretation of norm, and the supporting relationship between the data/evidence and the fact.

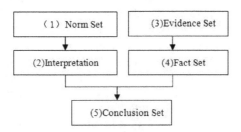

In this five-part structure, the conclusion is not one statement, but a set of statements, and the basic reasons/premises are divided into two parts: a set of norms or rules, and a set of evidences. In a legal process, any party, such as the suitor, the respondent or the trier, can and should make use of the above argument model. However, for the suitor and respondent, the set of norms may be empty, but the set of evidence cannot (at least in the Chinese legal system). And for the trier, none of these sets can be empty.

Furthermore, when presenting arguments with this model, no elements should contradict any of the other elements in the same set, because it is impossible for an acceptable conclusion to be logically inferred from some inconsistent set of premises. And each element of the conclusion set should follow from the interpretation set and the fact set. Moreover, seen from the perspective of set theory, the interpretation set should be a subset of the norm set. In other words, each element of the later set should imply or infer some element of the former set. If some element of the interpretation set is not inferred from any element of the norm set, this means the interpretation is illogical, so it becomes unacceptable in the legal process. However, it is unnecessary for the fact set to be a proper set of the evidence set, because there could be some fact which needs no proof. At last, the conclusion set does not follow directly from the norm set and the evidence set, it follows from the interpretation set and the fact set.

Thirdly, since a legal suit is the process and product of argument-based interaction between and among the suitor, the respondent and the trier, we cannot evaluate the three parties' litigation arguments in a separate manner. Therefore, litigation arguments must be located back into a framework of argument-based game, i.e., litigation argumentation, between and among the three parties. Accordingly, we propose to understand the structure of litigation argumentation in the following way:

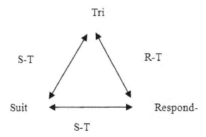

In a litigation argumentation, there are three sub-games which are dialectical and dialogical: the S-T game between the suitor and the trier, the S-R game between the suitor and the trier, and the R-T game between the respondent and the trier. Moreover, when a judicial decision is making, the evidence set of the judge's litigation argument should be based on both the suitor's and the respondent's evidence sets. The intersection between the evidence sets of the suitor and the respondent must be a proper set of that of the judge. And the admissibility of some evidence not to belong to this intersection will depend on the relevant legal procedure rules in a specific legal system.

8 Conclusion

Legal logic is used to explain the logical foundation for legal reasoning, argument and argumentation. It has undergone a course of development from zero-agent, one agent, and two-agent logic to three-agent logic. Meanwhile, this development also illustrates a shift from formal logic, which focuses only on semantic and syntactic elements, to informal logic or argumentation theory, which considers not only semantic and syntactic but also pragmatic dimensions.

As one kind of applied logic, legal logic always depends on the mainstream logic endorsed by legal philosophers in different period. Klug has conceived of legal logic based on symbolic logic, since formal deductive logic or symbolic logic dominated the whole field of logic at his time. Nevertheless, it is not for sure that the mainstream logic understood by legal philosophers is always the same one approved by coeval logicians. For instance, [13] has explained the logical reasonableness of legal reasoning through propositional logic developed by Stoic, which is obviously not a mainstream logic at that time. But we think it would be better and more desirable if legal reasoning could be logically justified by the mainstream logic of the time. Nowadays, we believe the dynamic logic could be deployed to describe litigation argumentation, because the multi-agents interactions are fully considered in this kind of logic. However, this approach to legal logic is not enough for evaluating litigation arguments, and a more comprehensive logical approach still needs to be developed. It should be a kind of *informal-formal* logical approach. That is, we should integrate both formal and informal perspectives by starting with investigations from the informal perspective, and ending with characterizations and modeling with formal tools. It is quite different from the formal-informal approach taken in the past which proceeds just in the reverse way.

Acknowledgements: The work in this paper is supported by the Chinese MOE Project of Key Research Institute of Humanities and Social Sciences at Universities (12JJD720006), and the National Social Science Fund of China (13AZX0017).

BIBLIOGRAPHY

[1] Alder, Mortimer J.: Legal Certainty, *Columbia Law Review* 31, 91-108 (1931).
[2] Alexy, Robert: *A Theory of Legal Argumentation: The Theory of Rational Discourse as Theory of Legal Justification*, Oxford University Press (1989).
[3] Feteris, Eveline T.: *Fundamentals of Legal Argumentation: A Survey of Theories on the Justification of Judicial Decisions*, Kluwer Academic Publishers (1999).
[4] Fogelin, Robert J., Sinnott-Armstrong, Walter: *Understanding Arguments: An Introduction to Informal Logic*, Wadsworth/Thomson (2001).

[5]   Gordon, Thomas F.: *The Pleadings Game: An Artificial Intelligence Model of Procedural Justice*, Kluwer Academic Publishers (1995).
[6]   Grossi, D., Rotolo, A: Logic in the Law: A Concise Overview, in A. Gupta and J. van Benthem (eds.), *Logic and Philosophy Today*, Studies in Logic, vol. 30, pp.251-274, College Publications (2011)
[7]   Hage, Jaap C.: *Reasoning with Rules*, Kluwer Academic Publishers(1997).
[8]   Holmes, Jr. Oliver Wendell: *The Common Law*, New York: Dover Publications (1881).
[9]   Holmes, Jr.Oliver Wendell: The Path of the Law, *Harvard Law Review* 10, 465-468 (1897).
[10] Horovitz, Joseph: *Law and Logic: A Critical Account of Legal Argument*, Springer-Verlag, (1972).
[11] Johnson, Ralph H. Blair, J. Anthony: Informal Logic: An Overview, *Informal Logic* 20, 93-107 (2000).
[12] Lodder, Arno R.: *DiaLaw: on Legal Justification and Dialogical Models of Argumentation*, Kluwer Academic Publishers (2001).
[13] MacCormick: *Legal Reasoning and Legal Theory*, Clarendon Press (1978).
[14] Perelman, Ch.: *The Realm of Rhetoric*, University of Notre Dame Press (1982).
[15] Prakken, Henry: *Logical Tools for Modeling Legal Argument: A Study of Defeasible Reasoning in Law*, Kluwer Academic Publishers (1997).
[16] Sartor, Giovanni: A Formal Model of Legal Argumentation, *Ratio Juris* 7, 177–211 (1997).
[17] Toulmin, Stephen, Rieke, Richard, and Janik, Allan: *An Introduction to Reasoning*, Macmillan Publishing Co., Inc. (1984).
[18] Toulmin, Stephen: *The Uses of Argument*, Cambridge University Press(1958).
[19] Van Benthem, Johan: A Mini-guide to Logic in Action, *Philosophical Research*, Supplement, 21-30 (2003).
[20] Van Eemeren, Frans, Grootendorst, Rob: *Speech Acts in Argumentative Discussions: A Theoretical Model for the Analysis of Discussions Directed towards Solving Conflicts of Opinion*, De Gruyter Mouton (1984).
[21] Van Eemeren, Frans, Grootendorst, Rob, Henkemans, Francisca Snoeck: *Argumentation: Analysis, Evaluation, Presentation*, Lawrence Erlbaum Associates (2002).
[22] Verheij, Bart: *Virtual Argument: On the Design of Argument Assistants for Lawyers and Other Arguers*, T.M.C.Asser Press (2005).
[23] Walton, Douglas: What is Reasoning? What is an Argument? *Journal of Philosophy* 87, 399-419 (1990).
[24] Xiong, Minghui: *Litigation Argumentation: A Logical Perspective of Litigation Games*, China University of Political Science and Law Press (2010).
[25] Xiong, Minghui: On the Inference Rules in Legal Logic, *Social Sciences in China* 30, 58-75 (2009).
[26] Kanger, Stig: Law and Logic, *Theoria* 38, 105-132 (1972).
[27] Grossi, Davide, Rotolo, Antonino: Logic in the Law: A Concise Overview, in: A. Gupta and J. van Benthem (eds.), *Logic and Philosophy Today*, Studies in Logic, Vol. 30, pp.251-274, College Publications London(2011).

Comments on Xiong and Xie

DAVIDE GROSSI

In *Logics for Litigation Argumentation* Minghui Xiong and Yun Xie outline an interesting interpretation of a century-old research tradition at the interface between logic and law and, informed by such overview, they set out to advocate specific directions of future research in the area. Their focus is the understanding of legal decision-making as it occurs in adversarial processes where two parties, viz. a plaintiff or prosecutor vs. a defendant, engage in debates in front of a court or judge to decide a lawsuit.

By reviewing a few landmark contributions to the field of logic and law, the authors identify two rather telling trends. First, a trend that from the development of logics where agency does not play any role brings to logics, or logic-based symbolic systems as developed in the AI literature, involving the interaction of at least two agents. Second, a trend of a more methodological nature from the use of 'formal' to the use of 'informal' logic for understanding legal processes like litigation. I see these are two very fruitful "dimensions" from which to look at the research on logic and law, and in particular at the analysis of argumentation as it occurs in legal settings. I will use them to give structure to my brief commentary.

Decision-making processes and games

Only recently, and thanks to its interaction with disciplines like computer science and artificial intelligence, has logic turned its attention to the modeling of multi-agent phenomena. But there is another formal discipline—with strong relations with logic—that can offer solid ground for the analysis of legal argumentation: game theory.

At the level of abstraction argued for by Minghui Xiong and Yun Xie, legal litigation can be viewed as a *multi-agent* adversarial decision-making process: two parties argue for and against a given claim, possibly in front of a judge or court. The winner determines the outcome of the process thereby establishing the claim or its negation.

The study of argumentation as an adversarial process—i.e., as a *game*—has already an established tradition, with contributions within logic [12, 7] and in particular within artificial intelligence, starting with [3]. Up till now, however, these games all model an idealized argumentative setting in the form of a two-player win-lose game with perfect information. Much has still to be understood about more complex argument games demanding for a more sophisticated game-theoretic analysis. In this view, Minghui Xiong and Yun Xie stress the importance of understanding 3-players argument games—where two parties argue in front of a judge or audience. Recent work has touched upon this issue in game theory [5, 10] and in artificial itelligenge [8], but much work still lies ahead.

"Formal" vs. "informal" logic.

Minghui Xiong and Yun Xie conclude their paper with a statement I feel to fully subscribe to: "[...] we should integrate both formal and informal perspectives by starting with investigations from the informal perspective, and ending with characterizations and modeling with formal tools". The opposition between 'formal' and 'informal' logic for the study of argumentation is only an apparent one.

As I see it, Minghui Xiong and Yun Xie's paper invites us to rethink and reinterpret the theses of Perelman and Toulmin in a modern light. Perelman and Toulmin in [11] pointed to the limitations of classical logic (eminently first-order logic) to study argumentation, rather than about the limitations of the mathematical method in general in comprehending argumentation: even though classical logic might not be the right tool for the analysis of argumentation, other *formal* logics could very well be fit for the job, and that is where researchers should direct their efforts if they wish to understand argumentation without derogating to the deployment of rigorous mathematical methods.

To further strengthen my point I want to conclude my commentary by quoting one of my favorite passages from [11]:

> "What features of our arguments should we expect to be field-invariant: which features will be field-dependent? We can get some hints, if we consider the parallel between the judicial process, by which the questions raised in a law court are settled, and the rational process, by which arguments are set out and produced in support of an initial assertion. [...] One broad distinction is fairly clear. The sorts of evidence relevant in cases of different kinds will naturally be very variable. [...] On the other hand there will be, within limits, certain broad similarities between the orders of proceedings adopted in the actual trial of different cases, even when these are concerned with issues of very different kinds. [...] When we turn from the judicial to the rational process, the same broad distinction can be drawn. Certain basic *similarities of pattern and procedure* [our emphasis] can be recognized, not only among legal arguments but among justificatory arguments in general, however widely different the fields of the arguments, the sort of evidence relevant, and the weight of the evidence may be." [11, pp.15-17]

Here Toulmin argues for an analysis of argumentation in terms of regularities and similarities of *pattern*—think of domain-independent argumentative patterns—and of *procedure*—think of the adversarial decision-making processes we addressed in the previous section. Patterns of argumentation, studied with the help of graph-theory, have obtained considerable attention since the work of [4][1] and so did argument games[2].

As shown in [6] these formal theories of argumentation are a natural playground for modal logic [2], and games—as well as argument games—could fruitfully be studied from a logical point of view[3]. These are just the beginnings of a research program Minghui Xiong and Yun Xie's paper strongly argues for: bridging the gap between the 'informal' and the 'formal' side of logic in the study of argumentation. 'Informal' logic points to a rich repertoire of research questions while 'formal' logic comes equipped with the tools to address them in a mathematical fashion. From this forward-looking point of view the 'formal-informal' opposition is just a deceptive appearance.

BIBLIOGRAPHY

[1] P. Baroni and M. Giacomin. Semantics of abstract argument systems. In I. Rahwan and G. R. Simari, editors, *Argumentation in Artifical Intelligence*. Springer, 2009.
[2] P. Blackburn, J. van Benthem, and F. Wolter, editors. *Handbook of Modal Logic*. Elsevier, 2006.
[3] P. M. Dung. Logic programming as dialogue games. Technical report, Division of Computer Science, Asian Institute of Technology, 1994.
[4] P. M. Dung. On the acceptability of arguments and its fundamental role in nonmonotonic reasoning, logic programming and n-person games. *Artificial Intelligence*, 77(2):321–358, 1995.
[5] J. Glazer and A. Rubinstein. Debates and decisions: On a rationale of argumentation rules. *Games and Economic Behavior*, 36(2):158–173, 2001.
[6] D. Grossi. Argumentation theory in the view of modal logic. In P. McBurney and I. Rahwan, editors, *Post-proceedings of the 7th International Workshop on Argumentation in Multi-Agent Systems*, number 6614 in LNAI, pages 190–208, 2011.
[7] D. Grossi. Abstract argument games via modal logic. *Synthese*, 2013.
[8] D. Grossi and W. van der Hoek. Audience-based uncertainty in abstract argument games. In F. Rossi and S. Thrun, editors, *Proceedings of IJCAI'13*. AAAI Press, 2013.
[9] S. Modgil and M. Caminada. Proof theories and algorithms for abstract argumentation frameworks. In I. Rahwan and G. Simari, editors, *Argumentation in Artificial Intelligence*, pages 105–132. Springer, 2009.
[10] A. Rubinstein. *Economics and Language: Five Essays*. The Churchill Lectures in Economic Theory. Cambridge University Press, 2000.
[11] S. Toulmin. *The Uses of Argument*. Cambridge University Press, 1958.
[12] J. van Benthem. Action and procedure in reasoning. *Cardozo Law Review*, 22:1575–1593, 2001.
[13] J. van Benthem. Extensive games as process models. *Journal of Logic, Language and Information*, 11:289–313, 2002.

---

[1] See [1] for a fairly recent overview.
[2] See [9] for an overview
[3] Cf. [13], and in particular [12] for an analysis directly targeting the application of logic games to law.

Comments on Xiong and Xie

HENRY PRAKKEN

This paper argues that a "legal logic for litigation argumentation" should have the following features:

1. The logic should be a "three-agent" logic, taking into account two opposing parties (plaintiff and defendant) and an adjudicator (a judge or jury).

2. The logic should be developed by first giving an informal analysis of legal argumentation and then formalising the informal account.

3. The logic should, like any legal logic, be based on the current "mainstream" logic, which according to the authors is dynamic logic.

In my commentary I want to argue that the thus proposed research programme is in fact already being carried out by the field of artificial intelligence (AI) and law. This field has been developing formal and computational models of legal argument since the mid 1980's. Its models have been based on mainstream AI formalisms for nonmonotonic logic and argumentation, but have also contributed to developing this mainstream.

But let me first briefly state my opinion on the three above claims of the authors. With the first claim I agree, although in my opinion such a three-agent logic should still have an 'unpersonal' logic of argumentation as a component, since otherwise the three agents have no rational criteria for how to state and evaluate their and each other's arguments. With the second point I could not agree more. When applying logic to the law, we ultimately aim at improving actual legal reasoning and legal decision making. However, much logical research is formalism-driven, which makes its relevance for actual reasoning sometimes hard to see (although this research may well have many other benefits). This is also one reason why I disagree with the authors' last point. In my opinion, the mainstream logic on which a legal logic should be based, should not be dynamic logic but an argumentation logic, since these logics have generally been the result of detailed analyses of actual argumentation.

Very briefly, argumentation logics model inference as the construction and comparison of alternative arguments for and against conclusions. The inference steps in an argument apply inference rules defined by the logic. One class of argumentation logics (e.g.[14, 10]) incorporates John Pollock's [12]'s distinction between two classes of inference rules: deductive rules, whose premises guarantee their conclusion, and defeasible rules, whose premises only create a presumption in favour of their conclusion. In argumentation logics with both deductive and defeasible inference rules, arguments can be attacked not only on their premises but also on their defeasible inferences, either by attacking their conclusion (Pollock's rebutting attacks) or by attacking the inference itself (Pollock's undercutting attacks). Conflicting arguments can be compared in terms of preferences, to see which attackers defeat their target. In epistemic reasoning such preferences may, for example, express degrees of uncertainty or reliability, in practical reasoning they may express subjective value judgements. The result of this is a so-called abstract argumentation framework, i.e., a set of arguments plus a binary relation of defeat. Then Dung's [4] theory of such frameworks can be used to assess which arguments are acceptable. An important aspect here is *reinstatement*: it may be that an argument A that is defeated by an argument B is still acceptable since B is in turn defeated by an argument C. Dung's formal theory is an excellent formal tool for capturing such, often complex defeat interactions between arguments (and this is all it can do).

Let me return to the authors' proposed research program and how AI (and law) has been carrying it out. As the authors mention, legal logic has been studied both by formal and informal means. Much informal research is inspired by Toulmin [15] and Perelman (e.g. [11]). Both Toulmin and Perelman argued that formal logicians of their time made the mistake of thinking that formal logic, which was developed for mathematical reasoning, would also apply to reasoning outside mathematics. A merit of their work is that they showed with careful analysis of actual reasoning that reasoning outside mathematics has many features not covered by the formal-logical tools of their days. However, we now know that they went a step too far in blaming formal methods; after more than 35 years of research in areas such as nonmonotonic logic, belief revision and computational argument we know that many features of non-mathematical reasoning that Toulmin and Perelman analysed can be analysed and modelled formally (see also [3]).

One such feature is the defeasibility of many non-mathematical arguments, captured by Toulmin's famous argument scheme, where data are connected to a claim by a warrant, which in turn has a backing. This scheme also has a place for rebuttals of the claim based on exceptions to the warrant. The fields of informal logic and argumentation theory, which emerged from Toulmin's and Perelman's work, generalised Toulmin's notion of a warrant into rich classifications of argument schemes, and they generalised Toulmin's notion of a rebuttal into lists of critical questions attached to argument schemes. In AI (and law) modern argumentation logics such as DefLog [16], Carneades [6] and $ASPIC^+$ [14, 10] are now being used to formalise argument schemes as defeasible inference rules and critical questions as pointers to counterarguments. These logics thus formalise the defeasible nature of non-mathematical arguments. Moreover, they provide general frameworks for modelling reasoning with argument schemes and thus respect that different legal issues are argued in different ways. For example, issues of fact are argued with argument schemes for epistemic reasoning, such as commonsense generalisations, witness or expert testimony, passage of time, or abductive reasoning, while issues of interpretation are argued with argument schemes for practical reasoning, such as appeal to precedent, purpose or value.

Another aspect of non-mathematical arguments is that they are meant to persuade an audience. Perelman stressed that whether a 'real-life' argument is good does not depend on its logical form but on whether it is capable of persuading the addressed audience. And an argument is more persuasive the more it takes the audience's "values" into account (for example, an argument that governments should not tap internet communications of their citizens since this infringes on their privacy is not very persuasive to an audience that values security over privacy.) In AI (and law) this idea was taken up by Trevor Bench-Capon [2]. Applying abstract argumentation frameworks to practical reasoning, he attached to each argument about what to do a value promoted by the proposed action and he made the evaluation of conflicting arguments dependent on subjective value orderings corresponding to different audiences. He later further developed this idea with Katie Atkinson in a formal account of reasoning with argument schemes for practical reasoning (e.g. [1]).

Like Perelman, Toulmin also argued that outside mathematics the validity of an argument does not depend on its logical form, but he did not refer to the notion of an audience. Instead, his criterion was whether an argument can be defended in a rational dispute. He pointed at legal procedure as a paradigmatic example of such procedural evaluation of arguments. Informal logic and argumentation theory developed this into the idea that arguments have to be evaluated in their context of use, namely, that of a dialogue. Among other things, this gave a new, dialogical account of fallacies. Traditional 'fallacies' such as appeal to authority or ad hominem arguments were recast as sensible presumptive argument schemes, which use can still be rational if their critical questions are adequately answered in the dialogue. Argumentation logics can formalise this new account of fallacies. For example, the argument scheme from expert opinion (traditionally the fallacy of appeal to authority) can be formalised as a defeasible inference rule, and an argument that the expert cannot be believed since he is biased (traditionally an ad hominem fallacy) is a rational counterargument applying a critical question of the expert testimony scheme.

But argumentation logics on their own do not yet model Toulmin's idea of procedural rationality, since they model neither dynamics nor communication between different agents: they just specify the nonmonotonic-logical consequences from a given knowledge base, in terms of the arguments that can be constructed from the knowledge base and their defeat relations as determined by given preferences. Yet these logics can very well be embedded in models of dynamics and communication, and AI and law has done so in formalised dialogue games of legal procedure (e.g. [7, 5, 8]), based on philosophical dialogue models of argumentation such as [9]. In the AI-and-law games, the players can formulate their own and attack each other's arguments, based on old or new information. They can also make other argumentation-related utterances, such as challenging, conceding or retracting claims or premises, or arguing that evidence is inadmissible. The players' utterances are regulated by a protocol, which is the formal counterpart of legal rules of procedure. In some such models, e.g. my own [13], a neutral adjudicator can allocate burdens of proof, decide on issues of admissibility and evaluate the strength of the adversaries' arguments. If the protocols for such games embody rational principles of interaction and investigation, then they can be seen as formal accounts of Toulmin's idea of procedural rationality, and then they come close to what the present authors call a legal logic of litigation argumentation.

The resulting picture of a logic for legal litigation argumentation is this. Two adversaries and an adjudicator (a judge or jury) state arguments and make other argumentative locutions within the procedural bounds defined by the protocol, thus moving from one information state to another. The players are driven by their goals (the adversaries want to win, the adjudicator wants to decide the case in a legal and just way) but also by the theory of argument schemes (their premises and critical questions point at evidence which, if found, gives rise to arguments or counterarguments). At each stage of the dispute, an argumentation logic can be used to check the well-formedness of the arguments and to evaluate the 'current' state of the dispute. The participants can do this privately, relative to their own estimates and preferences, to see whether they are likely winning, and the adjudicator can do this, for example, in verifying whether a so-called burden of production (a burden to provide credible evidence on an issue) has been met. In the closing stages the adjudicator assesses the internal strength of arguments and decides which attacking arguments defeat their targets, all this according to the applicable standards of proof. Then the argumentation logic determines which arguments in the end prevail.

I believe that this combined logical-procedural account of legal argumentation as developed in AI and law largely agrees with the authors' idea of a legal logic for litigation argumentation. This does not mean that the research program they propose has become obsolete. On the contrary, the AI-and-law research that I have briefly sketched above is still in full development, and many research issues are still unresolved. Therefore, contributions from other areas, such as logic and argumentation theory, would be extremely valuable.

BIBLIOGRAPHY

[1] K. Atkinson, T.J.M. Bench-Capon, and P. McBurney. Computational representation of persuasive argument. *Synthese*, 152:157–206, 2006.

[2] T.J.M. Bench-Capon. Persuasion in practical argument using value-based argumentation frameworks. *Journal of Logic and Computation*, 13:429–448, 2003.

[3] J.F.A.K. van Benthem. One logician's perspective on argumentation. *Cogency*, 2:13–25, 2009.

[4] P.M. Dung. On the acceptability of arguments and its fundamental role in nonmonotonic reasoning, logic programming, and n–person games. *Artificial Intelligence*, 77:321–357, 1995.

[5] T.F. Gordon. *The Pleadings Game. An Artificial Intelligence Model of Procedural Justice*. Kluwer Academic Publishers, Dordrecht/Boston/London, 1995.

[6] T.F. Gordon, H. Prakken, and D.N. Walton. The Carneades model of argument and burden of proof. *Artificial Intelligence*, 171:875–896, 2007.

[7] J.C. Hage, R.E. Leenes, and A.R. Lodder. Hard cases: a procedural approach. *Artificial Intelligence and Law*, 2:113–166, 1993.

[8] A.R. Lodder. *DiaLaw. On Legal Justification and Dialogical Models of Argumentation*. Law and Philosophy Library. Kluwer Academic Publishers, Dordrecht/Boston/London, 1999.

[9] J.D. Mackenzie. Question-begging in non-cumulative systems. *Journal of Philosophical Logic*, 8:117–133, 1979.

[10] S. Modgil and H. Prakken. A general account of argumentation with preferences. *Artificial Intelligence*, 195:361–397, 2013.

[11] Ch. Perelman and L. Olbrechts-Tyteca. *The New Rhetoric. A Treatise on Argumentation*. University of Notre Dame Press, Notre Dame, Indiana, 1969.

[12] J.L. Pollock. Defeasible reasoning. *Cognitive Science*, 11:481–518, 1987.

[13] H. Prakken. A formal model of adjudication dialogues. *Artificial Intelligence and Law*, 16:305–328, 2008.

[14] H. Prakken. An abstract framework for argumentation with structured arguments. *Argument and Computation*, 1:93–124, 2010.

[15] S.E. Toulmin. *The Uses of Argument*. Cambridge University Press, Cambridge, 1958.

[16] B. Verheij. DefLog: on the logical interpretation of prima facie justified assumptions. *Journal of Logic and Computation*, 13:319–346, 2003.

Response to Grossi and Prakken

MINGHUI XIONG AND YUN XIE

We would like to thank Davide Grossi and Henry Prakken, for their careful reading of our paper, and for their insightful comments. We are happy to see that both of them share the same assessment with us of the field of legal logic, and that they agree on most of the conclusions we have reached in this paper. Here we just give a brief response, clarifying some points raised in their comments.

The gist of our paper is to argue for a promising line of future research in the field of legal logic, that should take into account all significant features of legal argumentation in its practical context. We believe that it can only be realized by taking an informal-formal logical approach, that is, studies in legal logic should start with investigations from the informal perspective, and end with characterizations and modeling by formal methodology and tools. Therefore, we try to develop a five-part structure of legal argument, we propose to understand litigation argumentation as a three-agent game, and we urge bringing more advanced formal logics to bear, such as dynamic logic.

We agree with Davide Grossi that "the opposition between 'formal' and 'informal' logic for the study of argumentation is only an apparent one," and probably "from the forward-looking point of view... it is just a deceptive appearance". The opposition may well become illusory when the respective roles of formal and informal logic, and their proper interactions, are well defined, just as Davide Grossi has suggested in his commentary. It is common sense in contemporary argumentation theory that real arguments are not simply the embodiments of some formal, mathematical relationships between (sets of) static propositions. Any adequate theory of argumentation needs to capture its real-life nature first, such as the context of arguing, the roles of arguers, the practical rationality at work, etc. This point has been stressed early in 1950s by Toulmin and Perelman, who take pains to point out the limitations of the mathematical formalisms of their time. Nowadays, the lesson has been well learned in argumentation studies, and the deployment of rigorous mathematical methods, or tools of formal logic, has been carefully balanced by, and restricted to, real questions and rich features revealed by our informal investigations on argumentative practice. We believe that it is only in this way that an informal-formal logical approach in the studies of legal argumentation, and argumentation theory in general, can really be promising and fruitful. Otherwise our logical study of argumentation would be just heading to an unwelcome end, since, as Henry Prakken has observed, 'much logical research is formalism-driven, which makes its relevance for actual reasoning sometimes hard to see'.

Henry Prakken claims in his commentary that the research program we propose 'is in fact already being carried out by the field of artificial intelligence and law', which has developed since the 1980s, using rich formal models based on non-monotonic logic and argumentation logic. And he disagrees with us on the claim that, as one of the mainstream logics at the present time, 'dynamic logic could be deployed to describe litigation argumentation'. According to him, 'the mainstream logic on which a legal logic should be based, should not be dynamic logic, but an argumentation logic, since these logics have generally been the result of detailed analyses of actual argumentation'.

We are fully aware of work in the field of artificial intelligence and law, and we truly appreciate the brilliant achievements accomplished up to now. We mentioned this work in our paper, though we think most of it is trying to characterize legal argumentation in a two-agent model, which has, to some extent, simplified the real, more complex three-agent game structure of legal argumentation. However, in this paper we are not proposing a future research direction of legal logic which is radically opposite to artificial intelligence and law, but we are only urging some possible improvements on that basis. So the real point we would like to make is this: research in the field of artificial intelligence and law is surely relevant and promising, but still not enough to provide a sufficient and comprehensive characterization of legal argumentation. As Dr. Henry Prakken himself has admitted, 'many research issues are still unresolved'.

We might be guilty of making a misleading claim in this paper if readers get the impression that dynamic logic is the only logic that should be deployed for future studies on legal logic. But that is not our position. We are just proposing to borrow more advanced, and more suitable, formal methodology and logical tools for studying legal argumentation, with a hope to best capture its subtle features and complex structures. Argumentation logic has been, and still is, a useful mainstream logic in legal studies; its virtues are manifested in recent achievements in artificial intelligence and law. However, based on our proposed understanding of legal argumentation as a complex three-agent game, we believe that dynamic logic could be deployed as an additional tool in future studies in legal argumentation, since multi-agent interactions are at the heart of this kind of logic. With the insights and advanced tools provided in this way, we can better describe dynamics and formalities within legal argumentation, for example, the various kinds of acts of agents, the mechanism of information flow, and the changes of attitudes, beliefs, and preferences in relevant interactions among agents. In our view, legal logic needs both old and new logical perspectives, and we hope to show the utility of this in our ongoing and future work.

Part V

EVENING LECTURES: ABSTRACTS

The two evening lecturers at the Tsinghua conference are prominent representatives of active logicians working at interfaces with industry and society at large. We include their abstracts here.

From Aristotle to the iPhone

Moshe Y. Vardi

Logic started as a branch of philosophy, going back to Greeks, who loved debates, in the classical period. Computers are relatively young, dating back to World War II, in the middle of the 20th century. This talk tells the story of how logic begat computing, tracing the surprising path from Aristotle to the iPhone. This is a story full of both intellectual drama, as well as real-life drama, with most of the characters dying young, miserable, or both.

Additional reading
Martin Davis, *The Universal Computer: The Road from Leibniz to Turing*, A K Peters/CRC Press, 2011.

Bubbles[1]

VINCENT F. HENDRICKS

The more you can create that magic bubble, that suspension of disbelief, for a while, the better.

– Edward Norton

The term "bubble" has traditionally been associated with a particular situation occurring on financial markets:

"A bubble is considered to have developed when assets trade at prices that are far in excess of an estimate of the fundamental value of the asset, as determined from discounted expected future cash flows using current interest rates and typical long-run risk premiums associated with asset class. Speculators, in such circumstances, are more interested in profiting from trading the asset than in its use or earnings capacity or true value". ([12])

Textbook examples of bubbles include the Dutch tulip bulbs frenzy in the 1600s, the South Sea and Mississippi excesses about a century later, the US stock market as of 1929, the Japanese real estate and equity markets of the 1980s, the dot.com period and Internet stock boom of the 1990s, and of course the balloons, frenzies and speculative mania in the world economy leading to the global financial crisis of 2008 of which we are still in the midst of the aftermath.

In wake of the current crisis there have been many suggestions as to why financial bubbles occur, most of them composites in terms of explanatory factors involving different mixing ratios of bubble-hospitable market configurations and social psychological features of human nature and informational phenomena like *Infostorms* ([5]).

One seemingly paradoxical hypothesis suggests that too much liquidity is actually poisonous rather than beneficial for a financial market ([1]). Monetary liquidity in excess stimulated by easy access to credit, large disposable incomes and lax lending standards combined with expansionary monetary policies of lowering

---

[1] Joint work with David Budtz Pedersen, Pelle G. Hansen and Rasmus K. Rendsvig.

interests by banks and advantageous tax breaks and bars by the state, flush the market with capital. This extra liquidity leaves financial markets vulnerable to volatile asset price inflation the cause of which is to be found in short-term and possibly leveraged speculation by investors.

The situation becomes that too much money chases too few assets, good as well as bad, both of which in return are elevated well beyond their fundamental value to a level of general unsustainability. Pair up too much liquidity with robustly demonstrated socio-psychological features of human nature like boom-thinking, group-thinking, herding, informational cascades and other aggregated phenomena of social proof, it becomes a matter of time before the bubbles start to burst ([6]) at least in finance.

However, behind every financial bubble, crash and subsequent crisis "lurks a political bubble policy biases that foster market behaviors leading to financial instability" ([7]) with reference to the 2008 financial crunch. Thus there are political bubbles too and other sorts as well. There are stock, real-estate and other bubbles associated with financial markets but also filter bubbles, opinion bubbles, political bubbles, science bubbles, social bubbles, status bubbles, fashion bubbles, art bubbles all pushing collectives of agents in the same (often unfortunate) direction; not only buying the same stock or real estate but also thinking the same thing, holding the same opinions, appreciating the same art, "liking" the same posts on social media, purchasing the same brand names, subscribing to the same research program in science etc.

Internet activist Eli Pariser coined the term "filter bubble" ([9]) to refer to selective information acquisition by website algorithms (in search-engines, news feeds, flash messages, tweets) personalizing search results for users based on past search history, click behavior and location accordingly filtering away information in conflict with user interest, viewpoint or opinion. An automated but personalized information selection process in line with polarization mechanics isolating users in their cultural, political, ideological or religious bubbles. Filter bubbles may stimulate individual narrow-mindedness but are also potentially harmful to the general society undermining informed civic or public deliberation, debate and discourse making citizens ever more susceptible to propaganda and manipulation:

> "A world constructed from the familiar is a world in which there's nothing to learn ... (since there is) invisible autopropaganda, indoctrinating us with our own ideas." ([9])

Harvesting or filtering information in a particular way is part of aggregating opinion. One may invest an opinion on the free market place of ideas and a certain idea or stance, whether political, religious or otherwise, may at a certain point gain popularity or prominence and become an asset by the number of people apparently subscribing to it in terms of likes, upvotes, clicks or similar endorsements of minimum personal investment. Public opinion tends to shift depending on a variety of factors ranging from zeitgeist, new facts, current interests to premiums of social imprimatur. Opinion bubbles may accordingly suddenly go bust or gradually deflate depending.

Everyday personal opinions can serve as intellectual liquidity chasing assets of political or cultural ideas. But scientific inquiry may also be geared with too much intellectual liquidity in terms of explanatory expectations and available funding, paired up with boom-thinking in the scientific community. The short-term and possibly leveraged speculation by scientist may exactly occur in the way characterizing a ballooning market and science bubbles emerge ([2]). The modern commercialization of science and research has even been compared to downright Ponzi-schemes only surviving as long as you can steal from Peter to pay Paul scientifically so to speak ([8]).

Fashion in particular rely on getting everybody, or a selective few, to trend the same way – that's the point of the entire enterprise besides the occasional claim to artistic diligence. But even the art scene is tangibly ridden with bubbles: "The bubble that is Con Art blew up, like the sub-prime mortgage business, in the smoke-and-mirrors world of financial markets, where fortunes have been made on nothing" says Julian Spalding to *The Independent* (March 26, 2012), famous British gallery owner commenting on his recent book *Con Art – Why you ought to sell your Damien Hirsts while you can* (2012).

The concept of bubbles appears in seemingly different spheres. Perhaps it is more than just terminological coincidence – across spheres bubbles share similar structure and dynamics – from science to society. Irrational group behavior fuels bubbles. For instance, individual scientists may have doubts about the merits of bibliometric evaluation or excessive publishing practices much in vogue these days. However a strong public signal aggregated by the previous actions and endorsements of colleagues and institutions suggesting an aggressive publication strategy and abiding to the regulatory rules of evaluation and funding schemes may suppress the personal doubt of the individual scientist. But when personal information gets suppressed in favor of a public signal regulating individual behavior it may in turn initialize a lemming-effect, an informational cascade. Now, informational cascades have proven robust features in the generation of financial bubbles, where "individuals choose to ignore or downplay their private information and instead jump the bandwagon by mimicking the actions of individuals acting previously" ([12]).

Informational cascades may thus be considered pivotal to building bubbles – in science and elsewhere, and by using modern formal logic we have the means for uncovering their logical structure and dynamics independently of their realm of reign – and that's exactly what we are going to do ([3]), ([10]) and ([4]).

BIBLIOGRAPHY

[1] Buchanan, M. 2008. Why Economic Theory is Out of Whack. *New Scientist*: 07-19-2008.
[2] Budtz Pedersen, D. & Hendricks, V.F. (2013). Science Bubbles, *Philosophy & Technology* (in press).
[3] Hansen, P.G., Hendricks, V.F. & Rendsvig, R.K. (2013). Infostorms, *Metaphilosophy*, vol. 44(3), April: 301-326.
[4] Hansen, P.G. & Hendricks, V.F. (2014). *Infostorms: How to take Information Punches and Save Democracy*. New York Copernicus Books.
[5] Hendricks, V.F. & Rasmussen, J. Lundorff (2012). *Nedtur! Finanskrisen forstet filosofisk*. Copenhagen: Gyldendal Business 2012.
[6] Lee, I.H. (1998). Market Crashes and Informational Avalanches. *Review of Economic Studies* 65: 741-759.
[7] McCarty, N., Poole, K.T.& Rosenthal, H. (2013). *Political Bubbles: Financial Crises and the Failure of American Democracy*. Princeton: Princeton University Press.
[8] Mirowski, P. (2013). The Modern Commercialization of Science is a Passel of Ponzi Schemes. *Social Epistemology* 26 (4): 285-310.
[9] Pariser, E. (2011). *The Filter Bubble: What the Internet is Hiding from You*. New York: Penguin Books
[10] Rendsvig, R.K. & Hendricks, V.F. (2014). Social Proof in Extensive Games, submitted for publication.
[11] Spalding, J. (2012). *Con Art – Why you ought to sell your Damien Hirsts while you can*. CreateSpace Independent Publishing Platform.
[12] Vogel, H.L. 2010. *Financial Market Bubbles and Crashes*. New York: Cambridge University Press.

Contributors

Samson Abramsky
Department of Computer Science, University of Oxford
Email: samson@cs.ox.ac.uk

Thomas Ågotnes
Department of Information Science and Media Studies, University of Bergen
Email: Thomas.Agotnes@infomedia.uib.no

Sergei N. Artemov
Graduate Center, City University of New York
Email: sartemov@gc.cuny.edu

Alexandru Baltag
Institute for Logic, Language and Computation, University of Amsterdam
Email: A.Baltag@uva.nl

Arno Bastenhof
Department of Philosophy, University of Utrecht
Email: arnobastenhof@gmail.com

Johan van Benthem
Institute for Logic, Language and Computation, University of Amsterdam
Department of Philosophy, Stanford University
Email: johan.vanbenthem@uva.nl

Cristina Bicchieri
Department of Philosophy, University of Pennsylvania, Philadelphia
Email: cb36@sas.upenn.edu

Peter D. Bruza
Faculty of Science and Engineering, Queensland University of Technology
Email: p.bruza@qut.edu.au

Shushan Cai
Department of Psychology, Tsinghua University
Email: sscai@mail.tsinghua.edu.cn

Balder ten Cate
Department of Computer Science, University of California, Santa Cruz
Email: balder.tencate@gmail.com

Zoé Christoff
Institute for Logic, Language and Computation, University of Amsterdam
Email: zoe.christoff@uva.nl

Jianying Cui
Institute of Logic and Cognition, Sun Yat-sen University, Guangzhou
Email: cuijiany@mail.sysu.edu.cn

Franz Dietrich
CNRS; University of East Anglia
Email: fd@franzdietrich.net

Paul Égré
Institut Jean-Nicod (CNRS, ENS, EHESS)
Department of Philosophy, New York University
Email: paulegre@gmail.com

Patrick Girard
Department of Philosophy, University of Auckland
Email: p.girard@auckland.ac.nz

Noah D. Goodman
Department of Psychology, Stanford University
Email: ngoodman@stanford.edu

Davide Grossi
Department of Computer Science, University of Liverpool
Email: d.grossi@liverpool.ac.uk

Meiyun Guo
Institute of Logic and Intelligence, Southwest University, Chongqing
Email: guomy007@swu.edu.cn

Jens Ulrik Hansen
Department of Philosophy, Lund University
Email: Jens_Ulrik.Hansen@fil.lu.se

Vincent F. Hendricks
Department of Media, Cognition and Communication,
University of Copenhagen
Email: vincent@hum.ku.dk

Wesley H. Holliday
Department of Philosophy, University of California, Berkeley
Email: wesholliday@berkeley.edu

Huaxin Huang
Center for the Study of Language and Cognition, Zhejiang University, Hangzhou
Email: rw211@zju.edu.cn

Xiaoxi Huang
Institute of Cognition and Computing Intelligence, College of Computer
Science, Hangzhou Dianzi University, Hangzhou; Center for the Study of
Language and Cognition, Zhejiang University, Hangzhou
Email: huangxx@hdu.edu.cn

Thomas F. Icard III
Department of Philosophy, Stanford University
Email: icard@stanford.edu

Gerhard Jäger
Institute of Linguistics, University of Tübingen
Email: gerhard.jaeger@uni-tuebingen.de

Fengkui Ju
College of Philosophy and Sociology, Beijing Normal University
Email: fengkui.ju@bnu.edu.cn

Kevin T. Kelly
Department of Philosophy, Carnegie Mellon University
Email: kk3n@andrew.cmu.edu

Kohei Kishida
Department of Computer Science, University of Oxford
Email: kishidakohei@gmail.com

Hannes Leitgeb
Munich Center for Mathematical Philosophy, LMU Munich
Email: Hannes.Leitgeb@lrz.uni-muenchen.de

Beishui Liao
Center for the Study of Language and Cognition, Zhejiang University, Hangzhou
Email: bs3506@gmail.com

Hanti Lin
Department of Philosophy, Carnegie Mellon University
Email: hanti.lin@anu.edu.au

Christian List
Departments of Government and Philosophy, London School of Economics
Email: C.List@lse.ac.uk

Chanjuan Liu
School of Electronics Engineering and Computer Science, Peking University
Email: chanjuan.pkucs@gmail.com

Fenrong Liu
Department of Philosophy, Tsinghua University
Email: fenrong@tsinghua.edu.cn

Hu Liu
Insitutute of Logic and Cognition, Sun Yat-sen University, Guangzhou
Email: liuhu2@mail.sysu.edu.cn

Emiliano Lorini
CNRS and IRIT, Toulouse
Email: Emiliano.Lorini@irit.fr

Lawrence S. Moss
Department if Mathematics, Indiana University
Email: lsm@cs.indiana.edu

Sebastian Müller
Department of Computer Science, University of Toronto
Email: muller@karlin.mff.cuni.cz

Hiroakira Ono
Japan Advanced Institute of Science and Technology, Japan
Email: ono@jaist.ac.jp

Eric Pacuit
Department of Philosophy, University of Maryland
Email: epacuit@umd.edu

Sylvain Pogodalla
INRIA Lorraine, LORIA Lab, Nancy
Email: Sylvain.Pogodalla@inria.fr

Henry Prakken
Department of Information and Computing Sciences, Utrecht University
Faculty of Law, University of Groningen
Email: H.Prakken@uu.nl

Carlo Proietti
Department of Philosophy, Lund University
Email: Carlo.Proietti@fil.lu.se

Ramaswamy Ramanujam
National Institute of Mathematical Sciences, Chennai
Email: jam@imsc.res.in

Bryan Renne
Institute for Logic, Language and Computation, University of Amsterdam
Email: brenne@gmail.com

Alexander Reutlinger
Department of Philosophy, University of Cologne,
Munich Center for Mathematical Philosophy, LMU Munich
Email: Alexander.Reutlinger@uni-koeln.de

Robert van Rooij
Institute for Logic, Language and Computation, University of Amsterdam
Email: R.A.M.vanRooij@uva.nl

Olivier Roy
Munich Center for Mathematical Philosophy, LMU Munich
Email: Olivier.Roy@lrz.uni-muenchen.de

Katsuhiko Sano
School of Information Science, Japan Advanced Institute of Science and Technology, Japan
Email: v-sano@jaist.ac.jp

Jeremy Seligman
Department of Philosophy, University of Auckland
Email: j.seligman@auckland.ac.nz

Sonja Smets
Institute for Logic, Language and Computation, University of Amsterdam
Email: S.J.L.Smets@uva.nl

Kaile Su
School of Electronics Engineering and Computer Science, Peking University; Institute for Integrated and Intelligent Systems, Griffith University
Email: kailepku@gmail.com

Jakub Szymanik
Institute for Logic Language and Computation, University of Amsterdam
Email: J.K.Szymanik@uva.nl

Paolo Turrini
Department of Computing, Imperial College London
Email: p.turrini@imperial.ac.uk

Iddo Tzameret
The Institute for Interdisciplinary Information Sciences,
Tsinghua University
Email: tzameret@tsinghua.edu.cn

Jouko Väänänen
Department of Mathematics, University of Helsinki
Institute for Logic, Language and Computation, University of Amsterdam
Email: jouko.vaananen@helsinki.fi

Moshe Y. Vardi
Department of Computer Science, Rice University
Email: vardi@cs.rice.edu

Frank Veltman
Institute for Logic Language and Computation, University of Amsterdam
Email: f.j.m.m.veltman@uva.nl

Yde Venema
Institute for Logic Language and Computation, University of Amsterdam
Email: Y.Venema@uva.nl

Wen-fang Wang
Institute of Philosophy of Mind and Cognition,
Chung Cheng University, Taiwan
Email: wenfwang88@gmail.com

Wei Wang
Institute of Science, Technology and Society, Tsinghua University
Email: wangwei@tsinghua.edu.cn

Dag Westerståhl
Department of Philosophy, Stockholm University
Email: dag.westerstahl@philosophy.su.se

Shaun White
Department of Mathematics, University of Auckland
Email: s.white@math.auckland.ac.nz

John Woods
Department of Philosophy, University of British Columbia, Vancouver
Email: john.woods@ubc.ca

Yun Xie
Institute of Logic and Cognition, Department of Philosophy,
Sun Yat-sen University, Guangzhou
Email: xieyun6@mail.sysu.edu.cn

Minghui Xiong
Institute of Logic and Cognition, Sun Yat-sen University, Guangzhou
Email: hssxmh@mail.sysu.edu.cn

Cihua Xu
Center for the Study of Language and Cognition, Zhejiang University, Hangzhou
Email: xuch@zju.edu.cn

Difei Xu
Department of Philosophy, Renmin University of China, Beijing
Email: difeixu@163.com

Zhaoqing Xu
Department of Philosophy, Sichuan University, Chengdu
Email: zhaoqingxu@gmail.com

Congjun Yao
Department of Politics, Hunan University of Sicence and Engineering
Email: yaocongjun@126.com

Feng Ye
Department of Philosophy, Capital Normal University, Beijing
Email: fengye63@gmail.com

Liying Zhang
Institute of Modern Logic, Central University of Finance and Economics, Beijing
Email: clearliying@126.com

Xinshun Zhao
Institute of Logic and Cognition, Sun Yat-sen University, Guangzhou
Email: hsszxs@mail.sysu.edu.cn

Beihai Zhou
Department of Philosophy, Peking University
Email: zhoubh@phil.pku.edu.cn

Yuncheng Zhou
Department of Foreign Languages and Literatures, Tsinghua University
Email: yczhou@mail.tsinghua.edu.cn

Chongli Zou
Institute of Philosophy, Chinese Academy of Social Sciences
Email: chlizou@263.net

www.ingramcontent.com/pod-product-compliance
Lightning Source LLC
Chambersburg PA
CBHW071058050326
40690CB00008B/1058